Paradigms Lost: Learning from Environmental Mistakes, Mishaps, and Misdeeds

Paradigms Lost: Learning from Environmental Mistakes, Mishaps, and Misdeeds

Daniel A. Vallero, Ph.D.

ELSEVIER

AMSTERDAM • BOSTON • HEIDELBERG • LONDON
NEW YORK • OXFORD • PARIS • SAN DIEGO
SAN FRANCISCO • SINGAPORE • SYDNEY • TOKYO
Butterworth-Heinemann is an imprint of Elsevier

BH

BUTTERWORTH
HEINEMANN
ELSEVIER

30 Corporate Drive, Suite 400, Burlington, MA 01803, USA
Linacre House, Jordan Hill, Oxford OX2 8DP, UK

Library of Congress Cataloging-in-Publication Data
Vallero, Daniel A.
Paradigms lost: learning from environmental mistakes, mishaps, and misdeeds / Daniel A. Vallero.
p. cm.
Includes index.
ISBN 0-7506-7888-7 (harc cover : alk. paper)
1. Environmental education—History. 2. Cumulative effects assessment (Environmental assessment) I. Title
GE70.V35 2006
363.7—dc22

2005024537

British Library Cataloguing-in-Publication Data
A catalogue record for this book is available from the British Library.

ISBN 13: 978-0-75-067888-9
ISBN 10: 07-50-678887

For information on all Elsevier Butterworth–Heinemann publications visit our Web site at www.books.elsevier.com

Printed in the United States of America
06 07 08 09 10 11 10 9 8 7 6 5 4 3 2 1

For

Amelia and Michael

and

Daniel and Elise,

in their shared pursuit of new paradigms.

Eastern Box Turtle—*Terrapene carolina Carolina*.
Photo credit: Brookhaven National Laboratory and U.S. Fish and Wildlife Service, Upton Ecological and Research Reserve: http://www.bnl.gov/esd/reserve/turtles.htm.

Table of Contents

	Preface and Introduction	**xiii**
	Structure and Emphasis	xiv
	Quality Control	xxi
	Acknowledgments	xxx
	Notes and Commentary	xxx
Part I: New Science and New Paradigms		**1**
1.	**Lessons Learned: A Case Approach to Environmental Problems**	**13**
	MTBE and Cross-Media Transfer	18
	The Incremental Effect	19
	Failure and Blame	21
	A Lesson from the Medical Community	31
	Professional Accountability	33
	Villain and Victim Status	34
	Other Lessons: Risk and Reliability	37
	Environmental Ethics and a New Environmental Ethic	38
	Sensitivity	42
	Notes and Commentary	50
2.	**Pollution Revisited**	**55**
	DDT versus Eco-Colonialism: Trading Risks	62
	Reliability	65
	Characterizing Pollutants	69
	Partitioning to Solids—Sorption	75
	Partitioning to the Liquid Phase—Dissolution	80
	Partitioning to the Gas Phase—Volatilization	85
	Solubility as a Physical and Chemical Phenomenon	87
	Partitioning to Organic Tissue	89

Emissions, Effluents, Releases, Leaks, and Spills 91
Notes and Commentary 92

Part II: Key Environmental Events by Media 95
Fluids in the Environment: A Brief Introduction 96
Three Major Media 107

3. Something in the Air 109
London Air Pollution and the Industrial Revolution 109
Contaminants of Concern: Sulfur and Nitrogen Compounds 111
Notorious Air Pollution Cases of the Twentieth Century 118
The Meuse Valley Acid Fog 119
Contaminants of Concern: Particulate Matter 123
Donora, Pennsylvania 138
Poza Rica, Mexico 140
Contaminant of Concern: Hydrogen Sulfide 140
London, England 142
New York City 144
Toxic Clouds 144
The Bhopal Tragedy 144
Preparing for Intentional Toxic Clouds 150
Airshed in the Developing World: Mexico City 152
Lessons Learned 153
Contaminant of Concern: Photochemical Oxidant Smog 154
Notes and Commentary 157

4. Watershed Events 163
The Death of Lake Erie: The Price of Progress? 163
Eutrophication 164
Cuyahoga River Fire 168
Lesson Learned: The Need for Regional Environmental Planning 168
Spills: Immediate Problem with Long-Term Consequences 174
Solubility 176
Torrey Canyon Tanker Spill 179
Santa Barbara Oil Spill 183
Exxon Valdez Spill: Disaster Experienced and Disaster Avoided 185
Prestige Oil Spill 188
Lessons Learned: Two-Edged Swords 191
Pfiesteria piscicida: Nature Out of Sync 191
Lesson Being Learned 193
Notes and Commentary 194

5. Landmark Cases **197**
Love Canal, New York 198
Hazardous Waste Cleanup 200
A Fire that Sparked Controversy: Chester, Pennsylvania 204
Dioxin Contamination of Times Beach 213
A Terrifying Discovery: Valley of the Drums 215
Stringfellow Acid Pits 216
The March Continues 219
Lessons Learned 219
Failure to Grasp the Land Ethic 219
Disasters: Real and Perceived 226
"Cancer Alley" and Vinyl Chloride 232
Bioaccumulation and Its Influence on Risk 234
The Kepone Tragedy 235
Biological Response 237
Organic versus Inorganic Toxicants 240
Pesticides and Sterility 241
Jersey City Chromium 244
Radioisotopes 246
Radiation Poisoning in Goiania, Brazil 247
Factors of Safety 248
Small Numbers and Rare Events 250
Exposure Estimation 255
Risk-Based Cleanup Standards 263
The Drake Chemical Company Superfund Site: A Risk-Based Case 265
Risk Assessment: The First Step 268
Notes and Commentary 269

6. By Way of Introduction **275**
Asian Shore Crab 286
Zebra Mussel Invasion of the Great Lakes 290
Lesson Learned: Need for Meaningful Ecological Risk Assessments 292
Notes and Commentary 295

7. Environmental Swords of Damocles **297**
The Tragedy of the Commons 298
Global Climate Change 299
The Greenhouse Effect 300
Persistent, Bioaccumulating Toxicants 302
The Inuit and Persistent Organic Pollutants 302
Extrinsic Factors 305
Persistence 314
Endocrine Disrupting Compounds 316
Lake Apopka: A Natural Experiment 316

Genetic Engineering 320
Nuclear Fission 320
Meltdown at Chernobyl 321
Terrorism 325
Ecosystem Habitat Destruction 326
Lessons Learned 327
The Butterfly Effect 332
Notes and Commentary 334

Part III: Other Paradigms **337**

8. Dropping Acid and Heavy Metal Reactions **339**
Case of the Negative pH: Iron Mountain, California 340
Acid Mine Drainage 346
Acid Precipitation 348
Lead: The Ubiquitous Element 351
Coeur d'Alene Valley and the Bunker Hill Lead Smelter 352
Mercury: Lessons from Minamata 353
Arsenic Tragedy in Bangledesh 356
Asbestos in Australia 360
Notes and Commentary 364

9. Spaceship Earth **367**
Changes in the Global Climate 368
Carbon Dioxide 375
Methane 377
Nitrous Oxide 377
Halocarbons and Other Gases 378
Land Use and Forestry 378
Threats to the Stratospheric Ozone Layer 379
Coral Reef Destruction 380
Syllogisms for Coral Reef Destruction 383
Notes and Commentary 385

10. Myths and Ideology: Perception versus Reality **389**
Solid Waste: Is It Taking over the Planet? 389
Alar and Apples 392
Parent versus Progeny 393
Agent Orange: Important If True 404
The Snail Darter: A Threat to the Endangered Species Act? 407
Seveso Plant Disaster 410
Poverty and Pollution 413
Notes and Commentary 415

11. Just Environmental Decisions, Please **419**
Environmental Justice 420
How Can Engineers Best Manage Risks in a Changing Environment? 425
Optimization in Environmental Risk Management 427
Precautionary Principle and Factors of Safety in Risk Management 444
Market versus Non-Market Valuation: Uncle Joe the Junk Man 446
The Warren County, North Carolina, PCB Landfill 452
The Orange County, North Carolina, Landfill 459
If It Does Occur, It Is Not Bad 463
If It Does Occur and It Is Bad, It Is Not Racially Motivated 464
Is Environmentalism a Middle-Class Value? 465
Habitat for Humanity 467
Carver Terrace, Texas 469
West Dallas Lead Smelter 471
Lessons Applied: The Environmental Justice Movement 471
Environmental Justice and the Catalytic Converter 473
Notes and Commentary 475

Part IV: What Is Next? **481**

12. Bottom Lines and Top of the Head Guesses **483**
The Future of Environmental Science and Engineering 487
The Systematic Approach 487
New Thinking 491
The Morning Shows the Day 498
Notes and Commentary 498

Appendix 1: Equilibrium **503**
Appendix 2: Government Reorganizations Creating the U.S. Environmental Protection Agency and the National Oceanic and Atmospheric Administration **511**
Appendix 3: Reliability in Environmental Decision Making **517**
Appendix 4: Principles of Environmental Persistence **521**
Appendix 5: Cancer Slope Factors **527**
Appendix 6: Equations for Calculating Lifetime Average Daily Dose (LADD) for Various Routes of Exposure **535**
Appendix 7: Characterizing Environmental Risk **539**
Appendix 8: Risk-Based Contaminant Cleanup Example **543**
Appendix 9: Shannon Weiner Index Example **547**
Appendix 10: Useful Conversions in Atmospheric Chemistry **551**

Index **553**

Preface and Introduction

Awake, arise, or be forever fallen!

John Milton (1608–1674), *Paradise Lost*. Book 1, Line 330

Granted, Milton is a questionable choice to quote at the beginning of any scientific text, even one that considers mistakes, mishaps, and misdeeds. Having been engaged in the practice and the teaching of environmental science and engineering during their formative periods, I frequently have drawn upon the lessons learned from key cases. Certainly, the cases in this book are predominantly those with negative outcomes. But there is also much about which to be optimistic. Engineers and scientists have made great progress in advancing the understanding of the principles underlying environmental quality and public health. When asked, in fact, my students often have labeled me a technological optimist. However, our contemporary understanding has all too often come at a great cost. And, what makes this even more tragic is that society and the scientific community so often forget or do not learn the lessons that should have been learned. Paying attention to the past instructs us about the future.

Our experiences are collected into a set of shared values, which are incorporated into paradigms of acceptable norms (positive paradigms) and malevolent behavior (negative paradigms). Such paradigms instruct us on standards and laws, including those that instruct us on how to care for the environment and what happens when we fail to do so.

Societies become comfortable with their paradigms. Even slight shifts are met with resistance. The twentieth-century paradigm of almost unbridled avarice and the expectation that the air, water, and soil could absorb whatever manner of wastes we introduced had to be revisited and revised. We have slowly come to accept that the *paradise* of a diverse and sustainable life support system here on earth was in jeopardy. Our own ignorance of the vulnerability and delicate balances of our natural resources and environment was putting us at risk.

Thomas S. Kuhn (1922–1996), the noted physicist and philosopher of science, is recognized as having been among the first to show that scientists are reticent to change their ways of thinking.[1] It is probably fair to extend this reluctance more generally to human nature. But scientists and engineers are supposed to be, in fact are paid to be, objective! The modern concept of objective science grew out of the Renaissance, when Robert Boyle and other leading scientists of the Royal Society of London required that scientific investigation always include experimentation (*a posteriori* knowledge),[2] publication of methods and results (literary technology), and peer review (witnesses). Kuhn grew to see science as it is practiced in contemporary times often to be void of reason. This is ironic in light of the so-called scientific method, which is built upon objectivity and reason. Scientific ways of seeing the universe—paradigms—only change after incremental evidence forces us to change. This book highlights some of this evidence (i.e., cases) that pushes us toward a new environmental ethic and awareness.

Structure and Emphasis

This book blends the historical case perspective with credible and sound scientific explanations of key environmental disasters and problems. Scientific, engineering, technological, and managerial concepts are introduced using real-life incidents. Famous, infamous, and not-so-famous but important cases are explained using narrative, photographs, figures, and tables, as appropriate. In some instances, flowcharts and event trees show how the result came to be, as well as demonstrate alternative approaches, including preventive measures and contingency plans that could have ameliorated or even prevented the disaster.

If you were to ask my students to describe my pedagogical approach, they may tell you that it is Socratic. They may also describe it as anachronistic. Some may say it is eclectic. I would have to say that it is all those things. My approach to teaching has evolved into a journey of sorts. And, journeys require storytelling; storytelling requires real-world cases. The Socratic approach allows the class to relive events and along the way to learn through the students' own inquisitiveness. The questioning and doubt about certainties to elicit the truth are ideally suited to environmental science and engineering subject matter.

Environmental problems usually have no unique solution. Environmental consequences are the result of highly complex contingencies. The contingent probabilities of a particular outcome in a specific situation at a particular time, to use an engineering concept, are miniscule. But that specific outcome did in fact occur, so we need to discover why. Anachronisms are also valuable teaching devices. When considering problems of the industrial revolution, why not discuss contemporary lyrics or poetry? No single teaching device works in every situation, so an eclectic approach using group projects, case studies, lectures, seminar discussions, and any number of graphical and presentation techniques is more useful than force-fitting a favorite approach. I have blended the lessons learned from these approaches into this book. I do not shy away from highly technical and scientific discussions in my classes, nor do I in this book. Sometimes, the best way to introduce a very technical concept is to "sneak it" into a discussion that students would be having anyway. I am a true believer in teachable moments.[3] When they occurred, every one of the cases in this book provided such a teachable moment. The trick is to bring these teachable moments back to the present. The style and delivery of this book are quite similar to my pedagogy, so depending on the subject at hand the best approach will vary.

The lessons learned go beyond the typical environmental science and environmental engineering format. Indeed, these will be a part of the explanation of what occurred and what can be done to prevent the problems. In addition, process engineering, risk assessment and management, and practical solutions are considered, where appropriate.

Each case gives a platform to discuss larger, more widely applicable concepts that are important to engineers, planners, and decision makers. For example, Love Canal is an interesting and important case in its own right, but it also provides larger lessons about the importance of managers requiring contingency plans, the need to consider all possible effects from all options, and the need to coordinate public health responses and epidemiology once a problem begins to emerge. Such lessons apply to hazardous waste siting, landfill decisions, and health and public works services worldwide. Also, considering some of the nearly forgotten lessons learned from history provides insights into ways to address current problems. For example, were the deaths from the soot and smoke incidents of London and Pennsylvania in the 1950s all that different from those in developing countries now? The answer is open to debate, but at least some parallels and similarities seem apparent. And can we revisit steps taken and opportunities missed the past 50 years as lessons from which to advise those vulnerable populations today? The answer is clearly "yes."

The book is unabashedly technical, yet understandable to most readers. It is annotated with side bars and discussion boxes to keep the reader's interest and to help to extend the lessons beyond each case. As in my previous books, any technical term is introduced with a full explanation, including the generous use of examples. Each case is described in a way that it can stand on its own, alleviating the need for cross-referencing with other cases in the book or needing to refer to other sources. This makes for a better teaching device, as instructors may choose to begin with cases in a different order than that of the book.

There is much value in discussing the general lessons learned from the totality of the cases. So, each chapter ends with a litany of these lessons

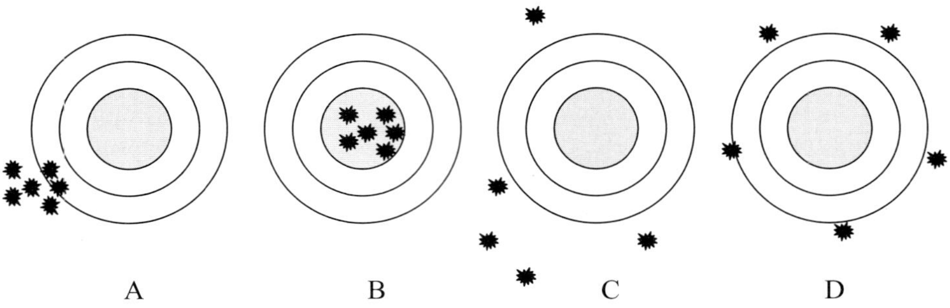

FIGURE P.1. Precision and accuracy. The bull's eye represents the true value. Targets A and B demonstrate data sets that are precise; Targets B and D, data sets that are accurate; and Targets C and D, data sets that are imprecise. Target B is the ideal data set, which is precise and accurate.

specific to that chapter, as well as insights as to the consequences of ignoring or adhering to these lessons. Environmental endeavors are always interconnected and integrated, so even though each case will be treated thoroughly, collective lessons from the myriad cases are considered.

Of course, like all things in the physical sciences and engineering, such predictions are always accompanied by uncertainties. Uncertainties are brought about both by variability and error.[4] Variability is ever-present in space and time. Every case has a unique set of factors, dependent variables, situations, and scenarios, so that what occurred will never be completely repeated again. Every cubic centimeter of soil is different from every other cubic centimeter. The same goes for a sample of water, sediment, air, and organic tissue. And these all change with time. Taking a sample in the winter is different from that in the summer. Conditions in 1975 are different in so many ways from conditions in 2005. And, of course, there are errors. Some are random in that the conditions that led to the cases in this book are partially explained by chance and things that are neither predictable nor correctable, although we can explain (or at least try to explain) them statistically, for example, with normal distributions.

Other error is systematic, such as those of my own bias. I see things through a prism different from anyone else's. This prism, like yours, is the result of my own experiences and expertise. This prism is my perception of what is real and what is important. My bias is heavily weighted in sound science, or at least what I believe to be sound science (as opposed to "junk science").[5] Sound science requires sufficient precision and accuracy in presenting the facts. *Precision* describes how refined and repeatable an operation is, such as the exactness in the instruments and methods used to obtain a result. It is an indication of the uniformity or reproducibility of a result. This can be likened to shooting arrows,[6] with each arrow representing a data point. Targets A and B in Figure P.1 show equal precision. Assuming that the center of the target, the bull's eye, is the "true value," data set B is more accurate than A. If we are consistently missing the bull's eye in the same direction at the same distance, this is an example of bias or systematic error. The good news is that if we are aware that we are missing the bull's eye (e.g., by comparing our results to those of known standards when using our analytical equipment), we can calibrate and adjust the equipment. To stay with our archery analogy, the archer would move her sight up and to the right.

Thus, *accuracy* is an expression of how well a study conforms to some defined standard (the true value). So, accuracy expresses the quality of what we find, and precision expresses the quality of the operation by which we obtained our finding. So, the two other scenarios of data quality are shown in Targets C and D. Thus, the four possibilities are that our data is precise but inaccurate (Target A), precise and accurate (Target B), imprecise and inaccurate (Target C), and imprecise and accurate (Target D).

At first blush, Target D may seem unlikely, but it is really not all that uncommon. The difference between Targets B and D are simply that D has more "spread" in the data. For example, the variance and standard deviation of D is much larger than that of B. However, their measures of central tendency, the means, are nearly the same. So, both data sets are giving us the right answer, but almost all the data points in B are near the true value. None of the data points in D are near the true value, but the mean (average location) is near the center of the bull's eye, so it has the same accuracy as Target B, but with much less precision. The key is that precision and accuracy of the facts surrounding a case must be known.

I recognize that science is a crucial part of any case analysis, but so are other factors. To wit, philosophers tell us that the only way to make a valid argument is to follow the structure of the syllogism:

1. Factual Premise
2. Connecting Premise (i.e., factual to evaluative)
3. Evaluative Premise
4. Moral Conclusion

For example, the facts may show that exposing people to a chemical at a certain dosage (e.g., one part per million) leads to cancer in one in every ten thousand people. We also know that, from a public health perspective, allowing people to contract cancer as a result of some human activity is morally wrong. Thus, the syllogism would be:

1. Factual Premise: Exposure to chemical X at 1 ppm leads to cancer.
2. Connecting Premise: Release of 10 kg per day of chemical X leads to 1 ppm exposure to people living near an industrial plant.

3. Evaluative Premise: Decisions that allow industrial releases that lead to cancer are morally wrong.
4. Moral Conclusion: Therefore, corporate executives who decide to release 10 or more kilograms of chemical X from their plants are morally wrong.

Upon examination, the syllogism is not as straightforward as it may first appear. In fact, the exact meanings of the premises and moral conclusions have led to very vigorous debates (and lawsuits). For example, all parties may agree with the evaluative premise, that releases should not lead to cancer, but they strongly disagree on the facts, such as whether the data really show that these dosages "cause" cancer or whether they are just coincidental associations. Or, they may agree that they cause cancer, but not at the rate estimated by scientists. Or, they may disagree with the measurements and models that project the concentrations of chemical X to which people would be exposed (e.g., a conservative model may show high exposures and another model, with less protective algorithms, such as faster deposition rates, may show very low exposures). Or, they may argue that measurements are not representative of real exposures. There are even arguments about the level of protection. For example, should public health be protected so that only one additional cancer would be expected in a population of a million or one in ten thousand? If the former (10^{-6} cancer risk) were required, the plant would have to lower emissions of chemical X far below the levels that would be required for the latter (10^{-4} cancer risk). This is actually an argument about the value of life. Believe it or not, there are "price tags" placed quite frequently on a prototypical human life, or even expected remaining lifetimes. These are commonly addressed in actuarial and legal circles. For example, Paul Schlosser in his discussion paper, "Risk Assessment: The Two-Edged Sword" states:

> The processes of risk assessment, risk management, and the setting of environmental policy have tended to carefully avoid any direct consideration of the value of human life. A criticism is that if we allow some level of risk to persist in return for economic benefits, this is putting a value on human life (or at least health) and that this is inappropriate because a human life is invaluable—its value is infinite. The criticism is indeed valid; these processes sometimes do implicitly put a finite, if unstated, value on human life.
>
> A bit of reflection, however, reveals that in fact we put a finite value on human life in many aspects of our society. One example is the automobile. Each year, hundreds or thousands of U.S. citizens are killed in car accidents. This is a significant risk. Yet we allow the risk to continue, although it could be substantially reduced or eliminated by banning cars or through strict, nation-wide speed limits of 15 or 20 mph. But we do not ban cars and allow speeds of

TABLE P.1
Regulation cost of saving one life (in U.S. dollars).

Activity	*Cost ($ US)*
Auto passive restraint/seat belt standards	100,000.00
Aircraft seat cushion flammability standard	400,000.00
Alcohol and drug control standards	400,000.00
Auto side door support standards	800,000.00
Trenching and excavation standards	1,500,000.00
Asbestos occupational exposure limit	8,300,000.00
Hazardous waste listing for petroleum refining sludge	27,600,000.00
Cover/remove uranium mill tailings (inactive sites)	31,7000,000.00
Asbestos ban	110,700,000.00
Diethylstilbestrol (DES) cattle feed ban	124,800,000.00
Municipal solid waste landfill standards (proposed)	19,107,000,000.00
Atrazine/Alachlor drinking water standard	92,069,700,000.00
Hazardous waste listing for wood preserving chemicals	5,700,000,000,000.00 (This is not a typo.)

Source: P.M. Schlosser, 1997. "Risk Assessment: The Two-Edged Sword": http://pw2.netcom.com/~drpauls/just.html; accessed April 12, 2005.

> 65 mph on major highways because we derive benefits, largely economic, from doing so. Hence, our car "policy" sets a finite value on human life.
>
> You can take issue with my car analogy because, when it comes to cars, it is the driver who is taking the risk for his or her own benefit, while in the case of chemical exposure, risk is imposed on some people for the benefit of others. This position, however, is different from saying that a human life has infinite value. This position says that a finite value is acceptable if the individual in question derives a direct benefit from that valuation. In other words, the question is then one of equity in the risk-benefit trade-off, and the fact that we do place a finite value on life is not of issue.

Another way to address this question is to ask, "How much are we willing to spend to save a human life?" Table P.1 provides one group's estimates of the costs to save one human life. From what I can gather from the group that maintains the Web site sharing this information, they are opposed to much of the environmentalist agenda, and their bias colors these data. However, their method of calculating the amount of money is fairly straightforward. If nothing else, the amounts engender discussions about possible risk trade-offs since the money may otherwise be put to more productive use.

Schlosser asks "How much is realistic?" He argues that a line must be drawn between realistic and absurd expenditures. He states:

> In some cases, risk assessment is not used for a risk-benefit analysis, but for comparative risk analysis. For example, in the case of water treatment one can ask: is the risk of cancer from chlorination by-products greater than the risk of death by cholera if we do not chlorinate? Similar, if a government agency has only enough funds to clean up one of two toxic waste sites in the near future, it would be prudent to clean up the site which poses the greatest risk. In both of these cases, one is seeking the course of action which will save the greatest number of lives, so this does not implicitly place a finite value on human life. (In the second example, the allocation of finite funds to the government agency does represent a finite valuation, but the use of risk assessment on how to use those funds does not.)[7]

We, as fallible human beings, are not the best assessors or predictors of value. We can rationalize the elimination of a "problem." Humans are very good at that. So, how do moral arguments about where to place value and the arguments made by Schlosser and others (such as the concept of willingness to pay) fit with moral theories, such as duty-based ethics (i.e., deontology), consequence-based ethics (teleology), or social contract theory (contractarianism)? Where do concepts like John Stuart Mill's harm principle, John Rawls' veil of ignorance, and Immanuel Kant's categorical imperative come into play? How do such concepts fit with the code in one's chosen profession? How do teleological, deontological, contractarian, and rational models hold up this scrutiny?

One method for testing our ethics is to try to look back from a hundred years hence, such as we can do now with slavery, women's rights, and so forth. What would you expect the future societies to think of what we are doing with those in our society with the weakest voices? As I mentioned, even though I continue to be strongly utilitarian in my support for animal testing, I fear that through the prism of future perspective, I may be found lacking. . . . I have seen every one of these arguments in environmental situations. Some are valid, some are not.

Syllogisms are not specifically drawn in most of the cases, but they are there just the same. Whenever we draw a moral conclusion—that the behavior of certain groups was improper, unacceptable, or downright immoral—we have intuitively drawn a syllogism. Intuitive syllogisms are present every time we give credit or place blame. The best we can hope for is that we have thoroughly addressed the most important variables and with wisdom may prevent similar problems in the future.

I have learned that syllogisms can easily be inverted to fit the perception and needs of those applying them. That is, people already have a conclusion in mind and go searching for facts to support it. The general public expects that its professionals understand the science and that any arguments being made are based in first principles. We must be careful that this "advocacy science" or, as some might call it, "junk science" does not find its way into environmental engineering. There is a canon that is common in most engineering codes that tells us we need to be "faithful agents." This, coupled with an expectation of competency, requires us to be faithful to the first principles of science. In a way, I fear that because of pressures from clients and political or ideological correctness, the next generation of engineers will be tempted to "repeal Newton's laws" in the interest of certain influential groups! This is not to say that engineers will have the luxury to ignore the wishes of such groups, but since we are the ones with our careers riding on these decisions, we must clearly state when an approach is scientifically unjustifiable. We must be good listeners, but honest arbiters.

Unfortunately, many scientific bases for decisions are not nearly as clear as Newton's laws. They are far removed from first principles. For example, we know how fluids move through conduits (with thanks to Bernoulli *et al.*), but other factors come into play when we estimate how a contaminant moves through very small vessels (e.g., intercellular transport). The combination of synergies and antagonisms at the molecular and cellular scales make for uncertainty. Combine this with uncertainties about the effects of enzymes and other catalysts in the cell, and we propagate even greater uncertainties. So, the engineer operating at the meso-scale (e.g., a wastewater treatment plant) can be fairly confident about the application of first principles of contaminant transport, but the biomechanical engineer looking at the same contaminant at the nano-scale is not so confident. That is where junk science sometimes is able to raise its ugly head. In the void of certainty, for example at the molecular scale, some crazy arguments are made about what does or does not happen. This is the stuff of infomercials! The new engineer had better be prepared for some off-the-wall ideas of how the world works. New hypotheses for causes of cancer, or even etiologies of cancer cells, will be put forward. Most of these will be completely unjustifiable by physical and biological principles, but they will sound sufficiently plausible to the unscientific. The challenge of the new engineer will be to sort through this morass without becoming closed-minded. After all, many scientific breakthroughs have been considered crazy when first proposed (recalling Copernicus, Einstein, Bohr, and Hawking, to name a few). But even more really were wrong and unsupportable upon scientific scrutiny.

Quality Control

The case-based approach to environmental problems does have the disadvantages of uncertainty and representativeness. We often are not sure of the physical scientific facts fundamental to a case, let alone the social science, humanities, and political subtleties. For example, I have attempted to choose cases that reflect the environmental paradigm shifts. This means that some important cases have been omitted, probably more than a few that you would have expected to see. As part of my quality control in this matter, after completing my manuscript, I inquired of a number of experts in various environmental disciplines such as science, engineering, and policy, as to what they considered to be important cases. The good news is that most of the cases they expected have been included. The not-so-good news is that some important cases are not directly addressed. Those identified that are either not covered or only mentioned in reference to other cases are:

1. The near meltdown of the nuclear reactor core at the Three Mile Island power facility near Harrisburg, Pennsylvania.
2. The Kuwaiti oil fires and eco-terrorism at the end of the first Gulf War.
3. The eco-disaster in the Danube basin resulting from the Iron Gates Dam project.
4. Rainforest destruction.
5. The ecosystem destruction wrought by introduced plant species.
6. The cadmium poisoning of miners in Japan.
7. Recent concerns about mercury, especially from fossil fuel combustion.
8. Exposure to asbestos, especially vermiculite and the Libby, Montana, mine.

To assuage my guilt for not directly addressing these eight issues as individual cases, allow me to discuss them briefly here. I also address them, with links to Web resources in the companion Web site to this book (http://books.elsevier.com/companions/0750678887).

I chose to address the Chernobyl nuclear disaster as a "sword of Damocles" in Chapter 7 rather than Three Mile Island because the consequences of the Ukrainian meltdown demonstrated failure at so many levels—design, implementation, oversight, regulatory, and emergency response. The 1979 accident at Three Mile Island did release radiation, especially the radioactive isotope iodine-131, which is formed after uranium undergoes fission. More importantly, the accident was an omen of what *could* happen and in fact did happen at Chernobyl. Our failure to heed the lessons of both nuclear disasters would be folly.

The 1991 Kuwait oil spills and fires do represent an important case in terms of intentional environmental destruction. I chose to discuss terrorism and environmental vulnerability, especially following the attacks on the Pentagon and the World Trade Center towers. However, every war and international conflict extracts an ecological and public health toll. There is no question that Iraq committed ecological terrorism in Kuwait by deliberately spilling millions of barrels of oil into the Persian Gulf and igniting, via sabotage, 500 Kuwaiti oil wells, storage tanks, and refineries. In fact, the oil spill was the largest ever: an estimated six million barrels of oil, 25 times larger than the 250,000 barrels from the *Exxon Valdez* in Alaska's Prince William Sound. The oil fires started in mid-February were the worst the world has ever suffered, releasing as much as six million barrels of oil residue in the plume per day at their peak. The thick, black clouds reached thousands of meters, eclipsing the sunlight, so that Kuwait City and Saudi Arabian cities just south of the border experienced almost constant night. The EPA Administrator at the time, William K. Reilly, said "If Hell had a national park, it would be those burning oil fires," and "I have never seen any one place before where there was so much compressed environmental degradation."[8] Indeed, it does represent an important case.

The Iron Gate Dam illustrates the importance of small things and a systematic approach. As such, it would fit nicely into the discussions in Chapter 12. It clearly represents the huge ecological price that must be paid when biodiversity is destroyed. The case is very interesting in that something that we do not ordinarily consider to be a limiting factor, silicates, led to major problems. The Black Sea is the largest enclosed catchment basin, receiving freshwater and sediment inputs from rivers draining half of Europe and parts of Asia. As such, the sea is highly sensitive to eutrophication (see Chapter 4) and has changed numerous times in recent decades. The Danube River receives effluents from eight European countries, flows into the Black Sea, and is the largest source of stream-borne nutrients. In less than a decade, the system changed from an extremely biodiverse one to a system dominated by jellyfish (*Aurelia* and the combjelly *Mnemiopsi*).[9] These invaders were unintentionally introduced in the mid-1980s, culminating in the fisheries almost completely vanishing by the early 1990s. This collapse was first attributed to unpalatable carnivores that fed on plankton, roe, and larvae. Subsequently, however, the jellyfish takeover was found to result from human perturbations in the coastal ecosystems and in the drainage basins of the rivers, including changing the hydrologic character of out-flowing rivers. The biggest of these was the damming of the Danube in 1972 by the Iron Gates, approximately 1,000 km upstream from the Black Sea. In addition, urban and industrial development, heavy use of commercial fertilizers, over-fishing, and the introduction of exotic, invasive organisms (e.g., *Mnemiopsi*) contributed to the problem. After 1970, this change in nutrient concentrations induced phytoplankton blooms during the warm months and changed the dominance to nonsiliceous species that were not a first choice as food for meso-zooplankton. The decreased fish stocks further increased the dominance of the jellyfish, since they competed better than the game fish for the same food. Ironically, since the mid-1990s, the ecosystems have begun to improve, mainly due to increased nutrient (phosphorus and nitrogen) loading. In most situations, we are looking to decrease this loading, to prevent eutrophication. But in this system, the added nutrients have allowed certain plankton and benthic (bottom dwelling) organisms to recolonize. The abundance of jellyfish has also stabilized, with a concomitant increase in anchovy eggs and larvae.

Nutrient limitation occurs when the presence of a chemical, such as phosphorus or nitrogen, is insufficient to sustain the growth of community or species. Usually, marine systems are nitrogen limited whereas freshwater plankton systems are phosphorus limited. Numerous freshwater organisms can "fix" atmospheric nitrogen but, with minor exceptions, the nitrogen is impeded in marine water. The nutrient requirements differ by species. A disturbance in the ratio of nitrogen, phosphorus, silica, and even iron changes the biotic composition of a particular plankton community. Often, all four nutrients can be considered as limiting. For instance, the lack of silica limits diatoms. This was observed first in natural blooms off Cape Mendocino in the United States and subsequently observed in the northwest part of the Black Sea, after closing the Iron Gates dam. The case also demonstrates that economics is crucial, since the marine ecosystem improvement directly corresponds to the decline of the economies of Central and Eastern European nations in the 1990s.

Rainforest destruction is certainly an important problem for numerous reasons, including the loss of irreplaceable habitat and the endangerment of species, the loss of "oxygen factories" as photosynthesis is reduced, and the loss of sinks to store carbon in both its oxidized forms (carbon dioxide) and reduced forms (methane). Both carbon dioxide and methane are principal greenhouse gases. This is touched on briefly in Chapter 9 when the major greenhouse gases are described and in the brief discussions on forestlands. In a sense, rainforest destruction is probably most akin to the coral reef destruction discussed in Chapter 9, since it is an example of resources that are almost impossible to recover. Public concern is increased in situations where the consequences are irreversible. The potential irreversibility means that what we are doing now will adversely affect future generations and is also evidence that we lack control and are uncertain about what the damage means. People want to prevent catastrophes or at least to catch problems before they become large and irreversible. The rates of rainforest losses are staggering; some estimates put the losses at 1 hectare per hour or about 31 million hectares per year, which is about the area of the country of Poland![10] Along with the sheer land numbers, about 50,000 rainforest species are becoming extinct each year.[11] Indeed, the problem is large and, given geopolitical realities, seemingly intractable.

Introduced plant species is a widespread problem. In fact, Table 6.1 includes a number of plants. The two species addressed in Chapter 6 (shore crab and zebra mussel), both aquatic, allow for comparisons and contrasts in the ways that the species are introduced and how they colonize. However, plants are certainly important. For example, numerous invasive plants have been introduced intentionally with good intentions; they represent an all-too-common problem of doing the wrong thing for the right reasons. This brings back memories of my father and Uncle Louie vigorously digging up the tough little multiflora rose (*Rosa multiflora* (Thunb.

ex Murr.)) seedlings that had popped up in my uncle's pastures in Collinsville, Illinois, near St. Louis (see Figure P.2).

The idea of using natural barriers to control livestock and to provide other agricultural barriers seemed brilliant at its conception. Instead of fences and barriers of steel, wood, or rock that were difficult to construct and in constant need of maintenance, why not "use nature" to keep animals from wandering off? And why not choose a plant that is beautiful on the one hand and akin to razor wire on the other? The seemingly perfect solution was the multiflora rose.[12] But, instead, the rose took over entire pastures. I remember hunting for mushrooms some years back in what was pastureland 10 years before, having to crawl through the thorns of these noxious, albeit pretty weeds. Needless to say, the cattle had long ago found "greener pastures." Since moving to North Carolina, I have seen large wooded areas completely covered in Kudzu (*Pueraria* spp.), as shown in Figure 6.1. The losses to arable land and destruction of sensitive habitat as a result of invasive plant species have been enormous.

Toxic heavy metals are addressed in detail in Chapter 8. I chose to emphasize lead, mercury, and the metalloid arsenic. However, the metal cadmium is not specifically addressed. I certainly agree that cadmium is highly toxic, even carcinogenic, and that its history is revealing in terms of the evolution of environmental protection. For example, one of the first documented cases of "industrial disease" is that of *Itai-Itai* (roughly translated from Japanese to mean "ouch-ouch"). *Itai-Itai* is a serious bone malady, the painful result of chronic cadmium poisoning from mining wastes that found their way into the Jinzu River basin in Toyama Prefecture. The case demonstrates the complexity of exposure pathways. For example, it appears that the exposures were predominantly by ingestion of rice, which was contaminated by the river water and the cadmium transposed to the plant tissues. The exposures could also have been by direct consumption of the water or by residues of cadmium on the plant materials. Many of the sufferers experienced extreme bone demineralization. Exposure to high concentrations of cadmium also causes other health problems including kidney damage that could also be responsible for the bone loss. So, the problem may be direct, such as cadmium's replacement of calcium in the calcium-phosphorous bone complexes, or it could be indirect where the bone diseases are due to nephrotoxicity (i.e., kidney damage). Likely, both processes are occurring. The cadmium poisoning is somewhat similar to the Minamata mercury case in Chapter 8, but the mining company appeared to be more of a victim of ignorance than the chemical company in the Minamata case. The similarities include a very vulnerable exposed population of farmers and anglers, about 200 severely affected patients in each case, thousands more with less severe effects, and their dependence on large industrial interests for economic development. In this sense, both cases may also be early examples of environmental injustices.

FIGURE P.2. Multiflora rose (*Rosa multiflora* [Thumb ex Murr.]). Top photo: Near the St. Peter and Paul Cemetery in Collinsville, Illinois, about 10 miles east of St. Louis, Missouri. The rose, in the middle of the photo, has likely colonized the area from bird droppings. Bottom photo: *Rosa multiflora* invading a garden east of downtown Collinsville.

Much attention is given in Chapter 8 to the chemistry of metal exposure and risk, including discussions about mercury. No other metal demonstrates the importance of chemical speciation better than mercury. For example, its persistence, its movement, and its toxicity are largely determined by its valence state and the compounds it forms. Dimethyl mercury, for example, is one of the most acutely toxic substances in the environment, or in the laboratory for that matter. Recently, much attention has been given to mercury emissions from coal-fired power plants. Mercury emitted from power plant stacks and other sources is carried by winds through the air and subsequently deposited to water and land. The actual distance traveled depends on the chemical form in which it is emitted, the height at which it is released and atmospheric conditions. Usually, mercury concentrations in the air are low and of little direct concern, but upon entering the water, microbes and other organisms transform the inorganic mercury into methyl mercury, a highly toxic form of mercury that bioaccumulates in fish and piscavores (i.e., animals that eat fish). Thus, the mercury increases in concentration as it moves up the food chain.

Human exposure to mercury occurs primarily through consumption of contaminated saltwater or freshwater fish; for example, large, predatory fish can be thousands of times higher in mercury than concentrations in the water. Low doses of mercury over time can damage the central and peripheral nervous systems. The greatest concerns are *in utero* (exposure of yet to be born children), and in babies and young children, whose nervous systems are developing. Highest exposure subpopulations are subsistence anglers and some Native Americans dependent upon fish and piscavores as a large part of their food supplies. Children of women exposed to relatively high levels of methyl mercury during pregnancy have exhibited a variety of abnormalities, including delayed onset of walking and talking, cerebral palsy, and reduced neurological test scores. Children exposed to far lower levels of methyl mercury in the womb have exhibited delays and deficits in learning ability. In addition, children exposed after birth potentially are more sensitive to the toxic effects of methyl mercury than adults. Thus, Minamata set the stage, but mercury emissions are now not only an engineering problem, they are a public health issue. An optimistic aspect of mercury is that it can now be detected at very low concentrations. For example, I recall working on a lake cleanup in St. Louis in the 1970s, when the health standard for mercury was below our level of detection, which was in the part per million (ppm) range. We can now not only detect mercury at several of orders of magnitude better (low parts per billion range), but we can quantify each chemical form, that is, its speciation.

Asbestos is addressed in both Chapter 3 in the discussion on particle matter and in Chapter 8 in a discussion on quantifying the unquantifiable, the value of a human life. However, several recent events have occurred that are changing the public perception of asbestos risks; most notably, epidemiological studies of asbestos exposure in vermiculite miners and other workers at the Libby, Montana, site, and recent findings of asbestos in soils where people live, work, and go to school. Vermiculite has been widely used as a soil additive and for insulation, as well as other uses. The mineral was discovered in Libby, Montana, in 1881 by gold miners, and then in 1919, Edward Alley identified its unique properties. In the 1920s, the Zonolite Company formed and began mining vermiculite. In 1963, W.R. Grace bought the Zonolite mining operations, which closed in 1990. According to the Agency for Toxic Substances and Disease Registry, current airborne asbestos levels in Libby now appear to be low, but levels were certainly much higher during the many decades that vermiculite was actively mined, processed, and shipped. In fact, air concentrations up to 15 times the current occupational limits were once reported for downtown Libby in the past. During its operation, the vermiculite mine in Libby may have produced 80% of the world's supply of vermiculite. Regrettably, the vermiculite from the Libby mine was contaminated with a toxic form of naturally occurring asbestos called tremolite-actinolite asbestiform mineral fibers. As discussed in Chapter 3, exposures to these fibers have been associated with asbestosis, lung cancer, and mesothelioma.

The case is also an example of the litigious nature of environmental protection that has grown out of the events of the 1970s. On February 7, 2005, a federal grand jury in the District of Montana indicted W.R. Grace and seven current and former Grace executives for knowingly endangering residents of Libby, Montana, and concealing information about the health effects of its asbestos mining operations. According to the indictment, W.R. Grace and its executives, as far back as the 1970s, attempted to conceal information about the adverse health effects of the company's vermiculite mining operations and distribution of vermiculite in the Libby, Montana, community. The defendants also are accused of obstructing the government's cleanup efforts and wire fraud. To date, according to the indictment, approximately 1,200 residents of the Libby area have been identified as suffering from some kind of asbestos-related abnormality. The case is pending as of this writing, and an indictment is merely an accusation, so the defendants are presumed innocent until proven guilty at trial beyond a reasonable doubt.

The comments of Ross McKinney, emeritus at the University of Kansas, sums up the challenge of environmental case analysis in general and this book specifically:

> No single event currently affects how the public perceives environmental pollution control. The biggest problem I see affecting how the public views environmental pollution control in the United States today comes from misinformation in the media; i.e. newspapers, radio, TV, magazines, and the Internet. The sources of the misinformation are various government agencies at the federal, state, and local levels, industries, professional and scientific organizations, consulting engineering firms, analytical firms,

> environmental organizations, universities, and John Q. Public. The problem is and always will be the drive for money and personal recognition. The lack of ethics throughout society is creating serious problems that could destroy our way of life.
>
> For the majority of Americans, environmental pollution control has been pushed to the back burner. It will stay on the back burner until there is a serious emergency that hits the media. The periodic stories about sewage spills caused by clogged sewers keeps the back burner lit but does not create any voice for action. The public does not trust government to handle the environmental pollution problem, but there is no alternative available to the public.

McKinney is currently collecting information of how America recognized and began to solve its wastewater disposal problem. It is the story of a rapidly growing problem and the few people who developed the solution to the problem in a series of stages. Once the solution for the first stage was clearly demonstrated, it was almost universally replicated. Next someone figured out how to move to the next level and demonstrated the improvement. Once again, everyone copied the solution. This continues to be the pattern. As government agencies became more involved, it grew more difficult to move to the next level. According to McKinney, "When the lawyers took control of the government agencies, chaos reigned supreme and progress ceased since the lawyers could not recognize progress or even the problem. Fortunately, the seriousness of the problem had dropped to a very low level." However, I would add, environmental protection is too important to entrust only to attorneys and the legal system. Engineers and scientists must ensure that sound science underlies every environmental decision.

These recommendations reflect the eclectic nature of environmental science and engineering. Each colleague who considers these cases to be paramount to the evolution of the environmental ethos has a unique perspective. Among their other comments is the need to include cases that demonstrate the positive paradigms, those that demonstrate the progress we have made. The air and water in most of the developed world has been steadily improving, especially for conventional pollutants. Environmental cleanup technologies are advancing rapidly. Engineering and science is improving risk assessment and management approaches. This progress has been significant and does need to be documented. I plan to get right on it!

The book ends by considering the reader's journey and where the next sojourn can be expected to take us. My predictions from past cases are based upon sound science (at least as sound as is currently available). There is a good chance I am wrong. In fact, I can guarantee that my predictions will be incorrect in kind or degree, and quite likely both. There is nothing like time and experience to humble a scientist. Explaining the past, as I do in this book, is easier than predicting the future. As they say in financial investments, "past performance does not guarantee future results."

Acknowledgments

Much has been written about environmental problems in the past 30 years. I have benefited from these writings. Unfortunately, having been in the practice of environmental protection for three decades, it is impossible for me to give proper acknowledgment and attribution to all of these sources. Their shared knowledge has become incorporated into my work, teaching, writing, and even my psyche, so I thank them all anonymously.

The students in my Professional Ethics course at Duke University (EGR 108S) have given me much insight into these cases. They have constructed storyboards, drawn up negative and positive paradigms, developed event trees, flowcharts, and line drawings, and conducted net goodness analyses for dozens of cases. Many of these tools have caused me to consider these cases in ways I would not have thought. This book benefits greatly from their enthusiasm and insights.

Philip Korn and Christine Minihane of Elsevier provided excellent insights and ideas on the publication aspects of book, and Ruby Nell Carpenter of Duke's Civil and Environmental Engineering Department was tireless in her administrative support. I especially want to thank P. Aarne Vesilind, professor of civil engineering at Bucknell University. Aarne has provided ideas, insights, critiques, and encouragement on many matters. I am particularly grateful that he has permitted me to draw from his ideas and our shared discussions and work on our current project related to environmental justice. The following cases have risen from my collaborations with Aarne, including some from Aarne's students at Bucknell:

- Poza Rica, Mexico
- The Kepone Tragedy
- Pesticides and Sterility
- Jersey City Chromium
- Drake Chemical Company Superfund Site
- Case of the Negative pH
- The Orange County, North Carolina, Landfill
- Carver Terrace
- West Dallas Lead Smelter

All information gathering for this book was conducted independently of my federal government employment and has not been subject to any governmental or administrative review. Therefore, the conclusions and opinions drawn are solely my own and should not be construed to reflect the views of any federal agency or department.

Notes and Commentary

1. T.S. Kuhn, 1962. *The Structure of Scientific Revolutions*, 2e, Enlarged, The University of Chicago Press, Chicago.

2. Although *a posteriori* knowledge is almost universally accepted in modern science, there was strong debate in the seventeenth century, with strong arguments for *a priori* knowledge in scientific inquiry. One of the best accounts of these debates, presented as a dialogue between Boyle and Hobbs, is found in S. Shapin and S. Schaffer, 1985, *Leviathan and the Air-Pump: Hobbes, Boyle, and the Experimental Life*, Princeton University Press, Princeton, NJ. Shapin and Schaffer attempt to "deal with the historical circumstances in which experiment as a systematic means of generating natural knowledge arose, in which experiment practices became institutionalized, and in which experimentally produced matters of fact were made into the foundations of what counted as proper scientific knowledge." To do this, they analyze Boyle's paradigm of experimental approach in using his air pump.
3. See, for example, D.A. Vallero, 2003. "Teachable Moments and the Tyranny of the Syllabus: September 11 Case," *Journal of Professional Issues in Engineering Education*, 129 (2), 100–105. The September 11, 2001, terrorist attacks on the World Trade Center and the Pentagon presented unique teachable moments to engineering educators but with the competing demand to complete the course as designed and as dictated by the tyranny of the syllabus. I found that for my students at Duke University and North Carolina Central University the percentage of courses addressing the events was highest in the Fall 2001 semester, when the attacks occurred, falling in the Spring 2002 semester, but increasing in Fall 2002. Most respondents supported the use of the events as teachable moments even if the syllabus and course outline had to be adjusted. I believe the results indicate that engineering education must be open to opportunities to teach physical science and engineering concepts and to introduce the students to the social sciences and humanities.
4. Another way to look at uncertainty is that it is a function of variability and ignorance. This has been well articulated by L. Ginzburg in his review of ecological case studies in U.S. Environmental Protection Agency, 1994, Peer Review Workshop Report on Ecological Risk Assessment Issue Papers, Report Number EPA/630/R-94/008. According to Ginzburg, "variability includes stochasticity arising from temporal and spatial heterogeneity in environmental factors and among exposed individuals. Ignorance includes measurement error, indecision about the form of the mathematical model or appropriate level of abstraction." Thus, variability can be lessoned by increased attention, e.g., empirical evidence, and "translated into risk (i.e., probability) by the application of a probabilistic model," but ignorance cannot. Ignorance simply translates into confidence intervals, or "error bounds" on any statement of risk.
5. See, for example, Physical Principles of Unworkable Devices: http://www.lhup.edu/~dsimanek/museum/physgal.htm. Donald E. Simanek's humorous but informative site on why perpetual motion machines cannot work; their inventors assumed erroneous "principles." This site is instructive to environmental decision makers to beware of junk science. Sometimes a good way to learn why something works the way it does is to consider all the reasons that it fails to work.

6. My apologies to the originator of this analogy, who deserves much credit for this teaching device. The target is a widely used way to describe precision and accuracy.
7. For a different, even contrary view, see http://www.brown.edu/Administration/George_Street_Journal/value.html. Richard Morin gives a thoughtful outline of Allen Feldman's model and critique of the "willingness to pay" argument (very commonly used in valuation).
8. R. Popkin, 1991. "Responding to Eco-Terrorism," *EPA Journal*, July/August.
9. The sources for the Iron Gates discussion are Global Environmental Facility, 2005. *Project Brief/Danube Regional Project—Phase 1:* ANNEX 11 Causes and Effects of Eutrophication in the Black Sea; http://www.gefweb.org/Documents/Council_Documents/GEF_C17/Regional_Danube_Annex_II_Part_2.pdf; accessed April 27, 2005. C. Lancelot, J. Staneva, D. Van Eeckhout, J.-M. Beckers, and E. Stanev, 2002. "Modelling the Danube-influenced North-western Continental Shelf of the Black Sea. II: Ecosystem Response to Changes in Nutrient Delivery by the Danube River after its Damming in 1972," *Estuarine, Coastal and Shelf Science*, 54: 473–499.
10. N. Myers, 1989. Deforestation Rates in Tropical Forests and Their Climatic Implications. Friends of the Earth. Myers suggest that 142,200 km^2 per year are lost to deforestation alone, with an additional loss estimated due to forest degradation. This amount was updated by Myers in a 1994 letter to the Rainforest Action Network, accounting for a 2% annual increase (not compounded). Thus, Myers' mid-1994 figure was 155,000 km^2 per year for deforestation with expected overall global rainforest destruction remaining more or less double the rate of deforestation.
11. E.O. Wilson, 1992. *The Diversity of Life*, Harvard University Press, Cambridge, MA.
12. The botanical information source is Southeast Exotic Pest Plant Council, http://www.se-eppc.org/manual/multirose.html.

Part I

New Science and New Paradigms

In a span of just a few decades, advances and new environmental applications of science, engineering, and their associated technologies have coalesced into a whole new way to see the world. Science is the explanation of the physical world, whereas engineering encompasses applications of science to achieve results. Thus, what we have learned about the environment by trial and error has incrementally grown into what is now standard practice of environmental science and engineering. This heuristically attained knowledge has come at a great cost in terms of the loss of lives and diseases associated with mistakes, poor decisions (at least in retrospect), and the lack of appreciation of environmental effects.

It is the right time to consider those events that have affected the state of environmental science and engineering. Environmental awareness is certainly more "mainstream," and less a polarizing issue than it was in the 1970s and 1980s. There has been a steady march of advances in environmental science and engineering for several decades, as evidenced by the increasing number of Ph.D. dissertations and credible scientific journal articles addressing a myriad of environmental issues. Corporations and government agencies, even those whose missions are not considered to be "environmental," have established environmental programs.

Old Paradigm:
  Pollution is best controlled by rigidly enforced standards.
Paradigm Shift:
  Green approaches can achieve environmental results beyond command and control.

Recently companies and agencies have been looking beyond ways to treat pollution to find better processes to prevent the pollution in the first place. In fact, the adjective "green" has been showing up in front of many disciplines—for example, green chemistry and green engineering—as has the adjective "sustainable." These approaches are being linked to improved computational abilities (see Table I.1) and other tools that were not available at the outset of the environmental movement. Increasingly, companies have come to recognize that improved efficiencies save time, money, and other resources in the long run. Hence, companies are thinking systematically about the entire product stream in numerous ways:

- Applying sustainable development concepts, including the framework and foundations of "green" design and engineering models
- Applying the design process within the context of a sustainable framework, including considerations of commercial and institutional influences
- Considering practical problems and solutions from a comprehensive standpoint to achieve sustainable products and processes
- Characterizing waste streams resulting from designs
- Understanding how first principles of science, including thermodynamics, must be integral to sustainable designs in terms of mass and energy relationships, including reactors, heat exchangers, and separation processes
- Applying creativity and originality in group product and building design projects

New systematic approaches, like almost everything else in environmental protection, call for new acronyms. These include Design for the Environment (DFE), Design for Disassembly (DFD), and Design for Recycling (DFR).[i]

For example, the concept of a cap-and-trade has been tested and works well for some pollutants. This is a system where companies are allowed to place a "bubble" over a whole manufacturing complex or trade pollution credits with other companies in their industry instead of a stack-by-stack and pipe-by-pipe approach; that is, the so-called command-and-control approach. Such policy and regulatory innovations call for some improved technology-based approaches as well as better quality-based approaches, such as leveling out the pollutant loadings and using less expensive technologies to remove the first large bulk of pollutants, followed by higher operation and maintenance (O&M) technologies for the more difficult-to-treat stacks and pipes. But, the net effect can be a greater reduction of pollutant emissions and effluents than treating each stack or pipe as an independent entity. This is a foundation for most sustainable design approaches; that is, conducting a life-cycle analysis, prioritizing the most important problems, and matching the technologies and operations to address them. The problems will vary by size (e.g., pollutant loading), difficulty in treating, and feasibility. The easiest ones are the big ones that are easy to treat (so-called "low hanging fruit"). You can do these first with

TABLE I.1
Principles of green programs.

Principle	*Description*	*Example*	*Role of Computational Toxicology*
Waste prevention	Design chemical syntheses and select processes to prevent waste, leaving no waste to treat or clean up.	Use a water-based process instead of an organic solvent-based process.	Informatics and data mining can provide candidate syntheses and processes.
Safe design	Design products to be fully effective, yet have little or no toxicity.	Use microstructures, instead of toxic pigments, to give color to products. Microstructures bend, reflect, and absorb light in ways that allow for a full range of colors.	Systems biology and "omics" technologies can support predictions of cumulative risk from products used in various scenarios.
Low hazard chemical synthesis	Design syntheses to use and generate substances with little or no toxicity to humans and the environment.	Select chemical synthesis with toxicity of the reagents in mind up front. If a reagent ordinarily required in the synthesis is acutely or chronically toxic, find another reagent or new reaction with less toxic reagents.	Computational chemistry can help predict unintended product formation and reaction rates of optional reactions.
Renewable material use	Use raw materials and feedstocks that are renewable rather than those that deplete nonrenewable natural	Construction materials can be from renewable and depleting sources. Linoleum flooring, for example, is highly durable,	Systems biology, informatics, and "omics" technologies can provide insights into the possible chemical reactions and toxicity

TABLE I.1
Continued

Principle	*Description*	*Example*	*Role of Computational Toxicology*
	resources. Renewable feedstocks are often made from agricultural products or are the wastes of other processes; depleting feedstocks are made from fossil fuels (petroleum, natural gas, or coal) that must be extracted by mining.	can be maintained with nontoxic cleaning products, and is manufactured from renewable resources amenable to being recycled. Upon demolition or reflooring, the linoleum can be composted.	of the compounds produced when switching from depleting to renewable materials.
Catalysis	Minimize waste by using catalytic reactions. Catalysts are used in small amounts and can carry out a single reaction many times. They are preferable to stoichiometric reagents, which are used in excess and work only once.	The Brookhaven National Laboratory recently reported that it has found a "green catalyst" that works by removing one stage of the reaction, eliminating the need to use solvents in the process by which many organic compounds are synthesized. The catalyst dissolves into the reactants. Also, the catalyst has the unique ability of being easily removed and recycled because, at the end of the reaction, the catalyst precipitates out of products as	Computation chemistry can help to compare rates of chemical reactions using various catalysts.

		a solid material, allowing it to be separated from the products without using additional chemical solvents.[1]	
Avoiding chemical derivatives	Avoid using blocking or protecting groups or any temporary modifications if possible. Derivatives use additional reagents and generate waste.	Derivativization is a common analytical method in environmental chemistry; i.e., forming new compounds that can be detected by chromatography. However, chemists must be aware of possible toxic compounds formed, including leftover reagents that are inherently dangerous.	Computational methods and natural products chemistry can help scientists start with a better synthetic framework.
Atom economy	Design syntheses so that the final product contains the maximum proportion of the starting materials. There should be few, if any, wasted atoms.	Single atomic and molecular scale logic used to develop electronic devices that incorporate design for disassembly, design for recycling, and design for safe and environmentally optimized use.	The same amount of value, e.g., information storage and application, is available on a much smaller scale. Thus, devices are smarter and smaller, and more economical in the long-term. Computational toxicology enhances the ability to make product decisions with better predictions of possible adverse effects, based on the logic.

TABLE I.1
Continued

Principle	*Description*	*Example*	*Role of Computational Toxicology*
Nano-materials	Tailor-made materials and processes for specific designs and intent at the nanometer scale (≤100 nm).	Emissions, effluent, and other environmental controls; design for extremely long life cycles. Limits and provides better control of production and avoids over-production (i.e., "throwaway economy").	Improved, systematic catalysis in emission reductions; e.g., large sources like power plants and small sources like automobile exhaust systems. Zeolite and other sorbing materials used in hazardous waste and emergency response situations can be better designed by taking advantage of surface effects; this decreases the volume of material used.
Selection of safer solvents and reaction conditions	Avoid using solvents, separation agents, or other auxiliary chemicals. If these chemicals are necessary, use innocuous chemicals.	Supercritical chemistry and physics, especially that of carbon dioxide and other safer alternatives to halogenated solvents, are finding their way into the more mainstream processes, most notably dry cleaning.	To date, most of the progress has been the result of wet chemistry and bench research. Computational methods will streamline the process, including quicker "scale-up."
Improved energy efficiencies	Run chemical reactions and other processes at ambient temperature and pressure whenever possible.	To date, chemical engineering and other reactor-based systems have relied on "cheap" fuels and, thus, have optimized on the	Heat will always be important in reactions, but computational methods can help with relative economies of scale.

		basis of thermodynamics. Other factors, e.g., pressure, catalysis, photovoltaics, and fusion, also should be emphasized in reactor optimization protocols.	Computational models can test feasibility of new energy-efficient systems, including intrinsic and extrinsic hazards, e.g., to test certain scale-ups of hydrogen and other economies. Energy behaviors are scale-dependent. For example, recent measurements of H_2SO_4 bubbles when reacting with water have temperatures in a range of those found on the surface of the sun.[2]
Design for degradation	Design chemical products to break down to innocuous substances after use so that they do not accumulate in the environment.	Biopolymers, e.g., starch-based polymers, can replace styrene and other halogen-based polymers in many uses. Geopolymers, e.g., silane-based polymers, can provide inorganic alternatives to organic polymers in pigments, paints, etc. These substances, when returned to the environment, become their original parent form.	Computation approaches can simulate the degradation of substances as they enter various components of the environment. Computational science can be used to calculate the interplanar spaces within the polymer framework. This will help to predict persistence and to build environmentally friendly products, e.g., those where space is adequate for microbes to fit and biodegrade the substances.

TABLE I.1
Continued

Principle	*Description*	*Example*	*Role of Computational Toxicology*
Real-time analysis to prevent pollution	Include in-process real-time monitoring and control during syntheses to minimize or eliminate the formation of byproducts.	Remote sensing and satellite techniques can be linked to real-time data repositories to determine problems. The application to terrorism using nano-scale sensors is promising.	Real-time environmental mass spectrometry can be used to analyze whole products, obviating the need for any further sample preparation and analytical steps. Transgenic species, though controversial, can also serve as biological sentries, e.g., fish that change colors in the presence of toxic substances.
Accident prevention	Design processes using chemicals and their forms (solid, liquid, or gas) to minimize the potential for chemical accidents including explosions, fires, and releases to the environment.	Scenarios that increase the probability of accidents can be tested.	Rather than waiting for an accident to occur and conducting failure analyses, computational methods can be applied in prospective and predictive mode; that is, the conditions conducive to an accident can be characterized computationally.

[1] U.S. Department of Energy, *Research News*, http://www.eurekalert.org/features/doe/2004–05/dnl-brc050604.php. Accessed March 22, 2005.
[2] D.J. Flannigan and K.S. Suslick, 2005. "Plasma formation and temperature measurement during single-bubble cavitation," *Nature* 434: 52–55.
Source: First two columns, except "Nano-materials" adapted from U.S. Environmental Protection Agency, 2005, "Green Chemistry": http://www.epa.gov/greenchemistry/principles.html; accessed April 12, 2005. Other information from discussions with Michael Hays, U.S. EPA, National Risk Management Research Laboratory, April 28, 2005.

immediate gratification! However, the most intractable problems are often those that are small but very expensive and difficult to treat, that is, less feasible. Thus, the environmental science requires that expectations be managed from both a technical and an operational perspective, including the expectations of the client, the government, and oneself.

Looking at key incidents and milestones can remind us of important principles so that we do not repeat mistakes unnecessarily. The retrospective view also gives us information on what may yet occur in the future. Like many other trends of the late twentieth and early twenty-first century, many people have a top-ten list of the most crucial events that have shaped the environmental agenda. There is no consensus on which events should be on such lists. For example, the Internet encyclopedia, *Wikipedia*,[ii] chronologically lists the most important environmental disasters as:

1. *Torrey Canyon* tanker oil spill in the English Channel (March 18, 1967)
2. Love Canal hazardous waste site, Niagara Falls, New York (discovered in the 1970s)
3. Seveso, Italy, explosion disaster, release of dioxin (July 10, 1976)
4. Bhopal, India, methylisocyanate explosion and toxic cloud (December 3, 1984)
5. *Exxon Valdez* tanker oil spill, Prince William Sound, Alaska (March 24, 1989)
6. *Prestige* tanker oil spill, off the Spanish coast (November 13, 2002)

It would be difficult to argue against any of these disasters as being important, but they certainly do not represent all those that have had profound impacts on environmental science, engineering, policy, and regulation. For example, important nuclear events also have been extremely influential in our perception of pollution and threats to public health. Most notably, the cases of Three Mile Island, in Dauphin County, Pennsylvania (March 28, 1979), and the Chernobyl nuclear power-plant disaster in the Ukraine (April 26, 1986) have had an unquestionable impact on not only nuclear power, but aspects of environmental policy, such as community right-to-know and the importance of risk assessment, management, and communication.

Numerous defense and war-related incidents also have had a major influence on the public's perception of environmental safety. For example, the atomic bombings of Hiroshima and Nagasaki (August 6 and August 9, 1945, respectively) were the world's first entrees to the linkage of chronic illness and mortality (e.g., leukemia and radiation disease) that could be linked directly to radiation exposure. Similarly, the use of the defoliant Agent Orange during the Vietnam War (used between 1961 and 1970) has made us aware of the importance of the *latency period*, where possible effects may not be manifested until years or decades after pesticide exposure. The Agent Orange problem also illustrates the problem of uncertainty in characterizing and enumerating effects. There is no consensus on whether the symptoms and disorders suggested to be linked to Agent Orange are sufficiently strong and well documented—that is, provide weight of evidence—to support cause and effect.

Other important industrial accidents and events must also be added to our list, such as the mercury releases to Minamata Bay in Japan, the effect of cadmium exposure that led to Itai-Itai disease in many Japanese, and air pollution episodes in Europe and the United States. Also, new products that at first appear to be beneficial have all too often been found to be detrimental to public health and the environment.

There is little agreement on the criteria for ranking. For example, death toll and disease (e.g., cancer, asthma, or waterborne pathogenic disease) are often key criteria. Also, the larger the affected, area, the worse the disaster, such as the extent of an oil slick or the size of a toxic plume in the atmosphere. Even monetary and other values are used as benchmarks. Sometimes, however, timing may be the most important criterion. Even if an event does not lead to an extremely large number of deaths or diseases, or its spatial extent is not appreciably big, it may still be very important because of where and when the event occurs. For example, the contamination of Times Beach, Missouri, although affecting much of the town, was not the key reason for the national attention. The event occurred shortly after the Love Canal hazardous waste problem was identified, and people were wondering just how extensively dioxin and other persistent organic compounds were going to be found in the environment. Times Beach also occurred at a time when scientists and engineers were beginning to get a handle on how to measure and even how to treat (i.e., by incineration) contaminated soil and water. Other events also seem to have received greater attention due to their timing, such as the worries about DDT and its effect on eagles and other wildlife, cryptosporidium outbreaks, and Legionnaire's Disease.

Some environmental incidents are not well defined temporarily, but are important because of the pollutants themselves. We would be hard pressed to identify a single event that caused the public concern about the metal lead. In fact, numerous incremental steps brought the world to appreciate lead toxicity and risk. For example, studies following lead reductions in gasoline and paint showed marked improvements in blood lead levels in many children. Meanwhile, scientific and medical research was linking lead to numerous neurotoxic effects in the peripheral and central nervous systems, especially of children. Similar, stepwise progressions of knowledge of environmental risk occurred for polychlorinated biphenyls (PCBs), numerous organochlorine, organophosphate, and other pesticides, depletion of the stratospheric ozone layer by halogenated (especially chlorinated) compounds, and even the effect of releases of carbon dioxide, methane, and other "greenhouse gases" on global warming (more properly called global climate change).

Thus, this book uses all these approaches to describe and to analyze different types of events that have one thing in common—they have had a profound impact on the new environmental paradigm. Some cases are on everyone's top-ten lists, others are a bit more obscure. Some may not be considered to be cases at all, but better defined as issues. No matter, they are considered even if they do not fit well into prototypical case categories, so long as they provide lessons and help to advance the science, engineering, and management of environmental risks.

Notes and Commentary

i. See S.B. Billatos, 1997. *Green Technology and Design for the Environment*, Taylor & Francis, Washington, D.C.

Also see V. Allada, 2000. "Preparing Engineering Students to Meet the Ecological Challenges through Sustainable Product Design," *Proceedings of the 2000 International Conference on Engineering Education*, Taipei, Taiwan.

ii. See http://en.wikipedia.org/wiki/List_of_disasters#Environmental_disasters; accessed February 26, 2005.

CHAPTER 1

Lessons Learned: A Case Approach to Environmental Problems

Progress, far from consisting in change, depends on retentiveness. . . . Those who cannot remember the past are condemned to repeat it.

George Santayana, 1905, *The Life of Reason*, Volume 1

Santayana's quotation is often repeated because it is advice that makes so much sense, but is too often ignored. What we remember can save us in the long run. We forget important events at our own peril. It is one thing to fail but quite another not to learn from our failures. We must consider the reasons and events that led to the failure in hopes that corrective actions and preventive measures are put in place to avoid their reoccurrence. This is not easy and is almost always complicated. Every disaster or failure has a unique set of events. Often, seemingly identical situations lead to very different conclusions. In fact, the mathematics and statistics of failure analysis are some of the most complicated, relying on nonlinear and chaotic approaches and nontraditional statistical methods, such as Bayesian theory.[1] Having said this, identifying these challenges certainly is not meant to imply that we cannot apply the lessons learned from environmental disasters to ongoing decisions. We can and must, and certainly will throughout this book.

The reasons for failure vary widely. All of the three types highlighted in this book's subtitle, "mistakes, mishaps, and misdeeds," have caused environmental problems, but in very different ways. The terms all include the prefix mis-, which is derived from Old English, "to miss." This type of failure applies to numerous environmental problems and disasters. However, the prefix mis- can connote something that is done poorly; that is, a mistake. It may also mean that an act leads to an accident because the original expectations were overtaken by events; that is, a mishap. This is an all too common shortcoming of professionals; that is, not upholding the levels of technical competence called for by their field. Medical and engineering codes of ethics, for example, include tenets and principles related to competence, such as only working in one's area of competence or specialty. Finally, mis- can suggest that an act is immoral or ethically impermissible; that is, a misdeed. Interestingly, the theological derivation for the word *sin* (Greek: *hamartano*) means that when a person has missed the mark—the goal of moral goodness and ethical uprightness—that person has sinned or has behaved immorally by failing to abide by an ethical principle, such as honesty and justice. Environmental failures have come about by all three means. The lesson from Santayana is that we must learn from all of these past failures. Learning must be followed by new thinking and action, including the need to forsake what has not worked and shift toward what needs to be done. Throughout this book, we will reconsider what was at one time consensus of thought. We will also reconsider some consensuses of current thinking. Our first paradigm has to do with how society has viewed the environment; our *zeitgeist*, if you will.

Old Paradigm:
The environment is nearly infinite in its capacity to withstand human waste.
Paradigm Shift:
Environmental resources have very finite limits on elasticity, with some resources being extremely sensitive to very small changes.

The title of this book may sound dire or even pessimistic. It may be the former, but hopefully not the latter. The "environmental movement" is a relatively young one. The emblematic works of Rachel Carson, Barry Commoner, and others in the 1960s were seen by many as mere straws in the wind. The growing environmental awareness was certainly not limited to the academic and scientific communities. Popular culture was also coming to appreciate the concept of "spaceship earth," i.e., that our planet consisted of a finite life support system and that our air, water, food, soil, and ecosystems were not infinitely elastic in their ability to absorb humanity's willful disregard. The poetry and music of the time expressed these fears and called for a new respect for the environment. The environmental movement was not a unique enterprise, but was interwoven into growing protests about the war in Vietnam, civil rights, and a general discomfort with the "establishment." The petrochemical industry, the military, and capitalism were coming under increased scrutiny and skepticism. Following the tumultuous 1960s, the musical group Quicksilver Messenger Service summed up this malaise and dissatisfaction with unbridled commercialism and the seeming disregard for the environment in their 1970 song *What about Me*. The song laments that the earth's "sweet water" has been poisoned, its forests clear cut, and its air is "not good to breathe." The songwriters also extend Rachel Carson's fears that the food supply is being contaminated, linking diseases to food consumption (i.e., ". . . the food you fed my children was the cause of their disease").

These sentiments took hold and became less polarized (and eventually politically bipartisan for the most part) and grew to be an accepted part of contemporary culture. For example, the mind-set of *What about Me* is quite similar to that of the words of the 1982 song *Industrial Disease*, written by Mark Knopfler of the band Dire Straits, but with the added health concerns and fears from chemical spills, radioactive leaks, and toxic clouds produced by a growing litany of industrial accidents.

In poetic terms and lyrical form, Knopfler is characterizing the growing appreciation of occupational hazards and the perils of whistle blowing, e.g., the cognitive dissidence brought on by people torn between keeping their jobs and complaining about an unhealthy workplace (". . . Somebody blew the whistle and the walls came down . . ."). His words also appear to present a hypothesis about the connection between contaminant releases (known and unknown) and the onset of adverse effects in human populations (i.e., ". . . Some come out in sympathy, some come out in spots; Some blame the management, some the employees . . .").

Such a connection is now evident, but in the early 1980s, the concept of risk-based environmental decision making was still open to debate. These concerns were the outgrowth of media attention given to environmental disasters, such as those in Seveso, Italy, and Love Canal, New York (for example, could Knopfler's "some come out in spots" be a reference to the chloracne caused by dioxin exposure at Seveso and Times Beach, Missouri?); and the near disaster at the Three Mile Island nuclear power plant in Pennsylvania. But Knopfler's lyrics are particularly poignant, prescient, and portentous in light of the fact that he penned these words years before the most infamous accidents at Bhopal, India, and Chernobyl, Ukraine; both causing death, disease, and misery still apparent decades after the actual incidents ("Sociologists invent words that mean *industrial disease*").

The momentum of the petrochemical revolution following the Second World War was seemingly inviolable. However, much of the progress we now take as given was the result of those who agitated against the *status quo* and refused to accept the paradigms of their time. In fact, several of the cases in this book provided evidence of the validity of these early environmentalists' causes. A handful of cases were defining moments in the progress in protecting public health and the environment. It seems that every major piece of environmental legislation was preceded by an environmental disaster precipitated from mistakes, mishaps, and misdeeds. Amendments to the Clean Air Act resulted from the episodes at Donora and London. Hazardous waste legislation came about after public outcries concerning Love Canal. "Right-to-Know" legislation worldwide grew from the Bhopal disaster. Oil spill and waste contingency plans were strengthened following the *Exxon Valdez* spill. International energy policies changed, with growing anti-nuclear power sentiments, following the near disaster at Three Mile Island and the actual catastrophe at Chernobyl. Most recently, engineering and public health emergency response planning has been completely revamped in response to the events of September 11, 2001.

Certainly these can all be classified as "environmental" problems, but they represent new, societal paradigms as well. Contemporary society has a way of thrusting problems upon us. Ironically, society simultaneously demands the promotion of emerging technologies and the control of the consequences, sometimes by the very same technologies of concern. For example, advances in radioisotope technology are part of the arsenal to treat cancer, but radioactive wastes from hospitals can increase the risk of contracting cancer if these wastes are not properly disposed of and handled safely. Likewise, cleanup of polluted waters and sediments can benefit from combustion and incineration to break down some very persistent contaminants, but combustion in general is problematic in its release of products of complete combustion (carbon dioxide) or incomplete combustion (e.g., dioxins, furans, polycyclic aromatic hydrocarbons, and carbon monoxide). In almost every case in this book and elsewhere, the environmental problems have emerged as a byproduct of some useful, high-demand enterprise.

In his recent book, *Catastrophe: Risk and Response*, Richard Posner, a judge of the U.S. Court of Appeals for the Second Circuit, describes this dichotomy succinctly when he says that "modern science and technology have enormous potential for harm" yet are "bounteous sources of social benefits." Posner is particularly interested in how technology can prevent natural and anthropogenic calamities, "including the man-made catastrophes that technology itself enables or exacerbates."[2]

Posner gives the example of the looming threat of global climate change, caused in part by technological and industrial progress (mainly the internal combustion engine and energy production tied to fossil fuels). Emergent technologies can help to assuage these problems by using alternative sources of energy, such as wind and solar, to reduce global demand for fossil fuels. We will discuss other pending problems, such as the unknown territory of genetic engineering, like genetically modified organisms (GMOs) used to produce food. There is both a fear that the new organisms will carry with them unforeseen ruin, such as in some way affecting living cells' natural regulatory systems. An extreme viewpoint, as articulated by the renowned physicist Martin Rees, is the growing apprehension about nanotechnology, particularly its current trend toward producing "nanomachines." Biological systems, at the subcellular and molecular levels, could very efficiently produce proteins, as they already do for their own purposes. By tweaking some genetic material at a scale of a few angstroms, parts of the cell (e.g., the ribosome) that manufacture molecules could start producing myriad molecules designed by scientists, such as pharmaceuticals and nanoprocessors for computing. However, Rees is concerned that such assemblers could start self-replicating (like they always have), but without any "shut-off." Some have called this the "gray goo" scenario, i.e., accidentally creating an "extinction technology" from the cell's unchecked ability to exponentially replicate itself if part of their design is to be completely "omnivorous," using all matter as food! No other "life" on earth would exist if this "doomsday" scenario were to occur.[3] Certainly, this is the stuff of science fiction, but it calls attention to the need for vigilance, especially since our track record for becoming aware of the dangers of technologies is so frequently tardy. In environmental situations, messing with genetic materials may harm biodiversity, i.e., the delicate balance among species, including trophic states (producer-consumer-decomposer) and predator-prey relationships. Engineers and scientists are expected to push the envelopes of knowledge. We are rewarded for our eagerness and boldness. The Nobel Prize, for example, is not given to the chemist or physicist who has aptly calculated important scientific phenomena, with no new paradigms. It would be rare indeed for engineering societies to bestow awards only to the engineer who for an entire career used only proven technologies to design and build structures. This begins with our general approach to contemporary scientific research. We are rugged individualists in a quest to add new knowledge. For example, aspirants seeking Ph.D.s must endeavor to add knowledge to their specific scientific discipline. Scientific journals are unlikely to publish articles that do not at least contain some modicum of originality and newly found information.[4] We award and reward innovation. Unfortunately, there is not a lot of natural incentive for the innovators to stop what they are doing to "think about" possible ethical dilemmas propagated by their discoveries.[5]

Products that contain dangerous materials like asbestos, lead, mercury, polybrominated compounds, and polychlorinated biphenyls (PCBs) were once considered acceptable and were even required by law or policy to protect the public safety and health, such as asbestos-containing and polybrominated materials to prevent fires, DDT and other persistent pesticides to kill mosquitoes in an effort to prevent disease, and methyl tert-butyl ether (MTBE) as a fuel additive to prevent air pollution (see the discussion box, *MTBE and Cross-Media Transfer*). Subsequently, these products all were found to cause adverse environmental and health problems, although there is still much disagreement within the scientific community about the extent and severity of these and other contaminants. We must also consider the cases that are yet to be resolved and those where there is incomplete or nonexistent unanimity of thought as to their importance or even whether indeed they are problems, such as global climate change, acid rain, and depletion of the stratospheric ozone layer.

MTBE and Cross-Media Transfer

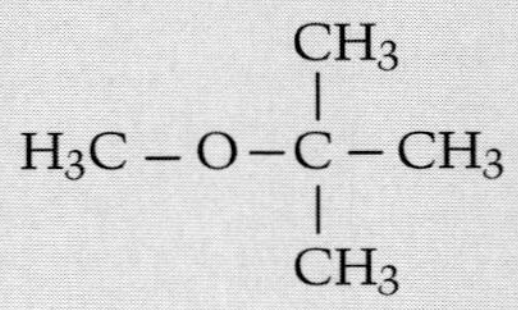

methyl *tert*-butyl ether (MTBE)

Automobiles generally rely on the internal combustion engine to supply power to the wheels.[6] Gasoline is the principal fuel source for most cars. The exhaust from automobiles is a large source of air pollution, especially in densely populated urban areas. To improve fuel efficiency and to provide a higher octane rating (for anti-knocking), most gasoline formulations have relied on additives. Up to relatively recently, the most common fuel additive to gasoline was tetraethyl-lead. But with the growing awareness of lead's neurotoxicity and other health effects, tetraethyl-lead has been banned in most parts of the world, so suitable substitutes were needed.

Methyl tertiary-butyl ether (MTBE) was one of the first replacement additives, first used to replace the lead additives in 1979. It is manufactured by reacting methanol and isobutylene and has been produced in very large quantities (more than 200,000 barrels per day in the United States in 1999). MTBE is a member of the chemical class of oxygenates. MTBE is quite volatile (vapor pressure = 27 kilopascals at 20°C), so that it is likely to evaporate readily. It also readily dissolves in water (aqueous solubility at 20°C = 42 grams per liter) and is very flammable (flash point = –30°C). In 1992, MTBE began to be used at higher concentrations in some gasoline to fulfill the oxygenate requirements set by the 1990 Clean Air Act Amendments. In addition, some cities, notably Denver, used MTBE at higher concentrations during the wintertime in the late 1980s.

The Clean Air Act called for greater use of oxygenates in an attempt help to reduce the emissions of carbon monoxide (CO), one of the most important air pollutants. Carbon monoxide toxicity results by interfering with the protein hemoglobin's ability to carry oxygen. Hemoglobin absorbs CO about 200 times faster than its absorption rate for oxygen. The CO-carrying protein is known as carboxyhemoglobin and when sufficiently high it can lead to acute and chronic effects. This is why smoking cigarettes leads to cardiovascular problems; the body has to work much harder because the normal concentration of oxygen in hemoglobin has been displaced by CO.

Carbon monoxide is also a contributing factor in the photochemistry that leads to elevated levels of ozone (O_3) in the troposphere. In addition, oxygenates decrease the emissions of volatile organic compounds (VOCs), which along with oxides of nitrogen are major precursors to the formation of tropospheric O_3. This is one of the most important roles of oxygenates, since unburned hydrocarbons can largely be emitted before catalytic converters start to work.

Looking at it from one perspective, the use of MTBE was a success by providing oxygen and helping gasoline burn more completely, resulting in less harmful exhaust from motor vehicles. The oxygen also dilutes or displaces compounds such as benzene and its derivatives (e.g., toluene, ethylbenzene, and xylene), as well as sulfur. The oxygen in the MTBE molecule also enhances combustion (recall that combustion is oxidation in the presence of heat). MTBE was not the only oxygenate, but it has very attractive blending characteristics and is relatively cheap compared to other available compounds. Another widely used oxygenate is ethanol.

The problem with MTBE is its suspected links to certain health effects, including cancer in some animal studies. In addition, MTBE has subsequently been found to pollute water, especially groundwater in aquifers. Some of the pollution comes from unburned MTBE emitted from tailpipes, some from fueling, but a large source is underground storage tanks at gasoline stations or other fuel operations (see Figure 1.1). A number of these tanks have leaked into the surrounding soil and unconsolidated media and have allowed the MTBE to migrate into the groundwater. Since it has such a high aqueous solubility, the MTBE is easily dissolved in the water.

When a pollutant moves from one environmental compartment (e.g., air) to another (e.g., water) as it has for MTBE, this is known as cross-media transfer. The problem has not really been eliminated, just relocated. It is also an example of a risk trade-off. The risks posed by the air pollution have been traded by the new risks from exposure to MTBE-contaminated waters.

The Incremental Effect

Sometimes, it is not the largest event that, in the long run, has the most profound effect on society, affects the health of the most people, changes environmental conditions, or has the most extensive damage in time and space. It is also often not even the most highly publicized cases that have had the most profound impact on environmental awareness and public policy. The incremental effects of a number of small and not very

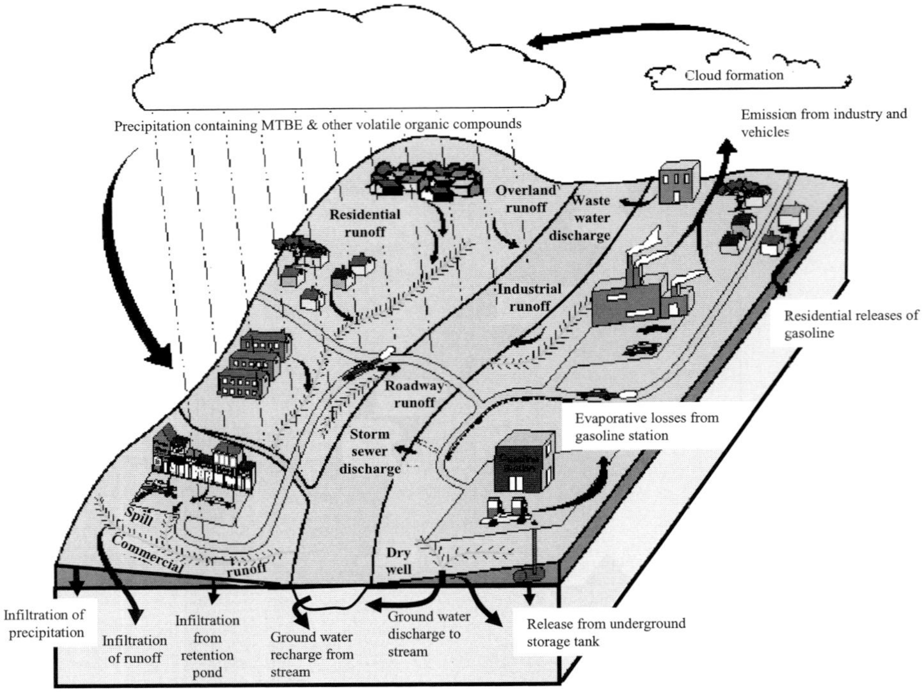

FIGURE 1.1. Migration of MTBE in the environment.
Source: G.C. Delzer, J.S. Zogorski, T.J. Lopes, and R.L. Bosshart, 1996. U.S. Geological Survey, "Occurrence of the Gasoline Oxygenate MTBE and BTEX in Urban Stormwater in the United States, 1991–95," Water Resources Investigation Report 96-4145, Washington, D.C.

noticeable cases in their own right have changed the landscape of environmental awareness. For example, federal projects, such as dams and highways, have caused incremental but dramatic changes to the environment. Growing levels of concern about escalating detrimental environmental effects eventually led to the passage of the National Environmental Policy Act and its creation of environmental impact statements (EISs) for every major federal action that could affect the quality of the environment. Also, social awareness has begun to coalesce recently with environmental movements. For example, the Warren County, North Carolina, landfill that was accepting PCBs became emblematic of what at the time was called environmental racism. Many similar cases, including some mentioned in this book, led to what is now the environmental justice movement. Sometimes, the state of the science and adequate information was insufficient to recognize large environmental and public health patterns until very recently, such as the epidemiological evidence showing lung cancer and mesothelioma in people exposed to asbestos, or the measurement of toxic compounds, like PCBs and pesticides, in people's fatty tissues. Therefore, this book discusses such incrementally important events and cases along with the more high-profile infamous ones.

The environmental revolution occurred simultaneously with the technological revolution, the petrochemical revolution, the nuclear revolution, and further entrenchment of energy, transportation, land use, and military demands of the twentieth century. This is not happenstance. Environmental awareness on the scale we know today came about as the direct result of the assaults and pressures coming from these other societal factions. Environmental problems can occur from combinations of benign and malignant trends in the marketplace and public policy initiatives, such as the noble efforts to bring electricity to all homes, rural and urban, which unfortunately increased the demand for reliable insulating and dielectric fluids, the PCBs. Since transformers and other electricity distribution equipment became ubiquitous across the twentieth century landscape, so were PCBs. When we became aware of the potential health and environmental effects from leaking PCBs, it became painfully obvious that there was a growing need to identify locations where these and other toxic substances could be stored, disposed, and destroyed. We also realized that new technologies beyond the state-of-the-science of the time would be needed. Thus, the public and private sectors had to reduce potential exposures to PCBs and their ilk. But, the scientific community was issuing scary information about the persistence, bioaccumulation, and toxicity of PCBs at the same time we were calling for more and better engineered landfills and PCB incinerators. This interaction confused and frightened the public so much that almost any recommendation dealing with PCBs met with disdain nearly everywhere they were (and are) proposed. When a locale was suggested for such technologies, the outrage and outcry were immediate. Thus another revolution was born: NIMBY, or not in my backyard.

Failure and Blame

Failure is also not completely attributed to science and engineering mistakes, mishaps, and misdeeds. Sometimes these misses are in the assessment and management of the risks brought on by environmental insults. Quite commonly, to paraphrase *Cool Hand Luke,* "we have a failure to communicate." Failed communication may not be the only or even the principal failure in many of these cases, but in almost every case considered in this book events were worsened and protracted because of poor risk communications, whether intentional (misdeed) or unintentional (mistakes). The good news is that environmental professionals are becoming more skillful in communications, bolstered by courses in engineering curricula and continuing education. But we still have much to do in this area. Thus, we have another paradigm shift.

Old Paradigm:
  Environmental problems are solved with sound science and engineering approaches.
Paradigm Shift:
  A sound risk-based approach to solving environmental problems requires credible risk assessment, risk management, and risk communication.

The cases in this book are examples of human failure, coupled with or complicated by physical realities. Water flows downhill. Air moves in the direction of high to low pressure. A rock in a landslide falls at an increasing rate of speed. Do we blame the water, the air, or the rock? Generally, no; we blame the engineer, the ship captain, the government inspector, or whomever we consider to have been responsible. And we hold them accountable. If we ignore physical, chemical, and biological principles, we have failed in a fundamental component of risk assessment. Risk assessment must be based in sound science. If we fail to apply these principles within the societal context and political, legal, economic, and policy milieu, we have failed in a fundamental component of risk management. And, if we fail to share information and include all affected parties in every aspect of environmental decisions, we have failed in a fundamental component of risk communication. Thus, environmental decisions can be likened to a three-legged stool. If any of the legs is missing or weak, our decision is questionable, no matter how strong the other two legs are.

Failure analysis is an important role of every engineering discipline, including environmental engineering. When there is a problem, especially if it is considered to be a disaster, considerable attention is given to the reasons that damages occurred. This is primarily an exercise in what historians refer to as "deconstruction" of the steps leading to the negative outcomes, or what engineers call a *critical path*. We turn back time to see which physical, chemical, and biological principles dictated the outcomes. Science and engineering are ready-made for such retrospective analyses. Factor A (e.g., gravity) can be quantified as to its effect on Factor B (e.g., stress on a particular material with a specified strength), which leads to Outcome C (e.g., a hole in the stern of a ship).

The severity of the outcome of an environmental event also affects the actual and perceived failure (see Table 1.1); that is, the greater the severity of the consequences, the more intense the blame for those expected to be responsible. The people thought to have caused it will assume more blame if they are professionals, for example, engineers and physicians.

Engineered solutions to environmental problems change the consequences from the outcome expected from the *status quo*. If a site is contaminated, the engineer can select from numerous interventions, all with

TABLE 1.1
Risk matrix comparing frequency to consequences of a failure event.

Risk matrix (example)

	Consequence				
Frequency	*Disastrous*	*Severe*	*Serious*	*Considerable*	*Insignificant*
Very likely	Unacceptable	Unacceptable	Unacceptable	Unwanted	Unwanted
Likely	Unacceptable	Unacceptable	Unwanted	Unwanted	Acceptable
Occasional	Unacceptable	Unwanted	Unwanted	Acceptable	Acceptable
Unlikely	Unwanted	Unwanted	Acceptable	Acceptable	Negligible
Very unlikely	Unwanted	Acceptable[a]	Acceptable	Negligible	Negligible

[a] Depending on the wording of the risk objectives it may be argued that risk reduction shall be considered for all risks with a consequence assessed to be "severe," and thus be classified as "unwanted" risks even for a very low assessed frequency.

Source: S.D. Eskesen, P. Tengborg, J. Kampmann, and T.H. Veicherts, 2004. "Guidelines for Tunnelling Risk Management: International Tunnelling Association, Working Group No. 2," Tunnelling and Underground Space Technology, 19: 217–237.

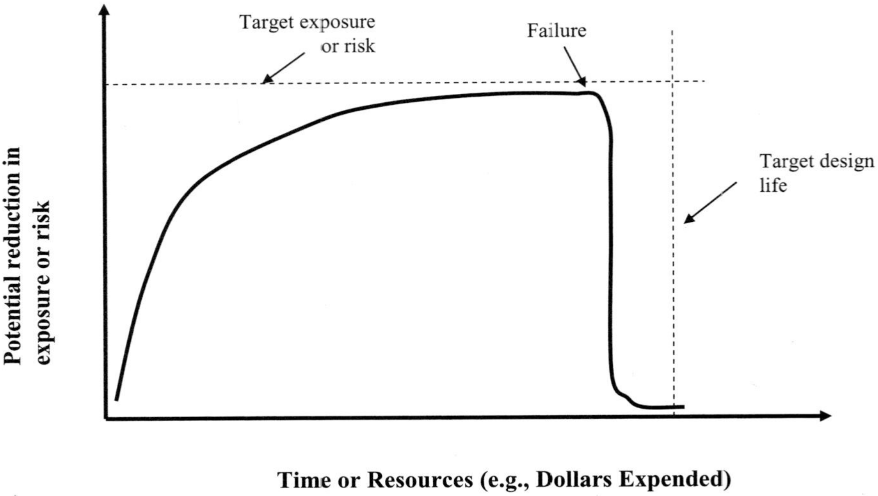

FIGURE 1.2. Hypothetical change in risk in relation to time and resources expended for exposure reduction (e.g., landfill capping or construction of a containment facility) without any reduction in the amount (mass) of the contaminant. Since the pollutant is still present, there is the potential for a catastrophic failure, followed by an increase in contaminant exposure and elevated risk.
Source: Adapted from National Research Council, 2003, Environmental Cleanup at Navy Facilities: Adaptive Site Management, Committee on Environmental Remediation at Naval Facilities, The National Academies Press, Washington, D.C.

different outcomes. A prototypical curve for an engineered facility that contains or caps the pollutant may reduce exposure to contaminants and, therefore, reduce health risks in manner similar to the curve in Figure 1.2, with relatively high risk reduction early or with the initial expenditure of resources and diminishing returns thereafter. The exposure or risk reduction is a measure of engineering effectiveness. The figure also depicts a catastrophic failure. This failure does not necessarily have to occur all at once, but could be an incremental series of failures that lead to a disaster, such as the containment and capping of hazardous wastes at the Love Canal, New York, site. Actions, including some environmentally irresponsible ones, were taken. This included burial of wastes and capping the landfill. Eventually, however, the failures of these engineered systems became obvious in terms of health endpoints—for example, birth defects, cancer, and other diseases—as well as measurements of contamination in the air, water, and soil. Whether or not the facilities reach catastrophic failure, the curve becomes asymptotic; that is, virtually no additional risk reduction with increased costs. The target design life for persistent chemical and

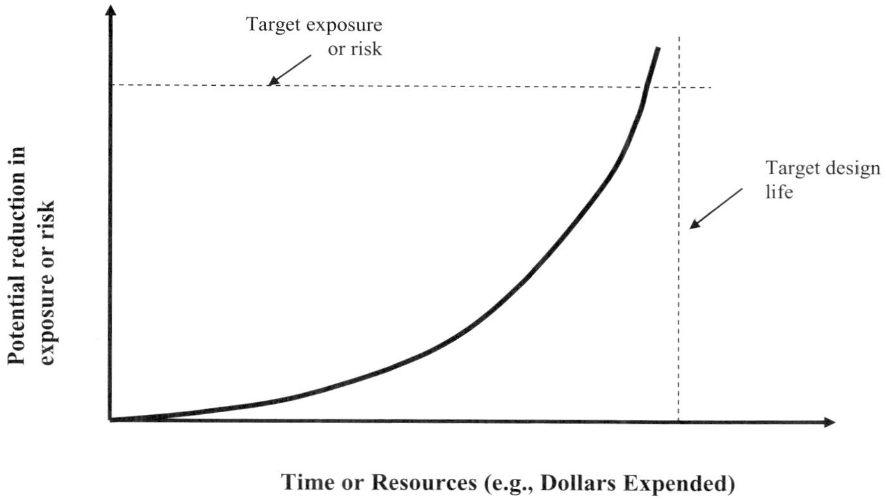

FIGURE 1.3. Hypothetical change in risk in relation to time and resources expended for exposure reduction from an aggressive cleanup action. In this depiction, the cleanup achieves the targeted risk reduction (e.g., less than one additional cancer per 10,000 population; i.e. cancer risk = 10^{-4}), within the specified project life (i.e., target cleanup date).
Source: Adapted from National Research Council, 2003. Environmental Cleanup at Navy Facilities: Adaptive Site Management, Committee on Environmental Remediation at Naval Facilities, The National Academies Press, Washington, D.C.

nuclear wastes can be many decades, centuries, even millennia. Any failure before this target is a design failure.

Another possible situation is where aggressive measures are taken, such as treating contaminants where they are found (i.e., *in situ*), including pump and treat for VOCs or chemical oxidation of dense nonaqueous phase liquids (DNAPLs) like PCBs (see Figure 1.3). The actual relationship of risk reduction with time and expended resources varies according to a number of factors, such as recalcitrance of the contaminant; ability to access the pollutant (e.g., in sediment or groundwater); matching the treatment technology to the pollution, microbial, and other biological factors; and natural variability, such as variability in meteorological and hydrological conditions (see Curves A and B in Figure 1.4).

Problems can result if the life of a project is shorter than what is required by the environmental situation. For example, "high maintenance" engineering solutions may provide short-term benefits, that is, rapid exposure reduction, but when the project moves to the operation and maintenance (O&M) stage, new risks are introduced (see Curve D in Figure 1.4). This is particularly problematic when designing environmental solutions

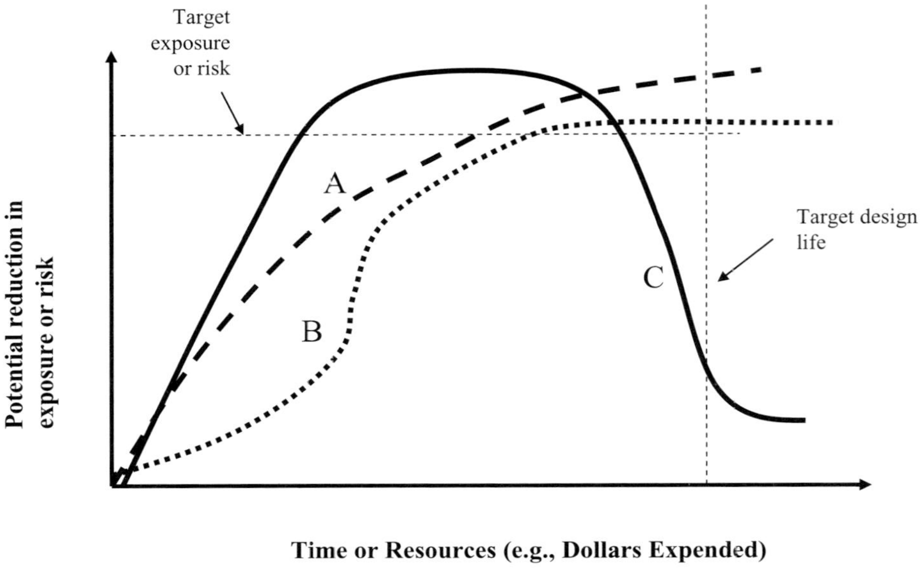

FIGURE 1.4. Hypothetical change in risk in relation to time and resources expended for exposure reduction from various actions, including natural attenuation (i.e., allowing the microbial populations to acclimate themselves to the pollutant and, with time, degrading the contaminations). For example, Curve A could represent an *in situ* treatment process. Curve B may represent natural attenuation, which lags the engineered approach, but the rate of biodegradation increases as the microbial populations become acclimated. Curve D is a situation where controls are effective up to a point in time. Thereafter, the risk increases either because of the treatment itself, for example, in pump and treat operations that pull in water from other aquifers that may be polluted, or when treatment technologies are high maintenance.
Source: Adapted from National Research Council, 2003. Environmental Cleanup at Navy Facilities: Adaptive Site Management, Committee on Environmental Remediation at Naval Facilities, The National Academies Press, Washington, D.C.

in developing countries or even in local jurisdictions with little technical capacity. For example, the local entities must retain expensive human resources or high-tech programs to achieve environmental and public health protection; there is a strong likelihood that these systems will not achieve the planned results and may even be abandoned once the initial incentives are gone. Certain engineering and environmental *pro bono* enterprises have recognized this and encourage low-tech systems that can easily be adopted by local people.

Engineering analyses not only require knowing how to solve problems, but also having the wisdom in deciding when conditions warrant one solution over another and where one solution is workable and another is not.

For example, the engineer is called upon to foresee which, if any, of the curves in Figure 1.4 applies to the situation at hand. Intuition has always been an asset for environmental engineers, and its value is increasing. The term *intuition* is widely used in a number of ways, so it needs to be defined here so that we are clear about what we mean by intuition, and more importantly, what engineering intuition is not. One of the things that sets apart engineers from most other scientists is the way that engineers process information. There are two ways of looking at data to derive information and, we hope, to gain knowledge. These are deductive and inductive reasoning. When we deduce, we use a general principle or fact to give us information about a more specific situation. This is the nature of scientific inquiry. We use general theories, laws, and experiential information to provide accurate information about the problem or the situation we are addressing. A classic example in environmental engineering is deducing from a cause to the effect. Low dissolved oxygen levels in a stream will not support certain fish species, so we reason that the fish kill is the result of low O_2. This demonstrates a product of deductive reasoning, i.e., synthesis.

Engineers also engage in inductive reasoning or analysis. When we induce, we move from the specific to the general and from the effect to the cause. We attribute the fish kill to the low dissolved oxygen levels in a stream that results from the presence of certain substances that feed microbes that, in turn, use up the O_2. We conduct experiments in microcosms that allow us to understand certain well-defined and well-controlled aspects of a system. We induce from these observations, so we gain larger principles beyond our specific study. The peril of induction is that any conclusion must be limited.[7] For example, our experiment may show a direct relationship between an independent and dependent variable, but we do not know just how far to extend the relationship beyond the controlled environment of the laboratory. We may show that increasing X results in growth of Y, but what happens in the presence of A, B, C, and Z? Engineers realize this and must be arbiters of what is useful and what will happen in real-world settings.

So, like other scientists, engineers build up a body of information and knowledge from deductive and inductive reasoning. They must rigorously apply scientific theory (deduction) and extend specific laboratory and field results (induction). Over time, the engineer's comfort level increases. Observing the decision making of a seasoned engineer might well lead to the conclusion that the engineer is using a lot of intuition. Engineers learn about how their designs and plans will work in two ways:

1. Their formal and continuing education; i.e., what others tell them.
2. What they have experienced personally.

The engineer learns both subject matter, that is, content, and processes, that is, rules. The scientific and practical content is what each engineer has learned about the world. Facts and information about matter and energy and the relationships between them are the content of engineering. Rules are the sets of instructions that each engineer has written (literally and figuratively) over time of how to do things.[8]

The accumulation of content and rules over our academic experience and professional practice leads to intuition. Thus, intuition can be explained as the lack of awareness of why or how professional judgments have come to be. Kenneth Hammond,[9] a psychologist who has investigated intuitive processes, says that intuition is, in fact, "a cognitive process that somehow produces an answer, solution, or idea without the use of a conscious, logically defensible step-by-step process." So, intuition is an example of something that we know occurs, and probably quite frequently, but it is not deliberative, nor can it be explained explicitly after it occurs. I argue that it is really a collective memory of the many deductive and inductive lessons learned (content), using a system to pull these together, sort out differences, synthesize, analyze, and come to conclusions (rules). The more we practice, the more content is gathered, and the more refined and tested the rules become.

Thus, the right solution in one instance may be downright dangerous in another. Or as the National Academy of Engineering puts it, "engineering is a profoundly creative process."[10] However, engineers must always design solutions to problems within constraints and tolerances called for by the problem at hand. For environmental engineers, this is a balance between natural and artificial systems. This balance depends on data from many sources. Good data makes for reliable information. Reliable information adds to scientific and societal knowledge. Knowledge, with time and experience, leads to wisdom. Environmental assessment and protection need to include every step in the "wisdom cascade" (see Figure 1.5).

Building a structure such as a hazardous waste treatment facility or an incinerator is part of the solution. At all times, the solution calls for a process that may or may not require the design and construction of a structure. Certainly, when a structure is called for, the operation and maintenance (O&M) and life-cycle analysis (LCA) are needed for the structure. However, the process may represent the entire solution to the environmental problem, such as instituting recycling or pollution prevention based entirely on virtual systems like waste clearinghouses. This thinking has gained currency in that it is a vital part of sustainable design, which applies to all engineering disciplines, not just environmental engineering. Standard practice in civil and mechanical engineering now embodies sustainable design; for example, we now expect engineers to design for the environment (DFE), design for recycling (DFR), and design for disassembly (DFD), as well as to consider ways to reduce the need for toxic chemicals and substances and to minimize the generation of wastes when they conceive of new products and processes.[11] Environmental engineering seldom, if ever, can rely exclusively on a single scientific solution, but is always a choice among

Concerns and Interests

↓

Data

↓

Information

↓

Knowledge

↓

Wisdom

FIGURE 1.5. Value-added chain from data to knowledge and, with experience, professional wisdom.
Source: D.A. Vallero, 2004. *Environmental Contaminants: Assessment and Control*, Elsevier Academic Press, Burlington, MA.

many possible solutions dictated by the particular environmental conditions. Thus, designing environmental solutions calls for the application of all the physical sciences, as well as the social sciences.

Throughout the first half of the twentieth century, when the field predominantly was considered sanitary engineering, structural considerations were paramount. However, even then, operational conditions had to include chemistry and biology, as well as fluid mechanics and other physical considerations. This amalgam of science grew more complex as we earned the designation of environmental engineering.

All engineers apply physical principles. Most also apply ample amounts of chemistry to their respective engineering disciplines. But, environmental and biomedical engineers must also account for biology. In the case of environmental engineering, our concern for biology ranges across all kingdoms, phyla, and species. Engineers use biological principles and concepts to solve problems (e.g., bacteria and fungi adapted to treat wastes; macrophytic flora to extract contaminants, that is, phytoremediation, and to restore wetlands; and benthic organisms to help to clean contaminated sediments). We use them as indicators of levels of contamination (e.g., algal blooms, species diversity, and abundance of top predators and other so-called sentry species) and act as our "canaries in the coal mine" to give us early warning about stresses to ecosystems and public health problems. And, arguably most important, we study organisms as endpoints in themselves. We care principally about human health. This particular area of biology that is so important to environmental engineers is known as *toxicology*, which deals with the harmful effects of substances on living organisms. Usually, toxicology that is not further specified deals with the harmful effects of substances on human beings, but there are subdisciplines, such as ecotoxicology, which address harm to components of ecosystems, and even more specific fields, such as aquatic toxicology, which is concerned with harm to those organisms living in water.

Scientists strive to understand and add to the knowledge of nature. This entails making decisions about what needs to be studied. In this way, science is a social enterprise. The reason we know more about many aspects of the environment today is that science has decided or been forced to decide to give attention to these matters.[12] Engineers have devoted entire lifetimes to ascertaining how a specific scientific or mathematical principle should be applied to a given event (e.g., why compound X evaporates more quickly, but compound Z under the same conditions remains on the surface). Such research is more than academic. For example, once we know why something does or does not occur, we can use it to prevent disasters (e.g., choosing the right materials and designing a ship hull correctly) as well as to respond to disasters after they occur. For example, compound X may not be as problematic in a spill as compound Z if the latter does not evaporate in a reasonable time, but compound X may be very dangerous if it is toxic and people nearby are breathing air that it has contaminated. Also, these factors affect what the Coast Guard, fire departments, and other first responders should do when they encounter these compounds. The release of volatile compound X may call for an immediate evacuation of human beings; whereas a spill of compound Z may be a bigger problem for fish and wildlife (it stays in the ocean or lake and makes contact with plants and animals). Thus, when deconvoluting a failure to determine responsibility and to hold the right people accountable, we must look at several compartments.

Arguably, the compartment that the majority of engineers and scientists are most comfortable with is the physical compartment. This is the one we know the most about. We know how to measure things. We can even use models to extrapolate what we find. We can also fill in the blanks between the places where we take measurements (what we call interpolations). So, we can assign values of important scientific features and extend the meaning of what we find in space and time. For example, if we use sound methods and apply statistics correctly, measuring the amount of crude oil on a few ducks can tell us a lot about the extent of an oil spill's impact on waterfowl in general. And good models can even give us an idea of how the environment will change with time (e.g., is the oil likely to be broken down by microbes and, if so, how fast?). This is not to say that the physical compartment is easy to deal with. It is often very complex and fraught with uncertainty. But it is our domain. Missions of government agencies, such as the Office of Homeland Security, the U.S. Environmental Protection Agency, the Agency for Toxic Substances and Disease Registry, the National Institutes of Health, the Food and Drug Administration, and the U.S. Public Health Service, devote considerable effort in just getting the science right. Universities and research institutes are collectively adding to the knowledge base to improve the science and engineering that underpins the physical principles that underpin public health and environmental consequences from contaminants, whether these be intentional or by happenstance.

Another important compartment in the factors that lead to a disaster is the anthropogenic compartment. This is a fancy word that scientists often use to denote the human component of an event (*anthropo* denotes human and *genic* denotes origin). This compartment includes the gestalt of humanity, taking into account all the factors that society imposes down to the things that drive an individual or group. For example, the anthropogenic compartment would include the factors that led to a ship captain's failure to stay awake. However, it must also include why the fail-safe mechanisms did not kick in. These failures do have physical factors that drive them, for example, a release valve may have rusted shut or the alarm clock's quartz mechanism failed because of a power outage, but there is also an arguably more important human failure in each. For example, one common theme in many disasters is that the safety procedures are often adequate in and of themselves, but the implementation of these procedures was insufficient. Often, failures have shown that the safety manuals and data sheets were properly written and available and contingency plans were adequate, but the workforce was not properly trained and inspectors failed in at least some crucial aspects of their jobs, leading to horrible consequences.

A Lesson from the Medical Community

To paraphrase Aristotle, an understanding of the physical factors is necessary to understand a disaster, but most certainly not sufficient. In this age of specialization in technical professions, one negative side effect is the increased likelihood that no single person can understand all the physical and human factors needed to prevent a disaster.

This book applies case analysis techniques. The engineering profession and to some extent the environmental science community does employ case analysis, particularly following an ethical or practical failure. However, case analysis is arguably a more familiar device to the medical profession, occurring at all levels, from the individual physician's review of similar cases when diagnosing a possible disease to the hospital review of a case to ensure that properly informed consent was given prior to a medical procedure to the American Medical Association review of cases to elicit ethical and practical lessons for the entire practice. An example of such a case review of failure is the Santillan case. Although the subject matter (i.e., surgical procedures) is outside the domain of this book, the case provides some important lessons.

Duke University is blessed with some of the world's best physicians and medical personnel. As a research institute, it often receives some of the most challenging medical cases, as was the case for Jesica Santillan, a teenager in need of a heart transplant. Although the surgeon in charge had an impeccable record and the hospital is world renowned for such a surgery, something went terribly wrong. The heart that was transplanted was of a different blood type than that of the patient. The heart was rejected, and even after another heart was located and transplanted, Jesica died due to the complications brought on by the initial rejection. The logical question is how could something so vital and crucial and so easy to know—blood type—be overlooked? It appears to be a systematic error. The system of checks and balances failed. And, the professional (the surgeon) is ultimately responsible for this or any other failure on his watch.

What can we learn from the Santillan case? One lesson is that a system is only as good as the rigor and vigilance given to it. There is really no such thing as "auto pilot" when it comes to systems. Aristotle helps us again here. He contended that the whole is greater than the sum of its parts. This is painfully true in many public health disasters. Each person or group may be doing an adequate or even superlative job, but there is no guarantee that simply adding each of the parts will lead to success. The old adage that things "fall through the cracks" is a vivid metaphor. The first mate may be doing a great job in open waters, but may not be sufficiently trained in dire straights when the captain is away from the bridge. A first response team may be adequately trained for forest fires (where water is a very good substance for firefighting), but may not be properly suited for a spill of an oxidizing agent (where applying water can make matters considerably more dangerous). Without someone with a global view to oversee the whole response, perfectly adequate and even exemplary personnel may contribute to the failure.

Systems often are needed and these systems must be tested and inspected continuously. Every step in the critical path that leads to failure is important. In fact, the more seemingly mundane the task, the less likely people are to think a lot about it. So, these small details may be the largest areas of vulnerability. We can liken this to the so-called "butterfly effect" of chaos theory, where the flapping of a butterfly's wings under the right conditions in a certain part of the world can lead to a hurricane. One of the adages of the environmental movement is that "everything is connected." A loss of a small habitat can lead to endangering a species and altering the entire diversity of an ecosystem. A seemingly safe reformulation of a pesticide can alter the molecule to make it toxic or even cancer-causing. Preventing an environmental disaster may rest on how well these details are handled. We must wonder how many meetings before the Santillan case had significant discussions on how to make sure that the blood type is properly labeled. We can venture that such a discussion occurs much more frequently now in pre-op meetings (as well as hospital board meetings) throughout the world.

Many of the cases in this book owe their origin or enlarged effect in part to a failure of fundamental checks and balances. Often, these requirements have been well documented, yet ignored.

A lesson going forward is the need to stay vigilant. One of the major challenges for safety and health units is that human beings tend to be alert to immediacy. If something has piqued their interest, they are more than happy to devote attention to it. However, their interest drops precipitously as they become separated from an event in space and time. Psychologists refer to this phenomenon as an extinction curve. For example, we may learn something, but if we have no application of what we have learned, we will forget it in a relatively short time. Even worse, if we have never experienced something (e.g., a real spill, a fire, or leak), we must further adapt our knowledge of a simulation to the actual event. We never know how well we will perform under actual emergency conditions.

Professional Accountability

The Santillan case provides a particularly noteworthy lesson for professional engineers and planners. One's area of responsibility and accountability is inclusive. The buck stops with the professional. The credo of the professional is *credat emptor*, let the client trust. Jesica's parents did not need to understand surgical procedure. Society delegated this responsibility exclusively to the surgeon. Likewise, environmental and public health professionals are charged with responsibilities to protect the public and ecosystems. When failures occur, the professionals are accountable. When a manufacturing, transportation, or other process works well, the professional can take pride in its success. The professional is responsible for the successful project. That is why we went to school and are highly trained in our fields. We accept the fact that we are accountable for a well-running system. Conversely, when things go wrong, we are also responsible and must account for every step, from the largest and seemingly most significant to those we perceive to be the most minuscule, in the system that was in place. Professional responsibility cannot be divorced from accountability. The Greeks called this *ethike areitai*, or skill of character. It is not enough to be excellent in technical competence. Such competence must be coupled with trust gained from ethical practice.

Villain and Victim Status

One of the difficult tasks in writing and thinking about failures is the temptation to assign status to key figures involved in the episodes. Most accounts in the media and even in the scientific literature readily assign roles of villains and victims. Sometimes, such assignments are straightforward and enjoy a consensus. However, often such classifications are premature and oversimplified. For example, in arguably the worst chemical disaster on record, the Bhopal toxic cloud killed thousands of people and left many more thousands injured, but there are still unresolved disagreements about which events leading up to the disaster were most critical. In addition, the incident was fraught with conflicts of interest that must be factored into any thoughtful analysis. In fact, there is no general consensus on exactly how many deaths can be attributed to the disaster, especially when trying to ascertain mortality from acute exposures *versus* long-term, chronic exposures. Certainly, virtually all the deaths that occurred within hours of the methylisocynate (MIC) release in nearby villages can be attributed to the Bhopal plant. However, with time, the linkages between deaths and debilitations to the release become increasingly indirect and more obscure. Also, lawyers, politicians, and businesspeople have reasons beyond good science for including and excluding deaths. Frequently, the best we can do is say that more deaths than those caused by the initial, short-term MIC exposure can be attributed to the toxic cloud. But just how many more is a matter of debate and speculation.

This brings us to the controversial topic of cause-and-effect, and the credible science needed to connect exposure to a risk and a negative outcome. Scientists frequently "punt" on this issue. We have learned from introductory statistics courses that association and causation are not synonymous. We are taught, for example, to look for the "third variable." Something other than what we are studying may be the reason for the relationship. In statistics classes, we are given simple examples of such occurrences:

Studies show that people who wear shorts in Illinois eat more ice cream.
Therefore, wearing shorts induces people to eat more ice cream.

The first statement is simply a measurement. It is stated correctly as an association. However, the second statement contains a causal link that is clearly wrong for most occurrences.[13] Something else is actually causing both variables, that is, the wearing of shorts and the eating of ice cream. For example, if we were to plot ambient average temperature and compare it to either the wearing of shorts or the eating of ice cream, we would see a direct relationship between the variables. That is, as temperature increases, so does shorts wearing and so does ice cream eating.

I said that we scientists often punt on causality. Punting is not a bad thing. (Ask the football coach who decides to go for the first down on fourth and inches and whose team comes up a half-inch short. He would have likely wished he had asked for a punt!) It is only troublesome when we use the association argument invariably. (The football coach who always punts on fourth and short might be considered to lack courage.) People want to know what our findings mean. Again the medical science community may help us deal with the causality challenge. The best that science usually can do in this regard is to provide enough weight-of-evidence to support or reject a suspicion that a substance causes a disease. The medical research and epidemiological communities use a number of criteria to determine the strength of an argument for causality, but the first well-articulated criteria were Hill's Causal Criteria[14] (see Table 1.2). Some of Hill's criteria are

TABLE 1.2
Hill's Criteria for Causality.

Factors to be considered in determining whether exposure to a chemical elicits an effect:
Criterion 1: Strength of Association. Strong associations between variables provide more certain evidence of causality than is provided by weak associations. Common epidemiological metrics used in association include risk ratio, odds ratio, and standardized mortality ratio.
Criterion 2: Consistency. If the chemical exposure is associated with an effect consistently under different studies using diverse methods of study of assorted populations under varying circumstances by different investigators, the link to causality is stronger. For example, the carcinogenic effects of Chemical X is found in mutagenicity studies, mouse and Rhesus monkey experiments, and human epidemiological studies; there is greater consistency between Chemical X and cancer than if only one of these studies showed the effect.
Criterion 3: Specificity. The specificity criterion holds that the cause should lead to only one disease and that the disease should result from only this single cause. This criterion appears to be based in the germ theory of microbiology, where a specific strain of bacteria and viruses elicits a specific disease. This is rarely the case in studying most chronic diseases, since a chemical can be associated with cancers in numerous organs, and the same chemical may elicit cancer, hormonal, immunological, and neural dysfunctions.
Criterion 4: Temporality. Timing of exposure is critical to causality. This criterion requires that exposure to the chemical must precede the effect. For example, in a retrospective study, the researcher must be certain that the manifestation of a disease was not already present before the exposure to the chemical. If the disease were present prior to the exposure, it may not mean that the chemical in question is not a cause, but it does mean that it is not the sole cause of the disease (see "Specificity" earlier).

TABLE 1.2
Continued

Factors to be considered in determining whether exposure to a chemical elicits an effect:
Criterion 5: Biologic Gradient. This is another essential criterion for chemical risks. In fact, this is known as the "dose-response" step in risk assessment. If the level, intensity, duration, or total level of chemical exposure is increased, a concomitant, progressive increase should occur in the toxic effect.
Criterion 6: Plausibility. Generally, an association needs to follow a well-defined explanation based on a known biological system. However, paradigm shifts in the understanding of key scientific concepts do change. A noteworthy example is the change in the latter part of the twentieth century of the understanding of how the endocrine, immune, and neural systems function, from the view that these are exclusive systems to today's perspective that in many ways they constitute an integrated chemical and electrical set of signals in an organism.[15]
Criterion 7: Coherence. The criterion of coherence suggests that all available evidence concerning the natural history and biology of the disease should "stick together" (cohere) to form a cohesive whole. By that, the proposed causal relationship should not conflict or contradict information from experimental, laboratory, epidemiologic, theory, or other knowledge sources.
Criterion 8: Experimentation. Experimental evidence in support of a causal hypothesis may come in the form of community and clinical trials, in vitro laboratory experiments, animal models, and natural experiments.
Criterion 9: Analogy. The term analogy implies a similarity in some respects among things that are otherwise different. It is thus considered one of the weaker forms of evidence.

more important than others. Interestingly, the first criterion is, in fact, association.

My Duke colleague, J. Jeffrey Peirce, is fond of saying that the right answer in engineering is usually "It depends." I believe he stresses this for future engineers because we are tempted to think that every solution can be found in a manual or handbook. He counsels students to consider elements beyond the "cookbook" answers. I contend that most of us, as we mature, find Peirce's advice to hold true in most human endeavors. The engineer, physician, accountant, attorney, clergy, parent, friend, or whatever role we take on benefits from a balanced view. It is not that the scientific principles are wrong, it is that we are missing some key information that, if it were available, would show that we should not use this equation or that the constant is not correct for this situation, or that an assumption that has been built into the formula was violated in our specific case.[16]

Other Lessons: Risk and Reliability

Returning to our discussion of the need to analyze cases from both physical and anthropogenic perspectives, it is important to point out that any technical analysis of why and how events occur must show the degree of certainty that we have in our assessment. Descriptive studies, for example, may simply provide a chronological step-by-step analysis of what happened. This may be enhanced by analytical studies that explain why such steps occurred.[17] Both types of studies, but particularly analytical studies, require documentation of the level of uncertainty. The data and information about what happened must be precise and accurate.

Scientists and engineers must apply quantitative methods when analyzing cases. This requires an assessment of the risks that occurred and the reliability of the findings. Both risk and reliability are engineering terms.

Risk is the likelihood of an adverse outcome. The likelihood is a mathematical expression, a probability. A probability must always range between 0 (no likelihood) and 1 (100% likelihood). Risk assessment is the application of scientifically sound methods to determine the contribution of each risk factor in the adverse outcome. In other words, a risk assessment is an effort to find what went wrong and to identify all the factors that led to the unfortunate outcome. The term "unfortunate" itself is to be used advisably. Its base, "fortune," is a synonym of "luck" and specifically in terms of risk it is "bad luck." Luck often has nothing to do with the outcome. In fact, a good failure engineer often takes an initial step of drawing an event tree or critical path that shows all the events and decisions that led to the ultimate outcome. Yes, many of the cases discussed in this book are the dreadful combinations of unlikely events followed by other unlikely events, but they can be explained. Sometimes, the scariest part of looking deeply at a failure is that we may wonder why other such failures had not already occurred elsewhere. Returning to the Santillan case, could such mislabeling and other weaknesses in the chain of custody of blood handling have already occurred in other hospitals, doctors' offices, or blood banks, but with less painful outcomes? Thus, this is another important reason to analyze public health and environmental failures sufficiently. We may find things that need to be fixed in other areas where the confluence of events that lead to tragic results have not yet occurred, but such a confluence remains all too likely.

So, risk is a statistical term, the probability (the statistical expression of an event's likelihood) of an adverse outcome. Risk is seldom used to denote a positive outcome. That would be reward. In fact, economists and financial experts often speak of risk/reward ratios, where the investor wants to minimize the former and maximize the latter. In virtually every case in this book, someone (maybe everyone involved) has their own conception of the risks and rewards of their particular role in what turned out to be a tragic outcome. For example, the pesticide DDT has the reward of eliminating mosquitoes that carry the malarial agent, but it simultaneously has the risk of eggshell thinning in top predatory birds and diseases in humans.

Reliability is a related term. Like risk, it is quantitative probability with values between 0 and 1. Whereas we want very small values for risk (e.g., less than 0.00001% or 10^{-6} for cancer risk), to understand acute toxicity, we strive to find concentrations that lead to risk values approaching 1 (e.g., 100% risk that a human will die after inhaling hydrogen cyanide, HCN, at a concentration of $100\,mg\,m^{-3}$ for a certain amount of time). In fact, that is the purpose of bioassays, where we use different species to see just how acutely toxic many compounds can be.

Besides risks and reliability associated with physical factors, the case analysis must include an analysis of the anthropogenic (human) factors associated with a case. However, the terms are used somewhat differently from how engineers and scientists generally apply them. For example, every night watchman has a certain degree of risk of falling asleep. Managers may institute measures to reduce this risk, such as the need to insert keys at various stations throughout the plant within specified time intervals. Likewise, reliability in management may be akin to predictability of outcomes. We consider both terms in greater detail in Chapters 2 and 5.

Environmental Ethics and a New Environmental Ethic

The shifting environmental paradigms have not only added emphasis to the need to be ethical about the environment, but also the need to form a whole new ethic. In many ways the former is much easier than the latter. Ethics, to paraphrase Socrates, is how we ought to live. So, environmental ethics is a set of expectations, rules of behavior if you will, of how we treat the planet's inhabitants, human and nonhuman. By extension, the only way to be ethical to all the earth's inhabitants is to take a reasoned approach to the living (biotic) and nonliving (abiotic) components. Ethics is the manner in which human beings govern their actions. Clearly, a number of environmental insults over the millennia resulted from unethical activities.

Although ethics is a part of an ethic, it does not completely describe it. In fact, ethics flow from an ethic. A society's ethic is a comprehensive view of what matters. For example, the well-known American "work ethic" is the view that society expects and gets a responsible workforce because the majority of North Americans hold fast to the belief that work is a moral good. Thus, although wages and other external measures are important, the quality of work and outputs are driven more by the view that work is "good." Likewise, the old paradigm of environmental resources was that they are to be exploited, even conquered. The desert was seen as wilderness that needed to be modified by irrigation systems. The swamps needed to be drained. The mountains needed to be toppled to release their treasures. The new ethic has come as a migration from the exploitation viewpoint to a greater appreciation of other values, such as the biodiversity of the desert and the swamp, and the risks of massive efforts to change the environment, as in the extraction of ores.

The ethic is even evidenced by the names we use to describe these resources; for example, swamps are a type of wetland. And old names have taken on new meanings, such as wilderness, which formerly invoked the need to change it to something "useful" to the present-day perception that wilderness is to be contrasted with the developed, urban, and "built" landscapes. In fact, to many, the negative and positive connotations have switched places. Wilderness is preferred by many in Western civilization to the built landscape. This is not the case throughout the world, however, since many developing countries wish to emulate the development of the West. This has presented special problems. For example, many of the most sensitive habitats (e.g., tropical rainforests) are in the developing world. These habitats are among the few resources available to these nations and they are being eliminated at an alarming rate. It is difficult to instill the West's newfound ethic in these countries, since it took us centuries to reach it ourselves. Plus, the developing countries have immediate and pressing needs, such as wood for heating and cooking. In large areas of the world, woodlands are burned to accommodate population shifts, to access minerals, and to provide for industrial expansion. The mangroves and other sensitive woodlands are thus at risk of large-scale destruction. Some of these problems may be mitigated as the countries develop and improve technologies, increase trade, and cooperate within the international community. The rainforests and wetlands are important for many reasons, but one of the most important is their production of molecular oxygen and the storage of carbon. Thus, the global atmospheric gas balances are quite dependent on the rainforests' steady-state conditions of oxygen (O_2) and carbon, especially the sequestration of carbon dioxide (CO_2) and methane (CH_4). However, the global value of leaving these resources intact and adopting the new environmental ethic is a bit much for many developing countries given their existing economic condition. One of the great challenges before the world is how to deal with the environmental problems brought on by increased economic development of the most populated countries in South America, southern Asia, and Africa. Interestingly, another challenge is the lack of development and the refraining of some in these same regions to adopt new, less resource-intensive practices (e.g., replacement of wood with other fuels). This challenge is ironic in that many of the most environmentally aware cultures have their cradles in these developing regions. Perhaps these traditions will provide a modicum of moderation against the trends of increasing exploitation.

The new environmental ethic is an outgrowth of the land ethic. Our contemporary understanding of environmental quality is often associated with physical, chemical, and biological contaminants, but in the formative years of the environmental movement, aesthetics and other "quality of life" considerations were essential parts of environmental quality. Most environmental impact statements, for example, began in the 1970s to address cultural and social factors in determining whether a federal project would have a significant effect on the environment. These included historic preservation, economics, psychology (e.g., open space, green areas, and crowding), aesthetics, urban renewal, and the so-called "land ethic." Aldo Leopold, in his famous essays, posthumously published as *A Sand County Almanac*, argued for a holistic approach:

> A thing is right when it tends to preserve the integrity, stability and of the biotic community. It is wrong when it tends otherwise.[18]

Laying out canons and principles for environmental professions to follow is a means of assuring that the individual practitioner approaches the environment from an ethical perspective. For example, engineers must find ways to reduce environmental risks to human populations and ecosystems. This is best articulated in the first canon of their codes of ethics; that is, engineers must hold paramount the health, safety, and welfare of the public. Environmental professionals spend their careers providing these public services and finding ways to ensure that what we design is safe and does not detract from the public good.

In addition to these new appreciations of environmental ethics, the new environmental ethic is also reflected in professionalism. For example the American Society of Civil Engineers (ASCE) recently modified its code of ethics to include a reference to "sustainable development." Recognizing this as a positive step in defining the responsibilities of engineers toward the environment, the First Fundamental Canon in the 1997 revisions of the ASCE Code of Ethics now reads:

> Engineers shall hold paramount the safety, health, and welfare of the public and shall strive to comply with the principles of sustainable development in the performance of their professional duties.

The term "sustainable development" was first popularized by the World Commission on Environment and Development (also known as the Brundtland Commission), sponsored by the United Nations. The report defines sustainable development as "development that meets the needs of the present without compromising the ability of future generations to meet their own needs."[19]

The land ethic causes us to consider the two- and three-dimensional aspects of pollution and environmental quality. Our view must be holistic. Every planned use of any resource must be seen as a life cycle. Even the good things we do for one part of the environment can have adverse consequences in another part, as the MTBE example aptly illustrates. And sustainability makes us consider what we do in light of future consequences, good and bad. So, our contemporary environmental ethic stretches environmental awareness in space and time.

The environmental ethic is quite inclusive. To achieve environmental success requires an eye toward numerous concerns. This was well articulated in the Stockholm Conference on Human Development held by the United Nations in 1972. In addition to establishing the UN Environmental Programme (UNEP), the conference reached a remarkable consensus on key principles that needed to be followed to begin to address the seemingly intractable environmental problems of the world (see Table 1.3). The principles are still guiding many international, regional, and even local

TABLE 1.3
Key principles of the United Nations Stockholm Conference on the Human Environment, June 1972.

1. Human rights must be asserted, apartheid and colonialism condemned.
2. Natural resources must be safeguarded.
3. The earth's capacity to produce renewable resources must be maintained.
4. Wildlife must be safeguarded.
5. Nonrenewable resources must be shared and not exhausted.
6. Pollution must not exceed the environment's capacity to clean itself.
7. Damaging oceanic pollution must be prevented.
8. Development is needed to improve the environment.
9. Developing countries therefore need assistance.
10. Developing countries need reasonable prices for exports to carry out environmental management.
11. Environment policy must not hamper development.
12. Developing countries need money to develop environmental safeguards.
13. Integrated development planning is needed.
14. Rational planning should resolve conflicts between environment and development.
15. Human settlements must be planned to eliminate environmental problems.
16. Governments should plan their own appropriate population policies.
17. National institutions must plan development of states' natural resources.
18. Science and technology must be used to improve the environment.
19. Environmental education is essential.
20. Environmental research must be promoted, particularly in developing countries.
21. States may exploit their resources as they wish but must not endanger others.
22. Compensation is due to states thus endangered.
23. Each nation must establish its own standards.
24. There must be cooperation on international issues.
25. International organizations should help to improve the environment.
26. Weapons of mass destruction must be eliminated.

decisions. Interestingly, the environmental principles are integrated in other important human endeavors, such as human rights, development, and security. Hints of concepts that were to emerge later, such as credible science, sustainability, and biodiversity, can be found in these principles.

This new environmental ethic is a strong indication of the how we have learned from the past. That is the good news. The not-so-good news is that we are only now learning how to implement measures that are comprehensive and sustainable. The cases in this book hold many lessons. They are discussed in a manner to comprise a rich resource from which we can explore new approaches and change our thinking so as not to be condemned to repeat our environmental mistakes, mishaps, and misdeeds.

Solving the current environmental problems and preventing future environmental pollution requires new ways of thinking. We have much to learn from our successes and failures, but we must shift from some of the old paradigms, enhance others, and produce new ones, where appropriate. The goal of this book is to place some of the many environmental events in a context in which they can be scrutinized objectively, systematically, and dispassionately.[20]

Sensitivity

The shifting paradigm recognizes that some environmental systems are highly sensitive to even very small insults. This brings up the very important scientific concept of sensitivity. Most environmental scientists have learned to rely on stoichiometry; that is, the quantities of substances entering into and produced by chemical reactions. We know that when methane combines with oxygen in complete combustion, 16 g of methane require 64 g of oxygen, and simultaneously 44 g of carbon dioxide and 36 g of water are produced by this reaction. We know that every chemical reaction requires that all elements in the reaction must be in specific proportions to one another. This is complicated when biology is added to the physics and chemistry, but biochemistry also adheres to the principles of stoichiometry. For example, empirical observations have led environmental engineers to understand the molar relationships in the oxidation of pollutants, such as benzene:

$$C_2H_6 + 7.5O_2 \rightarrow 6H_2O + 6CO_2 + \text{microbial biomass} \quad (1.1)$$

Environmental problems cannot ignore the stoichiometry. We can be quite certain that when benzene reacts, it will abide by the stoichiometry of Reaction 1.1. Yet, we know that in the real world, not all of the benzene reacts, even when it seems that there is plenty of oxygen. How then, do we begin to understand some of the other factors controlling the rate and extent of abiotic chemical and biological degradation processes?

Scientists are always concerned with how certain they need to be about data and information. An important means to determine, even quantify, the certainty of data and information needed to make a decision is the sensitivity analysis. Every chemical reaction or biological process in this book is, in fact, a model. A model is simply a system that represents another system, with the aim of helping to explain that target system. Reaction 1.1 explains just how many moles of oxygen are needed to react with benzene to form carbon dioxide and water. It is very robust and sensitive. Any change to the left-hand side will result in concomitant changes in the right-hand side. In fact, the way it is written, the only limits on how much CO_2, water, and microbes will be generated is the amount of oxygen and hydrocarbons (benzene) that are available. Of course, the model (reaction) does not show every variable influencing this reaction. Even if we were to pump large volumes of O_2 into an aquifer, it will speed up the degradation of the hydrocarbons but it will be immediate. Such a system has a surplus of oxygen; that is, it is not oxygen limited. Neither is it hydrocarbon limited. But, since these are the only two reactants, how can that be? So, other factors come into play. The + in the reaction indicates that the two reactants must contact one another, but does not show how abruptly or slowly this contact occurs. The system's scale is also important. Even if the overall environment is at an oxygen surplus, the place where the microbes live (e.g., the film around particles) may be oxygen deficient. There may also be discontinuities between individual particles, with some pockets of highly efficient biodegradation, but others isolated from water, oxygen, and substrate (including the benzene), not allowing the microbes to thrive. The actual environment can differ dramatically from the tightly controlled laboratory. The stoichiometric model in Reaction 1.1 simply expresses that biomass will also be produced, but is less specific about them than the abiotic parts of the model. The actual number and species of microbes will vary considerably from place to place. So, our reaction model is very good at expressing exactly how many moles will react and how many moles will be produced, but does not indicate many important conditions and variables outside of a strictly controlled laboratory.

It is possible to learn about the complexities and uncertainties in actual environmental problems by deconstructing some of the more complex models in use today. For example, if a model is being used to estimate the size of a plume of a contaminant in groundwater, a number of physical, chemical, and biological variables must be considered. Modelers refer to such variables as model parameters. So, an engineer or hydrologist interested in how far a plume extends and the concentrations of a contaminant within the plume must first identify hydrogeological parameters like aquifer thickness, porosity, transverse and longitudinal dispersivity,[21] source strength and type, recharge of the aquifer, as well as chemical parameters like sorption and degradation rates.

But not all parameters are of equal importance to the outcome of what is modeled; that is, some are more sensitive and others less sensitive. Some have a major influence on the result with even a slight change while others can change significantly with only a slight change in the result. In the former situation, the result is said to be highly sensitive to the parameter. In the latter, the result is considered to be nearly insensitive. If a result is completely insensitive, the parameter does not predict the outcome at all. This occurs when a parameter may be important for one set of microbes or one class of chemicals (i.e., sensitive), but when the model is used for another set of microbes or chemicals it is completely insensitive. For example, anaerobic bacteria may grow according to predictions of an oxygenation parameter in a model, but the same parameter is unimportant in predicting the growth of anaerobic bacteria. What the engineer and scientist want to find out is how much change is induced in a parameter per unit of perturbation. In other words, if the modeled results change 70% with a unit change to parameter A, but change only 7% with the same unit change to parameter B, we could characterize the model as being 10 times more sensitive to parameter A than to parameter B. This information is critical to solving environmental problems. For example, if we know which variables and parameters limit the change of contaminant concentrations, then we can optimize environmental cleanup.

Consider, for example, the natural attenuation prediction model, Bioplume III. This U.S. EPA model has been subjected to a sensitivity analysis for hydrogeological, physicochemical, and biological parameters. And, pertinent to this discussion, it has been subjected to tests to see just how sensitive benzene contamination is to changes in these parameters.[22] The Bioplume III sensitivity analysis evaluated five hydrogeological parameters and two chemical parameters:

1. Porosity of the soil or other media
2. Thickness of the aquifer
3. Transmissivity[23] of the aquifer
4. Longitudinal dispersivity
5. Horizontal dispersivity
6. Sorption (indirectly indicated by a retardation factor, Rf)[24]
7. Radioactive decay (as an analog to abiotic chemical half-life)

To test these parameters, the BIOPLUME model hypothesizes a base case with the characteristic shown in Table 1.4.

The parameters were manipulated to determine the difference in results (i.e., benzene concentrations in the plume) between the base case and other scenarios. The two most influential (i.e., sensitive) hydrogeological parameters on benzene concentrations in the plume appear to be thickness of the aquifer and transmissivity. Benzene concentrations appear to be sensitive to both of the chemical parameters (see Table 1.5).

TABLE 1.4
Base case conditions for BIOPLUME III natural attenuation model.

Characteristic	*Value*
Grid Size	9×10
Cell Size	900 ft × 900 ft
Aquifer Thickness	20 ft
Transmissivity	0.1 $ft^2 s^{-1}$
Porosity	30%
Longitudinal Dispersivity	100 ft
Transverse Dispersivity	30 ft
CELDIS	0.5
Simulation Time	2.5 yrs
Source and Loading of Contamination	1 injection well @ 0.1 cfs
Contaminant Concentration at Release	100 mg L^{-1}
Recharge	0 cfs
Boundary Conditions	Constant head, upgradient, and downgradient
Chemical Reactions	None
Biodegradation Reactions	None

Source: U.S. Environmental Protection Agency, 2003. Bioplume III Natural Attenuation Decision Support System, Users Manual, Version 1.0, Washington, D.C.

The benzene concentrations are very sensitive to biodegradation. Interestingly, however, the benzene concentrations were relatively insensitive to changes in molecular oxygen and only slightly sensitive to the electron acceptor concentrations (i.e., in addition to O_2 in aerobic systems, the model evaluates the anaerobic electron acceptors NO_3, Fe, SO_4, and CO_2). All other things being equal, microbes with the most efficient metabolic mechanisms grow at the fastest rate, so these organisms will overwhelm the growth of microbes with less efficient redox systems. Thus, if O_2 is available in surplus, this will be the preferred reaction in the model. Once a system becomes anaerobic, nitrate is the most preferred redox reaction, followed by solid phase ferric iron, sulfate, and carbon dioxide (the least preferred redox reaction).

A thermodynamically dictated system would give preference, even exclusivity, to the reaction that provides the most energy, so the model uses a sequential process that does not allow the microbes to use any other less preferred electron acceptor until the more preferred acceptor is depleted. However, in reality, when monitoring wells are analyzed near plumes undergoing natural attenuation (i.e., active biodegradation), they are seldom entirely depleted in one or more of these electron acceptors. There are seldom such "bright lines" in the field. For example, facultative aerobes, those that can shift from oxygen to anaerobic electron acceptors (especially nitrate), can change electron acceptors even when molecular oxygen is not

TABLE 1.5
Sensitivity of benzene concentrations of plume to hydrogeological and chemical parameters in the Bioplume III model.

Parameter	*Value (*base case)*	*Maximum Benzene Concentration in Plume* ($mg\,L^{-1}$)	*Plume Length (number of cells)*	*Plume Width (number of cells)*
Porosity	15%	75	6	5
	30%*	67	4	3
	45%	80	4	3
Aquifer Thickness (ft)	10	75	6	5
	20*	67	4	3
	40	47	2	2
Transmissivity ($ft^2\,s^{-1}$)	0.01	90	3	3
	0.1*	67	4	3
	0.2	57	5	3
Longitudinal Dispersivity (ft)	10	70	3	3
	50	69	4	3
	100*	67	4	3
Transverse Dispersivity (ft)	10	68	4	3
	30*	67	4	3
	60	66	4	3
Retardation Factor	1*	67	4	3
	2	49	3	2
	5	28	2	1
Abiotic Chemical Half-Life (s)	0*	67	4	3
	1×10^7	20	2	2
	2×10^7	33	2	3

Source: U.S. Environmental Protection Agency, 2003. Bioplume III Natural Attenuation Decision Support System, Users Manual, Version 1.0, Washington, D.C.

completely depleted. This can be attributed to the fact that redox potentials for oxygen and nitrate are not substantially different (at pH 7, O_2 = +820 volts and NO_3 = +740 volts, compared to CO_2 = −240 volts). Also, the apparent divergence from pure thermodynamics in the field may simply be a sampling artifact, which can be attributed to the way monitoring is conducted. For example, monitoring wells do not collect water from a "point." Rather, the screens (the perforated regions of underground piping where water enters) are set at 1.5- to 3-m intervals, so waters will mix from different vertical horizons. Thus, if different reactions are occurring with depth, these are actually aggregated into a single water sample.

When a contaminant degrades sequentially, the slowest degradation step has the greatest influence on the time it takes the chemical to break down. If this most sensitive step can be sped up, the whole process can be sped up. Conversely, if an engineer or scientist devotes much time and effort to one of the faster steps in the degradation sequence, little or no enhancement to the degradation process may occur. Thus, the model seems to point to the need to take care not to avoid, or at least not overgeneralize, the common assumption that a contamination plume is limited by oxygen or even other redox conditions. Adding iron to an anaerobic system or pumping air into an aerobic stratum of an aquifer will help, but only so much. Figure 1.6 demonstrates a way to apply microbial kinetics limits to redox.

Another difference between the lab and the field is the presence of confounding chemical mixtures in real contamination scenarios. For example, leaking underground storage tanks (LUSTs) are a widespread problem. It is tempting to think that since these tanks contain refined fuels, most spills will be similar. However, as we discuss throughout this text, each compound has specific physicochemical properties that will affect its reactivity and movement in the environment. As evidence, benzene, toluene, ethyl benzene, and xylenes (so-called BTEX) usually comprise only a small amount (ranging from about 15 to 26%) of the mole fraction of gasoline or jet fuel.[25] However, largely because the BTEX compounds have high aqueous solubilities (152 to 1780 $mg\,L^{-1}$) compared to the other organic constituents (0.004 to 1230 $mg\,L^{-1}$) in these fuels, they often account for more than two-thirds of the amount of the contaminants that migrate away from the LUST. Also, soils are seldom homogeneous, so even if the contaminant is well characterized, how it will react and move are largely affected by the media's characteristics, such as their potential to sorb pollutants.

Ease of implementation and sensitivity are both important considerations when deciding how to address environmental problems. In some situations, steps that are readily available may be relatively insensitive to the intended outcome. In other situations, immediate and relatively inexpensive measures can be taken that are sensitive, such as pumping air and water to speed up biodegradation in an aquifer that has already shown natural attenuation. This is analogous to the business world concept of "low hanging fruit." Managers are encouraged to make improvements that are relatively easy and that pay immediate dividends, before moving on to the more intractable problems. For example, if a survey shows that employees are unhappy and have low morale because a current policy does not allow them to eat their lunches at their desks, and no good reason can be found for this policy, a manager can simply change the policy at no cost to the company and reap immediate results. However, if the same survey showed that everyone in the organization needs to be retrained at considerable costs to the company, this would call for a more thoughtful and laborious correction pathway. The former improvement (i.e., eating at one's desk) may not greatly affect the bottom line, but it is easy to implement. Improved training may greatly influence the bottom line (i.e., profit is more sensitive

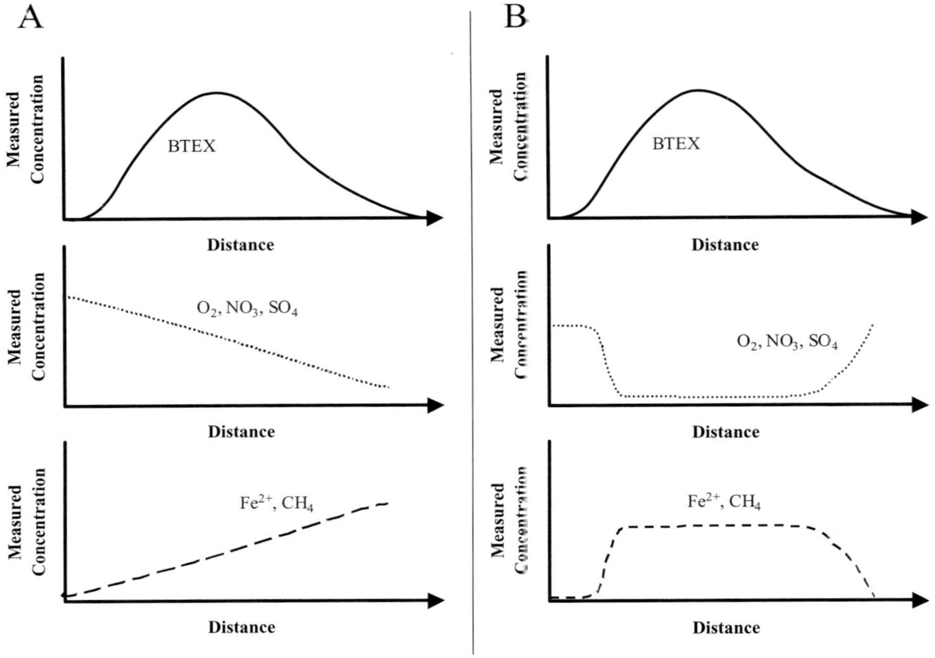

FIGURE 1.6. Two possible hypotheses for how microbes degrade benzene, toluene, ethyl benzene, and xylenes (BTEX): A. Rate of biodegration is limited by microbial kinetics. Concentrations of anaerobic electron acceptors (nitrate and sulfate) decrease at a constant rate downgradient from the pollutant source, with a concominant increase in the concentrations of the byproducts of these anaerobic reactions (ferrous iron and methane). B. Rate of biodegradation is relatively fast (days, not years, so compared to many groundwater replenishment rates, this can be characterized as instantaneous). Virtually all of the nitrate and sulfate anaerobic electron acceptors are depleted, and the iron and methane byproducts of these anaerobic reactions show the highest concentrations near the contaminant source. In both A and B the total concentrations of the byproducts are inversely related to the total concentrations of the principal electron acceptors in the anaerobic reactions overall. *Source*: Adapted from U.S. Environmental Protection Agency, 2003. Bioplume III Natural Attenuation Decision Support System, Users Manual, Version 1.0, Washington, D.C.

to a well-trained work force), but it is difficult to implement. Benzene degradation is highly sensitive to soil type, but there may be little that the engineer can do to manipulate this variable; that is, soil type is a sensitive parameter, but very difficult to change.

The challenge for the environmental professional is to understand the entire system. Based on this understanding, solutions to environmental problems can be developed. Ideally, improvements can be made by

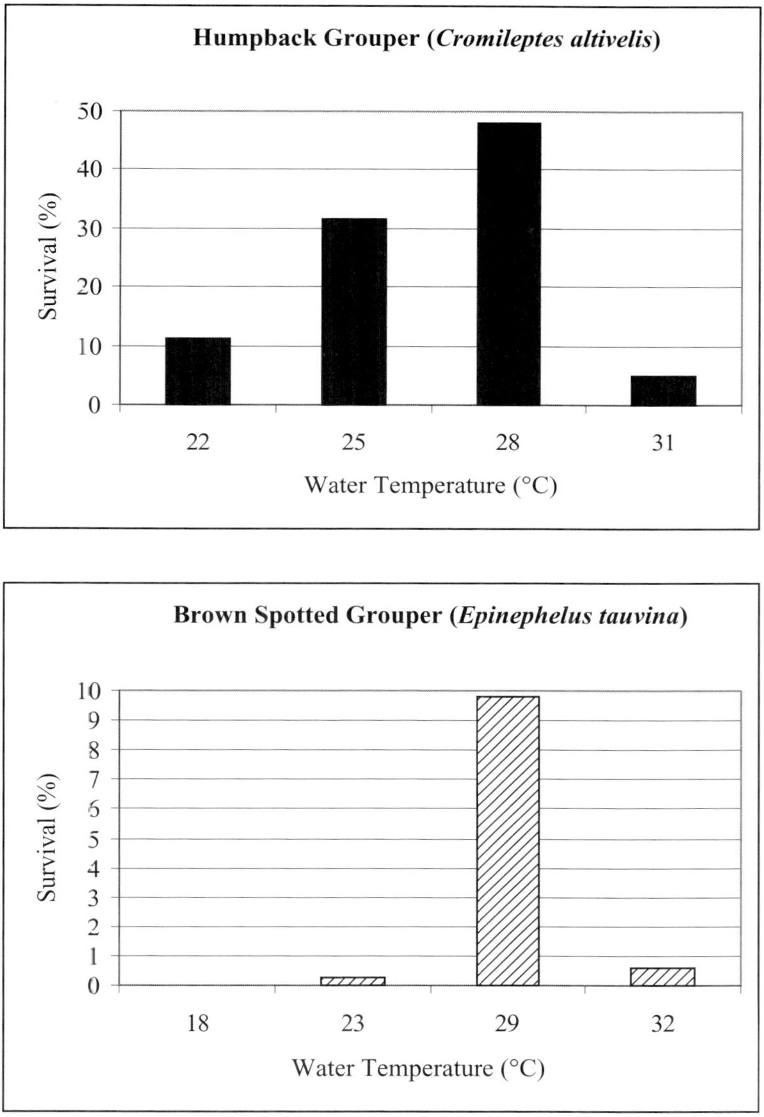

FIGURE 1.7. Effect of changes in temperature on the survival of the larvae of humpback grouper (*Cromileptes altivelis*) and brown spotted grouper (*Epinephelus tauvina*).
Source: Data for *Cromileptes altivelis* from K. Sugama, S. Trijoko, K. Ismi, and M. Setiawati, 2004. *Advances in Grouper Aquaculture,* M.A. Rimmer, S. McBride, and K.C. Williams, eds., ACIAR Monograph 110. Data for *Epinephelus tauvina* from S. Akatsu., K.M. Al-Abdul-Elah, and S.K. Teng, 1983. "Effects of Salinity and Water Temperature on the Survival and Growth of Brown Spotted Grouper Larvae (*Epinephelus tauvina*)," *Journal of the World Maricultural Society,* 14, 624–635.

focusing first on actions that bring about the most improvements; that is, where the environmental responses are most sensitive to changes. Unfortunately, this works the other way as well. That is, some parts of the environment are highly sensitive to small changes. Small changes in surface water temperature, pH, or dissolved oxygen and other essential factors can greatly affect survival. For example, larvae of certain grouper fish require a small range of water temperature in the range of about 28° to 29°C (see Figure 1.7). The figure also demonstrates the importance of interspecies variability.

Environmental systems are complex and complicated. Our understanding of pollution must consider many factors, even in seemingly straightforward instances of environmental degradation. It is prudent, in light of the burgeoning of environmental science and engineering in the past few decades, to reconsider what we mean by pollution.

Notes and Commentary

1. Thomas Bayes, English preacher and mathematician, argued that knowledge of prior events is needed to predict future events. Thus Bayes, like Santayana for political thought, advocated for the role of memory in statistics. Bayes' theorem, which was published two years after his death in 1761 in *An Essay Towards Solving a Problem in the Doctrine of Chances*, introduced the mathematical approach to predict, based on logic and history, the probability of an uncertain outcome. This is very valuable in science; it allows uncertainty to be quantified.
2. R.A. Posner, 2004. *Catastrophe: Risk and Response*, Oxford University Press, New York, NY.
3. M. Rees, 2003. *Our Final Hour: A Scientist's Warning: How Terror, Error, and Environmental Disaster Threaten Humankind's Future In This Century—On Earth and Beyond*, New York, NY.
4. Depending on the journal, this can contradict another tenet of scientific research; that is, the research should be able to be conducted by other researchers, following the methodology described in the article, and derive the same results. However, there is little incentive to replicate research if the likelihood of publication is low. That is, the research is no longer "new" because it was conducted by the original researcher, so the journal may well reject the second, replicate research.
5. However, the engineering profession is beginning to come to grips with this issue; for example, in emergent macroethical areas like nanotechnology, neurotechnology, and even sustainable design approaches. As evidence, see National Academy of Engineering, 2004. *Emerging Technologies and Ethical Issues in Engineering*, The National Academies Press, Washington, D.C.
6. The exception is electric cars, which represent a very small fraction of motorized vehicles; although a growing number of hybrid power supplies (i.e.,

electric systems charged by internal combustion engines) are becoming available.

7. Inductive reasoning is also called *abstraction*, because it starts with something concrete and forms a more abstract ideal. Philosophers have argued for centuries regarding the value of inductive reasoning. Induction is the process that takes specific facts, findings, or cases and then generally applies them to construct new concepts and ideas. Abstraction leaves out specific details, unifying them into a whole based on a defined principle. For example, a brown-feathered chicken, a white-feathered chicken, and a polka-dot-feathered chicken can all be integrated because each is a chicken, albeit with differences. The feather color, then, can be eliminated under the principle or criterion of being a chicken (i.e., chickenness); that is, color is not relevant. A brown chicken, brown bear, and brown paper bag can be integrated under the criteria of having brown color. The other aspects besides "brownness" of each item's characteristics are not relevant in this case, so they are omitted.

   In the eighteenth century, the Scottish philosopher, David Hume, postulated the so-called "problem of induction." To paraphrase, Hume was asking "Why should things that we may be observing on a regular basis continue to hold in the future?" In other words, there is no justification in using induction; because there is no reason that the conclusion of any inductive argument is valid. Like the scientific revolutionaries a couple of centuries earlier, Hume rejected *a priori* reason, since humans are incapable of fully and directly comprehending the laws of nature. This can be accomplished only *a posteriori*, through experience.

   Hume would have a problem with this inductive syllogism:

   Every time I add nickel to my activated sludge, the bacteria grow more rapidly. Therefore, the next time I add Ni to the sludge, my bacteria's growth rate will increase.

   Although engineers can think of many reasons why the Ni addition may not lead to increased growth (e.g., different strains may not have adapted an enzymatic need for Ni, temperature changes may induce changed behaviors that render the Ni ineffective, and incomplete mixing does not allow the microbes access to the Ni), we also know that under the regular (expected?) conditions in the plant that the fact it has worked every time is a strong indicator that it will work again. Mathematicians may have a harder time with this expectation, but is it really any different than pressing your brake pedal and expecting the car to stop? Yes, there is always a probability (hopefully very low) that a leak in the master cylinder or brake line could cause the hydraulics to fail and the car would not stop when the brake pedal is depressed, but such probabilities do not render, in my opinion, inductive reasoning useless.

8. The discussion on intuition draws upon R.M. Hogarth, 2001. *Educating Intuition*, University of Chicago Press, Chicago, IL.
9. *Ibid.* and K. Hammond, 1996. *Human Judgment and Social Policy: Irreducible Uncertainty, Inevitable Error, Unavoidable Injustice*, Oxford University Press, New York, NY.

10. National Academy of Engineering, 2004. *The Engineer of 2020: Visions of Engineering in the New Century*, The National Academies Press, Washington, D.C.
11. For example, see S.B. Billatos and N.A. Basaly, 1997. *Green Technology and Design for the Environment*, Taylor & Francis Group, London, UK.
12. For example, see D.E. Stokes, 1997. *Pasteur's Quadrant*, Brookings Institute Press, Washington, D.C. H. Brooks, 1979. "Basic and Applied Research," *Categories of Scientific Research*, National Academy Press, Washington, D.C, 14–18.
13. This is a typical way that scientists report information. In fact, there may be people who, if they put on shorts, will want to eat ice cream, even if the temperature is –30°. These are known as *outliers*. The term outlier is derived from the prototypical graph that plots the independent and dependent variables (i.e., the variable that we have control over and the one that is the outcome of the experiment, respectively). Outliers are those points that are furthest from the line of best fit that approximates this relationship. There is no standard for what constitutes an outlier, which is often defined by the scientists who conduct the research, although statistics and decision sciences give guidance in such assignments.
14. A. Bradford Hill, 1965. "The Environment and Disease: Association or Causation?" Proceedings of the Royal Society of Medicine, *Occupational Medicine* 58, p. 295.
15. For example, Candace B. Pert, a pioneer in endorphin research, has espoused the concept of mind/body, with all the systems interconnected, rather than separate and independent systems. C. Pert, 1999. *Molecules of Emotion: The Science Behind Mind-Body Medicine*, Scribner Book Company, New York, NY.
16. This is akin to the advice of St. Peter (Acts 24:25 and II Peter 1:6), who linked maturity with greater self-control or temperance (Greek *kratos* for strength). Interestingly, St. Peter's Epistle seems to argue that knowledge is a prerequisite for temperance. Thus, by extension to the professional point of view, it is logical to assume that he would argue that we can really only understand and appropriately apply scientific theory and principles after we practice them. This is actually the structure of most professions. For example, engineers who intend to practice must first submit to a rigorous curriculum (approved and accredited by the Accreditation Board for Engineering and Technology), then must sit for the Future Engineers examination. After some years in the profession (assuming tutelage by more seasoned professionals), the engineer has demonstrated the *kratos* (strength) to sit for the Professional Engineers (PE) exam. Only after passing the PE exam does the National Society for Professional Engineering certify that the engineer is a "professional engineer" and eligible to use the initials PE after his or her name. The engineer is, supposedly, now schooled beyond textbook knowledge and knows more about why in many problems the correct answer is "It depends."

    Likewise, Aristotle (384–322 B.C.) considered excellence (i.e., character and ethics) in one's endeavors to be a matter of practice: "We are what we repeatedly do. Excellence, then, is not an act, but a habit." (*Nicomachean Ethics*,

Book 2, Chapter 1). In fact, the ancient Greek term for habit is the same as its word for character—*ethos*—and the contemporary meaning of ethos is a set of beliefs held by a group. So, the ethos of the engineering profession is in part annunciated through the engineering codes of ethics (e.g., those of the National Society of Professional Engineers or specific engineering disciplines like the American Society of Civil Engineers). Likewise, the American Institute of Certified Planners (AICP) articulates codes of practice for city and regional planners. Environmental scientists like most other scientists, however, do not have such a codification representing their ethos. There has been some debate within the scientific community regarding the need for a scientific code of ethics. One of the arbitrators of these discussions has been *Sigma Xi*, the Scientific Research Society. In fact, it has published two valuable publications addressing ethical issues in research: Sigma Xi, 1997, *Honor in Science*, Research Triangle Park, NC; and, Sigma Xi, 1997, *The Responsible Researcher: Paths and Pitfalls*, Research Triangle Park, NC.

17. For example, epidemiologists who study diseases in populations often begin with descriptive epidemiological studies, such as migration studies showing the incidence and prevalence of diseases in a population that moves from one country to another. Such migration studies may be subsequently subjected to analytical epidemiological studies that look at various risk factors such as diet, lifestyles, and environmental conditions that may differ between the two studies. A prominent example is the description of differences in stomach cancer incidence in Japanese immigrants (i.e., higher in Japan, but lower in the United States) and intestinal cancer incidence in the same Japanese immigrants (i.e., lower in Japan, but higher in the United States). Following such descriptive studies, analytical epidemiology showed that diet changes and refrigeration differences in the two cultures may explain the change. Another example is the migration studies of Irish immigrants to tropical and subtropical climates (e.g., Australia) that showed increases in melanoma and other skin cancers in the next generation, subsequently explained by analytical studies linking increased ultraviolet light exposures to the increased skin cancers. The Japanese and Irish migration studies are also examples of studies from which scientists can generate and test hypotheses. Such extrapolations are also drawn from extreme cases, such as those considered in this book.
18. A. Leopold, 1949. *A Sand County Almanac*, Oxford University Press (1987), New York, NY.
19. World Commission on Environment, 1987. *Our Common Future: Report of the World Commission on Environment and Development*, Oxford University Press, Oxford, UK.
20. This should not be interpreted to mean that advocacy and passion are bad things. They have been extremely important in raising the consciousness and making the case for improving the environment. However, science must be objective. Conflicts of interest and perspective can damage good science. Science must be systematic. Research and other investigations must be able to be repeated and verified. And science must be dispassionate. The scientist must

be an honest arbiter of truth, whether the scientist likes the results or not. All engineers and scientists that I have had the pleasure of knowing have had the ability to compartmentalize their lives. They can be passionate Cardinal or Yankee fans at the ballpark and they can be as bureaucratic as any policy wonk when they are assigned as project officer on contracts or grants. But, in the laboratory or field of investigation, they must be completely objective, systematic, and dispassionate about the methods, approach, interpretations of data, and conclusions.

21. Dispersivity (D) is defined as the ratio of the hydrodynamic dispersion coefficient (*d*) to the pore water velocity (*v*); thus $D = \frac{d}{v}$.
22. For another excellent sensitivity analysis that illustrates the importance of numerous parameters, see J.E. Odencrantz, J.M. Farr, and C.E. Robinson, 1992. "Transport model parameter sensitivity for soil cleanup level determinations using SESOIL and AT123D in the context of the California Leaking Underground Fuel Tank Field Manual," *Journal of Soil Contamination*, 1(2): 159–182. The study found that benzene concentrations are most sensitive to biodegradation rate, climate, effective solubility, and soil organic carbon content.
23. Transmissivity is the rate at which water passes through a unit width of the aquifer under a unit hydraulic gradient. It is equal to the hydraulic conductivity multiplied by the thickness of the zone of saturation. It is expressed as volume per time per length such as gallons per day per foot ($\text{gal}\,\text{d}^{-1}\,\text{ft}^{-1}$) or liters per day per meter ($\text{L}\,\text{d}^{-1}\,\text{m}^{-1}$).
24. Retardation represents the extent to which a contaminant is slowed down compared to if it were moving entirely with the advective movement of the fluid (usually water). For example, if the water in an aquifer or vadose zone is moving at $1 \times 10^{-5}\,\text{cm}\,\text{s}^{-1}$, but due to sorption and other partitioning mechanisms the contaminant is only moving at $1 \times 10^{-6}\,\text{cm}\,\text{s}^{-1}$, the retardation factor (Rf) = 10, so an Rf of 10 means that the contaminant is moving at 1/10 the velocity of the water. Rf is a correction factor that accounts for the degree to which a contaminant's velocity is affected by sorption in the groundwater system. An Rf calculation must consider the bulk density of the media, porosity, and the distribution coefficient (K_d).
25. See, for example, P.C. Johnson, M.W. Kemblowski, and J.D. Colthart, 1990a. "Quantitative Analysis of Cleanup of Hydrocarbon-Contaminated Soils by In-Situ Soil Venting," *Ground Water*, Vol. 28, No. 3, May–June, 1990, pp. 413–429. P.C. Johnson, C.C. Stanley, M.W. Kemblowski, D.L. Byers, and J.D. Colthart, 1990b. "A Practical Approach to the Design, Operation, and Monitoring of In Site Soil-Venting Systems," *Ground Water Monitoring and Remediation*, Spring 1990, pp. 159–178. M.E. Stelljes and G.E. Watkin, 1993. "Comparison of Environmental Impacts Posed by Different Hydrocarbon Mixtures: A Need for Site Specific Composition Analysis," *Hydrocarbon Contaminated Soils and Groundwater*, Vol. 3, P.T. Kostecki and E.J. Calabrese, eds., Lewis Publishers, Boca Raton, p. 554.

CHAPTER 2

Pollution Revisited

Before discussing the shifting environmental paradigms, it is important to understand what the term *pollution* actually means. Over the past few decades the central feature of pollution has been its association with harm. The objects of the harm have received varying levels of interests. In the 1960s, harm to ecosystems, including threats to the very survival of certain biological species, was paramount. This concern was coupled with harm to humans, especially in terms of diseases, such as respiratory diseases associated with air pollutants and infectious diseases brought on by polluted drinking water.

There was also growing concern that sheer volumes of pollutants and diminishing resources like land and water would lead to major demographic and health problems in human populations. The need to close "open dumps" and replace them with engineered (sanitary) landfills, for example, and the construction of wastewater treatment plants completely changed the engineering profession. Sanitary engineering became environmental engineering. Also, the concern about the large volumes of municipal solid wastes was seen as a crisis. Other emerging concerns were also being apparent, including anxiety about nuclear power plants, particularly the possibilities of meltdown and the generation of cancer-causing nuclear wastes; petrochemical concerns, such as the increasing production and release of ominous-sounding chemicals like DDT and other pesticides; as well as spills of oil and other chemicals. These apprehensions would increase in the next decade, with the public's growing wariness about toxic chemicals added to the more familiar conventional pollutants like soot, carbon monoxide, and oxides of nitrogen and sulfur. The major concern about toxics was cancer. The next decades kept these concerns, but added new ones, including threats to hormonal systems in humans and wildlife, neurotoxicity (especially in children), and immune system disorders.

Growing numbers of studies in the last quarter of the twentieth century provided evidence linking disease and adverse effects to extremely low levels of certain particularly toxic substances. For example, exposure to dioxin at almost any level above what science could detect could be associated with numerous adverse effects in humans. During this time, other objects of pollution were identified, including loss of aquatic diversity in lakes due to deposition of acid rain. Acid deposition was also being associated with the corrosion of materials, including some of the most important human-made structures, such as the pyramids in Egypt and monuments to democracy in Washington, D.C. Somewhat later, global pollutants were identified, such as those that seemed to be destroying the stratospheric ozone layer or those that appeared to be affecting the global climate.

Old Paradigm:
Pollution can be absolutely defined.
Paradigm Shift:
Pollution can be defined only within its physical, chemical, and biological context.

This escalation of awareness of the multitude of pollutants complicated matters. For example, many pollutants under other circumstances would be "resources," such as compounds of nitrogen. In the air, these compounds can cause respiratory problems directly or, in combination with hydrocarbons and sunlight, indirectly can form ozone and smog. But, in the soil, nitrogen compounds are essential nutrients. So, it is not simply a matter of removing pollutants, but one of managing systems to ensure that optimal conditions for health and environmental quality exist.

> It isn't pollution that's harming the environment. It's the impurities in our air and water that are doing it. J. Danforth Quayle, U.S. Vice President[1]

Although the cases in this book vary considerably in scope and effect, they have at least one thing in common; each situation involves some type of environmental damage. At first blush, former Vice President Quayle's quote appears to be contradictive and even absurd, but upon closer examination it seems to point to a challenge for environmental professionals. When does something in our air, water, food, or soil change from being harmless or even beneficial to being harmful? Scientists often are asked to define pollution. A working definition can be found by turning Quayle's quote around; that is, impurities are common, but in excessive quantities and in the wrong places they become harmful. One of the most interesting definitional quandaries about pollution was stimulated by language in the Federal Water Pollution Control Act Amendments of 1972 (Public Law 92-500). The objective of this law is to restore and maintain the chemical, physical, and biological integrity of the nation's waters. To achieve this objective, the law set two goals: the elimination of the discharge of all pollutants into the navigable waters of the United States by 1985; and to provide an interim level of water quality to protect fish, shellfish, and wildlife and recreation by 1983.[2] Was Congress serious? Could they really mean that they had expected all sources that drained into U.S. lakes and rivers to be completely free of pollutants in 13 years? Or did this goal hinge upon the definition of pollutant? In other words, even toxic substances are not necessarily pollutants if they exist below a threshold of harm. In light of the fact that this same law established so-called effluent limitations, there is a strong likelihood that the definition called for in this goal was concentration-based.[3] More recently, the term *zero-emission* has been applied to vehicles (as the logical next step following low-emission vehicles (LEVs) and ultra-low-emission vehicles (ULEVs) in recent years). However, zero emissions of pollutants will not be likely for the foreseeable future, especially if we consider that even electric cars are not emission free, but actually emission trading, since the electricity is generated at a power plant that is emitting pollutants as it burns fossil fuels or has the problem of radioactive wastes if it is a nuclear power plant. Even hydrogen, solar, and wind systems are not completely pollution-free since the parts and assemblages require energy and materials that may even include hazardous substances.

These definitional uncertainties beg the question, then, of when does an impurity become a pollutant? Renaissance thinking may help us here. Paracelsus, the sixteenth century scientist, is famous for his contention that "dose alone makes a poison. . . . All substances are poisons; there is none which is not a poison. The right dose differentiates a poison and a remedy."[4] Paracelsus' quote illuminates a number of physical, chemical, and biological concepts important to understanding pollution. Let us consider two. First, the poisonous nature, or the toxicology, of a substance must be related to the circumstances of exposure. In other words, to understand a pollutant, we must appreciate its context. What is the physical, chemical, and biological nature of the agent to which the receptor (e.g., a person, an endangered species, or an entire population or ecosystem) is exposed? What is that person's existing health status? What is the condition of the ecosystem? What is the chemical composition and physical form of the contaminant? Is the agent part of a mixture, or is it a pure substance? How was the person or organism exposed—from food, drink, air, through the skin? These and other characterizations of a contaminant must be known to determine the extent and degree of harm.

The second concept highlighted by Paracelsus is that dose is related to response. This is what scientists refer to as a biological gradient, or a *dose-response* relationship. Under most conditions, the more poison to which we are exposed, the greater the harm.

The classification of harm is an expression of *hazard*, which is a component of risk. The terms hazard and risk are frequently used interchangeably in everyday parlance, but hazard is actually a component of risk, not synonymous with risk. A hazard is expressed as the potential of unacceptable outcome, and risk is the likelihood (i.e., probability) that such an adverse outcome will occur. A hazard can be expressed in numerous ways (see Tables 2.1 and 2.2). For chemical or biological agents, the most

TABLE 2.1
Four types of hazards important to hazardous wastes, as defined by the Resource Conservation and Recovery Act of 1976 (42 U.S.C. s/s 6901 *et seq.*).

Hazard Type	*Criteria*	*Physical/Chemical Classes in Definition*
Corrosivity	A substance with an ability to destroy tissue by chemical reactions.	Acids, bases, and salts of strong acids and strong bases. The waste dissolves metals, other materials, or burns the skin. Examples include rust removers, waste acid, alkaline cleaning fluids, and waste battery fluids. Corrosive wastes have a pH of <2.0 or >12.5. The U.S. EPA waste code for corrosive wastes is D002.
Ignitability	A substance that readily oxidizes by burning.	Any substance that spontaneously combusts at 54.3°C in air or at any temperature in water, or any strong oxidizer. Examples are paint and coating wastes, some degreasers, and other solvents. The U.S. EPA waste code for ignitable wastes is D001.
Reactivity	A substance that can react, detonate, or decompose explosively at environmental temperatures and pressures.	A reaction usually requires a strong initiator (e.g., an explosive like TNT, trinitrotoluene), confined heat (e.g., salt peter in gunpowder), or explosive reactions with water (e.g., Na). A reactive waste is unstable and can rapidly or violently react with water or other substances. Examples include wastes from cyanide-based plating operations, bleaches, waste oxidizers, and waste explosives. The U.S. EPA waste code for reactive wastes is D003.
Toxicity	A substance that causes harm to organisms. Acutely toxic substances elicit harm soon after exposure (e.g., highly toxic pesticides causing neurological damage within hours after exposure). Chronically toxic substances elicit harm after a long period of time of exposure (e.g., carcinogens, immunosuppressants, endocrine disruptors, and chronic neurotoxins).	Toxic chemicals include pesticides, heavy metals, and mobile or volatile compounds that migrate readily, as determined by the Toxicity Characteristic Leaching Procedure (TCLP), or a TC waste. TC wastes are designated with waste codes D004 through D043.

TABLE 2.2
Biologically-based classification criteria for hazardous waste.[5]

Criterion	*Description*
Bioconcentration	The process by which living organisms concentrate a chemical contaminant to levels exceeding the surrounding environmental media (e.g., water, air, soil, or sediment).
Lethal Dose (LD)	A dose of a contaminant calculated to expect a certain percentage of a population of an organism (e.g., minnow) exposed through a route other than respiration (dose units are mg [contaminant] kg^{-1} body weight). The most common metric from a bioassay is the lethal dose 50 (LD_{50}), wherein 50% of a population exposed to a contaminant is killed.
Lethal Concentration (LC)	A calculated concentration of a contaminant in the air that, when respired for four hours (i.e., exposure duration = 4 h) by a population of an organism (e.g., rat) will kill a certain percentage of that population. The most common metric from a bioassay is the lethal concentration 50 (LC_{50}), wherein 50% of a population exposed to a contaminant is killed. (Air concentration units are mg [contaminant] L^{-1} air.)

important hazard is the potential for disease or death (referred to in medical literature as morbidity and mortality, respectively). So, the hazards to human health are referred to collectively in the medical and environmental sciences as toxicity. Toxicology is chiefly concerned with these health outcomes and their potential causes.

To scientists and engineers, risk is a straightforward mathematical and quantifiable concept. Risk equals the probability of some adverse outcome. Any risk is a function of probability and consequence.[6] The consequence can take many forms. In environmental sciences, a consequence is called a hazard. Risk, then, is a function of the particular hazard and the chances of a person (or neighborhood or workplace or population) being exposed to the hazard. In environmental situations, this hazard often takes the form of toxicity, although other public health and environmental hazards abound.

To illustrate the difference between hazard and risk, consider two students in the same undergraduate genetics class. Amy has made A's in all her science and math courses that are prerequisites for the genetics course. She has taken abundant notes, has completed all homework assignments, and participates in study groups every Tuesday evening. Mike, on the other hand, has taken only one of the four prerequisite courses, receiving a D. He has completed less than half of his homework assignments and does not participate in study groups. Amy and Mike share the identical hazard, that is, flunking the genetics course. However, based upon the data given here, we would estimate that their individual risks of flunking are very different, with Mike's being much greater. Of course, this does not mean Mike *will* flunk genetics, or even that Amy *will* pass. It only means that the probability is more likely that Mike will fail the course. Even an A student has the slim chance of failing the course (e.g., may experience testing anxiety, have personal problems the week of the final, etc.), just as a failing student has a slim chance of passing the course (e.g., becomes motivated, catches up on homework, reaches a state of illumination, happens to guess the right genetic outcomes, etc.). So, risk assessment is seldom a sure thing, or 100% probability, but the risk difference between Amy and Mike can be very large, say 0.0001 for Amy and 0.85 for Mike.

The example also illustrates the concept of risk mitigation. For example, if Mike does begin to take actions, he can decrease the probability (i.e., risk). Perhaps, by participating in a study group, he decreases the risk of flunking to 0.5 (50%), and by also catching up on his homework, the risk drops to 0.2 (20%). Thus, implementing two risk abatement actions lowered his risk of flunking genetics by 65% (from 85% to 20%). Mike still has a greater risk of failure than Amy does, but now he is more likely to pass than to fail. Risk mitigation is very important for environmental problems. In fact, one of the lessons learned in every case discussed in this book is that adverse outcomes can be avoided or at least greatly reduced if mitigative or preventive measures are taken.

To illustrate further the difference between hazard and risk, let us consider an environmental example: a highly exposed individual versus an individual with a very low exposure. Jimmy works in a lead foundry, is removing lead-containing paint from his home walls, drinks from a private well with average lead concentrations of $10\,\text{mg}\,\text{L}^{-1}$, and, in his spare time, breaks down automobile batteries to remove the lead cores. Louie is the same sex and age as Jimmy, but Louie's only exposure to lead is from the public drinking water supply, which on average is $0.001\,\text{mg}\,\text{L}^{-1}$. Lead is well known to be neurotoxic, causing damage to the central and peripheral nervous systems of mammals, including humans. The principal hazard in this instance is neurotoxicity. The hazard is identical for Jimmy and Louie—central and peripheral nervous system disorders. But the neurotoxic risk to Jimmy is much higher than the neurotoxic risk to Louie. Environmental risks can quickly become complicated. For example, if Jimmy were an adult and Louie were an infant, the risks of neurotoxicity could actually be much higher for Louie, even if Louie's exposure is much lower than Jimmy's exposure to the lead. This is because of the physiological differences, such as the rate of tissue growth (very high for the infant and much lower for an older adult) and the longer time period that the infant will have to accumulate lead in his tissues. Other factors like sex and critical times in life that are more vulnerable to the effects of certain agents (e.g., hormonally active compounds), such as during gestation, infancy, puberty, and pregnancy can result in completely different risks of two individuals, even though the hazard and exposures are identical.

Note that chemical concentration is part of the risk equation, but this is influenced by a person's or a population's activities (e.g., working, touching, drinking, and breathing in different situations). For example, several of Jimmy's activities would be greater than the 99th exposure percentile. A source of information about such activities is the *Exposure Factors Handbook*,[7] which summarizes statistical data on the different factors needed to assess how people are exposed to contaminants. The factors include:

- Drinking water consumption
- Soil ingestion
- Inhalation rates
- Dermal factors, such as skin area and soil adherence factors
- Consumption of fruits and vegetables, fish, meats, dairy products, and homegrown foods
- Breast milk intake
- Human activity factors
- Consumer product use
- Residential characteristics

The handbook provides the recommended exposure values for the general population, as well as for highly exposed, susceptible, and sensitive subpopulations, which may have characteristics different from the general population. This is particularly important for environmental justice projects. Engineers are encouraged to calculate exposures and risks that are greater than average risks; for example, some standard deviations higher than measures of central tendency (mean, median, or mode), out in the tail of the distribution. After all, environmental justice communities, by definition, are exposed to contaminants disproportionately, compared to the general population.

Ameliorating one risk can, if we are lucky, also lessen another risk, such as when pollution control equipment removes particles and in the process also removes heavy metals that are sorbed to the particles. This means that not only are risks to heart and lung diseases reduced, but neurological risks are also reduced because of the decrease in exposures to lead (Pb), mercury (Hg), and other neurotoxic metals. Conversely, reducing one risk can introduce other risks, such as when solid waste is incinerated, eliminating the possibility of long-term risks from contaminated groundwater, but increasing the concentrations of products of incomplete combustion in the air, as well as creating bottom ash with very high concentrations of toxic metals. Another environmental challenge is how to avoid switching one exposed group for another. For example, to address the concern of possible exposures of building inhabitants to asbestos in building materials, we are likely to create occupational asbestos exposures to workers called in to remove the materials. In fact, for environmental justice situations, sometimes the overall population risk is lowered by moving contaminants to sparsely populated regions, but the risk to certain groups is in fact increased. So, we are often confronted with risk tradeoffs.[8] (See the Discussion Box, "DDT versus Eco-Colonialism: Trading Risks.")

DDT versus Eco-Colonialism: Trading Risks

The molecule 1,1,1-trichloro-2,2-bis-(4-chlorophenyl)-ethane is best known as DDT. Interestingly, when you were born is a major factor in what you think about DDT. In the United States, the World War II generation is more inclined to consider DDT in a rather positive light. I have recently asked our undergraduate students in ethics and environmental courses whether DDT is good or bad. The question is intentionally open-ended, and is designed to gauge attitudes and preconceived ideologies about environmental values. These younger respondents are much more likely to call DDT bad. They are generally likely to have read Rachel Carson's seminal work, *Silent Spring*,[9] which was emblematic of the negative change in thinking about organic pesticides in the 1960s, particularly that these synthesized molecules were threats to wildlife, especially birds (hence the "silent" spring), as well as to human health (particularly cancer).

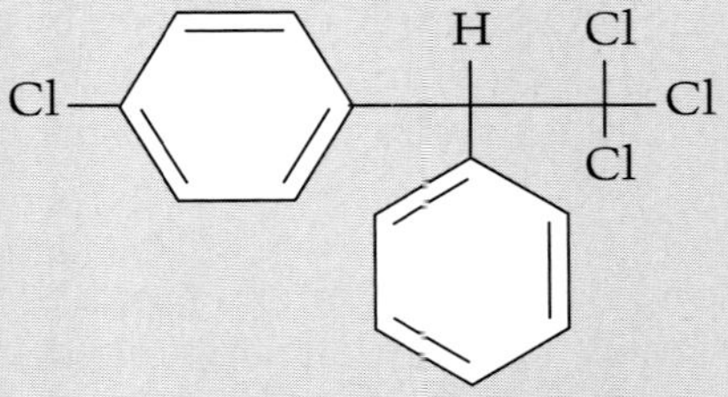

DDT

Conversely, the students are less aware that Allied troops were protected from malaria, typhus and other vector-borne diseases while stationed in tropical regions during wartime, and that the chemist Paul H. Müller won the 1948 Nobel Prize for Physiology or Medicine for synthesizing DDT. In his 1948 acceptance speech,

Müller was prescient in articulating the seven criteria for an ideal pesticide:[10]

1. Great insect toxicity.
2. Rapid onset of toxic action.
3. Little or no mammalian or plant toxicity.
4. No irritant effect and no or only a faint odor (in any case not an unpleasant one).
5. The range of action as wide as possible, covering as many Arthropoda as possible.
6. Long, persistent action; that is, good chemical stability.
7. Low price (= economic application).

Thus, the combination of efficaciousness and safety was, to Müller, the key. Disputes between the pros and cons of DDT are interesting in their own light. The environmental and public health risks *versus* the commercial benefits can be hotly debated. Our students rightfully are concerned that even though the use of a number of pesticides, including DDT, has been banned in Canada and the United States, we may still be exposed by importing food that has been grown where these pesticides are not banned. In fact, Western nations may still allow the pesticides be formulated at home, but do not allow their application and use. So, the pesticide comes back in the products we import; known as the "circle of poisons."

However, arguments of risks versus risks are arguably even more important. In other words, it is not simply a matter of taking an action (banning worldwide use of DDT), which leads to many benefits (less eggshell thinning of endangered birds and less cases of cancer). Instead, it sometimes comes down to trading off one risk for another. Since there are yet to be reliable substitutes for DDT in treating disease-bearing insects, policy makers must decide between ecological and wildlife risks and human disease risk. Also, since DDT has been linked to some chronic effects like cancer and endocrine disruption, how can these be balanced against expected increases in deaths from malaria and other diseases where DDT is part of the strategy for reducing outbreaks? Is it appropriate for economically developed nations to push for restrictions and bans on products that can cause major problems in the health of people living in developing countries? Some have even accused Western nations of eco-imperialism when they attempt to foist temperate climate solutions onto tropical, developing countries. That is, we are exporting fixes based upon our values (anti-cancer, ecological) that are incongruent with the values of other cultures (primacy of acute diseases over chronic effects, e.g., thousands of cases of malaria are more important to some than a few cases of cancer, and certainly more important than threats to the bald eagle from a global reservoir of persistent pesticides).

Finding substitutes for chemicals that work well on target pests can be very difficult. This is the case for DDT. In fact, the chemicals that have been formulated to replace it have either been found to be more dangerous (e.g., aldrin and dieldrin, which have also been subsequently banned) or much less effective in the developing world (e.g., pyrethroids). For example, spraying DDT in huts in tropical and subtropical environments, fewer mosquitoes are found compared to untreated huts. This likely has much to do with the staying power of DDT in mud structures compared to the higher chemical reactivity of pyrethroid pesticides.

Although the DDT dilemma represents a global issue, it has numerous lessons for us as we deal with local problems with risk tradeoffs. First, we must ensure that our recommendations are based upon sound science. This is not always easy. For example, a chemical that has been found to be effective may have an ominous sounding name, leading the community members to call for its removal. However, the chemical may have very low acute toxicity, has never been associated with cancer in any animal or human studies, and is not regulated by any agency. The engineer's job is not done by declaring that removal of the chemical is not necessary, but needs to provide clear information in a way that is understandable to the public.

Second, removal and remediation efforts are never risk-free in and of themselves. Sometimes, a spike in exposures is possible during the early stages of removal and treatment, as the chemical may have been in a place and form that made this less available until actions were taken. In fact, the concept of "natural attenuation" has recently gained greater acceptance within the environmental community. However, the engineer should expect some resistance from the local community when they are informed that the best solution is to do little or nothing but to allow nature (i.e., indigenous microbes) to take its course!

Third, the comparison of doing anything with doing nothing cannot always be captured with a benefit/cost ratio. Opportunity costs and risks are associated with taking no action (e.g., the community loses an opportunity to save a valuable wetland or enhance a shoreline). But the costs (time and money) are not the only reasons for avoiding an environmental action. Constructing the new wetland or adding sand to the shoreline could inadvertently attract tourists and other users who could end up presenting new and greater threats to the community's environment. So, it is not simply a matter of benefits versus cost, it is often one risk being traded for another.

Often, addressing contravening risk is a matter of optimization, which is a proven analytical tool in environmental engineering. However, the greater the number of contravening risks that are possible, the more complicated such optimization routines become.

Risk tradeoff is a very common phenomenon in everyday life. For example, local governments enforce building codes to protect health and safety. Oftentimes, these added protections are associated with indirect, countervailing risks. For example, the costs of construction may increase safety risks via income and stock effects. The income effect results from pulling money away from family income to pay the higher mortgages, making it more difficult for the family to buy other items or services that would have protected them. The stock effect results when the cost of the home is increased and families have to wait to purchase a new residence, so they are left in substandard housing longer.[11] Such countervailing risks are common in environmental decisions, such as the MTBE example in Chapter 1, where solving an air pollution problem created a different water pollution problem. Another example in the 1970s that continues today is that of exclusionary zoning and other community aesthetics measures and their effect on housing stock. The people arguing for major environmental standards were also arguing for increased risks from income and stock effects by imposing increased environmental controls, such as larger lot sizes. In fact, some were arguing against the housing development completely, meaning that stock effects would continue unabated until an alternate development is approved. Thus, the engineer frequently is asked to optimize two or more conflicting variables in environmental justice situations.

Reliability

Reliability is an engineering term that is important in understanding pollution. Like risk, reliability is an expression of likelihood, but rather than conveying something bad, it tells us the probability of a good outcome. Reliability is the extent to which something can be trusted. A system, process, or item is reliable to the extent that it performs the designated function under the specified conditions during a certain time period. Thus, reliability means that something will not fail prematurely. Or, stated more positively, reliability is expressed mathematically as the probability of success. Thus reliability is the probability that something that is in operation at time 0 (t_0) will still be operating until the designated life (time $t = (t_t)$) has expired. People in neighborhoods near the proposed location of a facility want to know if it will work and will not fail. This is especially true for those facilities that may affect the environment, such as landfills and power plants. Likewise, when environmental cleanup is being proposed, people want to know how certain the engineers are that the cleanup will be successful.

The probability of a failure per unit time is the hazard rate, a term familiar to environmental risk assessment, but many engineers may recognize it as a failure density, or $f(t)$. This is a function of the likelihood that an adverse outcome will occur, but note that it is not a function of the severity of the outcome. The $f(t)$ is not affected by whether the outcome is very severe (such as pancreatic cancer and loss of an entire species) or relatively benign (muscle soreness or minor leaf damage). The likelihood that something will fail at a given time interval can be found by integrating the hazard rate over a defined time interval:

$$P\{t_1 \le T_f \le t_2\} = \int_{t_1}^{t_2} f(t)dt \tag{2.1}$$

where T_f = time of failure.

Thus, the reliability function $R(t)$, of a system at time t is the cumulative probability that the system has not failed in the time interval from t_0 to t_t:

$$R(t) = P\{T_f \ge t\} = 1 - \int_{0}^{t} f(x)dx \tag{2.2}$$

One major point worth noting from the reliability equations is that everything we design will fail. Engineers can improve reliability by extending the time (increasing t_t). This is done by making the system more resistant to failure. For example, proper engineering design of a landfill barrier can decrease the flow of contaminated water between the contents of the landfill and the surrounding aquifer, for example, a velocity of a few microns per decade. However, the barrier does not completely eliminate failure (i.e., $R(t) = 0$); it simply protracts the time before the failure occurs (increases T_f). So, the failures noted in this book are those where the failure was unacceptable in time and space. Handling toxic materials is a part of contemporary society. Handling them properly is a function of reliable systems to ensure that the substances are not allowed to migrate or change in a matter that causes harm.

Equation 2.2 illustrates that if we have built-in vulnerabilities, such as unfair facility siting practices or the inclusion of inappropriate design criteria, like cultural bias, the time of failure is shortened. If we do not recognize these inefficiencies up front, we will pay by premature failures (e.g., law suits, unhappy clients, and a public that has not been well served in terms of our holding paramount their health, safety, and welfare).

A discipline within engineering, reliability engineering, looks at the expected or actual reliability of a process, system, or piece of equipment to identify the actions needed to reduce failures, and once a failure occurs, how to manage the expected effects from that failure. Thus, reliability is the mirror image of failure. Since risk is really the probability of failure (i.e., the probability that our system, process, or equipment will fail), risk and reliability are two sides of the same coin. A tank leaking chemicals into groundwater is an engineering failure, as is exposure of people to carcinogens in the air, water, and food. A system that protects one group of people at the expense of another is also a type of failure (i.e., environmental injustice), such as the location of a polychlorinated biphenyl (PCB) landfill in a historically African American neighborhood is Warren County, NC (see Chapter 11). So, if we are to have reliable engineering, we need to make sure that whatever we design, build, and operate is done so with fairness. Otherwise, these systems are, by definition, unreliable.

The most common graphical representation of engineering reliability is the bathtub curve (see Figure 2.1). The curve is U-shaped, meaning that failure is more likely to occur at the beginning (infant mortality) and near the end of the life of a system, process, or equipment. Actually, failure can occur even before infancy. In fact, many problems in environmental justice

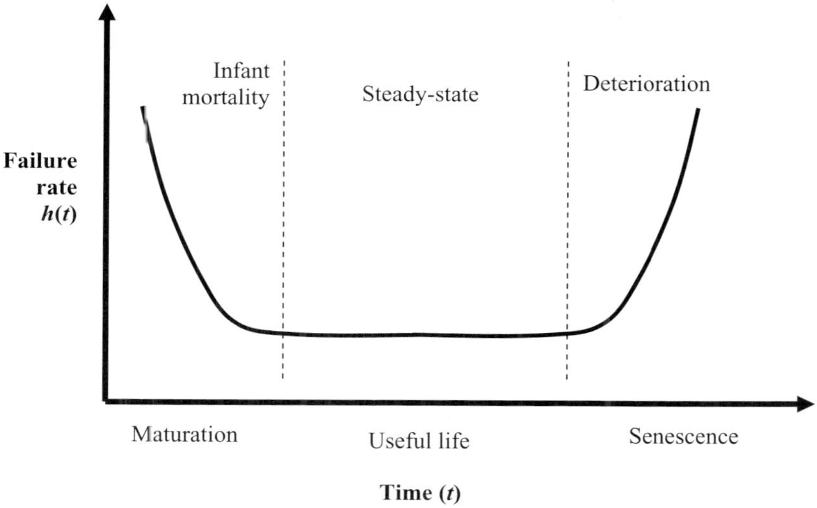

FIGURE 2.1. Prototypical reliability curve, i.e., the bathtub distribution. The highest rates of failure, $h(t)$, occur during the early stages of adoption (infant mortality) and when the systems, processes, or equipment become obsolete or begin to deteriorate. For well-designed systems, the steady-state period can be protracted, e.g., decades.

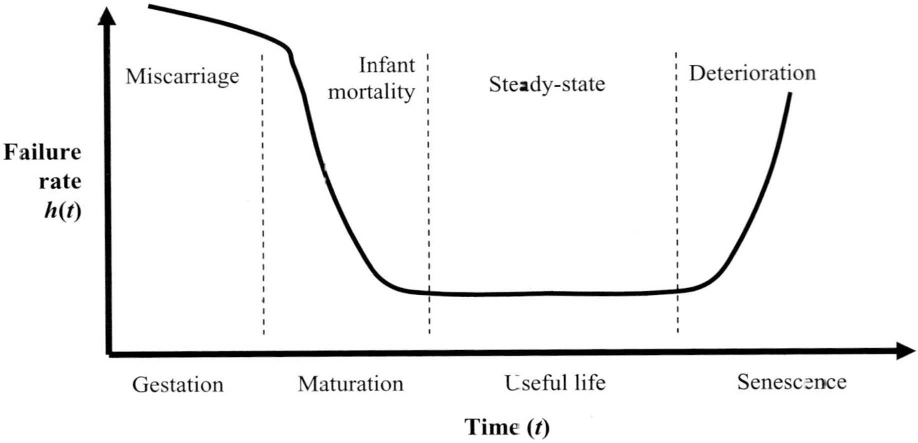

FIGURE 2.2. Prototypical reliability curve with a gestation (i.e., idea) stage. The highest rate of failure, $h(t)$, occurs even before the system, process, or equipment has been made a reality. Exclusion of people from decision making or failure to get input about key scientific or social variables can create a high hazard.

occur during the planning and idea stage. A great idea may be shot down before it is born.

Error can gestate even before the engineer becomes involved in the project. This "miscarriage of justice" follows the physiological metaphor closely. Certain groups of people historically have been excluded from preliminary discussions, so that if and when they do become involved they are well beyond the "power curve" and have to play catch up. The momentum of a project, often being pushed by the project engineers, makes participation very difficult from some groups. So, we can modify the bathtub distribution accordingly (see Figure 2.2).

Note that in environmental engineering and other empirical sciences there is another connotation of reliability, which is an indication of quality, especially for data derived from measurements, including environmental and health data. In this use, reliability is defined as the degree to which measured results are dependable and consistent with respect to the study objectives, for example, stream water quality. This specific connotation is sometimes called test reliability, in that it indicates how consistent measured values are over time, how these values, compare to other measured values, and how they differ when other tests are applied. Test reliability, like engineering reliability, is a matter of trust. As such, it is often paired with test validity; that is, just how near to the true value (as indicated by some type of known standard) the measured value is. The less reliable and valid the results, the less confidence scientists and engineers have in interpreting and using them. This is very important in engineering communications generally and risk communications specifically.

To solve environmental problems, scientists, engineers, and decision makers need to know how reliable and valid the data are. And this information must be properly communicated to those potentially or actually being affected. This includes candid and understandable ways to describe all uncertainties. Uncertainties are ubiquitous in risk assessment. The Chinese word for risk, *weij-ji*, is a combination of two characters, one representing danger and the other opportunity. *Weij-ji* indicates that risk is always an uncertain balance between benefit and cost; between gain and loss. The engineer should take care not to be overly optimistic, nor overly pessimistic about what is known and what needs to be done. Full disclosure is simply an honest rendering of what is known and what is lacking for those listening to make informed decisions. But, remember, a word or phrase can be taken many ways. Engineers should liken themselves to physicians writing prescriptions. Be completely clear, otherwise confusion may result and lead to unintended, negative consequences.

The concept of pollution is widely accepted today, but this has not always been the case. And, there are still raging battles over when something is a pollutant or simply an impurity. Benchmarks for environmental quality developed incrementally, often after "wakeup calls" in the form of pollution episodes that led to death and destruction. Let us consider some of the landmark events that helped to galvanize the public acceptance of the need for measures and protection against pollution, beginning with the landmark air pollution cases that helped to shape our contemporary environmental psyche and create a new ethos.

Characterizing Pollutants

A convenient way to categorize pollutants is by the ease with which they are able to change and move from one environmental compartment (surface waters) to another (sediment, soil, air, or even living tissues). During major leaks and spills, extremely high concentrations of very toxic pollutants have been released, but fortunately most pollutants generally exist at very low concentrations in the environment. These pollutant concentrations are the driving factor and constraint on how much of a pollutant will move from one compartment to another. In fact, the net amount that is transported is limited by equilibrium constraints which are quantified by partition coefficients related to the concentration of the gaining and losing compartments. So, a partition coefficient (K_{ij}) is defined as the ratio of the equilibrium concentration (C) of a pollutant in one environmental compartment (i) with respect to another environmental compartment (j):

$$K_{i,j} = \frac{C_i}{C_j} \tag{2.3}$$

These partition coefficients generally are derived from experiments where varying amounts of a pollutant are observed to see how much moves between the two compartments. For our purposes here (i.e., to understand environmental problems), we do not need to address the theoretical aspects of partitioning. Partitioning theory is covered in detail in the companion text, D.A. Vallero, 2004, *Environmental Contaminants: Assessment and Control* (Elsevier Academic Press, Burlington, MA). Pollutants eventually will reach equilibrium between two compartments. Equilibrium is both a physical and chemical concept. It is the state of a system where the energy and mass of that system are distributed in a statistically most probable manner, obeying the laws of conservation of mass, conservation of energy (first law of thermodynamics), and efficiency (second law of thermodynamics). So, if the reactants and products in a given reaction are in a constant ratio, that is, the forward reaction and the reverse reactions occur at the same rate, then that system is in equilibrium. Up to the point where the reactions are yet to reach equilibrium, the process is *kinetic*; i.e., the rates of particular reactions are considered (see Appendix 1).

Chemical kinetics is the description of the rate of a chemical reaction.[12] This is the rate at which the reactants are transformed into products. This may take place by abiotic or by biological systems, such as microbial metabolism. Since a rate is a change in quantity that occurs with time, the change we are most concerned with is the change in the concentration of our contaminants into new chemical compounds:

$$\text{Reaction rate} = \frac{\text{change in product concentration}}{\text{corresponding change in time}} \tag{2.4}$$

and

$$\text{Reaction rate} = \frac{\text{change in reactant concentration}}{\text{corresponding change in time}} \tag{2.5}$$

In environmental degradation, the change in product concentration will be decreasing proportionately with the reactant concentration, so, for contaminant X, the kinetics looks like:

$$\text{Rate} = -\frac{\Delta(X)}{\Delta t} \tag{2.6}$$

The negative sign denotes that the reactant concentration (the parent contaminant) is decreasing. It stands to reason then that the degradation product Y resulting from the concentration will be increasing in proportion to the decreasing concentration of the contaminant X, and the reaction rate for Y is:

$$\text{Rate} = \frac{\Delta(Y)}{\Delta t} \tag{2.7}$$

By convention, the concentration of the chemical is shown in parentheses to indicate that the system is not at equilibrium. $\Delta(X)$ is calculated as the difference between an initial concentration and a final concentration:

$$\Delta(X) = \Delta(X)_{\text{fiina}} - \Delta(X)_{\text{initial}} \tag{2.8}$$

So, if we were to observe the chemical transformation[13] of one isomer of the compound butane to different isomer over time, this would indicate the kinetics of the system, in this case the homogeneous gas phase reaction of *cis*-2-butene to *trans*-2-butene (see Figure 2.3) for the isomeric structures. The transformation is shown in Figure 2.4. The rate of reaction at any time is the negative of the slope of the tangent to the concentration curve at that specific time (see Figures 2.5 and 2.6).

For a reaction to occur, the molecules of the reactants must meet (collide). So, high concentrations of a contaminant are more likely to collide than low concentrations. Thus, the reaction rate must be a function of the concentrations of the reacting substances. The mathematical expression of this function is known as the rate law. The rate law can be determined experimentally for any contaminant. Varying the concentration of each reactant independently and then measuring the result will give a concentration curve. Each reactant has a unique rate law (this is one of a contaminant's physicochemical properties).

So, let us consider the reaction of reactants A and B, which yield C ($A + B \rightarrow C$), where the reaction rate increases in accord with the increasing concentration of either A or B. This means that if we triple the amount of A, the rate of this whole reaction triples. Thus, the rate law for such a reaction is:

$$\text{Rate} = k[\text{A}][\text{B}] \tag{2.9}$$

FIGURE 2.3. Two isomers of butane: *cis*-2-butene (left) and *trans*-2-butene (right).

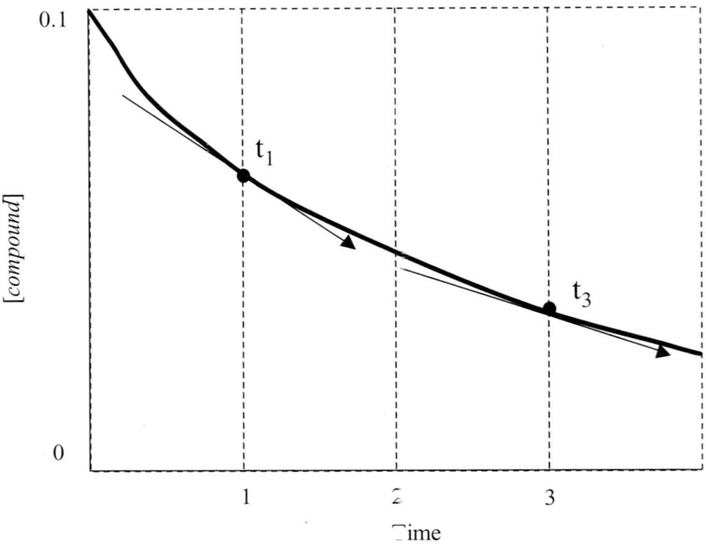

FIGURE 2.4. The kinetics of the transformation of a compound. The rate of reaction at any time is the negative of the slope of the tangent to the concentration curve at that time. The rate is higher at t_1 than at t_3. This rate is concentration-dependent (first-order).

However, let us consider another reaction X + Y → Z, in which the rate is increased only if the concentration of X is increased (changing the Y concentration has no effect on the rate law). In this reaction, the rate law must be:

$$\text{Rate} = k[\text{X}] \tag{2.10}$$

Thus, the concentrations in the rate law are the concentrations of reacting chemical species at any specific point in time during the reaction. The rate is the velocity of the reaction at that time. The constant k in the preceding equations is the *rate constant* which is unique for every chemical reaction and is a fundamental physical constant for a reaction, as defined by environmental conditions (e.g., pH, temperature, pressure, type of solvent). The rate constant is defined as the rate of the reaction when all reactants are present in a 1 molar (M) concentration, so the rate constant k is the rate of reaction under conditions standardized by a unit concentration. We can demonstrate the rate law by drawing a concentration curve for a contaminant that consists of an infinite number of points at each instant of time, then an instantaneous rate can be calculated along the concentration curve. At each point on the curve the rate of reaction is

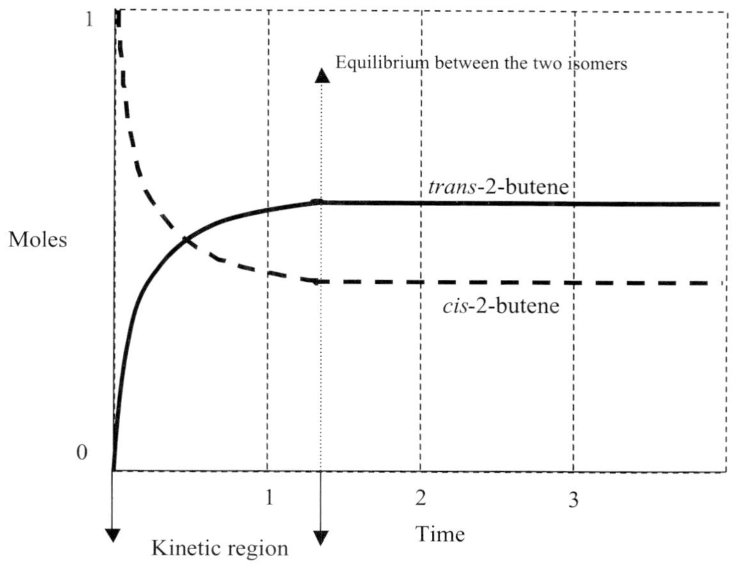

FIGURE 2.5. Change in respective moles of two butene isomers. Equilibrium at about 1.3 time units. The concentrations of the isomers depend on the initial concentration of the reactant (*cis*-2-butene). The actual time that equilibrium is reached depends upon environmental conditions, such as temperature and other compounds present; however, at a given temperature and conditions, the ratio of the equilibrium concentrations will be the same, no matter the amount of the reactant at the start. This is often described as the change of a "parent compound" into "chemical daughters" or "progeny." Pesticide kinetics often concerns itself with the change of the active ingredient in the pesticide to its degradation products (see Figure 2.7).

directly proportional to the concentration of the compound at that moment in time. This is a physical demonstration of *kinetic order*. The overall kinetic order is the sum of the exponents (powers) of all the concentrations in the rate law. So for the rate $k[A][B]$, the overall kinetic order is 2. Such a rate describes a second-order reaction because the rate depends on the concentration of the reactant raised to the second power. Other decomposition rates are like $k[X]$ and are first-order reactions because the rate depends on the concentration of the reactant raised to the first power.

The kinetic order of each reactant is the power that its concentration is raised in the rate law. So, $k[A][B]$ is first order for each reactant and $k[X]$ is first order for X and zero-order for Y. In a zero-order reaction, compounds degrade at a constant rate and are independent of reactant concentration. Further, if we plot the number of moles with respect to time, we would see the point at which kinetics ends and equilibrium begins. This simple example applies to any chemical kinetics process, but the kinetics is

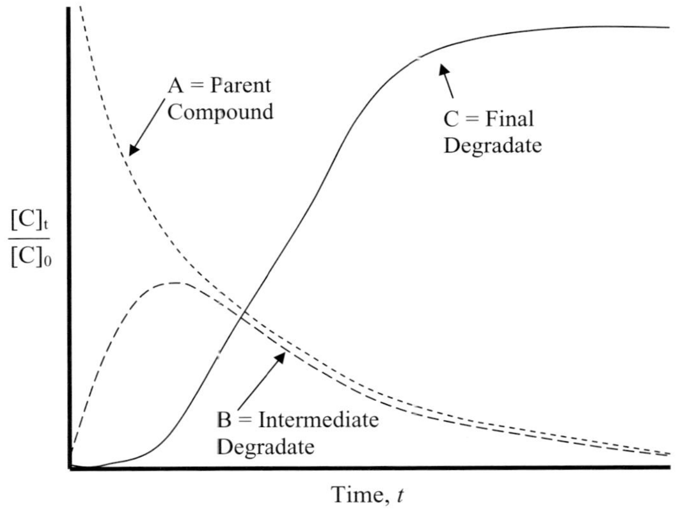

FIGURE 2.6. Distribution of chemical species for consecutive environmental reactions (first-order degradation of parent compound A). $[C]_t/[C]_0$ is the proportion of concentration of the compound at time t to the concentration at the time reaction begins. A persistent compound will require a relatively long time (i.e., a long half-life) for the parent compound to degrade. Also, intermediate degradation products (curve B) may themselves be persistent, such as the derivatives of DDT, e.g., DDA (2,2-bis-(4-chlorophenyl)acetic acid) and DDE (1,1′-(2,2-dichloroethenylidene)-bis[4-chlorobenzene]).
Source: Adapted from W.J. Weber, Jr. and F.A. DiGiano. 1996, *Process Dynamics in Environmental Systems*, John Wiley & Sons, New York, NY.

complicated in the real world by the ever-changing conditions of ecosystems, tissues, and human beings.

Specific partitioning relationships control the leaving and gaining of pollutants among particles, water, soil, and sediment surfaces, the atmosphere, and organic tissues. These relationships are sorption, solubility, volatilization, organic carbon-water partitioning, and bioconcentration, which are respectively expressed by coefficients of sorption (distribution coefficient, K_D, or solid-water partition coefficient, K_p), dissolution or solubility coefficients, air-water partitioning (and the Henry's Law (K_H) constant), organic carbon-water coefficient (K_{oc}), and bioconcentration factors (BCF).

The environment can be subdivided into finite compartments. The conservation laws dictate that the mass of the contaminant entering and the mass leaving a control volume must be balanced by what remains within the control volume. Likewise, within that control volume, each compartment may be a gainer or loser of the contaminant mass, but the overall mass must balance. The generally inclusive term for these compartmental changes is known as *fugacity* or the "fleeing potential" of a substance. It is the propensity of a chemical to escape from one type of environmental compartment to another. Combining the relationships between and among all of the partitioning terms give us the net chemical transport of a pollutant in the environment.[14] The simplest chemodynamic approach addresses each compartment where a contaminant is found in discrete phases of air, water, soil, sediment, and biota. However, this becomes complicated because even within a single compartment, a substance may exist in various phases (e.g., dissolved in water and sorbed to a particle in the solid phase). Within a compartment, a pollutant may remain unchanged (at least during the designated study period), or it may move physically, or it may be transformed chemically into another substance. In many instances all three mechanisms occur. Some of the pollutant will remain unmoved and unchanged. Another fraction remains unchanged but is transported to a different compartment. Another fraction becomes chemically transformed with all remaining products staying in the compartment where they were generated. And a fraction of the original contaminant is transformed and then moved to another compartment. For example, the octanol-water coefficient (K_{ow}) value is an indication of a compound's likelihood to exist in the organic versus aqueous phase. This means that if a substance is dissolved in water and the water comes into contact with another substance, for example, octanol, the substance will have a tendency to move from the water to the octanol. Its octanol-water partitioning coefficient reflects just how much of the substance will move until the aqueous and organic solvents (phases) will reach equilibrium. So, for example, in a spill of equal amounts of the polychlorinated biphenyl, decachlorobiphenyl (log K_{ow} of 8.23) and the pesticide chlordane (log K_{ow} of 2.78), the PCB has much greater affinity for the organic phases than does the chlordane (more than five orders of magnitude). This does not mean that a great amount of either compound is likely to stay in the water column, since they are both hydrophobic, but it does mean that they will vary in the time and mass of each contaminant moving between phases. The rate (kinetics) is different, so the time it takes for the PCB and chlordane to reach equilibrium will be different.

The cases in this book demonstrate the importance of a number of partitioning coefficients in environmental science and engineering. Understanding these coefficients will help to elucidate some of the scientific principles at work in these cases.

Partitioning to Solids—Sorption

Sorption is the process in which a contaminant or other solute becomes associated, physically or chemically, with a solid sorbent. Sorption is arguably the most important transfer process that determines how bioavailable or toxic a compound will be in surface waters and in contaminated sediments. The physicochemical transfer[15] of a chemical, A, from liquid to solid phase is expressed as:

$$A_{(solution)} + \text{solid} = \text{A-solid} \quad (2.11)$$

The interaction of the solute (i.e., the chemical being sorbed) with a solid surface can be complex and dependent upon the properties of the chemical and the water. Other fluids are often of such small concentrations that they do not determine the ultimate solid-liquid partitioning. Whereas it is often acceptable to consider net sorption, let us consider briefly the four basic types or mechanisms of sorption:

1. *Adsorption* is the process wherein the chemical in a solution attaches to a solid surface, which is a common sorption process in clay and organic constituents in soils. This simple adsorption mechanism can occur on clay particles where little carbon is available, such as in groundwater.
2. *Absorption* is the process that often occurs in porous materials so that the solute can diffuse into the particle and be sorbed onto the inside surfaces of the particle. This commonly results from short-range electrostatic interactions between the surface and the contaminant.
3. *Chemisorption* is the process of integrating a chemical into porous materials surface via chemical reaction. In soil, this is usually the result of a covalent reaction between a mineral surface and the contaminant.
4. *Ion exchange* is the process by which positively charged ions (cations) are attracted to negatively charged particle surfaces or negatively charged ions (anions) are attracted to positively charged particle surfaces, causing ions on the particle surfaces to be displaced. Particles undergoing ion exchange can include soils, sediment, airborne particulate matter, or even biota, such as pollen particles. Cation exchange has been characterized as being the second most important chemical process on earth, after photosynthesis. This is because the cation exchange capacity (CEC), and to a lesser degree anion exchange capacity (AEC) in tropical soils, is the means by which nutrients are made available to plant roots. Without this process, the atmospheric nutrients and the minerals in the soil would not come together to provide for the abundant plant life on planet earth.[16]

The first two sorption types are predominantly controlled by physical factors, and the second two are combinations of chemical reactions and physical processes. Generally, sorption reactions affect three processes[17] in environmental systems:

1. The chemical contaminant's transport in water due to distributions between the aqueous phase and particles.
2. The aggregation and transport of the contaminant as a result of electrostatic properties of suspended solids.
3. Surface reactions such as dissociation, surface-catalysis, and precipitation of the chemical contaminant.

When a contaminant enters a soil, some of the chemical remains in soil solution while the rest is adsorbed onto the surfaces of the soil particles. Sometimes this sorption is strong due to cations adsorbing to the negatively charged soil particles. In other cases the attraction is weak. Sorption of chemicals on solid surfaces needs to be understood because they hold onto contaminants, not allowing them to move freely with the pore water or the soil solution. Therefore sorption slows the rate at which contaminants move downward through the soil profile.

Contaminants eventually will establish a balance between the mass on the solid surfaces and the mass that is in solution. Molecules will migrate from one phase to another to maintain this balance. The properties of both the contaminant and the soil (or other matrix) will determine how and at what rates the molecules partition into the solid and liquid phases. These physicochemical relationships, known as *sorption isotherms*, are found experimentally. Figure 2.7 gives three isotherms for pyrene from experiments using different soils and sediments.

The x-axis in Figure 2.7 gives the concentration of pyrene dissolved in water, and the y-axis shows the concentration in the solid phase. Each line represents the relationship between these concentrations for a single soil or sediment. A straight-line segment through the origin represents the data well for the range of concentrations shown. Not all portions of an isotherm are linear, particularly at high concentrations of the contaminant. Linear chemical partitioning can be expressed as:

$$S = K_D C_W \tag{2.12}$$

where

S = concentration of contaminant in the solid phase (mass of solute per mass of soil or sediment)
C_W = concentration of contaminant in the liquid phase (mass of solute per volume of pore water)
K_D = partition coefficient (volume of pore water per mass of soil or sediment) for this contaminant in this soil or sediment

For many soils and chemicals, the partition coefficient can be estimated using:

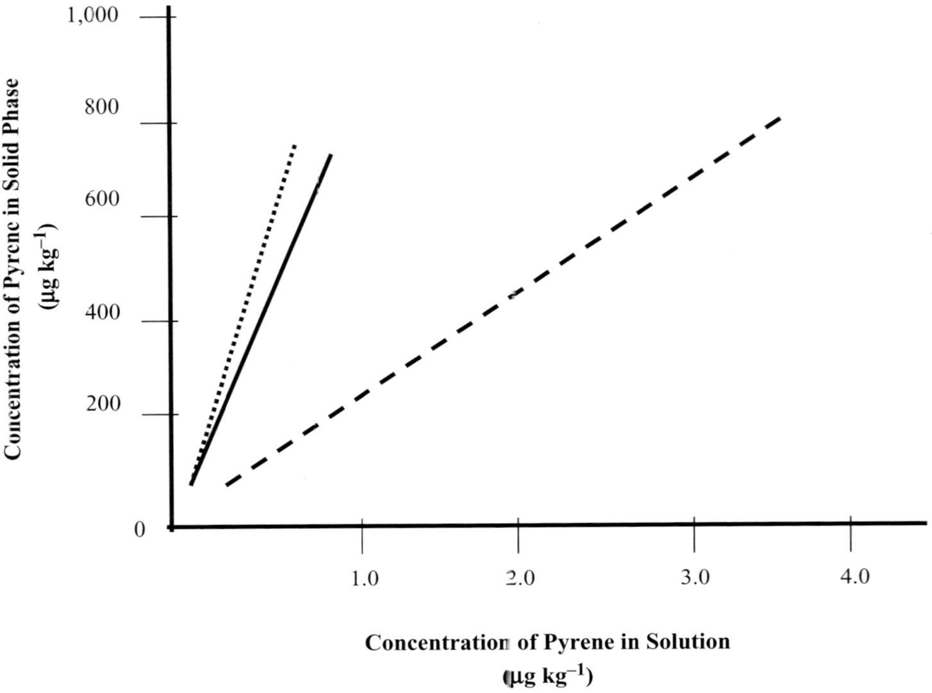

FIGURE 2.7. Three experimentally determined sorption isotherms for the polycyclic aromatic hydrocarbon, pyrene.
Source: J. Hassett and W. Banwart, 1989. "The sorption of nonpolar organics by soils and sediments," *Reactions and Movement of Organic Chemicals in Soils*, B. Sawhney and K. Brown, eds., Soil Science Society of America Special Publication 22. p. 35.

$$K_D = K_{OC}OC \tag{2.13}$$

where

K_{OC} = organic carbon partition coefficient (volume of pore water per mass of organic carbon)
OC = soil organic matter (mass of organic carbon per mass of soil)

This relationship is a very useful tool for estimating K_D from the known K_{OC} of the contaminant and the organic carbon content of the soil horizon of interest. The actual derivation of K_D is:

$$K_D = C_S(C_W)^{-1} \tag{2.14}$$

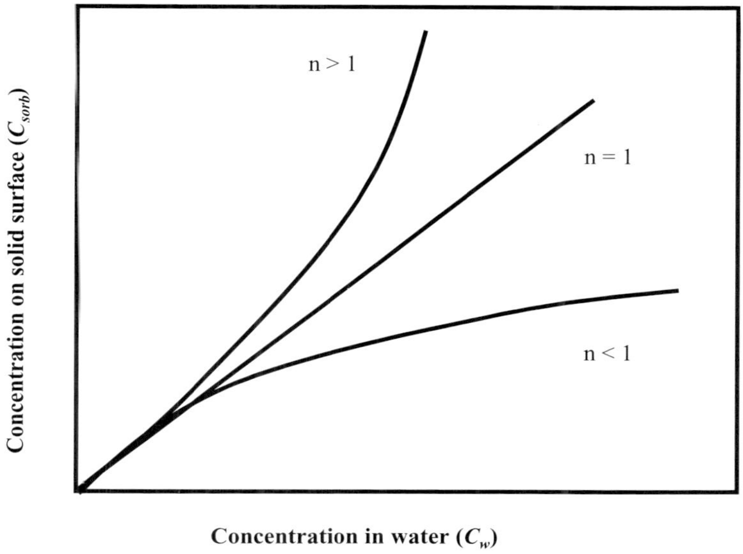

FIGURE 2.8. Hypothetical Freundlich isotherms with exponents (n) less than, equal to, and greater than 1, as applied to the equation $C_{sorb} = K_F C^n$.
Sources: R. Schwarzenbach, P. Gschwend, and D. Imboden, 1993. *Environmental Organic Chemistry*, John Wiley & Sons, Inc, New York, NY. H.F. Hemond and E.J. Fechner-Levy, 2000. *Chemical Fate and Transport in the Environment*, Academic Press, San Diego, CA.

where C_S is the equilibrium concentration of the solute in the solid phase and C_W is the equilibrium concentration of the solute in the water.

Therefore, K_D is a direct expression of the partitioning between the aqueous and solid (soil or sediment) phases. A strongly sorbed chemical like a dioxin or the banned pesticide DDT can have a K_D value exceeding 10^6. Conversely, a highly hydrophilic, miscible substance like ethanol, acetone, or vinyl chloride will have K_D values less than 1. This relationship between the two phases demonstrated by Equation 2.15 and Figure 2.8 is roughly what environmental scientists call the *Freundlich Sorption Isotherm*:

$$C_{sorb} = K_F C^n \tag{2.15}$$

where C_{sorb} is the concentration of the sorbed contaminant; that is, the mass sorbed at equilibrium per mass of sorbent, and K_F is the Freundlich isotherm constant. The exponent determines the linearity or order of the reaction. Thus, if $n = 1$, then the isotherm is linear; meaning the more of the contaminant in solution, the more would be expected to be sorbed to surfaces. For values of $n < 1$, the amount of sorption is in smaller proportion to the amount of solution and, conversely, for values of n > 1, a greater proportion of sorption occurs with less contaminant in solution. These three isotherms are shown in Figure 5.5. Also note that if n = 1, then Equation 5.6 and the Freundlich sorption isotherm are identical.

When organic matter content is elevated in soil and sediment, the amount of a contaminant that is sorbed is directly proportional to the soil/sediment organic matter content. This allows us to convert the K_D values from those that depend on specific soil or sediment conditions to those that are soil/sediment independent sorption constants, K_{OC}:

$$K_{OC} = K_D\left(\frac{K_D}{f_{OC}}\right) \tag{2.16}$$

where f_{OC} is the dimensionless weight fraction of organic carbon in the soil or sediment. The K_{OC} and K_D have units of mass per volume. Table 2.3 provides the log K_{OC} values that are calculated from chemical structure and those measured empirically for several organic compounds, and compares them to the respective K_{ow} values.

Partitioning to the Liquid Phase—Dissolution

Substances can be dissolved in numerous solvents. The most important solvent in environmental systems is water, but we must also consider solutions in organic solvents, such as dimethylsulfoxide (DMSO), ethanol, acetone, methanol, and toluene. A good resource for finding the solubility of numerous toxicants in water (aqueous solubility) or other solvents is the National Toxicology Program's *Chemical Solubility Compendium*[18] and the program's Health and Safety reports.[19]

The polarity of a molecule is its unevenness in charge. The water molecule's oxygen and two hydrogen atoms are aligned so that there is a slightly negative charge at the oxygen end and a slightly positive charge at the hydrogen ends. Since "like dissolves like," polar substances have an affinity to become dissolved in water, and nonpolar substances resist being dissolved in water. Consider the very polar water molecule (see Figure 2.9). The hydrogen atoms form an angle of 105° with the oxygen atom. The asymmetry of the water molecule leads to a dipole moment (see the discussion in the next session) in the symmetry plane pointed toward the more positive hydrogen atoms. Since the water molecule is highly polar, it will more readily dissolve other polar compounds than nonpolar compounds.

An element's ability to attract electrons toward itself is known as electronegativity. It is a measure of an atom's ability to attract shared electrons toward itself. The values for electronegativity range from 0 to 4, with fluorine (electronegativity = 4) being the most electronegative. Each atom is uniquely able to attract electrons to varying degrees owing to its size, the

TABLE 2.3
Calculated and experimental organic carbon coefficients (K_{ow}) for selected contaminants found at hazardous waste sites.

		Calculated		*Measured*	
Chemical	*log* K_{ow}	*log* K_{oc}	K_{oc}	*log* K_{ow}	K_{oc} *(geomean)*
Benzene	2.13	1.77	59	1.79	61.7
Bromoform	2.35	1.94	87	2.10	126
Carbon tetrachloride	2.73	2.24	174	2.18	152
Chlorobenzene	2.86	2.34	219	2.35	224
Chloroform	1.92	1.60	40	1.72	52.5
Dichlorobenzene, 1,2- (*o*)	3.43	2.79	617	2.58	379
Dichlorobenzene, 1,4- (*p*)	3.42	2.79	617	2.79	616
Dichloroethane, 1,1-	1.79	1.50	32	1.73	53.4
Dichloroethane, 1,2-	1.47	1.24	17	1.58	38.0
Dichloroethylene, 1,1-	2.13	1.77	59	1.81	65
Dichloroethylene, *trans* 1,2-	2.07	1.72	52	1.58	38
Dichloropropane, 1,2-	1.97	1.64	44	1.67	47.0
Dieldrin	5.37	4.33	21,380	4.41	25,546
Endosulfan	4.10	3.33	2,138	3.31	2,040
Endrin	5.06	4.09	12,303	4.03	10,811
Ethylbenzene	3.14	2.56	363	2.31	204
Hexachlorobenzene	5.89	4.74	54,954	4.90	80,000
Methyl bromide	1.19	1.02	10	0.95	9.0
Methyl chloride	0.91	0.80	6	0.78	6.0
Methylene chloride	1.25	1.07	12	1.00	10
Pentachlorobenzene	5.26	4.24	17,378	4.51	32,148
Tetrachloroethane, 1,1,2,2-	2.39	1.97	93	1.90	79.0
Tetrachloroethylene	2.67	2.19	155	2.42	265
Toluene	2.75	2.26	182	2.15	140
Trichlorobenzene, 1,2,4-	4.01	3.25	1,778	3.22	1,659
Trichloroethane, 1,1,1-	2.48	2.04	110	2.13	135
Trichloroethane, 1,1,2-	2.05	1.70	50	1.88	75.0
Trichloroethylene,	2.71	2.22	166	1.97	94.3
Xylene, *o*-	3.13	2.56	363	2.38	241
Xylene, *m*-	3.20	2.61	407	2.29	196
Xylene, *p*-	3.17	2.59	389	2.49	311

Source: U.S. Environmental Protection Agency, 1996, Soil Screening Program.

charge of its nucleus, and the number of core (i.e., nonvalent) electrons. Values vary with the element's position in the periodic table, with electronegativity increasing from left to right across a row and decreasing downward within each group. This is due to the fact that smaller atoms allow electrons to get closer to the positively charged nucleus. Thus, the higher the net charge of the combined nucleus plus the electrons of the filled inner shells (collectively referred to as the *kernel*), the greater

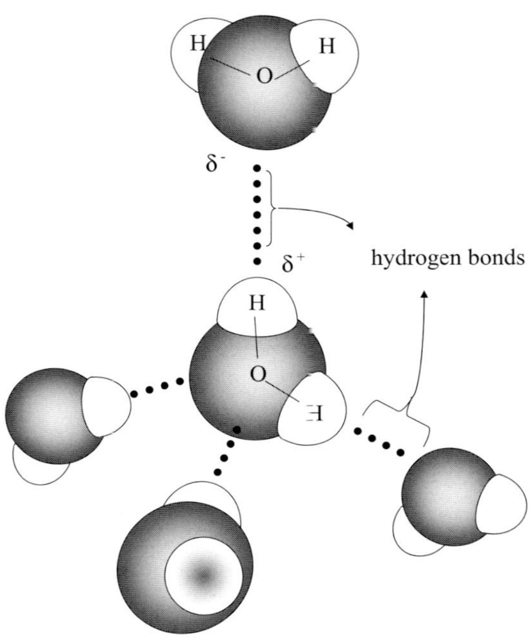

FIGURE 2.9. Configuration of the water molecule, showing the electronegativity (δ) at each end. The hydrogen atoms form an angle of 105° with the oxygen atom.

the electronegativity and the tendency of the atom to attract electrons (see Table 2.4).

The strength of a chemical bond in molecules is determined by the energy needed to hold the like and unlike atoms together with a covalent bond (i.e., a bond where electrons are shared between two or more atoms). The bond energy is expressed by the bond dissociation *enthalpy* (ΔH_{AB}). For a two-atom (i.e., *diatomic*) molecule the ΔH_{AB} is heat change of the gas phase reaction. That is, at constant temperature and pressure, ΔH_{AB} is:

$$A\text{-}B \rightarrow A\bullet + \bullet B \tag{2.17}$$

where A-B is the *educt* and A• and •B are the products of the reaction. The enthalpies and bond lengths for some of the bonds important in environmental engineering and science are given in Table 2.5.

Solubility is important and valuable information when considering whether a contaminant will move from one location to another (e.g., from the soil into the groundwater). If a compound is highly hydrophobic (i.e., not easily dissolved in water), we may be led to assume that it will not be found in the *water column*, in an environmental study. This is a reasonable

TABLE 2.4
Electronegativity of the elements.

	IA	IIA	IIIB	IVB	VB	V1B	VIIB	VIIIB			IB	IIB	IIIA	IVA	VA	VIA	VIIA	VIII
1	H 2.1																	He
		4																
2	Li 1.0	Be 1.5											B 2.0	C 2.5	N 3.0	O 3.5	F 4.0	Ne
3	Na 0.9	Mg 1.2											Al 1.5	Si 1.8	P 2.1	S 2.5	Cl 3.0	Ar
4	K 0.8	Ca 1.0	Sc 1.3	Ti 1.5	V 1.6	Cr 1.6	Mn 1.5	Fe 1.8	Co 1.8	Ni 1.8	Cu 1.9	Zn 1.6	Ga 1.6	Ge 1.8	As 2.0	Se 2.4	Br 2.8	Kr
5	Rb 0.8	Sr 1.0	Y 1.3	Zr 1.4	Nb 1.6	Mo 1.8	Tc 1.9	Ru 2.2	Rh 2.2	Pd 2.2	Ag 1.9	Cd 1.7	In 1.7	Sn 1.8	Sb 1.9	Te 2.1	I 2.5	Xe
6	Cs 0.7	Ba 0.9	La 1.1	Hf 1.3	Ta 1.5	W 1.7	Re 1.9	Os 2.2	Ir 2.2	Pt 2.2	Au 2.4	Hg 1.9	Tl 1.8	Pb 1.6	Bi 1.9	Po 2.0	At 2.2	Rn

TABLE 2.5
Bond lengths and enthalpies for bonds in molecules important in environmental studies.

Bond	*Bond Length (angstroms)*	*Enthalpy, ΔH_{AB}, ($kJ\,mol^{-1}$)*	*Notes*
Diatomic Molecules			
H—H	0.74	436	
H—F	0.92	566	
H—Cl	1.27	432	
H—Br	1.41	367	
H—I	1.60	298	
F—F	1.42	155	
Cl—Cl	1.99	243	
Br—Br	2.28	193	
I—I	2.67	152	
O═O	1.21	4.98	
N≡N	1.10	9.46	
Organic Compounds[20]			
H—C	1.11	415	
H—N	1.00	390	
H—O	0.96	465	
H—S	1.33	348	
C—C	1.54	348	
C—N	1.47	306	
C—O	1.41	360	
C—S	1.81	275	
C—F	1.38	486	
C—Cl	1.78	339	
C—Br	1.94	281	
C—I	2.14	216	
C═C	1.34	612	
C═N	1.28	608	
C═S	1.56	536	In carbon disulfide
C═O	1.20	737	In aldehydes
C═O	1.20	750	In ketones
C═O	1.16	804	In carbon dioxide
C≡C	1.20	838	
C≡N	1.16	888	

Source: R. Schwarzenbach, P. Gschwend, and D. Imboden, 1993. *Environmental Organic Chemistry*, John Wiley & Sons, Inc, New York, NY.

expectation theoretically and is based upon the expectation that the only solvent in water bodies is water. However, surface and groundwater is never completely devoid of other solvents. The process of *co-solvation* is a mechanism by which highly lipophilic (fat soluble) and hydrophobic compounds become dissolved in water. That is, if a compound is hydrophobic and nonpolar, but is easily dissolved in acetone or methanol, it may well end up in the water because these organic solvents are highly miscible in water. The organic solvent and water mix easily, and a hydrophobic compound will remain in the water column because it is dissolved in the organic solvent, which in turn has mixed with the water. Compounds like PCBs and dioxins may be transported as co-solutes in water by this means. So, the combination of hydrophobic compounds being sorbed to suspended materials and co-solvated in organic co-solvents that are miscible in water can mean that they are able to move in water bodies and receptors can be exposed through the water pathways.

The rate of dissolution is dependent upon the concentration of a contaminant being released to a water body (i.e., the volume of contaminant versus the volume of the receiving waters). However, concentrations of contaminants are usually at the ppm or lower level, so this is seldom a limiting factor in environmental situations. Other factors that influence dissolution are the turbulence of the water, temperature, ionic strength, dissolved organic matter present in the water body, the aqueous solubility of the contaminant, and the presence of co-solvents.[21]

Solubility is determined in the laboratory at a certain temperature, by adding as much of the solute to a solvent until the solvent can no longer dissolve the substance being added. So, if Compound A has a published solubility of $10\,\mathrm{mg\,L^{-1}}$ in water at 20°C, this means that one liter of water could dissolve only 10 mg of that substance. If, under identical conditions, Compound B has a published aqueous solubility of $20\,\mathrm{mg\,L^{-1}}$, this means that one liter of water could dissolve 20 mg of Compound B, and that Compound B has twice the aqueous solubility of Compound A. Actually, solutions are really in dynamic equilibrium because the solute is leaving and entering the solution at all times, but the average amount of solute in solution is the same. The functional groups on a molecule determine whether it will be more or less polar. So, compounds with hydroxyl groups are more likely to form H-bonds with water. Thus, methane is less soluble in water than methanol. Also, since water interacts strongly with ions, salts are usually quite hydrophilic.

Partitioning to the Gas Phase—Volatilization

The change of phase to a gas, *volatilization*, is a function of the concentration of a contaminant in solution and the contaminant's partial pressure. Henry's law states that the concentration of a dissolved gas is directly proportional to the partial pressure of that gas above the solution:

$$p_a = K_{\mathrm{H}}[c] \tag{2.18}$$

where

K_H = Henry's law constant
p_a = Partial pressure of the gas
$[c]$ = Molar concentration of the gas

or,

$$p_A = K_H C_W \tag{2.19}$$

where C_W is the concentration of gas in water.

So, for any chemical contaminant we can establish a proportionality between the solubility and vapor pressure. Henry's law is an expression of this proportionality between the concentration of a dissolved contaminant and its partial pressure in the headspace (including the open atmosphere) at equilibrium. A dimensionless version of the partitioning is similar to that of sorption, except that instead of the partitioning between solid and water phases, it is between the air and water phases (K_{AW}):

$$K_{AW} = \frac{C_A}{C_W} \tag{2.20}$$

where C_A is the concentration of gas A in the air.

The relationship between the air/water partition coefficient and Henry's law constant for a substance is:

$$K_{AW} = \frac{K_H}{RT} \tag{2.21}$$

where R is the gas constant (8.21×10^{-2} L atm mol^{-1} K^{-1}) and T is the temperature (°K). Henry's law relationships work well for most environmental conditions. It represents a limiting factor for systems where a substance's partial pressure is approaching zero. At very high partial pressures (e.g., 30 Pascals) or at very high contaminant concentrations (e.g., >1000 ppm), Henry's law assumptions cannot be met. Such vapor pressures and concentrations are seldom seen in ambient environmental situations, but may be seen in industrial and other source situations. Thus, in modeling and estimating the tendency for a substance's release in vapor form, Henry's law is a good metric and is often used in compartmental transport models to indicate the fugacity, or leaving potential, from the water to the atmosphere.

Henry's law constants are highly dependent upon temperature, since both vapor pressure and solubility are also temperature dependent. So, when using published K_H values, we must compare them isothermically. Also, when combining different partitioning coefficients in a model or study, it is important either to use only values derived at the same temperature (e.g., sorption, solubility, and volatilization all at 20°C), or to adjust them accordingly. A general adjustment is an increase of a factor of 2 in K_H for each 8°C temperature increase. Also, any sorbed or otherwise bound fraction of the contaminant will not exert a partial pressure, so this fraction should not be included in calculations of partitioning from water to air. For example, it is important to differentiate between the mass of the contaminant in solution (available for the K_{AW} calculation) and that in the suspended solids (unavailable for K_{AW} calculation). This is crucial for many hydrophobic organic contaminants, where they are most likely not to be dissolved in the water column (except as co-solutes), with the largest mass fraction in the water column being sorbed to particles.

The relationship between K_H and K_{ow} is also important. It is often used to estimate the *environmental* persistence, as reflected the chemical *half-life* ($T_{1/2}$) of a contaminant. However, many other variables determine the actual persistence of a compound after its release. Note in the table, for example, that benzene and chloroform have nearly identical values of K_H and K_{ow} yet benzene is far less persistent in the environment. We will consider these other factors in the next chapters, when we discuss abiotic chemical destruction and biodegradation.

The relative affinity for a substance to reside in air and water can be used to estimate the potential for the substance to partition not only between water and air, but more generally between the atmosphere and biosphere, especially when considering the long-range transport of contaminants.[22] Such long-range transport estimates make use of both atmospheric $T_{1/2}$ and K_H. Also, the relationship between octanol-water and air-water coefficients can also be an important part of predicting a contaminant's transport. For example, Figure 2.10 provides some general classifications according various substances' K_H and K_{ow} relationships. In general, chemicals in the upper left-hand group have a great affinity for the atmosphere, so unless there are contravening factors, this is where to look for them. Conversely, substances with relatively low K_H and K_{ow} values are less likely to be transported a long distance in the air.

Solubility as a Physical and Chemical Phenomenon

Usually, when scientists use the term *solubility* without any other attributes, they mean the measure of the amount of the solute in water, that is, aqueous solubility. Otherwise, the solubility will be listed along with the solvent, such as solubility in benzene, solubility in methanol, or solubility in hexane. Solubility may also be expressed in mass per mass or volume per volume, represented as parts per million (ppm), parts per billion (ppb), or parts per trillion (ppt).

Occasionally, solubility is expressed as a percent or in parts per thousand; however, this is uncommon for contaminants and usually is reserved for nutrients and essential gases (e.g., percent carbon dioxide in water or

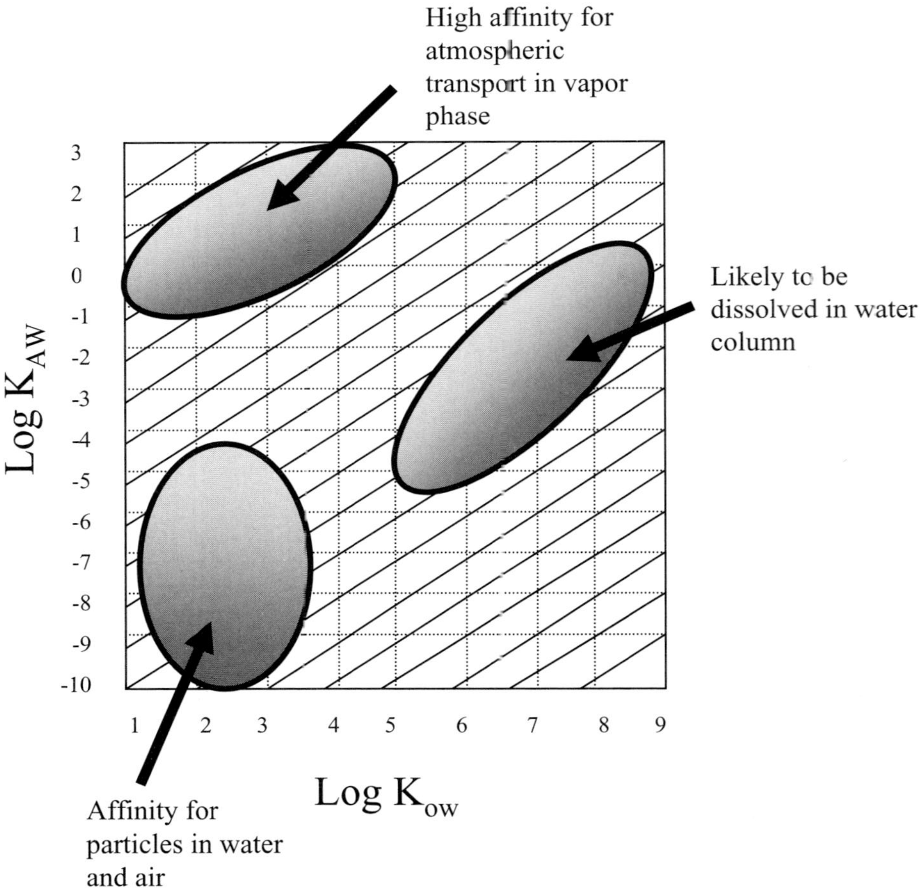

FIGURE 2.10. Relationship between air-water partitioning and octanol-water partitioning and affinity of classes of contaminants for certain environmental compartments.
Source: D. van de Meent, T. McKone, T. Parkerton, M. Matthies, M. Scheringer, F. Wania, R. Purdy, and D. Bennett, 1999. "Persistence and Transport Potential of Chemicals in a Multimedia Environment," *Proceedings of the SETAC Pellston Workshop on Criteria for Persistence and Long-Range Transport of Chemicals in the Environment, 14–19 July 1998, Fairmont Hot Springs, British Columbia, Canada,* Society of Environmental Toxicology and Chemistry, Pensacola, FL.

parts per thousand water vapor in the air). The solubility of a compound is very important to environmental transport. The diversity of solubilities in various solvents is a strong indication of where we are likely to find the compound. For example, the various solubilities of the most toxic form of dioxin, tetrachlorodibenzo-*para*-dioxin (TCDD) are provided in Table 2.6.

TABLE 2.6
Solubility of tetrachlorodibenzo-*para*-dioxin in water and organic solvents.

Solvent	*Solubility* ($mg\,L^{-1}$)	*Reference*
Water	1.93×10^{-5}	Podoll, *et al.* 1986. *Environmental Science and Technology* 20: 490–492
Water	6.90×10^{-4} (25°C)	Fiedler, *et al. Chemosphere* (20): 1597–1602
Methanol	10	International Agency for Research on Cancer[23] (IARC)
Lard oil	40	IARC
n-Octanol	50	IARC
Acetone	110	IARC
Chloroform	370	IARC
Benzene	570	IARC
Chlorobenzene	720	IARC
Orthochlorobenzene	1,400	IARC

From these solubilities, we would expect TCDD to have a much greater affinity for sediment, organic particles, and the organic fraction of soils. The low water solubilities indicate that dissolved TCDD in the water column should be at only extremely low concentrations.

Partitioning to Organic Tissue

Relatively hydrophobic substances frequently have a strong affinity for lipid-containing tissues (i.e., those containing high K_{ow} compounds). Therefore, such contaminants can be sequestered and can accumulate in organisms. Certain chemicals are very *bioavailable* to organisms that may readily take them up from the other compartments and store them. Bioavailability is an expression of the fraction of the total mass of a compound present in a compartment that has the potential of being absorbed by the organism. *Bioaccumulation* is the process of uptake into an organism from the abiotic compartments. *Bioconcentration* is the concentration of the pollutant within an organism above levels found in the compartment in which the organism lives. So, for a fish to bioaccumulate DDT, the levels found in the total fish or in certain organs (e.g., the liver) will be elevated above the levels measured in the ambient environment. In fact, DDT is known to bioconcentrate many orders of magnitude in fish. A surface water DDT concentration of 100 parts per trillion in water has been associated with 10 ppm in certain fish species (a concentration of 10,000 times!). Thus the straightforward equation for the *bioconcentration factor* (BCF) is the quotient of the concentration of the contaminant in the organism and the concentration of the contaminant in the host compartment. So, for a fish living in water, the BCF is:

$$\text{BCF} = \frac{C_{organism}}{C_w} \tag{2.22}$$

The BCF is applied to an individual organism that represents a genus or some other taxonomical group. However, considering the whole food chain and trophic transfer processes, in which a compound builds up as a result of predator/prey relationships, the term *biomagnification* is used. Some compounds that may not appreciably bioconcentrate within lower trophic state organisms may still become highly concentrated. For example, even if plankton have a small BCF (e.g., 10), if subsequently higher order organisms sequester the contaminant at a higher rate, in time, top predators (e.g., alligators, sharks, panthers, and humans) may suffer from the continuum of biomagnification, with levels many orders of magnitude higher than what is found in the abiotic compartments.

For a substance to bioaccumulate, bioconcentrate, and biomagnify, it must be somewhat persistent. If an organism's metabolic and detoxification processes are able to degrade the compound readily, it will not be present (at least in high concentrations) in the organism's tissues. However, if an organism's endogenous processes degrade a compound into a chemical species that is itself persistent, the metabolite or degradation product will bioaccumulate, and may bioconcentrate and biomagnify. Finally, cleansing or *depuration* will occur if the organism that has accumulated a contaminant enters an abiotic environment that no longer contains the contaminant. However, some tissues have such strong affinities for certain contaminants that the persistence within the organism will remain long after the source of the contaminant is removed. For example, the piscivorous birds, such as the Common Loon (*Gavia immer*), decrease the concentrations of the metal mercury in their bodies by translocating the metal to feathers and eggs. So, every time the birds molt or lay eggs they undergo mercury depuration. Unfortunately, when the birds continue to ingest mercury that has bioaccumulated in their prey (fish), they often have a net increase in tissue Hg concentrations because the bioaccumulation rate exceeds the depuration rate.[24]

Bioconcentration can vary considerably in the environment. The degree to which a contaminant builds up in an ecosystem, especially in biota and sediments, is related to the compound's persistence. For example, a highly persistent compound, if nothing else, lasts longer in the environment so there is a greater opportunity for uptake, all other factors being equal. In addition, persistent compounds often possess chemical structures that are also conducive to sequestration by fauna. Such compounds are generally quite often lipophilic, have high K_{ow} values, and usually low vapor pressures. This means that they may bind to the organic molecules in living tissues and may resist elimination and metabolic process, so that they build up over time. However, the bioaccumulation and bioconcentration can vary considerably, both among biota and within the same species of biota. For example, the pesticide mirex has been shown to exhibit bioconcentration factors of 2,600 and 51,400 in pink shrimp and fathead minnows, respectively. The pesticide endrin has shown an even larger interspecies variablility in BCF values, with factors ranging from 14 to 18,000 recorded in fish after continuous exposure. Intraspecies BCF ranges may also be high; for example, oysters exposed to very low concentrations of the organometallic compound, tributyl tin, exhibit BCF values ranging from 1,000 to 6,000.[25]

Even the same compound in a single medium, for example, a lake's water column or sediment, will show large BCF variability among species of fauna in that compartment. An example is the so-called "dirty dozen" compounds. This is a group of *persistent organic pollutants* (POPs) that largely have been banned, some for decades, but that are still found in environmental samples throughout the world. As might be expected from their partitioning coefficients, they are concentrated in sediment and biota. The worst combination of factors is when a compound is persistent in the environment, builds up in organic tissues, and is toxic. Such compounds are referred to as *persistent bioaccumulating toxic* substances (PBTs). Recently, the United Nations Environmental Programme (UNEP) reported on the concentrations of the persistent and toxic compounds. Each region of the world was evaluated for the presence of these compounds. For example, the North American report[26] includes scientific assessments of the nature and scale of environmental threats posed by persistent toxic compounds. Organometallic compounds, especially lead and its compounds, comprise the lion's share of PBTs in the United States. And the second largest quantity is represented by another metal, mercury, and its compounds.

The sources of PBTs are widely varied. Many are intentionally manufactured to serve some public need, such as the control of pests that destroy food and spread disease. Other PBTs are generated as unintended byproducts, such as the products of incomplete combustion. In either case, there are often measures and engineering controls available that can prevent PBT releases, rather than having to deal with them after they have found their way into the various environmental compartments.

Emissions, Effluents, Releases, Leaks, and Spills

Environmental problems are characterized differently, depending on who is doing the characterization. For example, an ongoing release of a contaminant into the air often is referred to as an *emission*. Regulatory agencies keep track of such emissions, often depending on self-reporting by the entity doing the emitting. These data are collected and published as emission inventories. Water programs generally refer to the same type of ongoing release as an *effluent* that is discharged. Again, the entity releasing the effluent reports the type and quantity of the released pollutant. The regulatory concept is similar to that of tax oversight by the Internal Revenue Service, with facilities randomly audited to ensure that the reported information is sufficiently precise and accurate and, if not, the facility is subject to civil and criminal penalties.

Other less predictable releases go by a number of names. In hazardous waste programs, such as the Leaking Underground Storage Tank (LUST) program, contaminant intrusions into groundwater are called *leaks*. In fact, new underground tanks are often required to have leak detection systems and alarms. In solid waste programs, such as landfill regulations, the intrusion may go by the name *leachate*. Landfills often are required to have leachate collection systems to protect adjacent aquifers and surface waters. *Spills* are generally liquid releases that occur suddenly, such as an oil spill. Air releases that occur suddenly are called *leaks*, such as chlorine or natural gas leaks. The general term for expected and unplanned environmental releases is just that, *releases*, such as those reported in the U.S. Environmental Protection Agency's Toxic Release Inventory (TRI).

All these terms are used in this book. Although consistency has been strived for, not every case fits neatly into an air, water, or other problem. In those cases, some or all of these terms apply.

Notes and Commentary

1. This quote may in fact be an urban legend. It is quite easy to find it in references and on the Internet, but I have been unable to find the actual citation. To add to the confusion, some of the Internet sites attribute the quote to another former vice president, Al Gore. Although possible, the use of the same quote by two so ideologically different people is highly improbable.
2. 33 U.S.C. 1251.
3. In fact, my own environmental career began shortly after the passage of this law, when it, along with the National Environmental Policy Act and the Clean Air Act of 1970, was establishing a new environmental policy benchmark for the United States. At the time environmentalists recited an axiom frequently: "Dilution is not the solution to pollution!" I recall using it on a regular basis myself. However, looking back over those three decades, it seems the adage was not completely true. Cleanup levels and other thresholds are concentration based, so if we do an adequate job in diluting the concentrations (e.g., dioxin concentrations below 1 part per billion), we have at least in part solved that particular pollution problem. Also, when it came to metal pollution, dilution was a preferred solution, since a metal is an element and cannot be destroyed. A sufficient amount of the metal wastes are removed from water or soil and moved to a permanent storage site. The only other engineering solution to metal pollution was to change its oxidation state and chemical species, which is not often preferable because when environmental conditions change,

so often do the oxidation states of the metals, allowing them to again become toxic and bioavailable.

4. W.C. Kreigher, 2001. "Paracelus: Dose Response," *Handbook of Pesticide Toxicology,* 2e, R. Kreiger, J. Doull and D. Ecobichon, eds., Elsevier Academic Press, San Diego, CA.
5. P. Aarne Vesilind, J. Jeffrey Peirce, and Ruth F. Weiner, 1993. *Environmental Engineering,* 3e, Butterworth-Heinemann: Boston, MA.
6. H.W. Lewis, 1990. *Technological Risk,* Chapter 5: The Assessment of Risk, W.W. Norton & Company, Inc., New York, NY.
7. U.S. Environmental Protection Agency, 1990. Exposure factors handbook. Report No. EPA/600/8–89/043, Washington, D.C.
8. J.D. Graham and J.B. Wiener, 1995. "Confronting Risk Tradeoffs," *Risk versus Risk: Tradeoffs in Protecting Health and the Environment,* J.D. Graham and J.B. Wiener, eds., Harvard University Press, Cambridge, MA.
9. R. Carson, 1962. *Silent Spring,* Houghton Mifflin, Boston, MA.
10. P.H. Müller, 1948. "Dichloro-diphenyl-trichloroethane and Newer Insecticides," Nobel Lecture, December 11, 1948, Stockholm, Sweden.
11. J.K. Hammitt, E.S. Belsky, J.I. Levy, and J.D. Graham, 1999. "Residential building codes, affordability, and health protection: A risk-tradeoff approach," *Risk Analysis,* 19 (6), 1037–1058.
12. Although "kinetics" in the physical sense and the chemical sense arguably can be shown to share many common attributes, for the purposes of this discussion, it is probably best to treat them as two separate entities. Physical kinetics is concerned with the dynamics of material bodies and the energy in a body owing to its motions. Chemical kinetics addresses rates of chemical reactions. The former is more concerned with mechanical dynamics, the latter with thermodynamics.
13. This example was taken from J. Spencer, G. Bodner, and L. Rickard, 2003. *Chemistry: Structure and Dynamics,* 2e, John Wiley & Sons, New York, NY.
14. Fugacity models are valuable in predicting the movement and fate of environmental contaminants within and among compartments. This discussion is based on work by one of the pioneers in this area, Don MacKay and his colleagues at the University of Toronto. See, for example, D. MacKay and S. Paterson, 1991. "Evaluating the fate of organic chemicals: A level III fugacity model," *Environmental Science and Technology,* Vol. 25: 427–436.
15. W. Lyman, 1995. "Transport and Transformation Processes," Chapter 15 in *Fundamentals of Aquatic Toxicology: Effects, Environmental Fate, and Risk Assessment,* 2e, G. Rand, ed., Taylor & Francis, Washington, D.C.
16. I credit Daniel Richter of Duke University's Nicholas School of the Environment for much of what I know about this topic.
17. J. Westfall, 1987. "Adsorption Mechanisms in Aquatic Surface Chemistry," *Aquatic Surface Chemistry,* Wiley-Interscience, New York, NY.
18. L. Keith and D. Walters, 1992. *National Toxicology Program's Chemical Solubility Compendium,* Lewis Publishers, Inc., Chelsea, MI.

19. See http://ntp-db.niehs.nih.gov/htdocs/Chem_Hs_Index.html.
20. The single bond lengths given are as if the partner atoms are not involved in double or triple bonds. If that were not the case, the bond lengths would be shorter.
21. W. Lyman, 1995. "Transport and Transformation Processes," Chapter 15 in *Fundamentals of Aquatic Toxicology: Effects, Environmental Fate, and Risk Assessment*, 2e, G. Rand, ed., Taylor & Francis, Washington, D.C.
22. D. Mackay, D. and F. Wania, 1995. "Transport of contaminants to the arctic: Partitioning, processes and models," *The Science of the Total Environment* 160/161:25–38.
23. Reference for all of the organic solvents: International Agency for Research on Cancer, 1977. *Monographs on the Evaluation of the Carcinogenic Risk of Chemicals to Man: 1972–Present*, World Health Organization, Geneva, Switzerland.
24. N. Schoch, N. and D. Evers, 2002. "Monitoring Mercury in Common Loons: New York Field Report," 1998–2000. Report BRI 2001-01 submitted to U.S. Fish & Wildlife Service and New York State Department of Environmental Conservation, BioDiversity Research Institute, Falmouth, ME.
25. United Nations Environmental Programme, 2002. Chemicals: North American Regional Report, Regionally Based Assessment of Persistent Toxic Substances, Global Environment Facility.
26. United Nations Environmental Programme, 2002.

Part II

Key Environmental Events by Media

Although not perfect reflections of the problems they need to solve, the missions of government agencies often reflect the public challenges of their time. With regard to environmental protection, the lineage of laws, rules, and court cases differs considerably by environmental media; for example, air quality, drinking water, surface water and groundwater protection, solid and hazardous waste, consumer products, pesticides, nuclear wastes, ecological risks, habitat loss, soil loss and contamination, sediment contamination, stratospheric ozone destruction, and global climate change. In many aspects, agencies that address this panoply of environmental problems are less organic entities and more confederations. For example, the U.S. Environmental Protection Agency (U.S. EPA) and the National Oceanic and Atmospheric Administration (NOAA), two of the most important environmental agencies in the United States, were created as a reorganization of parts of agencies in existence at the time of the reorganization (see Appendix 2), meaning that many of the remnants of these older agencies still exist today, hence the characterization of environmental confederations. This means that there can be little incentive to address cross-cutting issues, especially if such issues fall outside the environmental media of a particular office. The agencies that first addressed environmental problems differed considerably from each other, depending on the "media." That is, air programs tended to grow from a public health perspective, soil and pesticide programs from an agricultural prospective, toxic substance control programs from a consumer product perspective, and water and sediment programs from a natural resource (e.g., Department of the Interior) perspective. For example, the U.S. Environmental Protection Agency was not formed *de novo* in 1970 but merely from a governmental reorganization that transferred programs from various Cabinet-level departments.

The vestiges of previous programs are still very apparent from the structure and organization of the U.S. EPA. Further complicating the organizational structures, environmental agencies usually do not enjoy a single piece of enabling legislation, but numerous media-specific laws, such as the Clean Water Act; the Public Drinking Water Act; the Clean Air Act; the Resource Conservation Recovery Act (solid and hazardous waste); the Federal Insecticide, Fungicide, and Rodenticide Act (pesticides); and the Toxic Substances Control Act. Each of these and several other laws are administered by separate programs.

Old Paradigm #3:
Environmental problems occur within a single compartment.
Paradigm Shift:
Environmental problems can be understood only from a multimedia, multicompartmental perspective.

The policy and scientific inertia of the first half of the twentieth century led to a viewpoint that problems and events can be grouped by media—air, water, and land. This view is inconsistent with environmental science, however, which requires an appreciation for the interactions within and between environmental media. However, it is a convenient and common way to categorize environmental problems. Thus, the next three chapters will group problems media-specifically, but interactions with other media and the comprehensive environmental perspectives will also be discussed.

A convenient way to begin to approach environmental problems from a multimedia, multicompartmental perspective is to consider the properties and behavior of the principal environmental fluids, especially air and water.

Fluids in the Environment: A Brief Introduction

The movement of pollutants is known as *transport*. This is half of the often-cited couplet of environmental fate and transport. *Fate* is an expression of what a contaminant becomes after all the physical, chemical, and biological processes of the environment have acted (see Figure II.1). It is the ultimate site for a substance after it finds its way into the environment, that is, the fate of a substance is where it ends up after its release. Substances undergo numerous changes in place and form before reaching their fate. Throughout its journey a substance will be physically transported and will undergo simultaneous chemical processes, known as *transformations*, such as photochemical and biochemical reactions.[i]

Physical transport is a function of the mechanics of fluids, but it is also a chemical process, such as when and under what conditions transport and chemical transformation processes become steady-state or nearly steady-state; for example, sequestration and storage in the environment. Thus, transport and transformation depends on the characteristics of

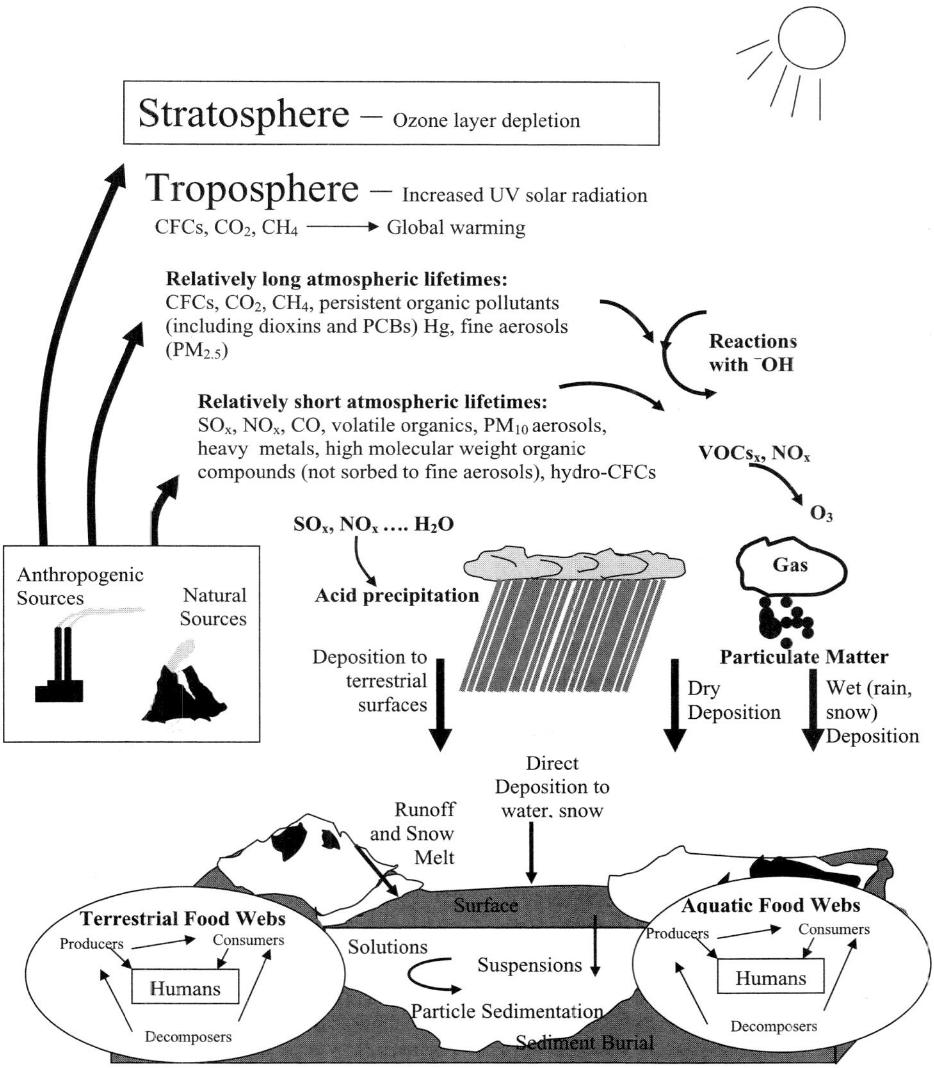

FIGURE II.1. The physical movement and accumulation of contaminants after release.

Sources: Commission for Environmental Cooperation of North America, 2002. The Sound Management of Chemicals (SMOC) Initiative of the Commission for Environmental Cooperation of North America: Overview and Update, Montreal, Canada. Adapted in D.A. Vallero, 2004. *Environmental Contaminants: Assessment and Control*, Elsevier Academic Press, Burlington, MA.

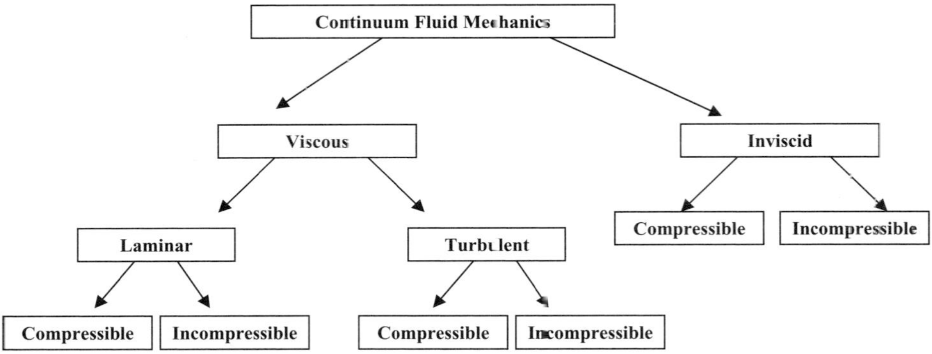

FIGURE II.2. Classification of fluids based on continuum fluid mechanics. *Source*: Research and Education Association 1987. *The Essentials of Fluid Mechanics and Dynamics I*, REA, Piscataway, NJ.

environmental fluids. A fluid is a collective term that includes all liquids and gases.[ii] A liquid is matter that is composed of molecules that move freely among themselves without separating from each other. A gas is matter composed of molecules that move freely and are infinitely able to occupy the space in which they are contained at a constant temperature. Engineers define a fluid as a substance that will deform continuously upon the application of a shear stress; That is a stress in which the material on one side of a surface pushes on the material on the other side of the surface with a force parallel to the surface.

Fluids can be classified according to observable physical characteristics of flow fields. A continuum fluid mechanics classification is shown in Figure II.2. Laminar flow is in layers, whereas turbulent flow has random movements of fluid particles in all directions. In incompressible flow, the variations in density are assumed to be constant, but the compressible flow has density variations that must be included in flow calculations. Viscous flows must account for viscosity whereas inviscid flows assume viscosity is zero.

The time rate of change a fluid particle's position in space is the fluid velocity (V). This is a vector field quantity. Speed (V) is the magnitude of the vector velocity V at some given point in the fluid, and average speed ($\overline{V}$) is the mean fluid speed through a control volume's surface. Therefore, velocity is a vector quantity (magnitude and direction), and speed is a scalar quantity (magnitude only). The standard units of velocity and speed are meter per second (m sec^{-1}).

Velocity is important to determine pollution, such as mixing rates after an effluent is discharged to a stream, how rapidly an aquifer will become contaminated, and the ability of liners to slow the movement of leachate from a landfill toward the groundwater. The distinction between velocity and speed is seldom made, even in technical discussion. Surface water flow is known as *stream discharge*, Q, with units of volume per time. Although the appropriate units are $m^3 sec^{-1}$, most stream discharge data in the United States is reported as number of cubic feet of water flowing past a point each second (cfs). Discharge is derived by measuring a stream's velocity at numerous points across the stream. Since heights (and volume of water) in a stream change with meteorological and other conditions, stream-stage/stream-discharge relationships are found by measuring stream discharge during different stream stages. The flow of a stream is estimated based upon many measurements. The mean of the flow measurements at all stage heights is reported as the estimated discharge. The calculation of discharge of the stream of width w_s is the sum of the products of mean depth, mean width, and mean velocity[iii]:

$$Q = \sum_{n=1}^{n} \tfrac{1}{2}(h_n + h_n - 1)(w_n + w_n - 1) \times \tfrac{1}{2}(v_n + v_n - 1)\tfrac{1}{2}(h_n + h_{n-1}) \qquad \text{(II.1)}$$

where

Q = Discharge ($m^3 sec^{-1}$)
w_n = n^{th} distance from baseline or initial point of measurement (m)
h_n = n^{th} water depth (m)
v_n = n^{th} velocity ($m\,sec^{-1}$) from velocity meter

Another important fluid property is pressure. A force per unit area is pressure (p):

$$p = \frac{F}{A} \qquad \text{(II.2)}$$

So, p is a type of stress that is exerted uniformly in all directions. It is common to use pressure instead of force to describe the factors that influence the behavior of fluids. The standard unit of pressure is the Pascal (P), which is equal to $1\,N\,m^{-2}$. The preferred pressure unit in this book is the kilopascal (kP), since the standard metric unit of pressure is the Pascal, which is quite small. Fluid pressure is measured against two references: zero pressure and atmospheric pressure. Absolute pressure is compared to true zero pressure and gage pressure is reported in reference to atmospheric pressure. To be able to tell which type of pressure is reported, the letter "a" and the letter "g" are added to units to designate whether the pressure is absolute or gage, respectively. So, it is common to see pounds per square inch designated as "psia" or inches of water as "in wg". If no letter is designated, the pressure can be assumed to be absolute pressure. When a gage measurement is taken, and the actual atmospheric pressure is known, absolute and gage pressure are related:

$$p_{absolute} = p_{gage} + p_{atmospheric} \tag{II.3}$$

Potential and kinetic energy discussions must consider the fluid acceleration due to gravity. In many ways, it seems that acceleration was a major reason for Isaac Newton's need to develop calculus.[iv] Known as the mathematics of change, calculus is the mathematical means of describing acceleration, and addressed Newton's need to express mathematically his new law of motion. Acceleration is the time rate of change in the velocity of a fluid particle. In terms of calculus, it is a second derivative. That is, it is the derivative of the velocity function. And a derivative of a function is itself a function, giving its rate of change. This explains why the second derivative must be a function showing the rate of change of the rate of change, which is readily apparent from the units of acceleration: length per time per time (m sec^{-2}).

The relationship between mass and volume is important in both environmental physics and chemistry and is a fundamental property of fluids. The density (ρ) of a fluid is defined as its mass per unit volume. Its metric units are kg m^{-3}. The density of an ideal gas is found using the specific gas constant and applying the ideal gas law:

$$\rho = p(RT)^{-1} \tag{II.4}$$

where

p = gas pressure
R = specific gas constant
T = absolute temperature

The specific gas constant must be known to calculate gas density. For example, the R for air is $287\,\text{J kg}^{-1}\,\text{K}^{-1}$. The specific gas constant for methane (R_{CH4}) is $518\,\text{J kg}^{-1}\,\text{K}^{-1}$.

Density is a very important fluid property for environmental situations. For example, a first responder must know the density of substances in an emergency situation. If a substance is burning, whether it is of greater or lesser density than water will be one of the factors on how to extinguish the fire. If the substance is less dense than water, the water will likely settle below the layer of the substance, making water a poor choice for fighting the fire. So, any flammable substance with a density less than water (see Table II.1), such as benzene or acetone, will require fire-extinguishing substances other than water. For substances heavier than water, like carbon disulfide, water may be a good choice.

Another important comparison in Table II.1 is that of pure water and seawater. The density difference between these two water types is important for marine and estuarine ecosystems. Saltwater contains a significantly

TABLE II.1
Densities of some important environmental fluids.

Fluid	*Density ($kg\,m^{-3}$) at 20°C unless otherwise noted*
Air at standard temperature and pressure (STP) = 0°C and 101.3 $N\,m^{-2}$	1.29
Air at 21°C	1.20
Ammonia	602
Diethyl ether	740
Ethanol	790
Acetone	791
Gasoline	700
Kerosene	820
Turpentine	870
Benzene	879
Pure water	1,000
Seawater	1,025
Carbon disulfide	1,274
Chloroform	1,489
Tetrachloromethane (carbon tetrachloride)	1,595
Lead (Pb)	11,340
Mercury (Hg)	13,600

greater mass of ions than does freshwater (see Table II.2). The denser saline water can wedge beneath freshwaters and pollute surface waters and groundwater (see Figure II.3). This phenomenon, known as saltwater intrusion, can significantly alter an ecosystem's structure and function, and threaten freshwater organisms. It can also pose a huge challenge to coastal communities who depend on aquifers for their water supply. Part of the problem and the solution to the problem can be found in dealing with the density differentials between fresh and saline waters.

The reciprocal of a substance's density is known as its specific volume (υ). This is the volume occupied by a unit mass of a fluid. The units of υ are reciprocal density units ($m^3\,kg^{-1}$). Stated mathematically, this is:

$$\upsilon = \rho^{-1} \qquad \text{(II.5)}$$

The weight of a fluid per its volume is known as specific weight (γ). Scientists and engineers sometimes use the term interchangeably with density. Geoscientists frequently refer to a substance's specific weight. A substance's γ is not an absolute fluid property because it depends upon the fluid itself and the local gravitational force:

TABLE II.2
Composition of freshwaters (river) and marine waters for some important ions.

Composition	*River Water*	*Saltwater*
pH	6–8	8
Ca^{2+}	4×10^{-5} M	1×10^{-2} M
Cl^-	2×10^{-4} M	6×10^{-1} M
HCO_3^-	1×10^{-4} M	2×10^{-3} M
K^+	6×10^{-5} M	1×10^{-2} M
Mg^{2+}	2×10^{-4} M	5×10^{-2} M
Na^+	4×10^{-4} M	5×10^{-1} M
SO_4^{2-}	1×10^{-4} M	3×10^{-2} M

Sources: K.A. Hunter, J.P. Kim, and M.R. Reid, 1999. Factors influencing the inorganic speciation of trace metal cations in freshwaters, *Marine Freshwater Research*, vol. 50, pp. 367–372. R.R. Schwarzenbach, P.M. Gschwend, and D.M. Imboden, 1993. *Environmental Organic Chemistry*, Wiley Interscience, New York, NY.

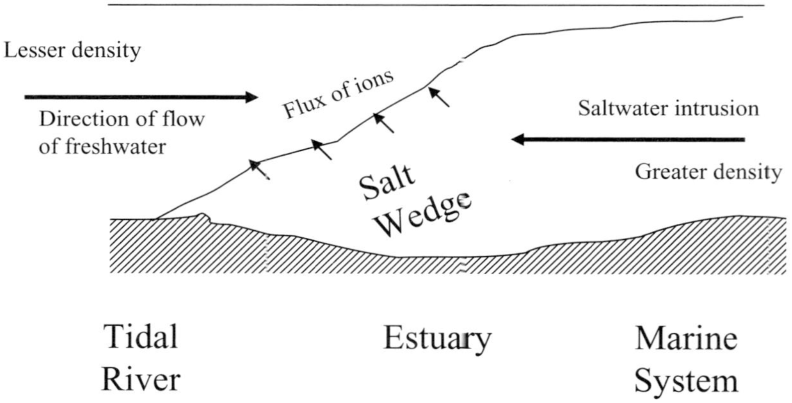

FIGURE II.3. Saltwater intrusion into a freshwater system. This denser saltwater submerges under the lighter freshwater system. The same phenomenon can occur in coastal aquifers.

$$\gamma = gp \tag{II.6}$$

Specific weight units are the same as those for density; for example, $kg\,m^{-3}$.

The fractional change in a fluid's volume per unit change in pressure at constant temperature is the fluid's coefficient of compressibility. Any fluid can be compressed in response to the application of pressure (p). For example, water's compressibility at 1 atm is $4.9 \times 10^{-5}\,\text{atm}^{-1}$. This compares to the lesser compressibility of mercury ($3.9 \times 10^{-6}\,\text{atm}^{-1}$) and the greater compressibility of hydrogen ($1.6 \times 10^{-3}\,\text{atm}^{-1}$). A fluid's bulk modulus, E, is a function of stress and strain on the fluid (see Figure II.4), and is a description of its compressibility and is defined according to the fluid volume (V):

$$E = \frac{stress}{strain} = -\frac{dp}{dV/V_1} \tag{II.7}$$

E is expressed in units of pressure (e.g., kP). Water is $E = 2.2 \times 10^6\,\text{kP}$ at 20°C.

Surface tension effects occur at liquid surfaces (interfaces of liquid-liquid, liquid-gas, liquid-solid). Surface tension, σ, is the force in the liquid surface normal to a line of unit length drawn in the surface. Surface tension decreases with temperature and depends on the contact fluid. Surface tension is involved in capillary rise and drop. Water has a very high σ value (approximately $0.07\,\text{N}\,\text{m}^{-2}$ at 20°C). Of the environmental fluids, only mercury has a higher σ (see Table II.3). The high surface tension creates a type of skin on a free surface, which is how an object that is denser than water (e.g., a steel needle) can "float" on a still water surface. It is the reason insects can sit comfortably on water surfaces. Surface tension is somewhat

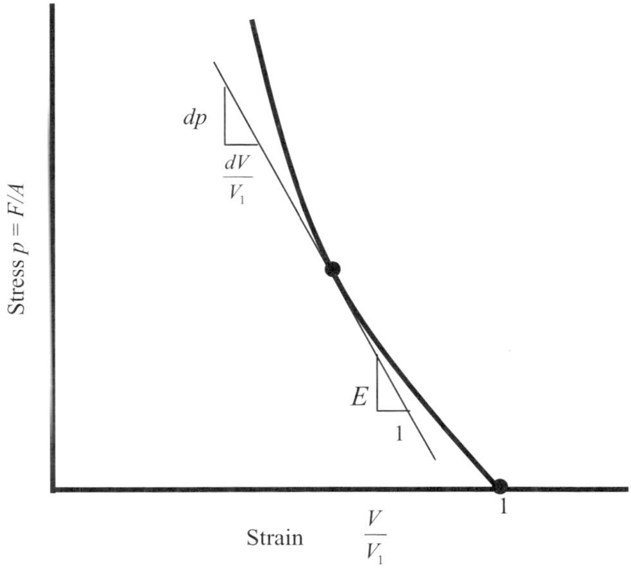

FIGURE II.4. Stress and strain on a fluid, and the bulk modulus of fluids.

TABLE II.3
Surface tension (contact with air) of selected environmental fluids.

Fluid	*Surface Tension, σ (Nm^{-1} at 20°C)*
Acetone	0.0236
Benzene	0.0289
Ethanol	0.0236
Glycerin	0.0631
Kerosene	0.0260
Mercury	0.519
n-Octane	0.0270
Tetrachloromethane	0.0236
Toluene	0.0285
Water	0.0728

dependent upon the gas that is contacting the free surface. If not indicated, it is usually safe to assume that the gas is the air in the troposphere.

Capillarity is a particularly important fluid property of groundwater flow and the movement of contaminants above the water table. In fact, the zone immediately above the water table is called the *capillary fringe*. Regardless of how densely soil particles are arranged, void spaces (i.e., pore spaces) will exist between the particles. By definition, the pore spaces below the water table are filled exclusively with water. However, above the water table, the spaces are filled with a mixture of air and water. As shown in Figure II.5, the spaces between unconsolidated material (e.g., gravel, sand, or clay) are interconnected, and behave like small conduits or pipes in their ability to distribute water. Depending on the grain size and density of packing, the conduits will vary in diameter, ranging from large pores (i.e., macropores), to medium pore sizes (i.e., mesopores), to extremely small pores (i.e., micropores).

Fluid pressures above the water table are negative with respect to atmospheric pressure, creating tension. Water rises for two reasons, its adhesion to a surface, plus the cohesion of water molecules to one another. Higher relative surface tension causes a fluid to rise in a tube (or a pore) and is indirectly proportional to the diameter of the tube. In other words, capillarity increases with decreasing diameter of a tube (e.g., tea will rise higher in a thin straw in your iced tea than in a fatter straw). The rise is limited by the weight of the fluid in the tube. The rise ($h_{capillary}$) of the fluid in a capillary is expressed as (Figure II.6 gives an example of the variables):

$$h_{capillary} = \frac{2\sigma \cos \lambda}{\rho_w g R} \quad \text{(II.8)}$$

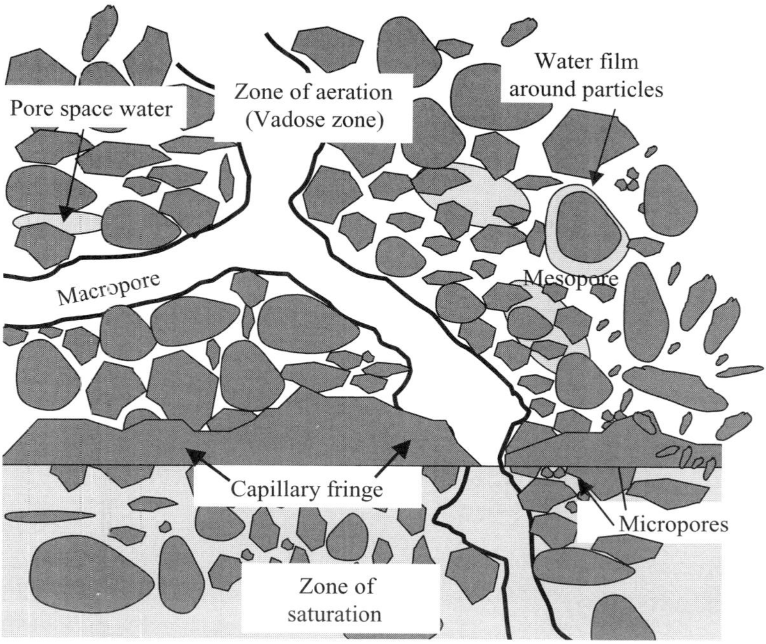

FIGURE II.5. Capillarity fringe above the water table of an aquifer.

where

σ = fluid surface tension ($g\,s^{-2}$)
λ = angle of meniscus (concavity of fluid) in capillary (degrees)
ρ_w = fluid density ($g\,cm^{-3}$)
g = gravitational acceleration ($cm\,sec^{-1}$)
R = radius of capillary (cm)

The contact angle indicates whether cohesive or adhesive forces are dominant in the capillarity. When λ values are greater than 90°, cohesive forces are dominant; when $\lambda < 90°$, adhesive forces dominate. Thus, λ is dependent upon both the type of fluid and the surface with which it comes into contact. For example, water-glass $\lambda = 0°$; ethanol-glass $\lambda = 0°$; glycerin-glass $\lambda = 19°$; kerosene-glass $\lambda = 26°$; water-paraffin $\lambda = 107°$; and mercury-glass $\lambda = 140°$. At the base of the capillary fringe the soil is saturated without regard to pore size. In the vadose zone, however, the capillary rise of water will be highest in the micropores, where relative surface tension and the effects of water cohesion are greatest.

Another property of environmental fluids is the mole fraction. If a composition of a fluid is made up of two or more substances (A, B, C, . . .),

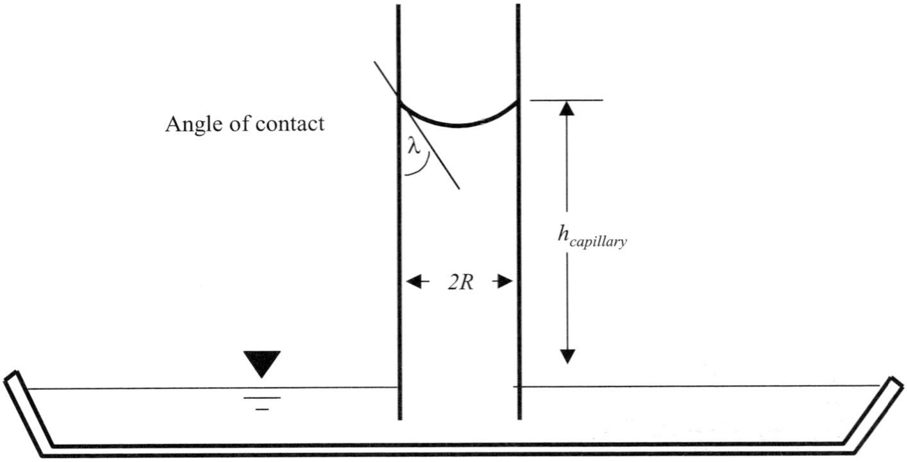

FIGURE II.6. Rise of a fluid in a capillary.

the mole fraction (x_A, x_B, x_C, . . .) is the number of moles of each substance divided by the total number of moles for the whole fluid:

$$x_A = \frac{n_A}{n_A + n_B + n_C + \cdots} \tag{II.9}$$

The mole fraction value is always between 0 and 1. The mole fraction may be converted to mole percent as:

$$x_{A\%} = x_A \times 100 \tag{II.10}$$

For gases, the mole fraction is the same as the volumetric fraction of each gas in a mixture of more than one gas.

The amount of resistance to flow when it is acted on by an external force, especially a pressure differential or gravity, is the fluid's viscosity. This a crucial fluid property used in numerous environmental applications, but particularly in air pollution plume characterization, sludge management, wastewater and drinking water treatment, and distribution systems.

Bernoulli's equation states that when fluid is flowing in a long, horizontal pipe with constant cross-sectional area, the pressure along the pipe must be constant. However, as the fluid moves in the pipe, there will be a pressure drop. A pressure difference is needed to push the fluid through the pipe to overcome the drag force exerted by the pipe walls on the layer of fluid that is making contact with the walls. Since the drag force is exerted by each successive layer of the fluid on each adjacent layer that is moving at its own velocity, then a pressure difference is needed (see Figure II.7). The

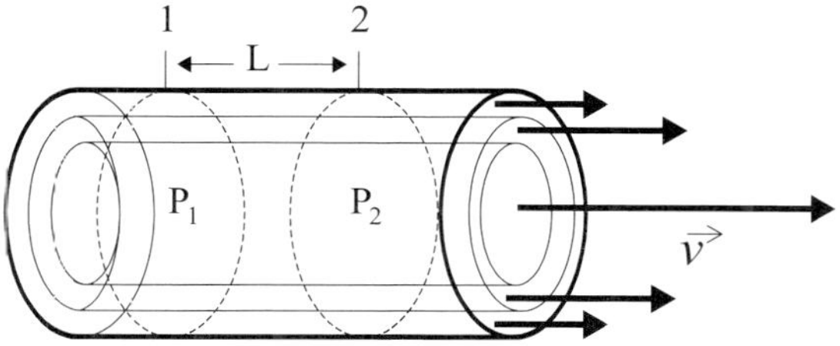

FIGURE II.7. Viscous flow through a horizontal pipe. The highest velocity is at the center of the pipe. As the fluid approaches the pipe wall, the velocity approaches zero.

drag forces are known as viscous forces. Thus, the fluid velocity is not constant across the pipe's diameter, owing to the viscous forces. The greatest velocity is at the center (furthest away from the walls), and the lowest velocity is found at the walls. In fact, at the point of contact with walls, the fluid velocity is zero.

So, if P_1 is the pressure at point 1, and P_2 is the pressure at point 2, with the two points separated by distance L, the pressure drop (ΔP) is proportional to the flow rate:

$$\Delta P = P_1 - P_2 \quad \text{(II.11)}$$

and,

$$\Delta P = P_1 - P_2 = I_v R \quad \text{(II.12)}$$

where I_v is volume flow rate and R is the proportionality constant representing the resistance to the flow. R depends on the length (L) of pipe section, the pipe's radius, and the fluid's viscosity.

Three Major Media

At the risk of hypocrisy, it is convenient to categorize environmental cases into three types: air, water, and land. The foregoing discussions should draw attention to the fact that, although we tend to place these cases in a single category, their causes and effects are often not so limited. For example, the pollutant loadings to the Great Lakes include those from water (feeding streams and direct discharges), air (dry and wet deposition), and land (nonpoint runoff). The effects from water pollution, such as polychlorinated biphenyl (PCB) contamination, can lead to air pollution when the PCBs volatilize from surface waters and, subsequently land pollution, when they are transported and are sequestered in soils (see Figure II.1).

Note and Commentary

i. Fate may also include some remediation reactions, such as thermal and microbial treatment, but in discussions of fate and transport, the reactions are usually those that occur in the ambient environment. The treatment and remediation processes usually fall under the category of environmental engineering.

ii. Even solids can be fluids at a very large scale. For example, in plate tectonics and other expansive geological processes, solid rock will flow, albeit very slowly.

iii. From C. Lee and S. Lin, eds., 1999. *Handbook of Environmental Engineering Calculations*, McGraw-Hill, New York, NY.

iv. Newton actually co-invented the calculus with Willhelm Leibniz in the seventeenth century. Both are credited with devising the symbolism and the system of rules for computing derivatives and integrals, but their notation and emphases differed. A debate rages on who did what first, but both of these giants had good reason to revise the language of science (i.e., mathematics) to explain motion.

CHAPTER 3

Something in the Air

As soon as I had gotten out of the heavy air of Rome, from the stink of the chimneys and the pestilence, vapors and soot of the air, I felt an alteration to my disposition.

Lucius Annaeus Seneca, the Younger, 61 A.D.

At least from the standpoint of how long we can last in its absence, air is the most critical of all human needs. We can survive for weeks without food, days without water, but only minutes without air. Like Seneca, the first century philosopher, we readily notice some acute effects of contaminated air. Unfortunately, however, some of the more toxic pollutants are not readily detectable with normal human senses. Interestingly, Seneca reported two of the major pollution types that are regulated commonly today, volatile compounds (i.e., vapors) and particulate matter (i.e., soot). The atmosphere has been affected by human activities for millennia, but only recently has the air been polluted on a scale and to a degree that natural processes have not be able to neutralize the widespread effects.

One of the first recorded episodes of air pollution was that of the village of Hit, west of ancient Babylon. In 900 B.C., the Egyptian King Tukulti described an offensive odor emanating from a bitumen mining operation that released high concentrations of sulfur dioxide (SO_2) and hydrogen sulfide (H_2S). The former is the result of oxidation and the latter is the reduction of the rather ubiquitous element sulfur. Even though these problems were for the most part confined to the most highly urbanized areas of ancient times, their effects became increasingly widespread.

London Air Pollution and the Industrial Revolution

London exemplified the transition from a predominantly agrarian Europe to a progressively more industrialized society. The air quality problem was accelerated after the dawn of the Industrial Revolution so that by the thirteenth century, urban air pollution had already come to be perceived as a public health problem. As evidence, in 1285, a special commission was formed in London to investigate the air pollution brought on by the increase in the combustion of coal, followed in 1307 with a law that prohibited using sea coal as the fuel for kilns and in blacksmith forging operations.[1] However, London and the other increasingly industrialized centers of Europe had to import increasing amounts of coal because wood had become unviable as an alternative fuel source due to deforestation. By the early seventeenth century, Great Britain's coal use had grown to greater than 50,000 tons per year. John Evelyn's 1661 *Fumifugium, or The Inconvenience of the Aer and the Smoak of London Dissipated,* railed against London's air pollution. He followed this with *A Character of England,* in which he described London as:

> . . . cloaked in such a cloud of sea-coal, as if there be a resemblance of hell upon earth, it is in this volcano in a foggy day: this pestilential smoke which corrodes the very iron, and spoils all the moveables, leaving a soot on all things that it lights; and so fatally seizing on the lungs of the inhabitants, that cough and consumption spare no man.

This interesting and vivid depiction includes both sources of pollutants (e.g., "sea-coal combustion") and receptors of the pollutants, corrosion of iron structures and respiratory effects, examples of welfare and health endpoints, respectively. The sea-coal reference points out that the pollution was pervasive well before the prominence of rail transport of coal to London. A map of the time (1746), drawn by cartographer John Rocque, depicts the wharf areas dedicated to unloading coal from barges along with timer yards along the Surrey bank of the River Thames.

In the next century, London had an aerosol problem. Visitors noted the presence of particle (soot) laden buildings and polluted air. This observation was recorded by the novelist Charles Dickens, who wrote in *Bleak House* (1852–1853):

> . . . fog everywhere, fog up the river where it flows among green aits and meadows—fog down the river, where it rolls defiled among the tiers of shipping and the waterside pollutions of a great (and dirty) city. Fog on the Essex marshes, fog on the Kentish heights. Fog creeping into the cabooses of collier-brigs; fog lying out in the yards, and hovering in the rigging of great ships . . . Fog in the eyes and throats of ancient Greenwich pensioners, wheezing by the firesides of their wards; fog in the stem and bowl of the afternoon pipe of the wrathful skipper, down in the close cabin; fog cruelly pinching the toes and fingers of his shivering little 'prentice boy on the deck.

Public officials began to take action in the latter part of the nineteenth century, writing legislation to control smoke, the Public Health (London)

Act of 1891. However, a major exemption in the law was that smoke reduction measures were not applicable to domestic chimney emissions. At that time, coal was the principal fuel for individual homes, so the sources were highly distributed spatially, making for a vexing challenge in terms of controlling the emissions of carbon and sulfur laden aerosols. It was a bit later that the term "smog" came into general use when, in 1905, the medical doctor H.A. Des Voeux, a member of the Coal Smoke Abatement Society coined the term to depict a mixture of smoke and fog.[2] In 1912, the *Lancet* estimated that 76,000 tons of soot fell on London every year. In 1926, the Public Health (Smoke Abatement) Act was passed, but once again ignored the problem of domestic fuel combustion.

During December 1952, unusual weather conditions created the ideal situations for the formation of a major fog. The fog caused numerous problems. A performance of La Traviata at Sadler's Wells was abruptly ended where the stage was no longer visible. Particles aggregated on people's skin and clothing, wind shields of cars were blackened by a slime of settled aerosols. Farm animals become ill (13 prize cattle of Earl's court had to be euthanized). About 12,000 people's deaths were the result of the fog. The subsequent outcry produced the Clean Air Act of 1956, which, by controlling domestic smoke output for the first time, got rid of most of the pea-souper fogs that had become synonymous with London.

Although we scientists are often reluctant to admit it, intuition may be the strongest approach in problem solving. In the case of aerosols, the chemical composition was yet to be known, but officials and the general public began to see the weight of evidence connecting soot and harm. For example, we are now able to ascertain the chemical composition of particles and gases in the atmosphere. A frequent culprit linked to health effects is sulfur. Exposure to numerous chemical species of sulfur can damage human health, harm ecosystems, and adversely affect welfare (e.g., economic costs by destroying materials, harm to quality of life from odors, especially from reduced sulfur species, like H_2S, and destruction of crops). Compounds of sulfur represent an important class of air pollutants today. Many compounds of sulfur are essential components of biological systems, but in the wrong place at the wrong time, these same compounds are hazardous to health, welfare, and the environment. (See the box "Contaminants of Concern: Sulfur and Nitrogen Compounds.")

Contaminants of Concern: Sulfur and Nitrogen Compounds

Talk to most farmers about the elements sulfur (S) and nitrogen (N), and they would quickly begin discussing the importance of fertilizers and the need for macro- and micronutrients to ensure productive crop yields. But talk to air quality experts, and they are likely to mention numerous sulfur and nitrogen compounds that can harm the health of humans, can adversely affect the environment, and can lead to welfare impacts, such as the corrosion of buildings and other structures and diminished visibility due to the formation of haze. So, S and N are important in all environmental media (air, water, soil, sediment, and biota). These nutrients also demonstrate the concept that pollution is often a resource that is simply in the wrong place.

The reason that sulfur and nitrogen pollutants are often lumped together may be that their oxidized species (e.g., sulfur dioxide (SO_2) and nitrogen dioxide (NO_2)) form acids when they react with water. The lowered pH is responsible for many environmental problems. Another reason may be that many sulfur and nitrogen pollutants result from combustion. Whatever the reasons, however, sulfur and nitrogen pollutants actually are very different in their sources and in the processes that lead to their emissions.

Sulfur is present in most fossil fuels, usually higher in coal than in crude oil. Prehistoric plant life is the source for most fossil fuels. Most plants contain S as a nutrient and as the plants become fossilized a fraction of the sulfur volatilizes (i.e., becomes a vapor) and is released. However, some sulfur remains in the fossil fuel and can be concentrated because much of the carbonaceous matter is driven off. Thus, the S-content of the coal is available to react with oxygen when the fossil fuel is combusted. In fact, the S-content of coal is an important characteristic in its economic worth; the higher the S-content, the less it is worth. So, the lower the sulfur content and volatile constituents and the higher the carbon content makes for a more valuable coal. Since combustion is the combination of a substance (fuel) with molecular oxygen (O_2) in the presence of heat, the reaction for complete or efficient combustion of a hydrocarbon results in the formation of carbon dioxide and water:

$$(CH)_x + O_2 \xrightarrow{\Delta} CO_2 + H_2O \tag{3.1}$$

However, the fossil fuel contains other elements, including sulfur, so the side reaction forms oxides of sulfur:

$$S + O_2 \xrightarrow{\Delta} SO_2 \tag{3.2}$$

Actually, many other oxidized forms of sulfur can form during combustion, so air pollution experts refer to them collectively as SO_x, which is commonly seen in the literature.

Likewise, nitrogen compounds also form during combustion, but their sources are very different from those of sulfur compounds. In fact, the atmosphere itself is the source of much of the nitrogen leading to the formation of oxides of nitrogen (NO_x). Molecular nitrogen (N_2) makes up most of the gases in the earth's atmosphere (79% by volume). Because N_2 is relatively nonreactive under most atmospheric conditions, it seldom enters into chemical reactions, but under pressure and at very high temperatures, it will react with O_2:

$$N_2 + O_2 \xrightarrow{\Delta} 2NO \tag{3.3}$$

Approximately 90–95% of the nitrogen oxides generated in combustion processes are in the form of nitric oxide (NO), but like the oxides of sulfur, other nitrogen oxides can form, especially nitrogen dioxide (NO_2), so air pollution experts refer to NO and NO_2 collectively as NO_x. In fact, in the atmosphere the emitted NO is quickly converted photochemically to nitrogen dioxide (NO_2). High temperature/high pressure conditions exist in internal combustion engines, like those in automobiles (known as mobile sources). Thus, NO_x is one of the major mobile source air pollutants. These conditions of high temperature and pressure can also exist in boilers such as those in power plants, so NO_x is also commonly found in high concentrations when leaving fossil fuel power generating stations.

In addition to the atmospheric nitrogen, other sources exist, particularly the nitrogen in fossil fuels. The nitrogen oxides generated from atmospheric nitrogen are known as "thermal NO_x" since they form at high temperatures, such as near burner flames in combustion chambers. Nitrogen oxides that form from the fuel or feedstock are called fuel NO_x. Unlike the sulfur compounds, a significant fraction of the fuel nitrogen remains in the bottom ash or in unburned aerosols in the gases leaving the combustion chamber, the fly ash. Nitrogen oxides can also be released from nitric acid plants and other types of industrial processes involving the generation and/or use of nitric acid (HNO_3).

Nitric oxide is a colorless, odorless gas and is essentially insoluble in water. Nitrogen dioxide has a pungent acid odor and is somewhat soluble in water. At low temperatures such as those often present in the ambient atmosphere, NO_2 can form the molecule NO_2-O_2N, or simply N_2O_4, that consists of two identical simpler NO_2 molecules. This is known as a *dimer*. The dimer N_2O_4 is distinctly reddish-brown and contributes to the brown haze that is often associated with photochemical smog incidents.

Both NO and NO_2 are harmful and toxic to humans, although atmospheric concentrations of nitrogen oxides are usually well below the concentrations expected to lead to adverse health effects. The low concentrations owe to the moderately rapid reactions that occur when NO and NO_2 are emitted into the atmosphere. Much of the concern for regulating NO_x emissions is to suppress the reactions in the atmosphere that generate the highly reactive molecule ozone (O_3). Nitrogen oxides play key roles as important reactants in O_3 formation. Ozone forms photochemically (i.e., the reaction is caused or accelerated by light energy) in the lowest level of the atmosphere, known as the troposphere, where people live. Nitrogen dioxide is the principal gas responsible for absorbing sunlight needed for these photochemical reactions. So, in the presence of sunlight, the NO_2 that forms from the NO incrementally stimulates the photochemical smog-forming reactions because nitrogen dioxide is very efficient at absorbing sunlight in the ultraviolet portion of its spectrum. This is why ozone episodes are more common in the summer and in areas with ample sunlight. Other chemical ingredients, that is, ozone precursors, in O_3 formation include volatile organic compounds (VOCs) and carbon monoxide (CO). Governments regulate the emissions of precursor compounds to diminish the rate at which O_3 forms.

Many compounds contain both nitrogen and sulfur along with the typical organic elements (carbon, hydrogen, and oxygen). The reaction for the combustion of such compounds, in general form, is:

$$C_aH_bO_cN_dS_e + 4a + b\text{-}2c \rightarrow aCO_2 + \left(\frac{b}{2}\right)H_2O + \left(\frac{d}{2}\right)N_2 + eS \quad (3.4)$$

Reaction 3.4 demonstrates the incremental complexity as additional elements enter the reaction. In the real world, pure reactions are rare. The environment is filled with mixtures. Reactions can occur in sequence, parallel, or both. For example, a feedstock to a municipal incinerator contains myriad types of wastes, from garbage to household chemicals to commercial wastes, and even small (and sometimes large) amounts of industrial wastes that may be illegally dumped. For example, the nitrogen content of typical cow manure is about 5 kg per metric ton (about 0.5%). If the fuel used to burn the waste also contains sulfur along with the organic matter, then the five elements will react according to the stoichiometry of Reaction 3.4.

Certainly, combustion specifically and oxidation generally are very important processes that lead to nitrogen and sulfur pollutants. But they are certainly not the only ones. In fact, we need to explain what oxidation really means. In the environment, oxidation *and* reduction occur. An oxidation-reduction (known as redox) reaction is the simultaneous loss of an electron (oxidation) by one substance joined by an electron gain (reduction) by another in the same reaction. In oxidation, an element or compound loses (i.e., donates) electrons. Oxidation also occurs when oxygen atoms are gained or when hydrogen atoms are lost. Conversely, in reduction, an element or compound gains (i.e., captures) electrons. Reduction also occurs when oxygen atoms are lost or when hydrogen atoms are gained. The nature of redox reactions means that each oxidation-reduction reaction is a pair of two simultaneously occurring "half-reactions." The formation of sulfur dioxide and nitric oxide by acidifying molecular sulfur is a redox reaction:

$$S(s) + NO_3^-(aq) \rightarrow SO_2(g) + NO(g) \quad (3.5)$$

The designations in parentheses give the physical phase of each reactant and product: "s" for solid; "aq" for aqueous; and "g" for gas.

The oxidation half-reactions for this reaction are:

$$S \rightarrow SO_2 \quad (3.6)$$

$$S + 2H_2O \rightarrow SO_2 + 4H^+ + 4e^- \quad (3.7)$$

The reduction half-reactions for this reaction are:

$$NO_3^- \rightarrow NO \quad (3.8)$$

$$NO_3^- + 4H^+ + 3e^- \rightarrow NO + 2H_2O \quad (3.9)$$

Therefore, the balanced oxidation-reduction reactions are:

$$4NO_3^- + 3S + 16H^+ + 6H_2O \rightarrow 3SO_2 + 16H^+ + 4NO + 8H_2O \quad (3.10)$$

$$4NO_3^- + 3S + 4H^+ \rightarrow 3SO_2 + 4NO + 2H_2 \quad (3.11)$$

Oxidation-reduction reactions are not only responsible for pollution, they are also very beneficial. Redox reactions are part of essential metabolic and respiratory processes. Redox is commonly used to treat wastes; for example, to ameliorate toxic substances and to treat wastes, by taking advantage of electron donating and accepting microbes, or by abiotic chemical redox reactions. For example, in drinking water treatment, a chemical oxidizing or reducing agent is added to the water under controlled pH. This reaction raises the valence of one reactant and lowers the valence of the other. Thus redox removes compounds that are "oxidizable," such as ammonia, cyanides, and certain metals like selenium, manganese, and iron. It also removes other "reducible" metals like mercury (Hg), chromium (Cr), lead, silver (Ag), cadmium (Cd), zinc (Zn), copper (Cu), and nickel (Ni). Oxidizing cyanide (CN^-) and reducing of Cr^{6+} to Cr^{3+} are two examples where the toxicity of inorganic contaminants can be greatly reduced by redox.*

A reduced form of sulfur that is highly toxic and an important pollutant is hydrogen sulfide (H_2S). Certain microbes, especially bacteria, reduce nitrogen and sulfur, using the N or S as energy sources through the acceptance of electrons. For example, sulfur-reducing bacteria can produce hydrogen sulfide (H_2S) by chemically changing oxidized forms of sulfur, especially sulfates (SO_4). To do so, the bacteria must have access to the sulfur; that is, it must be in the water, which can be in surface or groundwater, or the water in soil and sediment. These sulfur-reducers are often anaerobes; that is, bacteria that live in water where concentrations of molecular oxygen (O_2) are deficient. The bacteria remove the O_2 molecule from the sulfate leaving only the S, which in turn combines with hydrogen (H) to form gaseous H_2S. In groundwater, sediment, and soil water, H_2S is formed from the anaerobic or nearly anaerobic decomposition of deposits of organic matter, for example, plant residues. Thus, redox principles can be used to treat H_2S contamination; that is, the compound can be oxidized using a number of different oxidants (see Table 3.1). Strong oxidizers, like molecular oxygen and hydrogen peroxide, most effectively oxidize the reduced forms of S, N, or any reduced compound.

TABLE 3.1
Theoretical amounts of various agents required to oxidize $1\,mg\,L^{-1}$ of sulfide ion.

Oxidizing Agent	*Amount ($mg\,L^{-1}$) needed to oxidize $1\,mg\,L^{-1}$ of S_2^- based on practical observations*	*Theoretical stoichiometry ($mg\,L^{-1}$)*
Chlorine (Cl_2)	2.0 to 3.0	2.2
Chlorine dioxide (ClO_2)	7.2 to 10.8	4.2
Hydrogen peroxide (H_2O_2)	1.0 to 1.5	1.1
Potassium permanganate ($KMnO_4$)	4.0 to 6.0	3.3
Oxygen (O_2)	2.8 to 3.6	0.5
Ozone (O_3)	2.2 to 3.6	1.5

Source: Water Quality Association, 1999, Ozone Task Force Report, "Ozone for POU, POE & Small Water System Applications," Lisle, IL.

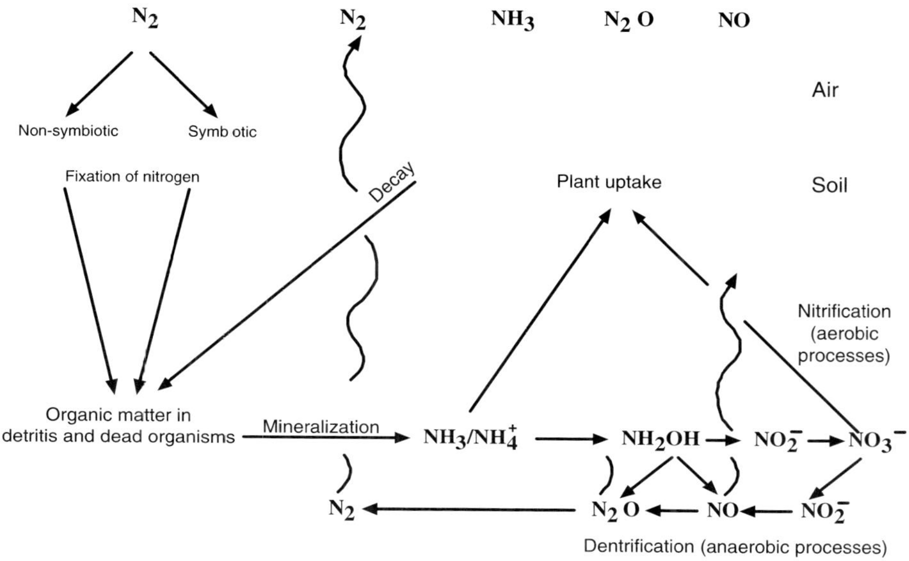

FIGURE 3.1. Biochemical nitrogen cycle.

Ionization is also important in environmental reactions. This is due to the configuration of electrons in an atom. The arrangement of the electrons in the atom's outermost shell (i.e., valence) determines the ultimate chemical behavior of the atom. The outer electrons become involved in transfer to and sharing with shells in other atoms; that is, forming new compounds and ions. An atom will gain or lose valence electrons to form a stable ion that has the same number of electrons as the noble gas nearest the atom's atomic number. For example, the nitrogen cycle (see Figure 3.1) includes three principal forms that are soluble in water under environmental conditions: the cation (positively charged ion) ammonium (NH_4^+), and the anions (negatively charged ions) nitrate (NO_3^-) and nitrite (NO_2^-). Nitrates and nitrites combine with various organic and inorganic compounds. Once taken into the body, NO_3^- is converted to NO_2^-. Since NO_3^- is soluble and readily available as a nitrogen source for plants (e.g., to form plant tissue such as amino acids and proteins), farmers are the biggest users of NO_3^- compounds in commercial fertilizers (although even manure can contain high levels of NO_3^-).

Ingesting high concentrations of nitrates (e.g., in drinking water) can cause serious short-term illness and even death. The serious illness in infants is due to the conversion of nitrate to nitrite by the body, which can interfere with the oxygen-carrying capacity of the blood, known as methemoglobinemia. Especially in small children, when nitrates compete successfully against molecular oxygen, the blood carries methemoglobin (as opposed to healthy hemoglobin), giving rise to clinical symptoms. At 15–20% methemoglobin, children can experience shortness of breath and blueness of the skin (i.e., clinical cyanosis). At 20–40% methemoglobin, hypoxia will result. This acute condition can deteriorate a child's health rapidly over a period of days, especially if the water source continues to be used. Long-term, elevated exposures to nitrates and nitrites can cause an increase in the kidneys' production of urine (diuresis), increased starchy deposits, and hemorrhaging of the spleen.[3]

Compounds of nitrogen and sulfur are important in every environmental medium. They are addressed throughout this book, as air pollutants, water pollutants, indicators of eutrophication (i.e., nutrient enrichment), ecological condition, and acid rain. They are some of the best examples of the need for a systematic viewpoint. Nutrients are valuable, but in the wrong place under the wrong conditions, they become pollutants.

* Redox reactions are controlled in closed reactors with rapid mix agitators. Oxidation-reduction probes are used to monitor reaction rates and product formation. The reactions are exothermic and can be very violent when the heat of reaction is released, so care must be taken to use only dilute concentrations, along with careful monitoring of batch processes.

Notorious Air Pollution Cases of the Twentieth Century

Characterizing air pollution requires an understanding of sources of pollutants, the means by which these pollutants move in the environment after they are released, as well the effects caused by these pollutants. The cases discussed in this section generally receive pollutants from two major source categories, natural and anthropogenic. Forest fires (although often started by human activities) are natural sources of particulate matter (PM), tars and polycyclic aromatic hydrocarbons (PAHs), and ozone (O_3). These plumes are often so large that they can be seen from orbiting spacecrafts and tracked using satellite imagery (see Figures 3.2 and 3.3). Anthropogenic sources include industrial, transportation, and energy production, as described in following cases.

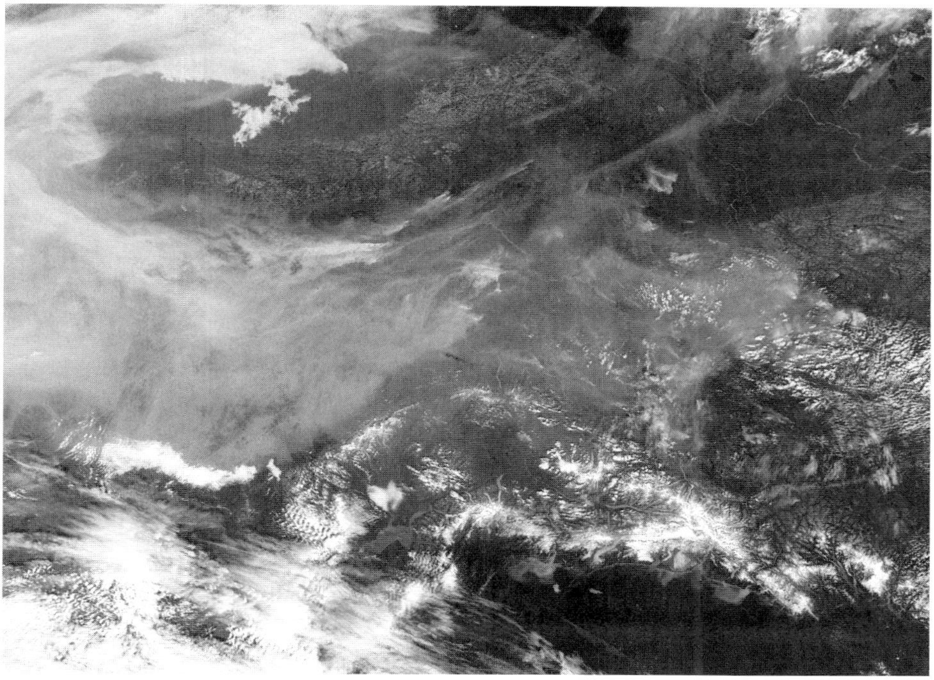

FIGURE 3.2. Airborne Visible and Infrared Imaging Spectrometer (AVIRIS) satellite image of a plume of particulate matter (PM) from a forest fire in Alaska and northern Canada (borders drawn), August 4, 2004.
Source: National Aeronautics and Space Administration, 2004, Moderate Resolution Imaging Spectroradiometer (MODIS) aboard the Terra (EOS AM) and Aqua (EOS PM) satellites.

The Meuse Valley Acid Fog

One of the first modern air pollution episodes took place in the Meuse Valley of Belgium on December 3–5, 1930. A thermal inversion (see Figure 3.4) trapped polluted air in the densely populated 15-mile valley, which had numerous sources, many related to metal production and refractory, such as steel mills, coke ovens, blast furnaces, zinc smelters, glass factories, and sulfuric acid plants. One of the principal contaminants in the trapped air resulting from these industrial emissions was sulfur dioxide (SO_2). The physicochemical properties of SO_2 are shown in Table 3.2. The SO_2 exposures probably triggered the death of 63 people and illnesses of an additional 600. Sulfur dioxide and fog droplets that form acid mists (i.e., acid aerosols) generate aerosols that can penetrate deeply into the respiratory system (see

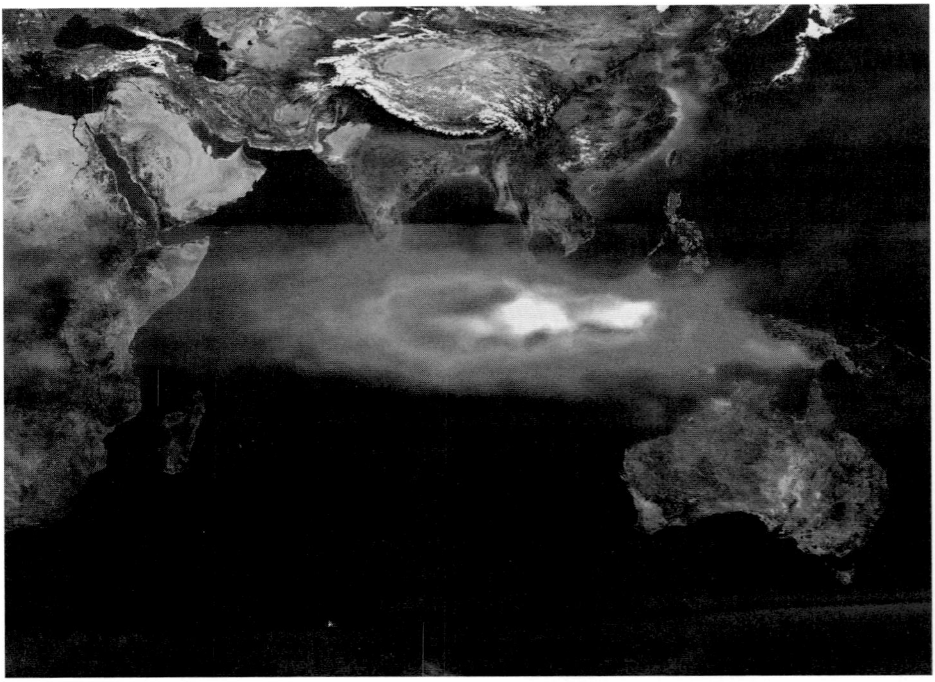

FIGURE 3.3. Tropospheric ozone and smoke over Indonesia on October 22, 1997. *Source*: National Aeronautics and Space Administration, 2001, Goddard Space Flight, Space Visualization Studio, Total Ozone Mapping Spectrometer (TOMS) project.

Figure 3.12 and the discussion box "Contaminant of Concern: Particle Matter"). Another possible cause of the deaths and illnesses is exposure to elevated concentrations of fluorine (F) compounds.[4] The most obvious symptom of those affected was dyspnea—shortness of breath—in severe cases, leading to cardiac failure.

A difference between the Meuse disaster and that of more recent air pollution episodes is that it took place in the winter. This may indicate the importance of home heating, especially the common use of coal to heat homes in Belgium at that time. With the widespread availability of air conditioning systems in North America and western Europe, emissions of particles, oxides of sulfur and oxides of nitrogen are now higher during the summer months, due in large part to the increased demand for electricity provided by coal-fired power plants. Also, ozone is a principal contaminant during summer months due to the greater amounts of sunlight available to produce photochemical oxidant smog.

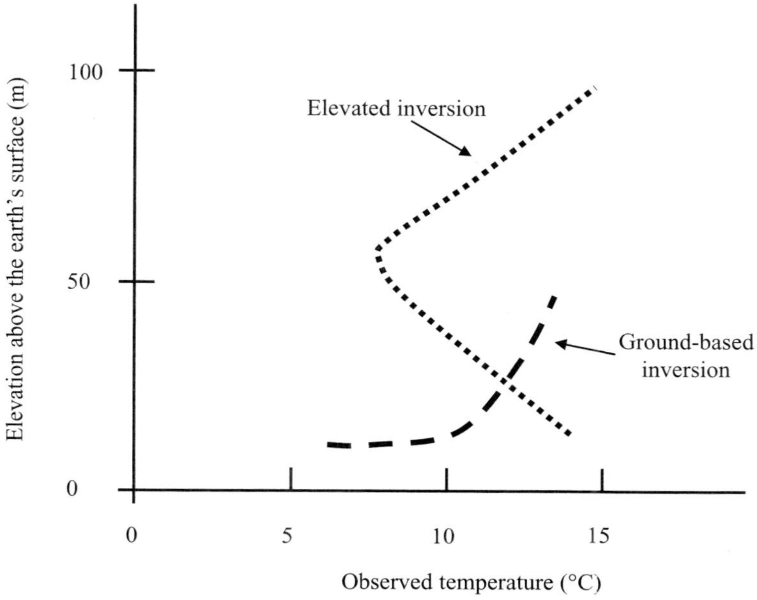

FIGURE 3.4. Thermal inversion. Under normal meteorological conditions, temperatures in the troposphere decrease with elevation, but under inversion conditions, the temperatures increase. Inversions cause the vertical air circulations to be trapped below a stagnant layer, so that pollutant concentrations build up until the inversion conditions dissipate.

Trapped by an inversion, pollutants accumulated in this steep-sided valley. Unfortunately no measurements were made during the 1930 episode, but first-hand accounts were documented. *Time Magazine's* December 15, 1930, article, "Poison Fog" recounted the episode quite vividly:

> During the winters of 1897, 1902, 1911, and last week Belgians experienced the dread phenomenon of "poison fog." In their Royal Palace at Brussels last week King Albert and Queen Elisabeth received dreadful tidings that men, women, animals (no children), were gasping, choking, dying in a fog which filled the valley of the River Meuse from Liege down through Namur. On the fourth day the fog lifted, on the fifth Queen Elisabeth motored through the stricken valley, where 67 human lives had been lost, was rousingly cheered. The Belgian Government officially announced that the deaths were due "solely to the cold fog," thus scotching rumors that War gases buried by the retreating German Armies had escaped. As

TABLE 3.2
Physical and chemical properties for sulfur dioxide.

Characteristic	*Value*
Molecular formula	SO_2
Synonyms	Sulfurous anhydride, sulfurous oxide, sulfur oxide, sulfurous acid anhydride
Molecular weight	64.07
Chemical Abstract Service (CAS) number	7446-09-5
Solubility	Soluble in water, alcohol, acetic acid, sulfuric acid, ether, and chloroform
Density	$2.811\,g\,L^{-1}$
Vapor pressure	3×10^{-3} mm Hg at 25°C
Saturated vapor pressure	$0.47\,lb/ft^3$ at 15°C
Melting point	−72°C
Boiling point	−10°C
Conversion factors in air, 1 atm	1 ppm = $2.6\,mg\,m^{-3}$ $1\,mg/m^3$ = 0.38 ppm

Source: National Academy of Sciences, 2002. *Review of Submarine Escape Action Levels for Selected Chemicals*, The National Academies Press, Washington, D.C.
Abbreviation: CAS, Chemical Abstracts Service.

on the three previous occasions when "poison" fogs have appeared, apparently no one in the panic stricken Meuse Valley thought to bottle a sample of the fog before it blew away. With nothing to work upon last week (for bereaved relatives, delayed attempts to obtain the bodies of fog-victims for autopsy), scientists could only guess what may have happened. Guesses: "Deadly gases from the tail of a dissipated comet."—Professor Victor Levine of Creighton University, Omaha, Neb. "Germs brought from the Near East by the winds which have carried dust from the Sahara Desert to Europe recently, producing muddy rains."—Colonel Joaquin Enrique Zanetti, Wartime poison gas expert, chemistry professor at Columbia University, Manhattan. "I did not allude to the Bubonic Plague in speaking of the Belgian fog. I said pneumonic plague. I meant . . . an acute respiratory infection attacking the lungs."—Famed J. B. S. Haldane, reader in biochemistry at Cambridge University, correcting worldwide reports that he had said Belgium was suffering from a return of the medieval "Black Death." Coincidence. Experts of the French Army were busy last week at Lille (80 mi. from the stricken Meuse Valley) producing enormous clouds of what they called "a cheap, harmless artificial fog made from chalk, sulphuric acid and tar products which will be extremely useful to hide the movements of troops in war time."

Contaminants of Concern: Particulate Matter

Although many contaminants of concern discussed in this book are best classified by their chemical composition, some contaminants are better classified according to their physical properties. Particulate matter (PM) is a common physical classification of particles found in the air, such as dust, dirt, soot, smoke, and liquid droplets.[5] Unlike other U.S. criteria pollutants subject to the National Ambient Air Quality Standards (ozone (O_3), carbon monoxide (CO), sulfur dioxide (SO_2), nitrogen dioxide (NO_2) and lead (Pb)), PM is not a specific chemical entity but is a mixture of particles from different sources and of different sizes, compositions, and properties. However, the chemical composition of PM is very important and highly variable. In fact, knowing what a particle is made of tells us much about its source; for example, receptor models use chemical composition and morphology of particles as a means to trace pollutants back to the source. The chemical composition of tropospheric particles includes inorganic ions, metallic compounds, elemental carbon, organic compounds, and crustal (e.g., carbonates and compounds, alkali and rare earth elementals) substances. For example, the mean 24-hour $PM_{2.5}$ concentrations measured near Baltimore, Maryland, in 1999 were composed of 38% sulfate, 13% ammonium, 2% nitrate, 36% organic carbon, 7% elemental carbon, and 4% crustal matter.[6] In addition, some atmospheric particles can be hygroscopic; that is, they contain particle-bound water. The organic fraction can be particularly difficult to characterize, since it often contains thousands of organic compounds.

The size of a particle results from how the particle is formed; for example, combustion can generate very small particles, and coarse particles are often formed by mechanical processes (see Figure 3.5). Particles, if they are small and have low mass, can be suspended in the air for long periods of time. Particles may be sufficiently large (e.g., >10 μm aerodynamic diameter) as to be seen as smoke or soot (see Figure 3.6), and others are very small (<2.5 μm). Sources of particles are highly variable; for example, emitted directly to the air from stationary sources, such as factories, power plants, and open burning, and from moving vehicles (known as mobile sources), especially those with internal combustion engines. Area or nonpoint sources of particles include construction, agricultural activities such as plowing and tilling, mining, and forest fires.

Particles may also form from gases that have been previously emitted, for example, when gases from burning fuels react with sunlight and water vapor. A common production of such secondary particles occurs when gases undergo chemical reactions in the

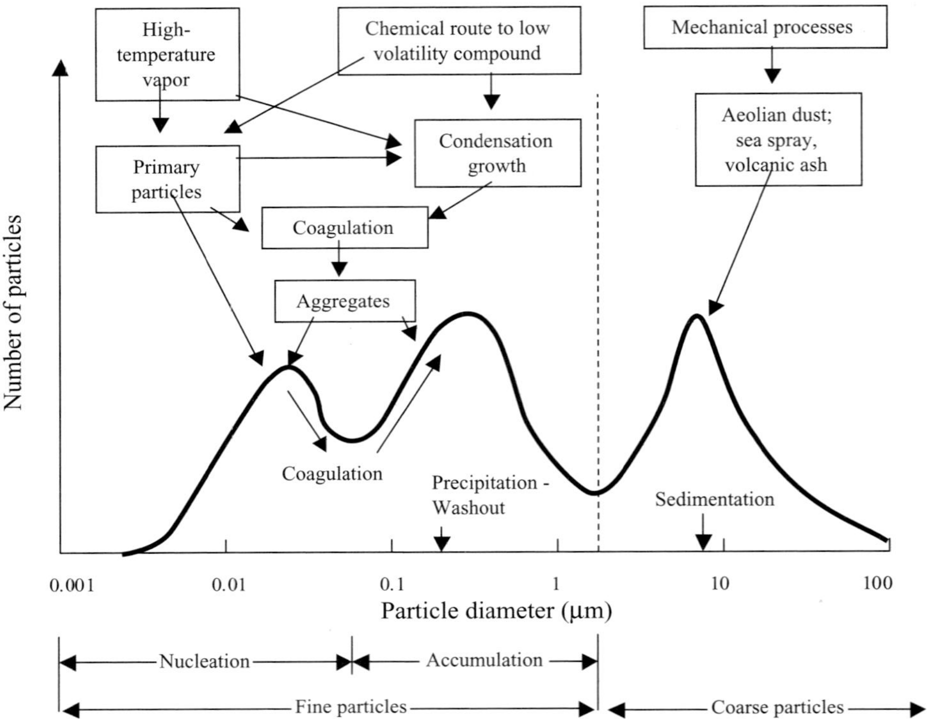

FIGURE 3.5. Prototypical size distribution of tropospheric particles with selected sources and pathways of how the particles are formed. Dashed line is approximately 2.5 μm diameter.
Source: Adapted from United Kingdom Department of Environment, Food, and Rural Affairs, Expert Panel on Air Quality Standards, 2004. *Airborne Particles: What Is the Appropriate Measurement on which to Base a Standard? A Discussion Document.*

atmosphere involving O_2 and water vapor (H_2O). Photochemistry can be an important step in secondary particle formation, resulting when chemical species like ozone (O_3) are involved in step reactions with radicals (e.g., the hydroxyl ($\cdot OH$) and nitrate ($\cdot NO_3$) radicals). Photochemistry also occurs in the presence of air pollutant gases like sulfur dioxide (SO_2), nitrogen oxides (NO_X), and organic gases emitted by anthropogenic and natural sources. In addition, nucleation of particles from low-vapor pressure gases emitted from sources or formed in the atmosphere, condensation of low-vapor pressure gases on aerosols

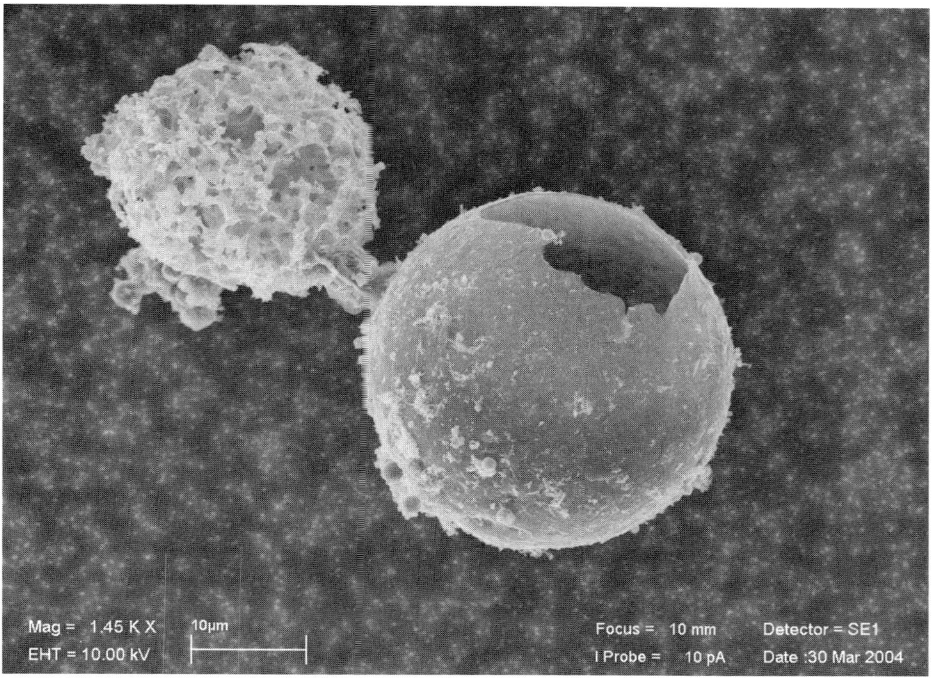

FIGURE 3.6. Scanning electron micrograph of coarse particles emitted from an oil-fired power plant. Diameters of the particles are greater than 20 µm optical diameter. Both particles are hollow, so their aerodynamic diameter is significantly smaller than if they were solid.
Source: Source characterization study by R. Stevens, M. Lynam, and D. Proffitt, 2004. Photo courtesy of R. Willis, Man Tech Environmental Technology, Inc., 2004; used with permission.

already present in the atmosphere, and coagulation of aerosols can contribute to the formation of particles. The chemical composition, transport, and fate of particles are directly associated with the characteristics of the surrounding gas. The term "aerosol" is often used synonymously with PM. An aerosol can be a suspension of solid or liquid particles in air, and an aerosol includes both the particles and all vapor or gas phase components of air.

Smaller particles are particularly problematic because they can travel longer distances and are associated with numerous health effects. Generally, the mass of PM falling in two size categories is measured: ≤2.5 micron diameter, and ≥2.5 micron ≤10 micron diameter. These measurements are taken by instruments (see Figure 3.7) with inlets using size exclusion mechanisms to segregate the mass of each size fraction (i.e., dichotomous samplers). Particles with diameters ≥10 microns are generally of less concern; however, they are occasionally measured if a large particulate emitting source (e.g., a coal mine) is nearby, since these particles rarely travel long distances.

Mass can be determined for a predominantly spherical particle by microscopy, either optical or electron, by light scattering and Mie theory, by the particle's electrical mobility, or by its aerodynamic behavior. However, since most particles are not spherical, PM diameters are often described using an equivalent diameter; that is, the diameter of a sphere that would have the same fluid properties. Another term, optical diameter, is the diameter of a spherical particle that has an identical refractive index as the particle. Optical diameters are used to calibrate the optical particle sizing instruments (e.g. nephelometers) which scatter the same amount of light into the solid angle measured. Diffusion and gravitational settling are also fundamental fluid phenomena used to estimate the efficiencies of PM transport, collection, and removal processes, such as in designing PM monitoring equipment and ascertaining the rates and mechanisms of how particles infiltrate and deposit in the respiratory tract.

Only for very small diameter particles is diffusion sufficiently important that the Stokes diameter is often used. The Stokes diameter for a particle is the diameter of a sphere with the same density and settling velocity as the particle. The Stokes diameter is derived from the aerodynamic drag force caused by the difference in velocity of the particle and the surrounding fluid. Thus, for smooth, spherical particles, the Stokes diameter is identical to the physical or actual diameter. The aerodynamic diameter (D_{pa}) for all particles greater than 0.5 micrometer can be approximated[7,8] as the product of the Stokes particle diameter (D_{ps}) and the square root of the particle density (ρ_p):

$$D_{pa} = D_{ps}\sqrt{\rho_p} \qquad (3.11)$$

If the units of the diameters are in μm, the units of density are g cm^{-3}.

Fine particles (<2.5μm) generally come from industrial combustion processes (see Figure 3.8) and from vehicle exhaust. This smaller sized fraction has been closely associated with increased respiratory disease, decreased lung functioning, and even premature death, probably due to their ability to bypass the body's trapping mechanisms, such as cilia in the lungs and nasal hair filtering. Some of the diseases

FIGURE 3.7. Sampler for measuring particles with aerodynamic diameters ≤2.5 microns. Each sampler has an inlet (top) that takes in particles ≤10 microns. An impacter downstream in the instrument cuts the size fraction to ≤2.5 microns, which is collected on a Teflon filter. The filter is weighed before and after collection. The Teflon construction allows for other analyses, e.g., X-ray fluorescence, to determine inorganic composition of the particles. Quartz filters would be used if any subsequent carbon analyses are needed since Teflon contains carbon. Photo courtesy of U.S. EPA.

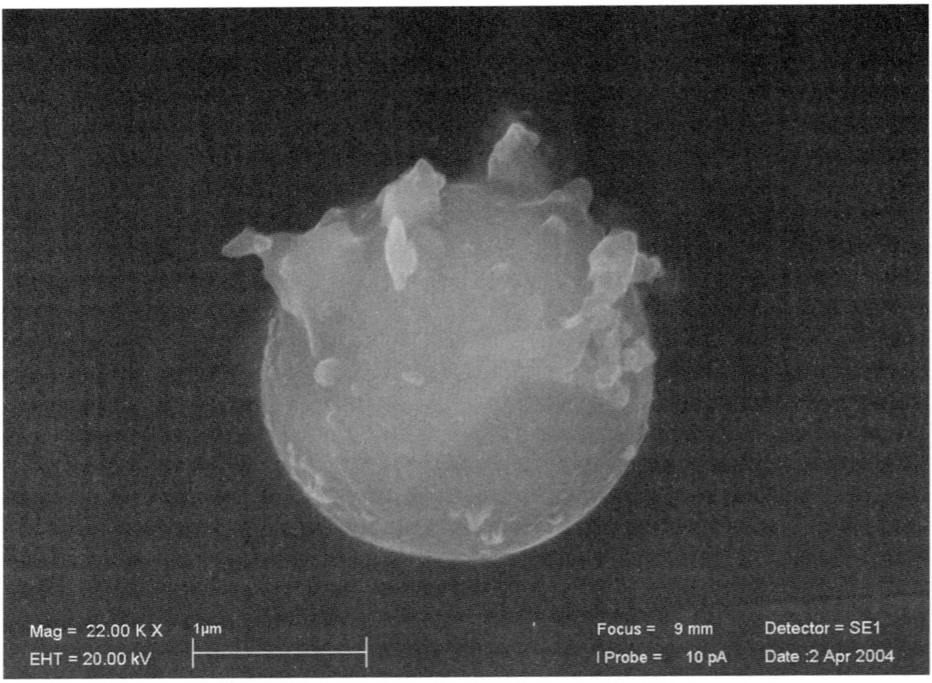

FIGURE 3.8. Scanning electron micrograph of spherical aluminosilicate fly ash particle emitted from an oil-fired power plant. Diameter of the particle is approximately 2.5 μm. Photo courtesy of R. Willis, Man Tech Environmental Technology, Inc., 2004; used with permission.

linked to PM exposure include aggravation of asthma, chronic bronchitis, and decreased lung function.

Particles aggravate bronchitis, asthma, and other respiratory diseases. Certain subpopulations of people are sensitive to PM effects, including those with asthma, cardiovascular or lung disease, as well as children and elderly people. Particles can also damage structures, harm vegetation, and reduce visibility. In 1987, the EPA changed the indicator for PM from TSP to PM_{10}; that is, particle matter ≤10 μm diameter.[9] The NAAQS for PM_{10} was a 24-hour average of $150 \mu g m^{-3}$ (not to exceed this level more than once per year), and an annual average of $50 \mu g m^{-3}$ arithmetic mean. However, even this change did not provide sufficient protection for people breathing PM contaminated air, since most of the particles that penetrate deeply into the air-blood exchange regions of the lung are quite small (see Figure 3.9). So, in 1997, the U.S. EPA added a new fine particle (diameters ≤2.5), known as $PM_{2.5}$.[10]

In addition to health impacts, PM is also a major contributor to reduced visibility, including near national parks and monuments. Also, particles can be transported long distances and serve as vehicles on which contaminants are able to reach water bodies and soils. Acid deposition, for example, can be dry or wet. Either way, particles play a part in acid rain. In the first, the dry particles enter ecosystems and potentially reduce the pH of receiving waters. In the latter, particles are washed out of the atmosphere and, in the process, lower the pH of the rain. The same transport and deposition mechanisms can also lead to exposures to persistent organic contaminants like dioxins and organochlorine pesticides and heavy metals like mercury that have sorbed in or on particles.

Routes of exposure for PM are similar to those of chemical contaminants. In fact, the lifetime average daily dose (LADD) equations include provisions for particle exposure. Also, particles can serve as vehicles for carrying chemical contaminants. For example, compounds that are highly sorptive (e.g., those with large K_{oc} partitioning coefficients discussed in Chapter 2) can use particles as a means for long-range transport. Also, charge differences between the particle and ions (particularly metal cations) will also make particles a means by which contaminants are transported.

Fibers

Generally, when environmental scientists discuss particles, they mean those that are somewhat spherical or angular, like soil particles. Particles that are highly elongated usually are differentiated as fibers. Such elongation is expressed as a particle's aspect ratio—the ratio of the length to width. Fibers generally have aspect ratios greater than 3:1. Environmentally important fibers include fiberglass, fabrics, and minerals (see Figures 3.10 and 3.11). Exposure to fiberglass and textile fibers are found most commonly in industrial settings, such as has been associated with the health problems of textile workers exposed to fibrous matter in high doses for many years. For example, chronic exposure to cotton fibers has led to the ailment, byssinosis, also referred to as "brown lung disease," which is characterized by the narrowing of the lung's airways. However, when discussing fibers, it is highly likely that the first contaminant to come to mind is asbestos, a group of

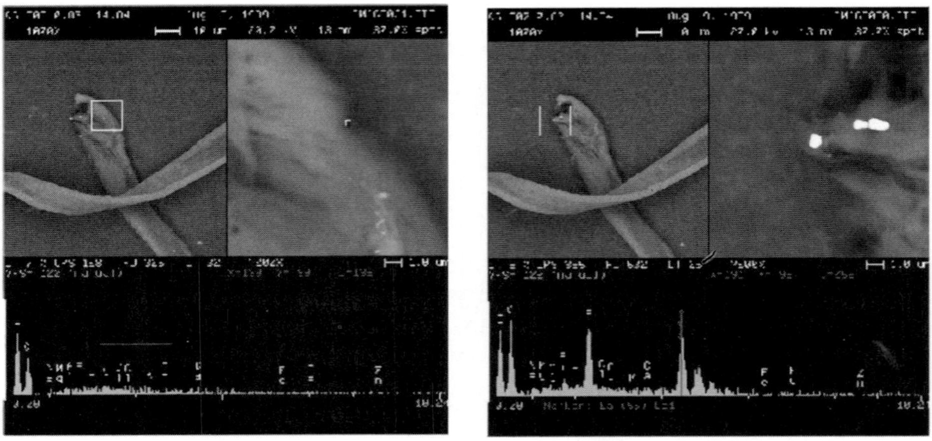

FIGURE 3.9. Scanning electron micrograph of cotton fibers. Acquired using an Aspex Instruments, Ltd., scanning electron microscope.
Source: U.S. Environmental Protection Agency, 2004. Photo courtesy of T. Conner, used with permission.

highly fibrous minerals with separable, long, and thin fibers. Separated asbestos fibers are strong enough and flexible enough to be spun and woven. Asbestos fibers are heat resistant, making them useful for many industrial purposes. Because of their durability, asbestos fibers that get into lung tissue will remain for long periods of time.

PM Control Technology

Controlling PM emissions tracks fairly closely the homeowner demands for devices to remove dust. Allow me to share a recent quest to purchase a vacuum cleaner as a gift for my daughter, Amelia. Although she is not an engineer or scientist, in her purchases and in other daily pursuits she is doggedly determined to achieve optimal designs. So, it came to me as no surprise when she articulated the kind of vacuum cleaner she needs, she articulated specifications and tolerances.

First, she specified a cyclone system, not a bag system. Allow me to digress at this point. All vacuum cleaners work on the principle of pressure differential. The laws of potentiality state that flow moves from high to low pressure. So, if pressure can be decreased signifi-

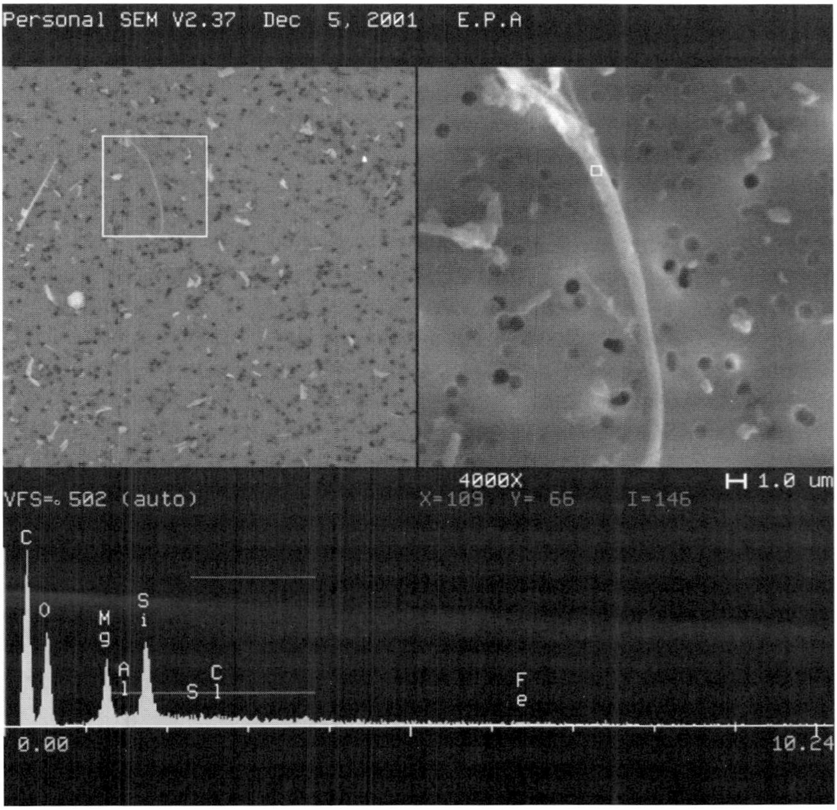

FIGURE 3.10. Scanning electron micrograph of fibers in dust collected near the World Trade Center, Manhattan, NY, in September 2001. Acquired using an Aspex Instruments, Ltd., scanning electron microscope. The bottom of the micrograph represents the elemental composition of the highlighted 15 μm-long fiber by energy dispersive spectroscopy (EDS). This composition (i.e., O, Si, Al, and Mg) and the morphology of the fibers indicate they are probably asbestos. The EDS carbon peak results from the dust being scanned on a polycarbonate filter.
Source: U.S. Environmental Protection Agency, 2004. Photo courtesy of T. Conner, used with permission.

cantly below that of the atmosphere, air will move to that pressure trough. If there is a big pressure difference between air outside and inside, the flow will be quite rapid. So, the "vacuum" is created inside the vacuum cleaner using an electric pump. When the air rushes to the low pressure region it carries particles with it. The greater the

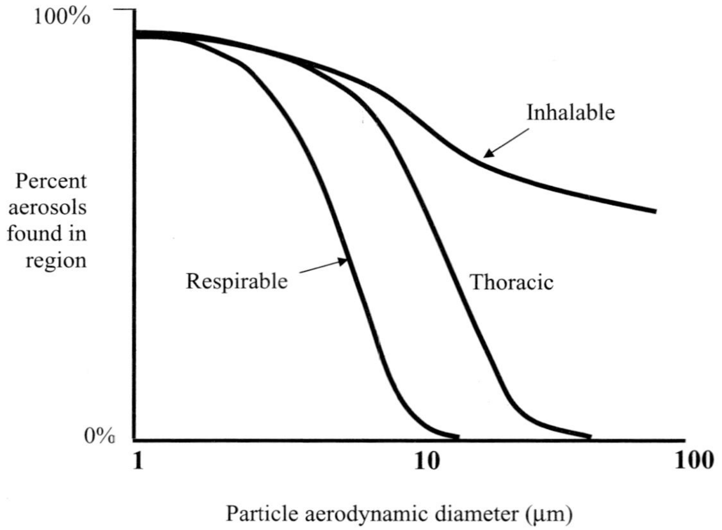

FIGURE 3.11. Three regions of the respiratory system where particle matter is deposited. The inhalable fraction remains in the mouth and head area. The thoracic fraction is the mass that penetrates the airways of the respiratory system, and the smallest fraction—the respirable particulates—are those that can infiltrate most deeply into the alveolar region.

velocity the more particles it can carry. This is the same principle as the "competence" of a stream, which is high (i.e., can carry heavier loads) in a flowing river, but the competence declines rapidly at the delta where stream velocity approaches zero. This causes the river to drop its particles in descending mass; that is, sedimentation of heavier particles first, but colloidal matter remaining suspended for much longer times.

I was introduced to cyclones and bag systems in a graduate air pollution engineering course. It is illustrative to consider the scientific and engineering principles of both technologies. The vacuum cleaner, the cyclone, and the bag (fabric filter) are all designed to remove particles. In the United States, the Clean Air Act established the national ambient air quality standards (NAAQS) for particulate matter (PM) in 1971. These standards were first directed at total suspended particulates (TSP) as measured by a high volume sampler—a device that collected a large range of sizes of particles (aerodynamic diameters up to 50 μm).

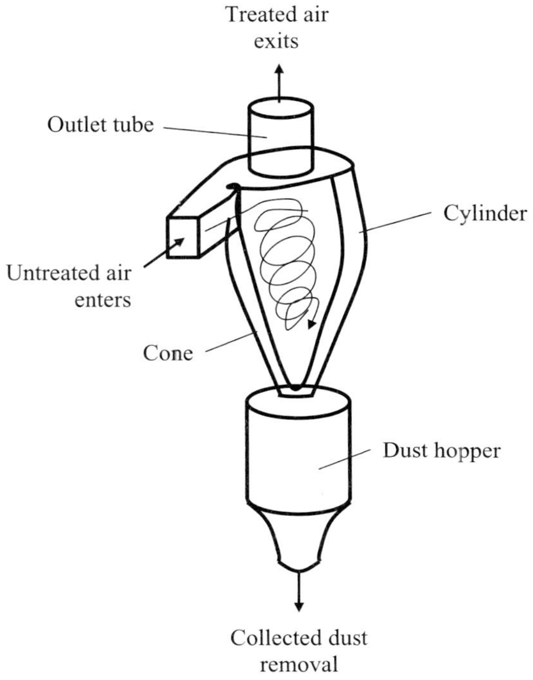

FIGURE 3.12. Schematic of a simple cyclone separator.
Source: U.S. Environmental Protection Agency, 2004. Air Pollution Control Orientation Course. http://www.epa.gov/air/oaqps/eog/course422/ce6.html; accessed September 20, 2004.

In a cyclone (see Figure 3.12), air is rapidly circulated causing suspended particles to change directions. Due to their inertia, the particles continue in their original direction and leave the air stream (see Figure 3.13). This works well for larger particles because of their relatively large masses (and greater inertia), but very fine particles are more likely to remain in the air stream and stay suspended. The dusty air is introduced in the cyclone from the top through the inlet pipe tangential to the cylindrical portion of the cyclone. The air whirls downward to form a peripheral vortex, which creates centrifugal forces. As a result, individual particles are hurled toward the cyclone wall and, after impact, fall downward where they are collected in a hopper. When the air reaches the end of the conical segment, it will change direction and move upward toward the outlet. This forms an inner vortex. The upward airflow against gravitation allows for addi-

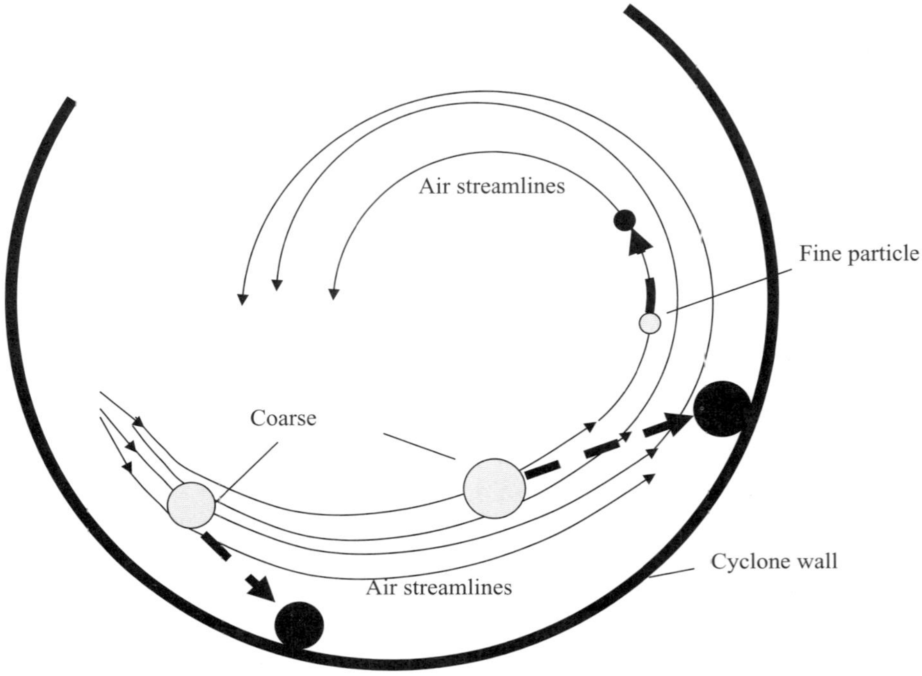

FIGURE 3.13. Inertial forces in a cyclone separator. Coarse (heavier) particles' inertial forces are large enough that they leave the air stream of the vortex and collide with the cyclone wall. Fine particles have smaller mass so that their inertial forces cannot overcome the air flow, so they remain suspended in the air stream.

tional separation of particles. The cyclone vacuum cleaner applies the same inertial principles, with the collected dust hitting the sides of the removable cyclone separator and falling to its bottom.

The other technology of vacuum cleaners is the bag. Actually, the engineering term for a "bag" is a fabric filter. Filtration is an important technology in every aspect of environmental engineering—air pollution, wastewater treatment, drinking water, and even hazardous waste and sediment cleanup. Basically, filtration consists of four mechanical processes: 1) diffusion, 2) interception, 3) interial impaction, and 4) electrostatics (see Figure 3.14). Diffusion is important only for very small particles ($\leq 0.1\,\mu m$ diameter) because the Brownian motion allows them to move in a "random walk" away from the air stream. Interception works mainly for particles with

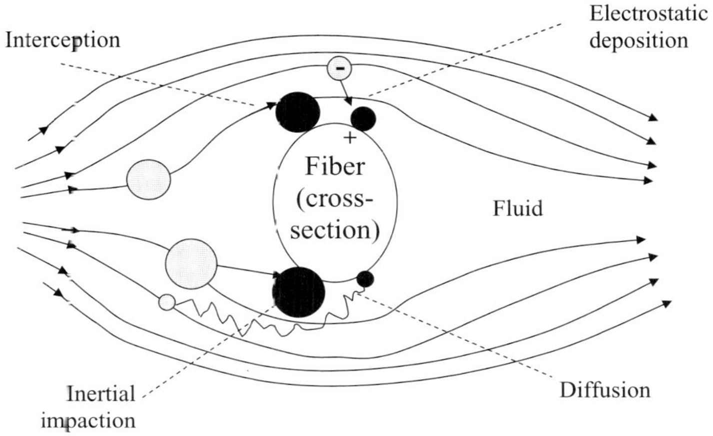

FIGURE 3.14. Mechanical processes important to filtration.
Source: Adapted from: K.L. Rubow, 2004. "Filtration: Fundamentals and Applications," Aerosol and Particle Measurement Short Course, University of Minnesota, August 16–18.

diameters between 0.1 and 1 μm. The particle does not leave the air stream but comes into contact with the filter medium (e.g., a strand of fiberglass). Inertial impaction, as explained in the cyclone discussion, collects particles sufficiently large to leave the air stream by inertia (diameters ≥1 μm). Electrostatics consist of electrical interactions between the atoms in the filter and those in the particle at the point of contact (Van der Waal's force), as well as electrostatic attraction (charge differences between particle and filter medium). These are the processes at work in large-scale *electrostatic precipitators* that are employed in coal-fired power plant stacks around the world for particle removal. Other important factors affecting filtration efficiencies include the thickness and pore diameter of the filter, the uniformity of particle diameters and pore sizes, the solid volume fraction, the rate of particle loading onto the filter (e.g., affecting particle bounce), the particle phase (liquid or solid), capillarity and surface tension (if either the particle or the filter media are coated with a liquid), and characteristics of air or other carrier gases, such as velocity, temperature, pressure, and viscosity.

Environmental engineers have been using filtration to treat air and water for several decades. Air pollution controls employing fabric filters (i.e., baghouses) remove particles from the air stream by passing the air through a porous fabric. The fabric filter is efficient at removing fine particles and can exceed efficiencies of 99%. Based solely upon an extrapolation of air pollution control equipment, a bag-type vacuum cleaner should be better than a cyclone-type vacuum cleaner. However, this does not take into account operational efficiencies and effectiveness, which are very important to the consumer and the engineer. Changing the bag and insuring that it does not exceed its capacity must be monitored closely by the user. Also, the efficiency of the vacuum cleaner is only as good as the materials being used. Does the bag filter allow a great deal of diffusion, interception, and inertial impaction? The cyclone only requires optimization for inertia.

Selecting the correct control device is a matter of optimizing efficiencies, which brings us to my daughter's second selection criterion—the vacuum must have a HEPA filter. I am not sure if Amelia knows that the acronym stands for "high efficiency particle air" filters, but she knows that it is needed to remove more of the "nasty" dust. In this case, HEPA filters are fitted to equipment to enhance efficiency. Efficiency is often expressed as a percentage; so a 99.99% HEPA filter is efficient enough to remove 99.99% particles from the air stream. This means that if 10,000 particles enter the filter, on average only one particle would pass all the way through the filter. This is exactly the same concept that we use for incinerator efficiency, but it is known as destruction removal efficiency (DRE). For example, in the United States, federal standards require that hazardous compounds can be incinerated only if the process is 99.99% efficient, and for the "nastier" compounds (i.e., extremely hazardous wastes) the so-called "rule of six nines" applies, DRE ≥99.9999%. The HEPA and DRE calculations are quite simple:

$$\text{DRE or HEPA efficiency} = \frac{M_{in} - M_{out}}{M_{in}} \times 100 \qquad (3.12)$$

Thus, if $10\,\text{mg}\,\text{min}^{-1}$ of a hazardous compound is fed into the incinerator, only $0.001\,\text{mg}\,\text{min}^{-1} = 1\,\mu\text{g}\,\text{min}^{-1}$ is allowed to exit the stack for a hazardous waste. If the waste is an extremely hazardous waste, only $0.00001\,\text{mg}\,\text{min}^{-1} = 0.01\,\mu\text{g}\,\text{min}^{-1} = 10\,\text{ng}\,\text{min}^{-1}$ is allowed to exit the stack. Actually, this same equation is used throughout the environmental engineering discipline to calculate treatment and removal efficiencies. For example, assume that raw wastewater enters a treatment facility with $300\,\text{mg}\,\text{L}^{-1}$ biochemical oxygen demand (BOD_5), $200\,\text{mg}\,\text{L}^{-1}$ suspended solids (SS), and $10\,\text{mg}\,\text{L}^{-1}$ phosphorous (P). The plant must meet effluent standards $\leq 10\,\text{mg}\,\text{L}^{-1}$ BOD_5, $\leq 10\,\text{mg}\,\text{L}^{-1}$ SS, and $\leq 1\,\text{mg}\,\text{L}^{-1}$ P, so using the efficiency equation we know that removal rates of these contaminants must be 97% for BOD_5, 95% for SS, and 90% for P. The pure efficiency values may be misleading because the ease of removal can vary significantly with each contaminant. In this case, gravitational settling in the primary stages of the treatment plant can remove most of the SS, and the secondary treatment stage removes most of the BOD, but more complicated, tertiary treatment is needed for removing most of the nutrient P.

As I shopped for the cyclone-HEPA vacuum cleaner, I came across some additional technical terminology, "allergen removal." I am not completely certain what this means, but I believe it is a filtration system that does not quite have the efficiency removal of a HEPA filter. However, this points out the somewhat chaotic nature of environmental engineering. Remember, the smaller particles are the ones that concern most health scientists. However, there has been a resurgence of interest in particles ranging between 2.5 and 10 μm aerodynamic diameter, referred to as coarse particles. These may consist of potentially toxic components; for example, resuspended road dust, brake lining residues, industrial byproducts, tire residues, heavy metals, and aerosols generated by organisms' (known as *bioaerosols*), such as tree pollen and mold spores. Figure 3.9 shows that a large fraction of these coarse particles may deposit to the upper airways, causing health scientists to link them to asthma. And, since asthma appears to be increasing in children, there may be a need to reconsider the importance of larger particles. In fact, the U.S. EPA is considering ways to develop a federal reference method (FRM)[11] for coarse particles to complement its $PM_{2.5}$ FRM.

Thus, my shopping trip was complicated by some fairly technical specifications. Efficiency is an important part of effectiveness, although the two terms are not synonymous. As we discussed, efficiency is simply a metric of what you get out of a system compared to what you put in. However, you can have a bunch of very efficient systems that may *en toto* be ineffective. They are all working well as designed but they may not be working on the right things, or their overall configuration is not optimal to solve the problem at hand. So, the correct control device is the one that gives not only optimal efficiency, but one that effectively addresses the pollution.

Asbestos: The Fiber of Concern

There are two general types of asbestos, *amphibole* and *chrysotile*. Some studies show that amphibole fibers stay in the lungs longer than chrysotile, and this tendency may account for their increased toxicity.

Generally, health regulations classify asbestos into six mineral types: chrysotile, a serpentine mineral with long and flexible fibers; and five amphiboles, which have brittle crystalline fibers. The amphiboles include actinolite asbestos, tremolite asbestos, anthophyllite asbestos, crocidolite asbestos, and amosite asbestos (see Figure 3.15).

Asbestos Routes of Exposure

Ambient air concentrations of asbestos fibers are about 10^{-5} to 10^{-4} fibers per milliliter (fibers mL^{-1}), depending on location. Human exposure to concentrations much higher than 10^{-4} fibers mL^{-1} is suspected of causing health effects.[12] Asbestos fibers are very persistent and resist chemical degradation (i.e., they are inert under most environmental conditions) so their vapor pressure is nearly zero meaning they do not evaporate, nor do they dissolve in water. However, segments of fibers do enter the air and water as asbestos-containing rocks and minerals are weathered naturally or when extracted during mining operations. One of the most important exposures is when manufactured products (e.g., pipe wrapping and fire-resistant materials) begin to wear down. Small diameter asbestos fibers may remain suspended in the air for a long time and be transported advectively by wind or water before sedimentation. Like more spherical particles, heavier fibers settle more quickly. Asbestos seldom moves substantially via soil. The fibers are generally not broken down to other compounds in the environment and will remain virtually unchanged over long periods. Although most asbestos is highly persistent, chrysotile, the most commonly encountered form, may break down slowly in acidic environments. Asbestos fibers may break into shorter strands and, therefore, increase the number of fibers, by mechanical processes such as grinding and pulverization. Inhaled fibers may get trapped in the lungs and with chronic exposures build up over time. Some fibers, especially chrysotile, can be removed from or degraded in the lung with time.

Donora, Pennsylvania

In 1948, the United States experienced its first major air pollution catastrophe in Donora, Pennsylvania. Effluents from a number of industries, including a sulfuric acid plant, a steel mill, and a zinc production plant,

FIGURE 3.15. Scanning electron micrograph of asbestos fibers (amphibole) from a former vermiculite-mining site near Libby, Montana.
Source: U.S. Geological Survey and U.S. Environmental Protection Agency, Region 8, Denver, Colorado.

became trapped in a valley by a temperature inversion and produced an unbreathable mixture of fog and pollution. Six thousand suffered illnesses ranging from sore throats to nausea. There were 20 deaths in three days. Sulfur dioxide (SO_2) was estimated to reach levels as high as $5{,}500\,\mu g\,m^{-3}$ (see the box, "Contaminants of Concern: Sulfur and Nitrogen Compounds"). For a town of only 14,000, the number of deaths in such a short time was unprecedented; in fact, the town did not have enough coffins to accommodate the burials.[13]

TABLE 3.3
Crude oil production by field.

Field	*Share of National Petroleum Production* 1940	1945	1949
Poza Rica	65%	53%	59%
Pánuco-Ebano	14%	12%	17%
Faja de Oro (Golden Lane)	9%	23%	13%
Isthmus de Tehuantepec	13%	12%	11%
Reynosa	0%	0%	0*%

* less than 1%

Source: Data provided by Professor Thayer Watkins, San Jose State Univeristy: http://www2.sjsu.edu/faculty/watkins/watkins.htm; accessed April 19, 2005.

The Donora incident is important because it was the first dramatic evidence that unchecked pollution was an American problem. Air pollution had moved from being a nuisance and an aesthetic problem to an urgent public health concern in North America.

Poza Rica, Mexico

Poza Rica, a town of 15,000 people on the Gulf of Mexico, was a dominant petrochemical center in the 1940s (see Table 3.3). The highly toxic chemical hydrogen sulfide (H_2S) was the culprit when it escaped from a pipeline in 1950 (see "Contaminant of Concern: Hydrogen Sulfide"). The disaster originated from an accident at one of the local factories that recovers sulfur from natural gas in part. The release of hydrogen sulfide into the ambient air lasted for only 25 minutes. However, because it is as dense as air, the spread of the gas under a shallow inversion with foggy and calm conditions killed 22 people and hospitalized 320.

Contaminant of Concern: Hydrogen Sulfide

Hydrogen sulfide (H_2S), a colorless gas, is often produced as an unintended byproduct of combustion, desulfurization processes, and chemical operations (see Table 3.4 for physicochemical properties). The compound also has industrial uses, such as a reagent and as an intermediate in the preparation of other reduced sulfur compounds. Some principal sources of H_2S include geothermal power plants (due to the reduction of sulfur in makeup water), petroleum refining (reduc-

TABLE 3.4
Physicochemical properties of hydrogen sulfide (H_2S).

- Molecular weight: 34.08 g/mol
- Melting point: −86°C
- Latent heat of fusion (1,013 bar, at triple point): 69.75 kJ kg^{-1}
- Liquid density (1.013 bar at boiling point): 914.9 kg m^{-3}
- Liquid/gas equivalent (1.013 bar and 15°C (59°F)): 638 vol vol^{-1}
- Boiling point (1.013 bar): −60.2°C
- Latent heat of vaporization (1.013 bar at boiling point): 547.58 kJ/kg
- Vapor pressure (at 21°C or 70°F): 18.2 bar
- Critical temperature: 100°C
- Critical pressure: 89.37 bar
- Gas density (1.013 bar at boiling point): 1.93 kg m^{-3}
- Gas density (1.013 bar and 15°C (59°F)): 1.45 kg m^{-3}
- Compressibility factor (Z) (1.013 bar and 15°C (59°F)): 0.9915
- Specific gravity (air = 1) (1.013 bar and 15°C (59°F)): 1.189
- Specific volume (1.013 bar and 21°C (70°F)): 0.699 m^3 kg^{-1}
- Heat capacity at constant pressure (Cp) (1 bar and 25°C (77°F)): 0.034 kJ $(mol.K)^{-1}$
- Viscosity (1.013 bar and 0°C (32°F)): 0.0001179 Poise
- Thermal conductivity (1.013 bar and 0°C (32°F)): 12.98 mW $(m.K)^{-1}$
- Solubility in water (1.013 bar and 0°C (32°F)): 4.67 vol vol^{-1}
- Autoignition temperature: 270

Source: Air Liquide: http://www.airliquide.com/en/business/products/gases/gasdata/index.asp?GasID=59; accessed April 20, 2005.

tion of sulfur in the crude oil), and gas releases from sewers (when plants go anaerobic or when pockets of anoxic conditions exist in the plants or in the sewer lines).

Hydrogen sulfide stinks! In 1969, California set an ambient air quality standard for hydrogen sulfide of 0.03 ppm (42 μg m^{-3}), averaged over a period of 1 hour and not to be equaled or exceeded, to protect against the substance's nuisance odor, the "rotten egg smell."[14] As new studies began to link H_2S exposures to health effects, the acceptable air levels dropped further. Based on a study demonstrating nasal histological changes in mice, the State of California lowered the acceptable concentration to 8 ppb (10 μg m^{-3}) as the chronic Reference Exposure Level (cREL) for use in evaluating long-term emissions from facilities in high emission areas.[15] The United States does not currently classify H_2S as either a criteria air pollutant or a Hazardous Air Pollutant (HAP). The U.S. Environmental Protection Agency has issued safety levels, known as reference concentrations (RfCs), for H_2S for long-term (chronic) effects of 0.001 mg m^{-3} (1 μg m^{-3}) for hydrogen sulfide.[16] The RfC is an estimate of a daily inhalation exposure of the human population that is likely to be without an appreciable risk of adverse effects during a person's expected lifetime (see Chapter 5 for more details on how the RfC is developed). Exposure to H_2S concentrations of 250 ppm irritates mucous membranes and can lead to conjunctivitis, photophobia, lacrimation, corneal opacity, rhinitis, bronchitis, cyanosis, and acute lung injury. Concentrations between 250 to 500 ppm can induce headache, nausea, vomiting, diarrhea, vertigo, amnesia, dizziness, apnea, palpitations, tachycardia, hypotension, muscle cramps, weakness, disorientation, and coma. Higher levels, greater than 700 ppm, may cause abrupt physical collapse, respiratory paralysis, asphyxial seizures, and death.[17]

Hydrogen sulfide is one of those compounds whose detection is directly related to olfactory sensitivity. The greater the concentration, the more likely people can identify H_2S (see Table 3.5), although the sense of smell is paralyzed (known as "extinction") at airborne concentrations greater than 150 ppm (or as low as 50 ppm). Perceiving the presence of a pollutant is not often the case in environmental pollution, where many very toxic compounds are imperceptible even at dangerously high concentrations.

Environmentally, H_2S converts abiotically to elemental sulfur in surface waters. In soil and sediment, microbes mediate oxidation-reduction reactions that oxidize hydrogen sulfide to elemental sulfur. Bacterium genera *Beggiatoa*, *Thioploca*, and *Thiotrix* exist in transition zones between aerobic and anaerobic conditions where both molecular oxygen and H_2S are present. A few photosynthetic bacteria also oxidize hydrogen sulfide to elemental sulfur. Purple sulfur bacteria (*Chlorobiaceae* and *Chromatiaceae*), which are phototropic aerobes, can live in high H_2S water. The interactions of these organisms with their environments are key components of the global sulfur cycle. Due to its abiotic and biological reactions, H_2S does not bioaccumulate.[18]

London, England

Known today as "the London Fog," from December 5–8, 1952, London experienced the worst ambient air pollution disaster ever before reported. With daily temperatures below average, fireplaces and industries supplied pollutants that combined with condensation in the air to form a dense fog. Concentrations of pollutants reached very high levels under these adverse conditions. The fog finally cleared away, but 4,000 Londoners had perished.

TABLE 3.5
Predicted effects of exposure to ambient concentrations of hydrogen sulfide (H_2S).

H_2S (ppb)	*% able to detect odor*[a]	*Perceived odor intensity*[b] *(ratio)*	*Median odor units*[c]	*% annoyed by odor*[d]
200	99	2.31	25	88
100	96	1.93	12	75
50	91	1.61	6.2	56
40	88	1.52	5.0	50
35	87	1.47	4.4	47
30	83	1.41	3.7	40
25	80	1.34	3.1	37
20	74	1.27	2.5	31
15	69	1.18	1.9	22
10	56	1.06	1.2	17
8	50	1.00	1.00	11
6	42	0.93	0.75	8
4	30	0.83	0.50	5
2	14	0.70	0.25	2
1	6	0.58	0.12	1
0.5	2	0.49	0.06	0

[a] Based on mean odor detection threshold of 8.0 ppb and standard deviation ± 2.0 binary steps.
[b] Based on intensity exponent of 0.26.
[c] H_2S concentration divided by mean odor detection threshold of 8 ppb.
[d] Based on assumption that mean annoyance threshold is 5× the mean odor detection threshold, and standard deviation ±2.0 binary steps.
Sources: California Air Resources Board, 2000, *Hydrogen Sulfide: Evaluation of Current California Air Quality Standards with Respect to Protection of Children*, Sacramento, CA; and J.E. Amoore, 1985, *The Perception of Hydrogen Sulfide Odor in Relation to Setting an Ambient Standard*, Olfacto-Labs, Berkeley, CA: prepared for the California Air Resources Board.

The meteorological conditions were ideal for a pollution event. Anticyclonic or high pressure weather with stagnating continental polar air masses trapped under subsidence inversions produced a shallow mixing layer with an almost complete absence of vertical and horizontal air motion. Fireplaces and industries supplied the hygroscopic condensation nuclei into the air to form dense fog. The daily temperatures were below the average. With such adverse conditions the concentrations of pollutants reached high values.

The elderly are often more sensitive to the effects of air pollutants than is the general population; that is, they are considered to be a "sensitive subpopulation." This was the case in the London fog incident. Deaths from bronchitis in the elderly increased by a factor of 10, influenza by 7, pneumonia by 5, tuberculosis by 4.5, other respiratory diseases by 6, heart diseases by 3, and lung cancer by 2. When a change in weather finally cleared the fog, 4,000 Londoners had perished in a toxic version of the city's famous "pea soup."

Subsequent air pollution episodes with comparably elevated pollution occurred from 1957 to 1958 and again from 1962 to 1963. But the 1952 incident showed much greater mortality and morbidity.

New York City

New York City experienced air pollution-related deaths in 1953, 1962–1963, and 1966, beyond what epidemiologist would have expected. New York had very high SO_2 atmospheric concentrations through much of the twentieth century, but its local meteorology often helped the city avert air pollution disasters. However, when these conditions change, as they did in December 1962, which experienced calm winds and the occurrence of shallow inversions, then the SO_2 and aerosol concentrations peaked. Total deaths increased to 269, which was in excess (more than three standard deviations greater than the mean) beyond the expected mortality for that week.

Toxic Clouds

We live in a world that is very different from even a couple of decades ago. Some would argue that September 11, 2001, marked a change in what many professions will be called upon to do, including environmental engineering. For example, the threats to environmental quality and human health were often seen as unfortunate and unplanned byproducts of something else, such as profit making and business decisions (e.g., chemical manufacturing) or lifestyle support and individual decisions (e.g., wastewater treatment). Now, we have to realistically consider the possibility of intentional harm to our environmental resources. Environmental engineers are increasingly being called upon to design and build facilities that are not as vulnerable to terrorist acts.[19]

The Bhopal Tragedy

Perhaps the biggest air pollution disaster of all time occurred in Bhopal, India, in 1984 when a toxic cloud drifted over the city from the Union Carbide pesticide plant. This gas leak managed to kill 20,000 people and permanently injure another 120,000. We often talk about a failure that results from not applying the science correctly (e.g., a mathematical error and an incorrect extrapolation of a physical principle). Another type of failure results from misjudgments of human systems. Bhopal had both.

Although the Union Carbide Company was headquartered in the United States, as of 1984, it operated in 38 different countries. It was quite

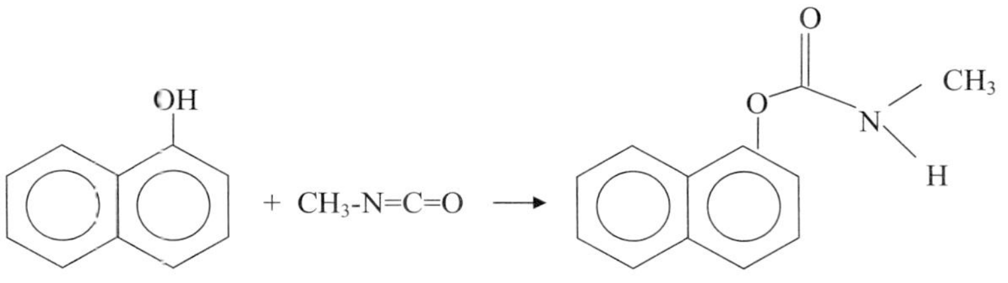

FIGURE 3.16. Chemical reaction producing methyl isocyanate at the Bhopal, India, Union Carbide plant.

large (35th largest U.S. company) and was involved in numerous types of manufacturing, most of which involved proprietary chemical processes. The pesticide manufacturing plant in Bhopal, India, had produced the insecticides Sevin and Cararyl since 1969, using the intermediate product, methyl isocyanate (MIC), in its gas phase. The MIC was produced by the reaction shown in Figure 3.16.[20] This process was highly cost-effective, involving only a single reaction step. The schematic of the MIC process is shown in Figure 3.17. MIC is highly water reactive (see Table 3.6); that is, it reacts violently with water, generating a very strong exothermic reaction that produces carbon dioxide. When MIC vaporizes it becomes a highly toxic gas that, when concentrated, is highly caustic and burns tissues. This can lead to scalding nasal and throat passages, blinding, and loss of limbs, as well as death.

On December 3, 1984, the Bhopal plant operators became concerned that a storage tank was showing signs of overheating and began to leak. The tank contained MIC. The leak rapidly increased in size, and within one hour of the first leakage, exploded and released approximately 80,000 lbs (4×10^4 kg) of MIC into the atmosphere. Introduction of water to the MIC storage tank resulted in a highly exothermic reaction generating CO_2, which would have led to a rapid increase in pressure that likely caused the release of 40 metric tons of MIC into the atmosphere. The release led to arguably the worst industrial disaster on record. The human exposure to MIC was widespread, with a half million people exposed. Ten years after the incident, 12,000 death claims had been filed, along with 870,000 personal injury claims. However, only $90 million of the Union Carbide settlement agreement had been paid out.

As of 2001, many victims did receive compensation, averaging about $600 each, although some claims are still outstanding.

So what caused the accident? Many factors were involved. The Indian government required that the plant be operated exclusively by Indian workers, so Union Carbide agreed to train them, including flying them to a sister plant in West Virginia for hands-on sessions. In addition, the company required that U.S. engineering teams make periodic on-site

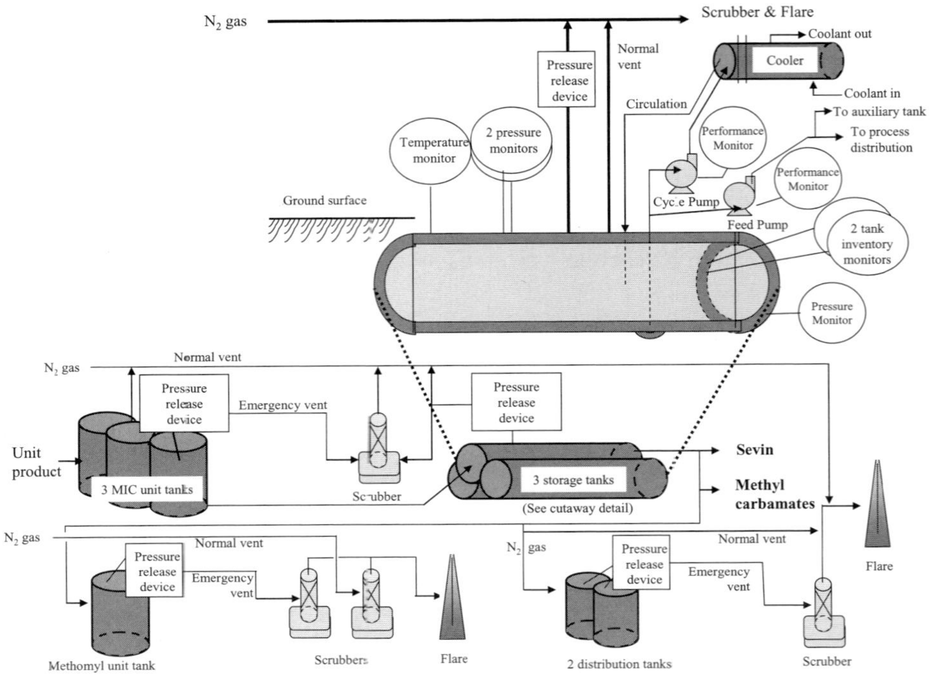

FIGURE 3.17. Schematic of methyl isocyanate processes at the Bhopal, India, plant (*circa* 1984). Based on information in: W. Worthy, 1985, Methyl isocyanate: The chemistry of a hazard, *Chemical Engineering News*, 63 (66), p. 29.

inspections for safety and quality control, but these ended in 1982 when the plant managers decided that these costs were too high. So, instead, the U.S. contingency was responsible only for budgetary and technical controls, but not safety. The last U.S. inspection in 1982 warned of many hazards, including a number that have since been implicated as contributing to the leak and release.

From 1982 to 1984, safety measures declined, attributed to high employee turnover, improper and inadequate training of new employees, and low technical savvy in the local workforce. On-the-job experiences were often substituted for reading and understanding safety manuals (remember, this was a pesticide plant). In fact workers would complain of typical acute symptoms of pesticid e exposure, such as shortness of breath, chest pains, headaches, and vomiting, yet they would typically refuse to wear protective clothing and equipment. The refusal in part stemmed from the lack of

TABLE 3.6
Properties of methyl isocyanate.

Common name	*isocyanic acid, methylester, and methyl carbylamine*
Molecular mass	57.1
Properties	Melting Point: –45°C; Boiling Point: 43–45°C Volatile liquid. Pungent odor. Reacts violently with water and is highly flammable. MIC vapor is denser than air and will collect and stay at low areas. The vapor mixes well with air and explosive mixtures are formed. May polymerize due to heating or under the influence of water and catalysts. Decomposes on heating and produces toxic gases like hydrogen cyanide, nitrogen oxides, and carbon monoxide.
Uses	Used in the production of synthetic rubber, adhesives, pesticides, and herbicide intermediates. It is also used for the conversion of aldoximes to nitriles.
Side effects	MIC is extremely toxic by inhalation, ingestion, and skin absorption. Inhalation of MIC causes cough, dizziness, shortness of breath, sore throat, and unconsciousness. It is corrosive to the skin and eyes. Short-term exposures also lead to death or adverse effects like pulmonary edema (respiratory inflammation), bronchitis, bronchial pneumonia, and reproductive effects. The Occupational Safety and Health Administration's permissible exposure limit to MIC over a normal 8-hour workday or a 40-hour workweek is 0.05 mg m^{-3}.

Sources: U.S. Chemical Safety and Hazards Board, http://www.chemsafety.gov/lib/bhopal.0.1.htr; Chapman and Hall, *Dictionary of Organic Chemistry*, Volume 4, 5e, Mack Printing Company, United States of America, 1982; and T.W. Graham, *Organic Chemistry*, 6e, John Wiley and Son, Inc., Canada, 1996.

air conditioning in this subtropical climate, where masks and gloves can be uncomfortable.

Indian, rather than the more stringent U.S., safety standards were generally applied at the plant after 1982. This likely contributed to overloaded MIC storage tanks (company manuals cite a maximum of 60% fill).

The release lasted about two hours, after which the entire quantity of MIC was released The highly reactive MIC arguably could have reacted and become diluted beyond a certain safe distance. However, over the years, tens of thousands of squatters had taken up residence just outside of the plant property, hoping to find work or at least take advantage of the plant's water and electricity. The squatters were not notified of hazards and risks associated with the pesticide manufacturing operations, except by a local journalist who posted signs saying: "Poison Gas. Thousands of Workers and Millions of Citizens are in Danger." This is a classic instance of a "confluence of events" that led to a disaster. More than a few mistakes were made. The failure analysis found the following:

- The tank that initiated the disaster was 75% full of MIC at the outset.
- A standby overflow tank for the storage tank contained a large amount of MIC at the time of the incident.
- A required refrigeration unit for the tank was shut down five months prior to the incident, leading to a three- to four-fold increase in tank temperatures over expected temperatures.
- One report stated that a disgruntled employee unscrewed a pressure gauge and inserted a hose into the opening (knowing that it would do damage, but probably not nearly the scale of what occurred).
- A new employee was told by a supervisor to clean out connectors to the storage tanks, so the worker closed the valves properly, but did not insert safety discs to prevent the valves from leaking. In fact, the worker knew the valves *were* leaking, but they were the responsibility of the maintenance staff. Also the second-shift supervisor position had been eliminated.
- When the gauges started to show unsafe pressures, and even when the leaking gases started to sting mucous membranes of the workers, they found that evacuation exits were not available. There had been no emergency drills or evacuation plans.
- The primary fail-safe mechanism against leaks was a vent-gas scrubber; normally, this release of MIC would have been sorbed and neutralized by sodium hydroxide (NaOH) in the exhaust lines, but on the day of the disaster, the scrubbers were not working. (The scrubbers were deemed unnecessary, since they had never been needed before.)
- A flare tower to burn off any escaping gas that would bypass the scrubber was not operating because a section of conduit connecting the tower to the MIC storage tank was under repair.
- Workers attempted to mediate the release by spraying water 100 ft high, but the release occurred at 120 feet.

Thus, according to the audit, many checks and balances were in place, but the cultural considerations were ignored or given low priority such as, when the plant was sited, the need to recognize the differences in land use planning and buffer zones in India compared to Western nations or the difference in training and oversight of personnel in safety programs. Every engineer and environmental professional needs to recognize that much of what we do is affected by geopolitical realities and that we work in a global economy. This means that we must understand how cultures differ in their expectations of environmental quality. We cannot assume that a model that works in one setting will necessarily work in another without adjusting for differing expectations. Bhopal demonstrated the consequences of ignoring these realities.

Smaller versions of the Bhopal incident are more likely to occur, but with more limited impacts. For example, two freight trains collided in Graniteville, SC, just before 3:00 A.M. on January 6, 2005, resulting in the derailment of three tanker cars carrying chlorine (Cl_2) gas and one tanker car carrying sodium hydroxide (NaOH) liquids. The highly toxic Cl_2 gas was released to the atmosphere. The wreck and gas release resulted in hundreds of injuries and eight deaths.

In February of 2005, the District of Columbia City Council banned large rail shipments of hazardous chemicals through the U.S. Capital, making it the first large metropolitan area in the United States to attempt to reroute trains carrying potentially dangerous materials. The CSX Railroad is opposing the restrictions, arguing that they violate Constitutional protections and interstate commerce legislation and rules. The Graniteville chlorine leak is a most recent example of rail-related exposure to hazardous wastes, and it serves as a reminder that roads and rails are in very close contact to where people live. And the incidents are not really all that rare. Seven months before the Graniteville incident, three people died after exposure to chlorine as a result of derailment in San Antonio, Texas. Fifty more people were hospitalized. Although one of the concerns is occupational safety—the engineer died in the San Antonio wreck—transportation also increases community exposures. The two other deaths and most of the hospitalized were people living in the neighborhood where the leak occurred.

Many metropolitan areas also have areas where rail, trucks, and automobiles meet, so there is an increased risk of accidents. Most industrialized urban areas have the problematic mix of high density population centers, multiple modes of transport, dense rail and road networks, and rail-to-rail and rail-to-truck exchange centers. Most cities are especially at risk if an accident were to involve hazardous chemicals, since they are major crossroads. And, rerouting trains is not feasible in many regions because transcontinental lines have run through most urban areas for many decades. So, other steps can be taken to reduce shipment risks from hazardous substances like chlorine and improvements in manifest reports are immediately available to first responders. At present, such information is not generally available. Even following the September 11, 2001, attacks, rail companies have been reticent to disclose what is being shipped. One local fire department spokesman has stated that one "could almost assume there are several cars of hazardous materials every time we see a train."[21]

This has an environmental justice aspect to it since many rail yards and spars where tank cars are parked are in low socioeconomic status neighborhoods. The potential for leaks and spills near the parked rail cars increases the potential for exposures to hazardous substances.

Preparing for Intentional Toxic Clouds

Robert Prieto, chairman of the board of the engineering firm Parsons Brinckenhoff, foresees the new engineering requirements to be "resistance, response, and recovery."[22] Prieto contends, and I agree, that infrastructures, including environmental facilities like drinking water plants, treatment plants, and hazardous waste management systems, must be able to withstand intentional attacks and the "catastrophic failure" that we saw at the Pentagon and the World Trade Center. It is particularly unsettling that these two facilities, by most accounts, had been designed with significant redundancies and safety factors, but they still failed.[23]

There are many improvements to be made in the design of facilities to decrease their vulnerabilities and the new engineer must include these as design criteria. I must add that in our attempt to build in resistance to malevolent acts, there is the possibility of "over-designing." Reaching for zero risk, of course, is impossible to attain because society is such a complex cacophony of values. This is further complicated by intentional malevolence. Like virtually everything else in environmental engineering, we are seeking a "sweet spot," that is, optimizing the most important variables (that is to say, choosing among "values"). If we go too far on the side of security, Prieto warns us that we should be ready "to live in an environment that resembles the complex caves of Afghanistan."

In addition to designing for resistance, Prieto tells us that we should be properly prepared as an engineering community to respond once something goes wrong. Although the structural aspects of this response have been highlighted in many accounts, perhaps the environmental response has not. In fact, however, the environmental emergency response activities in Manhattan and Washington, D.C., were significant. Many lessons can be drawn from the work of local, state, and federal agencies, as well as private sector firms, in responding. As a matter of fact, the actions by the Coast Guard, Federal Emergency Management Administration, Environmental Protection Agency, New York environmental and health agencies, the Port Authority, and others reflected a growing ability to address environmental emergencies. The good news is that the United States has built an intricate network of first responders for natural and human-induced disasters in a relatively short time since the 1980s.

Large efforts in emergency response go beyond, in both time and space, the specific area of the disaster. In the instance of the towers and the

Pentagon building, of course the fire and potential exposure to toxic substances at the immediate disaster site comprised the primary concern. However, many people intuitively knew when they saw the plume of dust surge away from Ground Zero, the potential harm would go well beyond the collapse site. In addition, people wanted the professional community to explain the long-term effects of short-term exposures and to give an assessment of how long these exposures to toxic substances like asbestos, lead, smoke, and organic contaminants would continue, albeit at lower doses than the initial exposure. Although first responders had asked themselves the same questions in various emergencies before, 9/11 was unique in its cause and its potential extent. Environmental systems are intricately connected to other systems, like medical triage, transportation, telephony, and energy distribution. On September 11, 2001, and for some extended time after, services were completely disrupted, creating logistic challenges for the response teams.

Unfortunately, when engineers design systems and structures, they must consider what will happen when environmental systems fail and how that failure will affect and be affected by other infrastructures. Engineers have a major role in recovery, and we must be realistic about that role. No matter what we do to resist and to respond to the increased challenges of environmental protection and security, systems will fail and we will need to recover from these failures. Again, the World Trade Center is illustrative. After the towers collapsed, rescue and recovery teams accomplished amazing feats. The first responders acted less on instinct than on what they were trained to do. The preparation of the fire fighters, police, hospital personnel, and rescue squads allowed many lives to be saved by evacuation and care that had been planned for years before. So then, what does the environmental engineering profession need to do to help in this process? First, we must see that environmental engineers are just as critical to emergency response as any other discipline. Once that is recognized, we must go about including emergency response in engineering curricula for undergraduates, as well as in continuing education of practicing professionals. And this must not be seen as solely a specialty area of just some engineers. I recommend that all engineers receive ample training in emergency response. I cannot think of any area of environmental engineering where response to failure, whether the result of natural or intentional causes, is not a core need. For example, if we design air pollution controls, failsafe considerations in air movement must be part of those controls (e.g., is the air mover vulnerable to sabotage?). Providing high-quality water and food, handling chemical and radioactive wastes, and protecting indoor and occupational settings all have obvious needs to design facilities that can resist failures, that have a need to respond to such failures, and in the event that these measures do not work, that have explicit plans for recovery.

Environmental engineers have always had to balance many perspectives to reach optimal solutions to problems. I believe this is becoming even more complicated, even without the exacerbation of terrorism. The September 11, 2001, attack on the World Trade Center in Manhattan, New York, actually resulted in at least three different toxic clouds. The first occurred immediately after the planes crashed into the buildings, leading to the discharge of large amounts of jet fuel, combustion products from the explosion, and building debris. Later, when the towers fell, large plumes of pollutants were released, consisting of building materials in addition to the fire that had begun burning after the plane crashes. Next, the longest lasting plume was that of the fire that continued to burn under the large piles of debris that remained after the towers imploded and collapsed. The contaminants varied, including volatile organic compounds like benzene, semivolatile organic compounds like the dioxins, and metals, including lead. The pollutants existed in both the gas and particle phases. Thus, a large emergency response effort should include provisions for sampling and analysis of a broad range of substances.

Airshed in the Developing World: Mexico City

Air quality in Mexico City, Mexico, illustrates many of the problems currently being encountered by policy makers and practitioners. The Metropolitan Area is quite large, encompassing 3,969 square kilometers (km^2), with one of the world's densest populations, about 15 million. It is also representative demographically for much of the developing world in the tropical and subtropical latitudes (19°26′ N, 99°08′ W), with the majority of the population younger than 25 years of age. Because of pressures to develop and remnant old technologies, the Mexico City airshed frequently violates Mexican air quality standards, as well as internationally recommended air quality levels, for a variety of contaminants. Like many other cities in both developed and developing nations, the contaminants come from both point sources (e.g., easily identifiable stacks, like the one shown in Figure 3.18) and areal sources, especially cars, trucks, and other transportation-related causes. Ozone (O_3) and its precursors, nitrogen oxides and hydrocarbons, and carbon monoxide are the principal pollutants of concern.

These conventional air pollutants are certainly not the only contaminants of concern, but like so many other highly urbanized areas of the world, they must be dealt with immediately, before other air toxics can be addressed. This demonstrates that the developing nations, to a great extent, follow a similar path as that experienced by the more highly developed countries. That is, the most pressing and dangerous cases are driving the need for environmental improvements, much like the Meuse and Donora incidents. As these are addressed, then other problems, such as air toxics, can be tackled more vigorously. In a way, this is similar to the logic behind the hierarchy of needs. A society must first address the primary needs before moving on to higher level needs. Thus, the important air pollution episodes (summarized in Table 3.7) varied widely in impact. However, they share many lessons.

FIGURE 3.18. Air pollutant point source.
Photo credit: U.S. Environmental Protection Agency.

Lessons Learned

A principal lesson learned from the twentieth century air pollution episodes is the importance of meteorological conditions. The same emissions in terms of absolute mass released in the same time period can result in very different air quality, depending on winds, ambient temperature, atmospheric pressure, and, most importantly, inversion conditions. This is particularly important when conditions lead to elevated inversions above

TABLE 3.7
Some important air pollution episodes and their impact.

Location	*Year*	*Deaths/Injuries*
Meuse Valley, Belgium	1930	63 died, 600 sick
Donora, Pennsylvania	1948	20 died, 6,000 sick
London, England	1952	4,000 died
Bhopal, India	1984	12,000 died, 120,000 injured
World Trade Center	2001	Unknown (nearly 3,000 died as result of attack, but long-term health effects not known)
Graniteville, South Carolina	2004	8 died, 240 injured

valleys, such as the Meuse River valley in Belgium and the Los Angeles Basin. Elevated inversions can cap pollutants in a stagnant air stratum when vertical air circulation occurs above the valley, so that pollution levels build up over days (see Figure 3.4). This indicates that pollution controls need to be targeted toward worst-case scenarios, rather than average meteorological conditions.

Another lesson is the importance of paying attention to evolving knowledge. The increasing awareness of the importance of certain conventional pollutants, like PM, SO_x, and NO_x, and their links to health effects should have been heeded and incorporated into city planning and pollution control decision making processes. Also, the toxic cloud episodes could have been expected from these more conventional episodes, given that toxic substances like methylsiocyanate are much more toxic than oxides of sulfur and nitrogen. Thus, a release even at low concentrations should have been provided for in contingency plans. Even worse, shortly after the Bhopal disaster, engineers and regulators were given a test on what they had learned about toxic releases and communicating risk when a release of another toxic gas (this time, aldicarb oxime) occurred at the Institute, West Virginia, pesticide plant. The same company implicated at Bhopal, Union Carbide, owned the Institute plant and, in the minds of many, once again failed the test. There should have been significant progress made, since in many ways the two plants and situations were so very similar, but some of the same weaknesses remained even after the horrible consequences of Bhopal.

The Bhopal disaster again reminds us that forgetting the past can be deadly.

Contaminant of Concern: Photochemical Oxidant Smog

The term *smog* is a shorthand combination of "smoke-fog." However, it is really the code word for photochemical oxidant smog, the brown haze that can be seen when flying into Los Angeles, St. Louis, Denver, and other metropolitan areas around the world. Smog is made up of at least three ingredients: light, hydrocarbons, and radical sources, such as the oxides of nitrogen. Therefore, smog is found most often in the warmer months of the year, not because of temperature, but because these are the months with greater amounts of sunlight. More sunlight is available for two reasons, both attributed to the earth's tilt on its axis. In the summer, the earth is tilted toward the sun, so the angle of inclination of sunlight is greater than when the sun is tipped away from the earth leading to more intensity of light per earth surface area. Also, the days are longer in the summer, so these two factors increase the light budget.

Hydrocarbons come from many sources, but the fact that internal combustion engines burn gasoline, diesel fuel, and other mixtures of hydrocarbons makes them a ready source. Complete combustion results in carbon dioxide and water, but anything short of combustion will be a source of hydrocarbons, including some of the original ones found in the fuels, as well as new ones formed during combustion. The compounds that become free radicals, like the oxides of nitrogen, are also readily available from internal combustion engines, since the ambient air is more than three-quarters molecular nitrogen (N_2). Although N_2 is relatively not chemically reactive, with the high temperature and pressure conditions in the engine, it does combine with the O_2 from the fuel/air mix and generates oxides that can provide electrons to the photochemical reactions.

The pollutant most closely associated with smog is ozone (O_3), which forms from the photochemical reactions just mentioned. In the early days of smog control efforts, O_3 was used more as a surrogate or *marker* for smog, since one could not really take a sample of smog. Later, O_3 became recognized as a pollutant in its own right since it was increasingly linked to respiratory diseases.

Cities that failed to achieve human health standards as required by the Clean Air Act's National Ambient Air Quality Standards (NAAQS) were required to reach attainment within six years of passage, although Los Angeles was given 20 years, since it was dealing with major challenges in reducing ozone concentrations. Almost 100 cities failed to achieve ozone standards and were ranked from *marginal* to *extreme*. The more severe the pollution, the more rigorous controls required, although additional time was given to those extreme cities to achieve the standard. Measures included new or enhanced inspection/maintenance (I/M) programs for autos; installation of vapor recovery systems at gas stations and other controls of hydrocarbon emissions from small sources; and new transportation controls to offset increases in the number of miles traveled by vehicles. Major stationary sources of nitrogen oxides also have to reduce emissions.

The ozone threshold value is 0.12 parts per million (ppm), measured as a one-hour average concentration. An area meets the ozone NAAQS if there is no more than one day per year when the highest hourly value exceeds the threshold. (If monitoring did not take place every day because of equipment malfunction or other operational problems, actual measurements are prorated for the missing days. The estimated total number of above-threshold days must be 1.0 or less.) To be in attainment, an area must meet the ozone NAAQS for three consecutive years.

Calculating compliance can be tricky. Air quality ozone value is estimated using a calculation usually based on the fourth highest monitored value with three complete years of data and is selected as the updated air quality value because the standard allows one exceedance for each year. It is important to note that the 1990 Clean Air Act Amendments required that ozone nonattainment areas be classified on the basis of the design value at the time the Amendments were passed; generally the 1987–1989 period was used.

The strong seasonality of O_3 levels makes it possible for areas to limit their O_3 monitoring to a certain portion of the year, termed the O_3 season. Peak O_3 concentrations typically occur during hot, dry, stagnant summertime conditions; that is, high temperature and strong solar insolation (i.e., incoming solar radiation). The length of the O_3 season varies from one area of the country to another. The months of May through October are typical, but states in the south and southwest may monitor the entire year. Northern states have shorter O_3 seasons, for example, May through September for North Dakota. This analysis uses these O_3 seasons to ensure that the data completeness requirements apply to the relevant portions of the year.

Children have higher health risks associated with exposure to ozone than do most adults. The average adult breathes 13,000 liters of air per day, but on an air-per-kilogram basis, children breathe even more air than do adults. Because children's respiratory systems are prolific and still developing, they are more susceptible than adults to many environmental threats. Children are outside playing and exercising during the summer months more frequently than in the less warm months. Unfortunately, this is also the time of year with elevated O_3. In addition, asthma is a growing threat to children and adults. Children make up 25 percent of the U.S. population, but comprise 40 percent of the asthma cases. The asthma death rate has increased three-fold in the past 20 years, and African Americans die at a rate six times that of Caucasions. Even moderately exercising healthy adults can experience 20% or greater reductions in lung function from exposure to low levels of ozone over several hours. These factors make smog an important public health concern.

The principal lesson from the history of air pollution episodes, as from every case in this book, is that the atmosphere is not infinite in its ability to absorb wastes. Although this appears obvious to the twenty-first century scientist, it is actually a fairly recent realization.

Notes and Commentary

1. The source of this discussion is the Vauxhall Society (http://www.vauxhallsociety.org.uk/), a civic society covering the northern part of the London's borough of Lambeth with adjacent areas of Southwark and Wandsworth. The Vauxhall Society's aims are:

   - To stimulate public interest in and care for the character, history, and future of the area and to maintain and improve the quality of life for all who live there.
   - To encourage high standards of architecture and civic design.
   - To encourage the conservation of both the natural and built environment and above all the living community of which we are all part.

2. The original connotation of the term smog has evolved. Today, most atmospheric scientists refer to smog as a short form of "photochemical oxidant smog." For example, the Northeast States for Coordinated Air Use Management (NASCAUM), an interstate association of air quality control divisions in the Northeast states in the United States, defines smog as:

   > . . . a mixture of *pollutants,* principally ground-level *ozone,* produced by chemical reactions in the air involving smog-forming chemicals. A major portion of smog-formers come from burning of petroleum-based fuels such as gasoline. Other smog-formers, *volatile organic compounds,* are found in products such as paints and solvents. Smog can harm health, damage the environment, and cause poor visibility. Major smog occurrences are often linked to heavy motor vehicle traffic, sunshine, high temperatures and calm winds, or *temperature inversion* (weather condition in which warm air is trapped close to the ground instead of rising). Smog is often worse away from the source of the smog-forming chemicals, since the chemical reactions that result in smog occur in the sky while the reacting chemicals are being blown away from their sources by winds.

Frequently, tropospheric ozone (O_3) is used as a surrogate for smog, because smog is actually the result of a series of photochemical reactions between hydrocarbons and oxides, especially oxides of nitrogen and sulfur. The rates of these reactions are limited by any of these components, that is, rate limits by sunlight, by oxides of nitrogen (NO_x limited), or volatile organic compounds (hydrocarbon limited). So, it is convenient and allows for standardization to measure O_3. In fact, O_3 is the U.S. criteria air pollutant under the Clean Air Act's National Ambient Air Quality Standard (NAAQS). Interestingly, although the principal, original importance of O_3 was that it represents smog, the toxicity of ozone itself has become well documented. In fact, since ozone first became regulated as a ground-level pollutant, numerous health studies linked health effects that occur at levels lower than the previous standard and that exposure times longer than one hour (reflected in the previous standard) became a particular concern. So tropospheric O_3 is now considered an air pollutant in its own right, irrespective of its value as an indication of smog. As a result, the U.S. EPA replaced the previous standard with an eight-hour standard set at 0.08 ppm; an area will attain the standard when the three-year average of the annual fourth-highest daily maximum eight-hour concentrations is less than or equal to 0.08 ppm.

See U.S. Environmental Protection Agency, 1996. External Draft, Air Quality Criteria for Ozone and Related Photochemical Oxidants, EPA report no. EPA/600/AP-93/004af-cf, Research Triangle Park, NC; and U.S. Environmental Protection Agency, 1996. Review of National Ambient Air Quality Standards for Ozone: Assessment of Scientific and Technical Information: Office of Air Quality Planning and Standards Staff Paper, EPA report no. EPA-452\R-96-007, Research Triangle Park, NC.

3. U.S. Environmental Protection Agency (EPA), 1999. National Primary Drinking Water Regulations: Technical Fact Sheets. Washington, D.C.: http://www.epa.gov/OGWDW/hfacts.html.
4. K. Roholm, 1937. "The Fog Disaster in the Meuse Valley, 1930: A Fluorine Intoxication," *Journal of Industrial Hygiene and Toxicology*, 19, 126–137.
5. United Kingdom Department of Environment, Food, and Rural Affairs, Expert Panel on Air Quality Standards, 2004. *Airborne Particles: What Is the Appropriate Measurement on Which to Base a Standard? A Discussion Document.*
6. G. Bonne, P. Mueller, L.W. Chen, B.G. Doddridge, W.A. Butler, P.A. Zawadzki, J.C. Chow, R.J. Tropp, and S. Kohl. *Proceeding of the PM2000: Particulate Matter and Health Conference*, "Composition of PM2.5 in the Baltimore-Washington Corridor," Air & Waste Management Association, Washington, D.C., pp. W17–18, Jan 2000.
7. Aerosol textbooks provide methods to determine the aerodynamic diameter of particles less than 0.5 micrometer. For larger particles gravitational settling is more important and the aerodynamic diameter is often used.
8. Particle matter is regulated in the United States to protect both health and welfare. Primary standards protect health, while secondary standards protect welfare, which includes damage to property, harm to ecosystem, and aesthetics

(e.g. visibility). Presently, the primary and secondary standards are the same for PM, but some groups are calling for more stringent secondary standards.

9. The diameter most often used for airborne particle measurements is the "aerodynamic diameter." The aerodynamic diameter (D_{pa}) for all particles greater than 0.5 micrometer can be approximated as the product of the Stokes particle diameter (D_{ps}) and the square root of the particle density (ρ_p):

$$D_{pa} = D_{ps}\sqrt{\rho_p} \tag{3.13}$$

If the units of the diameters are in μm, the units of density are $\mathrm{g\,cm^{-3}}$.

The Stokes diameter D_{ps} is the diameter of a sphere with the same density and settling velocity as the particle. The Stokes diameter is derived from the aerodynamic drag force caused by the difference in velocity of the particle and the surrounding fluid. Thus, for smooth, spherical particles, the Stokes diameter is identical to the physical or actual diameter.

10. For information regarding particle matter (PM) health effects and inhalable, thoracic, and respirable PM mass fractions, see U.S. Environmental Protection Agency, 1996. Air Quality Criteria for Particulate Matter, Technical Report No. EPA/600/P-95/001aF, Washington, D.C.

11. The federal reference method (FRM) is a certified reference analogous to the National Institute for Standards and Testing (NIST) standard reference material (SRM), which is a standard of known quality to be used by scientists and engineers, allowing them:

    a. To help to develop and validate representative and accurate methods of analysis.
    b. To verify that tests meet validated performance criteria.
    c. To calibrate systems of measurements.
    d. To establish acceptable quality assurance (QA) programs.
    e. To provide test materials for interlaboratory and interstudy comparisons and proficiency test programs.

    Rather than materials, however, the FRM applies to environmental measurement equipment against which all other equipment is compared. For example, in ambient monitoring the U.S. federal standard method dictates how to monitor a specific pollutant, such as aerosols or sulfur dioxide (SO_2). The FRM may be specified by technique (such as a particular physical principle, like a gravimetric or an optical technique for estimating mass) or by design (such as a particular system for collecting and weighing particles).

12. For more information on asbestos exposure, see the Public Health Statement on Asbestos: ATSDR, 2001, Public Health Statement for Asbestos, at http://www.atsdr.cdc.gov/toxprofiles/phs61.html.

13. D.L. Davis, 2000. "Air Pollution Risks to Children: A Global Environmental Health Problem," *Environmental Manager*, February issue: 31–37. H.H.

Schrenk, H. Heimann, G.D. Clayton, W.M. Gafafer, and H. Wexler, 1949. "Air Pollution in Donora, PA: Epidemiology of the Unusual Smog Episode of October 1948," Preliminary Report. Public Health Bulletin No. 306. U.S. Public Health Service, Washington, D.C.

14. California State Department of Public Health, 1969. Recommended Ambient Air Quality Standards. Statewide standards applicable to all California Air Basins, HS-3.
15. Canadian Office of Environmental Health Hazard Assessment, 2000. Air Toxics Hot Spots Program Risk Assessment Guidelines, Part III. Technical Support Document for the Determination of Noncancer Chronic Reference Exposure Levels. Available online at http://www.oehha.ca.gov.
16. U.S. Environmental Protection Agency, 1999. Integrated Risk Information System (IRIS) database: Reference concentration (RfC) for hydrogen sulfide. Available online at http://www.epa.gov/ngispgm3/iris/subst/index.html.
17. Air Products, Inc., 2005. Material Safety Data Sheet for Hydrogen Sulfide. http://avogadro.chem.iastate.edu/MSDS/hydrogen_sulfide.pdf; accessed April 20, 2005.
18. Ibid.
19. I am writing this the day after the commemoration of the third year after the September 11, 2001, attacks on the World Trade Center (WTC) and the Pentagon, so please view this editorial through the prism of the time and place of its writing. The environmental engineer today is in a very different world than what existed on September 10, 2001. Every decision in the public domain and most in the corporate world is tainted with the need to consider willful malevolence against what we design, build, and operate. In fact, if not thoroughly considered, these marvelous projects can be turned against us. Our drinking water supplies, air handling systems, structures, waste treatment, storage and disposal (TSD) systems, and transportation routes and security systems designed for the good of humankind must be revisited as to their potential to be used as instruments of harm. This goes beyond the very rational concern about interruption of service, which is a traditional criterion for any good operation and maintenance (O&M) program. Before, we were likely to consider inadvertent and even neglectful malfeasance in design and operation, such as violation of pretreatment standards when a company is trying to save money or avoid permit requirements. Nowadays, in addition to these concerns, we can have inadequate design and planning if we do not consider terrorism scenarios, like the deliberate contamination of our public life support systems.

    The role of the environmental engineer has also changed drastically in what we will be called upon to do in the event that terrorist actions occur. Our expectations now include emergency response (many environmental engineers were deployed following the WTC attacks to evaluate and recommend actions to reduce post-attack risks). Also, HAZMAT and other first responder teams will include environmental engineers to conduct failure analyses of what went wrong and to advise on how to prevent and to design against previously unforeseen weaknesses in design and O&M. This creates a new onus for the environmental engineers. I cannot help but see that engineers will, increasingly, be found lacking for what they have failed to foresee, that is, sins of omission. One of the lessons learned from the 9/11 Commission was that the intelligence community and others "lacked vision" insofar as we should have known much more clearly the degree of enmity that our terrorist enemies have toward the United States. They have been "at war" with us for some time, but we did not recognize it. Environmental engineers will be among the professionals that are no longer excused for their ignorance. "Fool me once, shame on you. Fool me twice, shame on me!"

Even the military must include environmental engineering in its antiterrorism deployments, as illustrated by the surgical-like special operations of the Navy SEALs in preventing a major ecological disaster at the beginning of the second Iraq War by preventing oil well and rig fires and crude oil spills that many believe Saddam Hussein's regime had planned in the event of an invasion.

20. The principal sources for this case are M.W. Martin and R. Schinzinger, 1996. *Ethics in Engineering*, 3e, McGraw-Hill, New York. Also C.B. Fledderman, 1999. *Engineering Ethics*, Prentice Hall, Upper Saddle River, NJ.
21. W. Moore, 2005. "Analysis: Chlorine Tankers Too Risky for Rails?" *Sacramento Bee*, February 20, 2005.
22. R. Prieto, 2002. "A 911 Call to the Engineering Profession," *The Bridge*, 32 (1).
23. See, for example: L. Robinson, 2002. "Reflections on the World Trade Center," *The Bridge*, 32 (1). Leslie Robinson was responsible for the structural design of the towers. The best way to express Robinson's anguish in trying to deconvolute and to deconstruct what, if anything, could have been foreseen by the engineers is to quote the last paragraph of his article:

> . . . the events of September 11 have profoundly affected the lives of countless millions of people. To the extent that the structural design of the World Trade Center contributed to the loss of life, the responsibility must surely rest with me. At the same time, the fact that the structures stood long enough for tens of thousands to escape is a tribute to the many talented men and women who spent endless hours toiling over the design and construction of the project . . . Surely, we have all learned the most important lesson—that the sanctity of human life rises far above all other values.

The difference between now and 9/10/2001 is that we now know better what we are up against.

CHAPTER 4

Watershed Events

Man is not an aquatic animal, but from the time we stand in youthful wonder beside a Spring brook till we sit in old age and watch the endless roll of the sea, we feel a strong kinship with the waters of this world.

Hal Borland (1900–1978), *Sundial of the Seasons*, 1964

Water, to humans, is more than dihydrogen monoxide (H_2O). It is, with air and blood, one of the indispensable fluids of life. Borland aptly places humanity in a strongly bound relationship with water at every scale, from cytoplasm to oceans and continental air masses. Water is a central theme of most faith traditions, including every account of creation. For example, in the book of Genesis of Judeo-Christian Scripture, the creation account declares that "the Spirit of God was moving over the water." And later, in Exodus, Chapter 7, one of the first miracles provided is a case of water pollution: "Aaron raised his stick and struck the surface of the river, and all the water in it was turned into blood. The fish in the river died, and it smelled so bad that the Egyptians could not drink from it." Water quality has enjoyed a primacy in human populations throughout history.

Water pollution is a problem that affects us at all scales in space and time, but also at the very essence of our being. Thus, we will consider some of the most noteworthy cases, not so much in terms of volume of contamination, but in how these cases have shaped our collective environmental psyches and ethos. To do so, we will consider cases in two different types of water pollution: incremental buildup, as represented by Lake Erie and the Cuyahoga River, and sudden releases, as represented by oil spills.

The Death of Lake Erie: The Price of Progress?

The demise of Lake Erie in the 1960s was emblematic of the environmental challenge growing out of the industrial and petrochemical revolutions of the nineteenth and twentieth centuries in the West. Companies, municipalities, and people in general had seemingly perceived water to be completely elastic in its ability to absorb any amount and type of pollution. Any waste that needed to be disposed of was directly discharged into surface waters, such as rivers, lakes, and oceans.

Lake Erie provides numerous lessons. For those of us who were learning the new language of environmental science, Lake Erie gave us a new paradigm, one of optimism where even highly polluted waters could be saved. For the decades that followed the 1960s, water bodies began to recover from even extremely polluted conditions. Although the science of limnology (the study of freshwater systems) had been well established within the hydrologic and biological science communities, the problems in Lake Erie helped to propel it into a much wider application. For example, eutrophication of freshwaters, especially ponds and lakes, became better understood in the context of Lake Erie (see the discussion box, "Eutrophication").

Eutrophication

Healthy water bodies contain sufficiently low amounts of contaminants and sufficiently high amounts of dissolved oxygen (DO) to support a balance of aquatic life. The DO concentrations in surface waters can be reduced by both natural and human factors. Evidence of a healthy water body is its trophic state. Every lake fits into a particular trophic state, according to its degree of eutrophication, and all lakes change their trophic status over time. All lakes, even the most pristine, are undergoing nutrient enrichment and filling. Lakes can be divided into three categories based on trophic state—oligotrophic, mesotrophic, and eutrophic. These categories reflect a lake's nutrient and clarity levels.

Limnologists refer to healthy water bodies as oligotrophic systems; that is, they contain little plant nutrients and are often continuously cool and clear. Oligotrophic waters have very low production of organic matter by photosynthesis and can support diverse animal life and collect optimal amounts of nutrients, mainly phosphorous and nitrogen, from natural sources, such as decomposing plant matter. When a water body becomes enriched in dissolved nutrients, especially phosphorous and nitrogen, they stimulate the growth of aquatic plant life, which can lead to the depletion of dissolved oxygen (DO). This is known as *eutrophication.*

Oligotrophic lakes (see Figure 4.1) are generally clear, deep, and free of weeds or large algae blooms. Though aesthetically appealing, they are low in nutrients and do not support large fish populations. Nutrient concentrations, such as phosphorous and nitrogen, are limiting, and aquatic macrophytes (large plants) and algae are less abun-

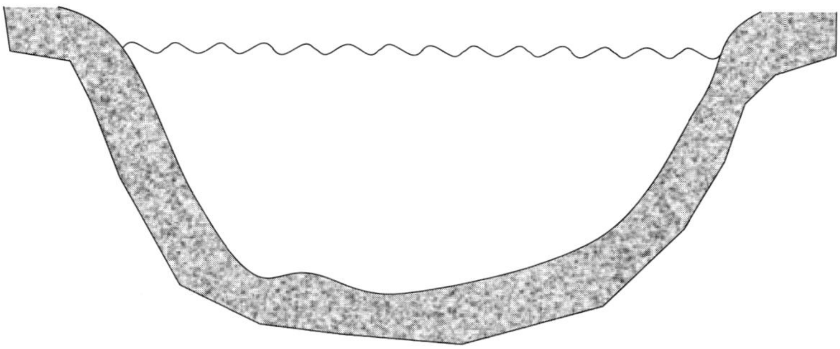

Oligotrophic lake:
•Low in nutrient concentrations (i.e., low enrichment)
•Low productivity, low turbidity (clear water)
•Highly desirable habitat for sensitive species, game fish

FIGURE 4.1. Oligotrophic lake system.

dant. Oligotrophic water bodies typically have accumulated little plant debris on the bottom over the years since aquatic macrophytes and algae are less abundant. They generally have water clarity greater than four meters (i.e., the distance one can see down into the water) since the amounts of free-floating algae are low, as well as the absence of presence of coloring agents in dissolved substances and low concentrations of suspended particles. Fish and wildlife populations generally will be small because food and habitat are often limited. Oligotrophic water bodies usually do not support abundant populations of sportfish such as large-mouth bass and bream, and it usually takes longer for individual fish to grow in size in oligotrophic waters. However, oligotrophic lakes often develop a food chain capable of sustaining a very desirable fishery of large game fish, but these conditions can deteriorate in a short amount of time due to fishing pressure increases.

A mesotrophic lake (see Figure 4.2) can support moderate populations of living organisms. These lakes have moderate concentrations of nutrients and moderate growth of plant life, such as algae and/or macrophytes, owing to the moderate concentrations of nutrients (especially N and P). There is evidence of slight sediment buildup and organic accumulations. Clarity is between 2 and 4 m, so a mesotrophic lake is usually "swimmable and fishable," to use a phrase made famous by the Federal Water Pollution Control Act of 1972.

Eutrophic waterbodies (see Figure 4.3) may be dominated by algal growth or by larger plant growth. If algae-dominated, the water may

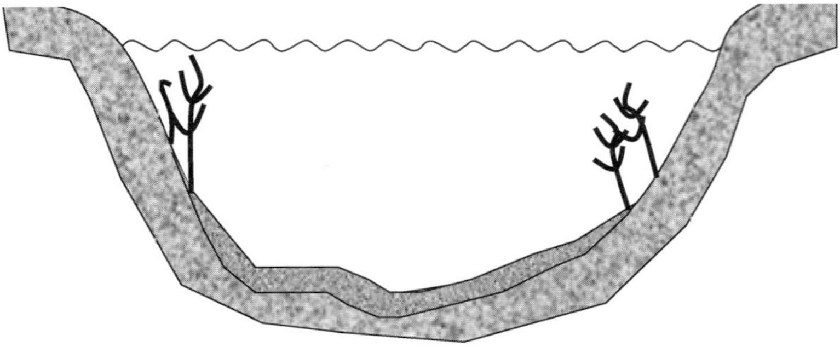

Mesotrophic lake:

- **Higher nutrient concentrations (i.e., increasing enrichment)**
- **Organic matter accumulating**
- **Increased sedimentation**
- **Good fishery**
- **Algal blooms occasionally**

FIGURE 4.2. Mesotrophic lake system.

Eutrophic lake:

- **Very high nutrient concentrations (i.e., highly enriched)**
- **Very productive (supports much plant life)**
- **Depletion of dissolved oxygen**
- **Poor fishery (rough fish common)**
- **Frequent algal blooms**

FIGURE 4.3. Eutrophic lake system.

have a green, cloudy appearance from the colonies of algae suspended or floating in the water. If plant-dominated, the submersed macrophytes will decrease the concentrations of the green pigment, chlorophyll, and use much of the nutrient concentrations, making for clearer water. Thus clarity as indicated by Secchi depth readings are higher than if the water body were an algae-dominated eutrophic system. This makes trophic state classification based on appearance very difficult.

Like almost everything in the environment, there is an optimal range between too little and too much nutrient loading. A minimal amount of nutrients is needed in any ecosystem, but when this amount is exceeded, as was the case for Lake Erie some decades ago, algal growth can become prolific. In the right balance, algae serve as food sources and are crucial to energy and mass balances in aquatic systems. Out of balance, however, the algae use up too much of the DO needed by fish and other aquatic organisms. The nutrients find their way into water bodies through numerous avenues, but the major categories are either point sources or nonpoint sources. As the name implies, point sources deliver nutrients and other pollutants to surface waters from a single point, such as a pipe, conduit, outfall structure, or ditch. Nonpoint pollutants are those that flow over broad expanses, such as runoff from agricultural practices, mining, roads, neighborhoods, and urbanized areas. Another nonpoint source of pollutants is the atmosphere. In fact, atmospheric deposition can be the largest source of many contaminants. The nutrients can take on many physical and chemical properties. For example, nitrogen can be in solid phase, such as in a commercial fertilizer, in liquid phase, such as when ammonia is dissolved in runoff water, or in gas phase, such as when ammonia or nitric acid is found in soil pores. The chemical forms can also be diverse, such as when conditions make for a reduced form, for example, ammonia, or in an oxidized form, such as nitrite or nitrate. Thus, the process of eutrophication includes elevated biological productivity resulting from increased input of nutrients or organic matter into aquatic systems. For lakes, which do not flow as rapidly as streams, such increased biological productivity usually leads to decreased lake volume because organic detritus accumulates. Natural eutrophication continues as aquatic systems fill in with organic matter. This is contrasted with cultural eutrophication, which is exacerbated by human activities and the consequent point and nonpoint pollution.

In retrospect, the death sentence to Lake Erie was premature. The assumptions of the time were that things could not change enough to return biodiversity to the lake ecosystem. Thankfully, these predictions were wrong. However, the Great Lakes are still vulnerable. In fact, some new problems have emerged, whereas others have been solved. To wit, the invasion of opportunistic species that are threatening the diversity of enormous regions; notably the appearance and proliferation of the zebra mussel (*Dreissenia polymorpha*) throughout much of the Great Lakes (see the discussion in Chapter 6).

Cuyahoga River Fire

June of 1969 is one of the most important months in the history of environmental protection. That is the month that pollution of surface water reached national prominence. Collectively, even people who had thought the problem of water pollution was overblown began to appreciate the magnitude and apparent irreversibility of the problem when the Cuyahoga River, flowing through Cleveland, Ohio, making its way to Lake Erie, caught fire. The river had such high concentrations of light (i.e., less dense than water), flammable contaminants that film crews recorded the unthinkable—burning water. Actually, the Cuyahoga River had been in flames previously, but this time the whole nation was watching. Many point to the fire as the seminal event that heightened awareness and began the call for greater water pollution controls, ultimately leading to the passage of the Great Lakes Water Quality Act and Clean Water Act in the 1970s.

Lesson Learned: The Need for Regional Environmental Planning[1]

It was becoming apparent in the 1960s that pollution was multifaceted. Its sources could be obvious, such as outfall structures and other "point sources" leaving factories and wastewater facilities. But the scientists' understanding of the complex riverine, lacustrine, and marine systems were also receiving significant contributions of pollutants from less obvious sources, the nonpoint and atmospheric sources. Thus, a significant lesson from Lake Erie and the Cuyahoga fire that we learned while watching these amazing recoveries was the importance of management. Even the best designs and engineering approaches were not nearly as effective on their own compared to a unified, comprehensive planning approach. The Clean Water Act was the first environmental legislation to articulate the need for wide-ranging environmental planning in the "areawide waste treatment management plan." Such plans were authorized under Section 208 of the Clean Water Act to develop a comprehensive program for the treatment of water and for controlling water pollution from all point and nonpoint sources in the geographic area. These plans, which came to be known as "208 plans" were first written in the 1970s, principally to ensure that wastewater treatment plants were sited and operated efficiently, especially those plants and facilities that were paid for under the multibillion dollar construction grant program (Section 201 of the Clean Water Act), which provided federal funds for the design and construction of sewage collection and treatment facilities.[2] Most states have continued to require the 208 plans, which are updated periodically by the responsible areawide planning agency or the state. Areawide planning agencies often are designated by the governor of the state under authority of Section 208 of the Clean Water Act, which has responsibilities for areawide waste treatment management planning within a specified area of a state. However, regional planning usually transcends the geographic boundaries of individual governmental units, because the whole region, made up of different municipalities and county jurisdiction, shares common social, economic, political, natural resource, and transportation characteristics.

Regional planning agencies can be either voluntary associations of local government or mandated or authorized by state legislation. The agencies are generally only advisory; they have little enforcement capacity. As such they provide technical assistance, share information, and train decision makers in matters important to a particular region. They also coordinate issues among the disparate governments represented, including ways to obtain and administer funding. Some regional planning agencies do have regulatory authority, if granted by the state, regarding land use controls, such as regional zoning and subdivision regulations.[3]

The first regional planning agency was formed in Boston, Massachusetts, the Boston Metropolitan Improvement Commission in 1902, followed by the Commercial Club of Chicago, a private organization, which spearheaded the Plan of Chicago, developed by the famous urban architects Daniel H. Burnham and Edward H. Bennett and their planning team.[4]

Interestingly, four pieces of legislation, none specifically environmental, pushed the regional planning perspective. The Housing and Community Development Act of 1965 allowed regional planning agencies to receive federal funds to develop and implement plans. The Public Works and Economic Development Act of 1965 gave funds to multicounty economic development districts. It also authorized the establishment of federal multistate economic development commissions. The Appalachian Regional Development Act established the multistate Appalachian Regional Commission, which worked through multicounty development districts. And, the Water Resources Planning Act of 1965 established federal multistate river basin missions.[5] Note that all these major laws were being written when Lake Erie was reaching its worst environmental conditions. In 1966, for example, the U.S. Congress supported regional approaches to governmental initiatives by requiring all recipient governments to coordinate federal programs with regional clearinghouses, often taking the form of regional planning councils.[6] This eventually led to one of the most important documents directing regional planning, the U.S. Office of Management and Budget's Circular A-95, entitled "Evaluation, Review and Coordination of Federal and Federally Assisted Programs and Projects," which gave regional agencies authority to review applications for federal assistance to ensure that these projected funds would be consistent with regional and local plans. Congress directed that regional clearinghouses that received federal planning grants prevent duplication of services and enhance the planning of federally assisted projects. Melvin Mogulof, an observer of regional clearinghouses, aptly stated that "the use of A-95 procedures represents the single, most potentially powerful device to affect the distribution of resources in a region according to some regional point of view."[7]

The growth and establishment of the regional environmental planning can be summed up by B.D. McDowell of the U.S. Advisory Commission on Intergovernmental Relations:

> This explosion of "areawide" regional councils and the multistate river basin and economic development regions occurred because of very intentional and systematic federal action which drew in the states as well as local governments. In the cases of the areawide councils, the federal actions included establishing 39 grant programs designed to require and fund regional planning, and direct appeal to the governors of all 50 states to establish statewide systems of substate districts to systematize the administration of the federal programs supporting regional councils. And many of the states did so.[8]

Lake Erie and the other seemingly intractable water pollution problems of the 1960s helped to establish a new focus on regional planning. The Great Lakes continues this tradition to date. For example, numerous Great Lakes Areas of Concern (AOCs) have been identified in the Great Lakes Basin. The AOCs are defined by the U.S.-Canada Great Lakes Water Quality Agreement (Annex 2 of the 1987 Protocol agreement between the two countries) as "geographic areas that fail to meet the general or specific objectives of the agreement where such failure has caused or is likely to cause impairment of beneficial use of the area's ability to support aquatic life." The two governments have identified 43 AOCs (see Figure 4.4); 26 in the United States and 17 in Canada (five are shared between the United States and Canada on connecting river systems). The two federal governments are collaborating with state and provincial governments to carry out Remedial Action Plans (RAPs) in each AOC. The RAPs are written to achieve and maintain 14 beneficial uses. An impaired beneficial use means a change in the chemical, physical, or biological integrity of surface waters. These include:

- Restrictions on fish and wildlife consumption
- Tainting of fish and wildlife flavor
- Degradation of fish and wildlife populations

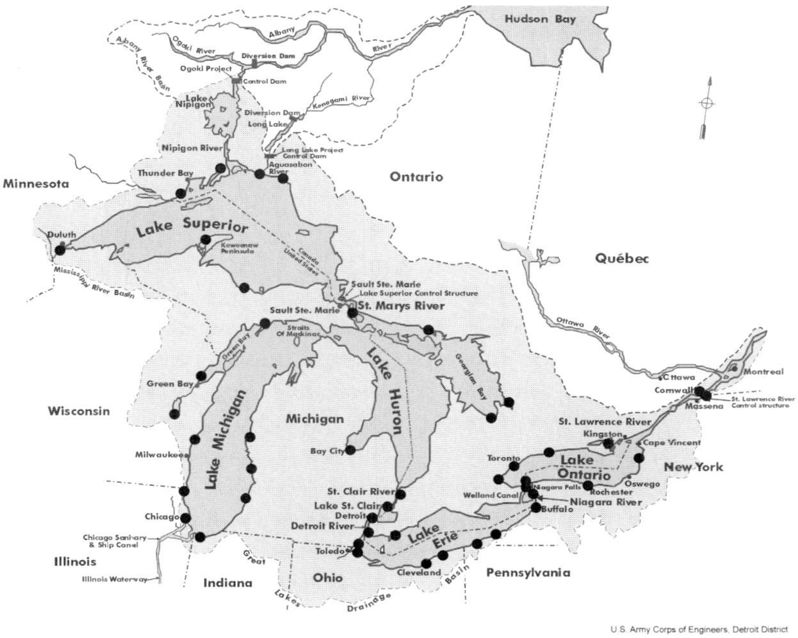

FIGURE 4.4. Areas of concern in the Great Lakes.
Source: U.S. Environmental Protection Agency; Base map from the U.S. Army Corps of Engineers.

- Fish tumors or other deformities
- Bird or animal deformities or reproduction problems
- Degradation of benthos
- Restrictions on dredging activities
- Eutrophication or undesirable algae
- Restrictions on drinking water consumption, or taste and odor problems
- Beach closings
- Degradation of aesthetics
- Added costs to agriculture or industry
- Degradation of phytoplankton and zooplankton populations
- Loss of fish and wildlife habitat

An example of a remedial action plan to address impaired uses is one for the Cuyahoga River, located in northeast Ohio. The river runs for about 100 miles from Geauga County, flowing south to Cuyahoga Falls where it turns sharply north until it empties into Lake Erie. The river drains 813 square miles of six counties.

The Cuyahoga AOC embodies the lower 45 miles of the river from the Ohio Edison Dam to the river's mouth, along with 10 miles of Lake Erie shoreline. The AOC also includes 22 miles of urbanized stream between Akron and Cleveland.

Waters have numerous uses. From an environmental perspective, a water body is not able to serve its purposes when it is contaminated. That is, its beneficial uses have become *impaired*.[9] For example, 10 of 14 use impairments have been identified in the Cuyahoga basin through the Remedial Action Plan (RAP) process. The environmental degradation resulted from nutrient loading, toxic substances (including PCBs and heavy metals), bacterial contamination, habitat change and loss, and sedimentation. Sources for these contaminants include municipal and industrial discharges, bank erosion, commercial/residential development, atmospheric deposition, hazardous waste disposal sites, urban stormwater runoff, combined sewer overflows (CSOs), and wastewater treatment plant bypasses.

Restrictions on Fish Consumption

In 1994, an advisory about eating fish was issued for Lake Erie and the Cuyahoga River AOC. The basis for the advisory was elevated PCB levels in fish tissue. The advisory restricted the consumption of white sucker, carp, brown bullhead, and yellow bullhead in the Cuyahoga River AOC, and walleye, freshwater drum, carp, steelhead trout, white perch, Coho salmon, Chinook salmon, small mouth bass, white bass, channel catfish, and lake trout in Lake Erie.

Degradation of Fish Populations

Beginning at the Ohio Edison Gorge and extending downstream to Lake Erie, measures of fish population conditions ranged from fair to very poor and were below applicable Ohio warm water habitat aquatic life use criteria. Although fish communities have recovered significantly compared to the historically depleted segments of the Cuyahoga River, pollution-tolerant species continue to compose the dominant fish population.

Degradation of Wildlife Populations

Anecdotal information indicates some recovery of Great Blue Heron nesting in the Cuyahoga River watershed. Resident populations of black-crowned night herons have been noted in the navigation channel. The RAP is seeking partners to undertake research in this area in order that an evaluation may be made.

Fish Tumors or Other Deformities

Although deformities like eroded fins, lesions and external tumors (DELT anomalies) have declined throughout the watershed, significant impairments continue to be found in the headwaters to the nearshore areas of Lake Erie.

Bird or Animal Deformities or Reproductive Problems
No data have been found to suggest this is impaired in the Cuyahoga River AOC. But, "no data" is not the same as "no problem."

Degradation of Benthos
Macroinvertebrate populations living at or near the bottom (i.e., benthic organisms) of the Cuyahoga River remain impaired at certain locations; however, there are indications of substantial recovery, ranging from good to marginally good throughout most free-flowing sections of the river. Some fair and even poor designations are still seen, however.

Restrictions on Dredging Activities
The U.S. Environmental Protection Agency restricts disposal of dredged sediment in most of the Cuyahoga AOC due to high concentrations of heavy metals. Only a small amount of the dredged material with contaminated sediments are transported and disposed of in a confined disposal facility in the Cleveland area.

Eutrophication or Undesirable Algae
The Cuyahoga navigation channel seems to be impaired due to extreme oxygen depletion during summer months. The oxygen demand of sediment is a factor.

Restrictions on Drinking Water Consumption, or Taste and Odor
The AOC contains no public drinking water sources, but contaminated aquifers and surface waters may still be sources for individual supplies and wells.

Beach Closings and Recreational Access
High bacterial counts following rain events periodically adversely affect the two beaches in the AOC. Swimming advisories are issued after a storm, or if microbial counts exceed certain thresholds.

Degradation of Aesthetics
Aesthetics are impaired throughout the AOC due to soil erosion, surface water contamination from debris, improperly operating septic systems, CSOs, and illegal dumping activities.

Degradation of Phytoplankton and Zooplankton Populations
According to some studies, phytoplankton populations in the AOC are impaired. No standards exist for zooplankton communities.

Added Cost to Agriculture and Industry
No registered water withdrawals for agricultural purposes are taking place in the AOC. Industry does not appear to be adversely affected.

Loss of Fish and Wildlife Habitat
Channelization, nonexistent riparian cover, silt, bank reinforcement with concrete and sheet piling, alterations of littoral areas and shorelines, and dredging are contributors to the impairment of fish and wildlife in the AOC.

As for most of the AOCs, planning and remediation in Lake Erie are ongoing and much progress has been made, but there are still a good deal of cleanup and environmental management actions remaining to be done.

Spills: Immediate Problem with Long-Term Consequences

The contamination of Lake Erie occurred progressively over decades. Another contamination scenario involves a sudden release of contaminants, invoking an emergency response. The escape of large quantities of oil and other toxic substances into waterways in the second half of the twentieth century was occurring with alarming frequency. Supertankers were capable of transporting huge volumes, so oil and petroleum product spills could cover many square kilometers.

Other types of spills have also caused devastating effects on the environment, such as a ruptured pipeline in Brazil in 2000 that released more than a million liters of heavy oil into Guanabara Bay, Rio de Janeiro, and the deliberate release of two million liters of crude oil into the Persian Gulf in 1991 by the Iraqi army.

In reverse chronological order, some of the most important water spills in recent decades include:

January 2000: A ruptured pipeline spewed more than 1 million liters of heavy oil into Guanabara Bay, Rio de Janeiro, Brazil.

December 1999: The tanker *Erika* spilled 13,000 tons of heavy diesel oil off the coast of Brittany, UK.

February 1996: The *Sea Empress* spilled about 72,000 tons of crude oil near the port of Milford Haven in Wales, UK.

January 1993: The *Braer* grounded off the Shetland Islands, UK, releasing 85,000 tons.

December 1992: A tanker in the Aegean Sea spilled 80,000 tons of crude near the port of La Coruna, Spain.

May 1991: The tanker *ABT Summer* leaked oil after an explosion off Angola, spilling 260,000 tons.

April 1991: The tanker *Haven* spilled more than 50,000 tons of oil off Genoa in Italy.

January 1991: During the first Gulf War, the Iraqi army released about two million liters of crude oil into the Persian Gulf.

March 1989: The *Exxon Valdez* was grounded and spilled 38,800 tons of crude oil into Prince William Sound in Alaska.

August 1983: A fire onboard the tanker *Castillo de Bellver* released over 250,000 tons of oil residue.

1979: The tanker *Atlantic Empress* spilled 160,000 tons of oil off Tobago.

1979: The IXTOC I exploratory well blew out in the Bay of Campeche off Ciudad del Carmen, Mexico. By the time the well was brought under control in 1980, about 600 million liters of oil had spilled into the bay.

1978: The wrecked tanker *Amoco Cadiz* spilled 120,000 tons of crude oil off the Brittany, UK, coast.

March 1967: The Torrey Canyon spilled 119,000 tons of crude off the Isles of Scilly, UK.

Similar to the toxic plumes in air discussed in Chapter 3, the adverse effects from these spills were expansive; they not only contaminated the receiving waters and their aquatic life, but also harmed shorelines where the contaminants washed ashore. Also like the air toxic plumes, the damage done by a particular spill is a function of more than the sheer volume of contaminants released. It is also determined by the toxicity and physicochemical characteristics of the substance that is spilled and the sensitivity of receiving waters, including the depth, flushing rate, and the amount and type of habitats supported by these waters and associated shorelines. Spills also can increase the risk of fires and explosions (see Figure 4.5). Thus, the spill's impact will be affected by existing meteorological conditions, such as wind, temperature, and precipitation; hydrological conditions, such as

FIGURE 4.5. In addition to the ecological damage of oil spills, there is a potential for oil to ignite or explode. The *Mega Borg* tanker spill is shown. Emergency response crews are reducing the danger by spraying down the tanker.
Source and photo credit: U.S. Environmental Protection Agency, http://www.epa.gov/superfund/programs/er/resource/d1_08.htm; accessed April 5, 2005.

depth, tides, and currents; and the damage is also a function of the characteristics of the spilled substance. Crude oil is a complex mixture of hydrocarbons, molecules in which hydrogen atoms are bonded to the carbon atoms. Refining the crude oil changes the molecular structure of the hydrocarbon (e.g., to gasoline mixtures). The compounds with fewer carbons generally make up the gasoline fraction, increased during refining by a process known as cracking, in which long carbon chains are shortened to form smaller hydrocarbon chains. For petroleum hydrocarbons these include (see discussion on fluid properties in the introduction to Part II):

- Density
- Vapor pressure and boiling range
- Viscosity
- Potential to create suspensions, especially to emulsify
- Aqueous solubility (see discussion box on Solubility)

Solubility

One of the first and most important characteristics of a contaminant is its solubility.

The extent to which a substance can be dissolved in water or some other fluid is an important aspect of its ability to move from one place to another, its toxicity to humans and wildlife, and its ability to be removed and treated. The most important solubility consideration is a substance's aqueous solubility; that is, how easily it is dissolved in water. Solubility is most often stated in terms of mass per volume (e.g., $mg L^{-1}$). The amount of a substance that can be held in solution at a given temperature and pressure is its solubility. For example, sucrose is very soluble in water. This means that under environmental conditions, water can hold a large amount of sugar. The sucrose in the solution is the solute and the water is the solvent. Conversely, many pollutants are not very soluble in water (see Table 3.8).

This brings us to the other fluids besides water. Chemical handbooks and manuals often show the solubility of a substance in water, as well as other fluids. These are usually organic solvents; that is, they are liquids under standard temperature and pressure (STP) and contain at least one covalent bond between carbon atoms or between a hydrogen atom and a carbon atom, such as ethanol (C_2H_6) or benzene (C_6H_6). Thus, PCB solubility in water is very low, but in organic solvents it is very high (see Table 3.8).

Actually, for air pollutants and meteorologically important substances (e.g., carbon dioxide and water vapor) solvents can also be gases, such as CO_2 dissolved in the air (mainly N_2 and O_2).

TABLE 3.8
Aqueous solubility and octanol-water coefficients (K_{ow}) for some important organic compounds.

Compound	*Number of Chlorine Atoms*	*Aqueous solubility ($mg\,L^{-1}$)*	*Log K_{ow}*
Benzene	0	1,780	2.13
Hexachlorobenzene	6	0.006	6.18
Phenol	0	82,000	1.45
Pentachlorophenol	5	14	3.7
Biphenyl	0	5.9–7.5	3.89
PCB 209	10	0.000004	8.23
Dibenzo-*p*-dioxin	0	0.842	4.3
2,3,7,8-tetrachlorodibenzo-*p*-dioxin (2,3,7,8-Cl_4DD)	4	0.000008	7
octachlorodibenzo-*p*-dioxin (Cl_8DD)	8	0.0000004	8.2

Source: UNEP, H. Fiedler, Polychlorinated Biphenyls (PCBs): Uses and Environmental Releases in Persistent Organic Pollutants; http://www.chem.unep.ch/pops/POPs_Inc/proceedings/bangkok/FIEDLER1.html; accessed April 22, 2005.

TABLE 4.1
Characteristics of typical residual fuel oils.

Parameter	*Residual Fuel Oils*
Density (@16°C)	0.904–1.02 $g\,cm^3$
API Gravity (@16°C)	7–25
Viscosity (Saybolt Universal sec @38°C)	45–18,000
Flash Point (°C)	66–121
Pour Point (°C) Sulfur Content (% by weight)	–18 or less

Source: National Oceanic and Atmospheric Adminstration.

The emulsification potential can consist of the substance's likelihood to form either an oil-in-water or a water-in-oil emulsion, which is determined by the substance's intrinsic properties and extrinsic circumstances like wind and wave conditions. The solubility of most spilled materials falls within a tight range, with aqueous solubility commonly below $0.1\,g\,L^{-1}$. Table 4.1 provides a few of the other important physicochemical characteristics for residual fuel oils.

Density is an extremely important factor. Since, under environmental conditions, pure water has a density of about $1\,g\,cm^{-1}$, it may be heavier or lighter than the densities of various petroleum hydrocarbon mixtures. In other words, depending on what type of petroleum product is being pumped, stored, or shipped, if it is released it may float on top of the water, sink toward the floor, or be dispersed atop and throughout the water column.

TABLE 4.2
Densities of some important environmental fluids.

Fluid	*Density ($g\,cm^{-3}$) at 20°C unless otherwise noted*
Air at standard temperature and pressure (STP) = 0°C and $101.3\,N\,m^{-2}$	0.000129
Air at 21°C	0.000120
Ammonia	0.602
Diethyl ether	0.740
Ethanol	0.790
Acetone	0.791
Gasoline	0.700
Kerosene	0.820
Turpentine	0.870
Benzene	0.879
Pure water	0.1000
Seawater	0.1025
Carbon disulfide	0.1274
Chloroform	0.1489
Tetrachloromethane (carbon tetrachloride)	0.1595
Lead (Pb)	0.11340
Mercury (Hg)	0.13600

In emergency situations, such as a spill or pipeline break, for example, the first responder must know the density of substances in an emergency situation. If a substance is burning, whether it is of greater or lesser density than water will be one of the factors on how to extinguish the fire. If the substance is less dense than water, the water will likely settle below the layer of water, making water a poor choice for fighting the fire. So, any flammable substance with a density less than water (see Table 4.2), such as benzene or acetone, will require fire-extinguishing substances other than water. For substances heavier than water, like carbon disulfide, water may be a good choice.

Another important comparison is that of pure water and seawater. The density difference between these two water types is important for marine and estuarine ecosystems. Saltwater contains a significantly greater mass of ions than does freshwater (see Table 4.3). The denser saline water can wedge beneath freshwaters and pollute surface waters and groundwater (see Figure II.3). This phenomenon, known as *saltwater intrusion*, can significantly alter an ecosystem's structure and function and threaten freshwater organisms. It can also pose a huge challenge to coastal communities, who depend on aquifers for their water supply. Part of the problem and the solution to the problem can be found in dealing with the density differentials between fresh and saline waters.

TABLE 4.3
Composition of freshwaters (river) and marine waters for some important ions.

Composition	*River Water*	*Saltwater*
pH	6–8	8
Ca^{2+}	4×10^{-5} M	1×10^{-2} M
Cl^{-}	2×10^{-4} M	6×10^{-1} M
HCO_3^{-}	1×10^{-4} M	2×10^{-3} M
K^{+}	6×10^{-5} M	1×10^{-2} M
Mg^{2+}	2×10^{-4} M	5×10^{-2} M
Na^{+}	4×10^{-4} M	5×10^{-1} M
SO_4^{2-}	1×10^{-4} M	3×10^{-2} M

Source: K.A. Hunter, J.P. Kim, and M.R. Reid, 1999, Factors influencing the inorganic speciation of trace metal cations in fresh waters, *Marine Freshwater Research*, vol. 50, pp. 367–372; and R.R. Schwarzenbach, P.M. Gschwend, and D.M. Imboden, 1993, *Environmental Organic Chemistry*, Wiley Interscience, New York, NY.

Torrey Canyon Tanker Spill

The *Torrey Canyon* oil spill is one of the most important spills since it was the first major tanker spill and because it released a large amount (119,000 tons) of crude oil. On March 18, the Liberian oil tanker *Torrey Canyon*, operated by an subsidiary of the Union Oil Company of California, ran aground between the Scilly Isles and the British coast (see Figure 4.6). The accident was precipitated by a navigational error causing the tanker to strike Pollard's Rock in the Seven Stones reef between the Scilly Isles and Land's End, England.

Heroic response efforts using dispersants and all available recovery means available ensued. Since the *Torrey Canyon* was the first major supertanker spill, no comprehensive plans were in place. Attempts to refloat the tanker were unsuccessful. A member of a Dutch salvage team was killed. Attempts to sink the wrecked ship and to combust the slick to reduce the leakage even included bombing by the Royal Navy. Also attempting to burn the slick, the Royal Air Force dropped gasoline and napalm. However, in spite of these efforts, oil slicks drifted in the English Channel, making their way to French and British shorelines. Remnants of damage from the spill remain today, including the lag in recovery in the diversity of bird populations, even after noble efforts to save them (see Figures 4.7 and 4.8).

The accident raised the awareness of Europeans to the risks associated with tankers and was a call to arms for new emergency response and water pollution prevention programs. Most notably, the United Kingdom government immediately formed the Coastguard Agency's Marine Pollution Control Unit (MPCU), to provide a command and control structure for deci-

FIGURE 4.6. *Torrey Canyon* tanker run aground at Scilly Isles and the British coast. *Source and photo credit*: National Oceanic and Atmospheric Administration and United Kingdom National Archives.

sion making and response following a shipping incident that causes, or threatens to cause, pollution in U.K. waters. The spill also subsequently led to a comprehensive National Contingency Plan (NCP) in the United Kingdom and internationally.[10] The NCP categorizes spills into an internationally adopted Tier system:

Tier 1: A small operational spill employing local resources during any cleanup.
Tier 2: A medium-sized spill, requiring regional assistance and resources.
Tier 3: A large spill, requiring national assistance and resources. The National Contingency Plan will be activated. (If the *Torrey Canyon* spill were to have occurred today, it would call for a Tier 3 response.)

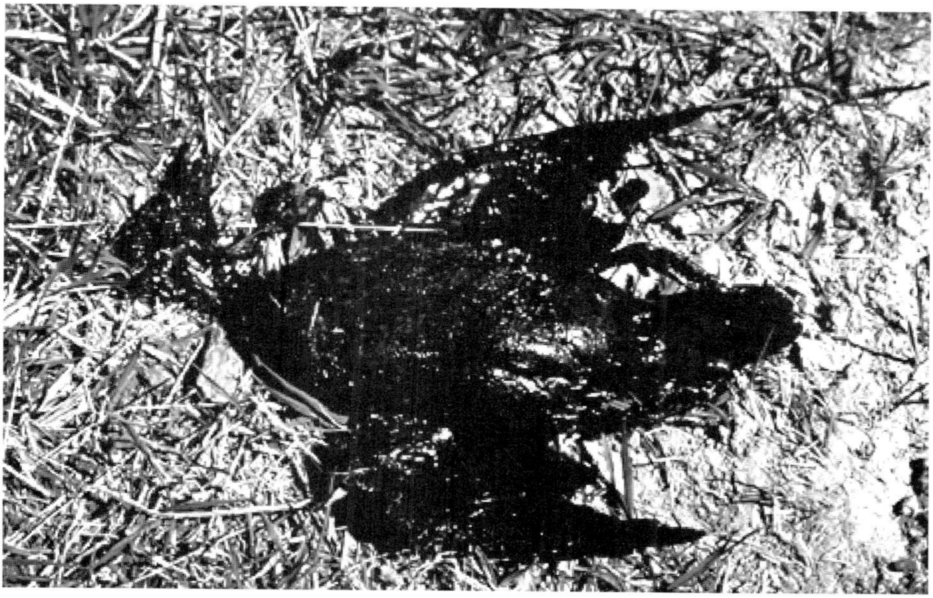

FIGURE 4.7. A bird that is completely covered in oil. Animals are very sensitive to the effects of oil and other lipophilic compounds. Birds die when they lose their insulation as a result of feathers being coated and/or their inability to fly and find food.
Source and photo credit: U.S. Environmental Protection Agency, http://www.epa.gov/superfund/programs/er/resource/d2_20.htm; accessed April 5, 2005.

A new procedure was required, "load on top," which reduced oil losses and helped prevent water pollution. The system collects washings resulting from tank cleaning by pumping them into a special tank. During the voyage back to the loading terminal the oil and water separate. The water at the bottom of the tank is pumped overboard and the terminal oil is pumped on to the oil left in the tank.

The spill led to substantial changes in international shipping conventions, with a number of important amendments to the International Convention for the Prevention of Pollution of the Sea by Oil, 1954 (OILPOL 1954).[11] In 1971, the size of cargo tanks was limited in all tankers ordered after 1972, so that in the event of a future spill, a limited and more manageable amount of oil would be leaked. In 1973, oil and other potentially hazardous cargo shipping rules were expanded and improved. The new provision specified requirements for continuous monitoring of oily water discharges and included the requirement for governments to provide shore reception and treatment facilities at oil terminals and ports. It also established a number of special areas in which more stringent discharge standards were applicable, including the Mediterranean, Red Sea and Gulf, and

FIGURE 4.8. Animal cleaning center set up in spill damaged areas. Following capture, birds and mammals are taken to a rehabilitation facility, checked for hypothermia and dehydration, and prepared for cleaning, including flushing oil from an animal's eyes and intestines, wiping feathers and fur with absorbent cloths to remove patches of oil, and examination to detect injuries. A detergent is used because it has both aqueous and fat solubility, allowing the lipophilic compounds to go into solution and be flushed. When they regain their health, birds are tested for their ability to fly, and mammals for their water-repellency and ability to float. If they meet these standards, they are returned to their natural habitat.
Source and photo credit: U.S. Environmental Protection Agency, http://www.epa.gov/superfund/programs/er/resource/d1_32.htm; accessed April 5, 2005.

Baltic Seas. These special areas would be implemented when the littoral states concerned had provided adequate reception facilities for dirty ballast and other oily residues.

In the United States, the national oil spill response strategy is based on the National Oil and Hazardous Substances Pollution Contingency Plan (National Contingency Plan, or NCP).[12] The first National Contingency Plan was issued in 1968, in part as a response to *Torrey Canyon*. The U.S. NCP provides the institutional framework to:

- Define responsibilities of federal, state and local governments
- Describe resources available for response
- Establish a hierarchy of response teams

- Specify a command structure to oversee spill response
- Require federal, regional, and area contingency plans
- Summarize state and local emergency planning requirements, as well as response priorities, phases, and procedures
- Provide procedures for the use of chemicals (e.g., dispersants, shoreline cleaning agents) in removing spilled hazardous materials

This general framework has been retained and periodically revised for the past 30 years.

Santa Barbara Oil Spill

The year 1969 was truly a "watershed year" (pun intended). The Cuyahoga caught fire; Congress was debating the need for a national environmental policy and passed the National Environmental Policy Act (NEPA); the United Nations Educational, Scientific, and Cultural Organization (UNESCO) held a revolutionary conference, "Man and His Environment: A View Towards Survival" in San Francisco; automobile manufacturers settled a lawsuit with the Justice Department regarding conspiracy to stifle the development of pollution-control devices started in the mid-1950s; and a major oil well off the coast of Santa Barbara, California, blew out, spilling almost a million liters of oil and depositing tar onto approximately 50 kilometers of beach.

On January 29, 1969, a Union Oil Company oil drilling platform (see Figure 4.9) about 10 kilometers off the Santa Barbara coast suffered a blowout. The problem resulted when riggers were pulling up pipe that had been dropped about 1,000 m below the ocean floor in an effort to replace a broken drill bit. The failure occurred because an insufficient amount of driller's mud was available to control pressure, leading to a natural gas blowout. After a successful capping, pressure increased to a point where the expansion of the capped material created five fissures in an ocean floor fault, allowing gas and oil to reach the surface.[13]

The cause of the spill was a rupture in Union Oil's Platform A due to an inadequate protective casing. The U.S. Geological Survey had given approval to operate the platform using casings that did not meet federal and California standards. Investigators would later determine that more steel pipe sheeting inside the drilling hole would have prevented the rupture.

Because the oil rig was beyond California's three-mile coastal zone, the rig did not have to comply with state standards. At the time, California drilling regulations were far more stringent than those required by the federal government.

For 11 days, oil workers struggled to cap the rupture. During that time, 800,000 liters of crude oil surfaced and spread into an 2,000-km^2 slick by winds and the ocean swells. Tides carried the thick tar onto beaches from Rincon Point to Goleta, damaging about 50 km of coastline. The slick also

FIGURE 4.9. Santa Barbara oil platform. Images Courtesy of the University of California—Santa Barbara Map & Image Library.
Photo credit: Santa Barbara Wildlife Care Network, 2005, http://www.sbwcn.org/spill.shtml; accessed April 15, 2005.

moved south, tarring Frenchy's Cove of Anacapa Island and beaches on the Santa Cruz, Santa Rosa, and San Miguel Islands.

The spill caused massive ecological and sea life damage. The thick oil clogged the blowholes of the dolphins, leading to lung hemorrhages. Terrestrial fauna that ingested the oil were acutely poisoned. Gray whales for months after altered their migratory routes to avoid the polluted channel. Shorebirds that feed on sand creatures fled. However, diving birds dependent on nourishment from the ocean water were soaked with tar in the feeding process. Less than 30% of the tarred birds treated by the wildlife response effort survived. Many other birds simply died on the beaches in search of sustenance. Even the cleanup was hazardous to wildlife. For example, detergents used to disperse the oil slick can be toxic to birds, because it removes the natural waterproofing depended on by seabirds to stay afloat. A total of nearly 4,000 birds were estimated to have died because of contact with oil.

The leak was not under control until eleven and a half days after the spill began. Cleanup consisted of pumping chemical mud down the 1,000 m shaft at a rate of 1,500 barrels per hour and capping it with a cement plug. Some gas continued to escape and another leak occurred in the following weeks, and residual oil was released for months after the spill. Skimmers scooped up oil from the surface of the ocean. In the air, planes dumped detergents on the tar-covered ocean in an attempt to break up the slick. On the beaches and harbors, straw was spread on oily patches of water and sand. The straw soaked up the black mess and was then raked up. Rocks were steam-cleaned, cooking marine life like limpets and mussels that attach themselves to coastal rocks.

The next spring, the first Earth Day was celebrated. The Santa Barbara spill was certainly a topic of discussion and a significant impetus to the environmental movement.

Exxon Valdez Spill: Disaster Experienced and Disaster Avoided

The *Exxon Valdez* oil spill has become emblematic of the perils of shipping potentially toxic substances in large quantities over great distances. It also heightened awareness of the delicacy of littoral ecosystems, the vulnerability due to human error and poor judgment, and the need for measures to prevent and respond to accidents.

The *Exxon Valdez* tanker's regular mission was to load oil from the trans-Alaska pipeline from the Valdez terminal and deliver it to West Coast states (see Figure 4.10). In March of 1989, oil was loaded onto the *Exxon Valdez* for shipment to Los Angeles/Long Beach, California. Shortly after leaving the Port of Valdez, on March 24, 1989, the tanker grounded on Bligh Reef, Alaska, releasing more than 40 million liters of oil into Prince William Sound (see Figure 4.11).

The spill is particularly tragic because it was the direct result of human error. At the time that the vessel ran aground on the reef, the Captain of the ship, Joe Hazelwood, was not at the wheel because he was allegedly in his bunk sleeping in an alcoholic stupor. As a result, the third mate was in

FIGURE 4.10. Valdez, Alaska, oil transport terminal.
Photo credit: National Oceanic and Atmospheric Administration.

FIGURE 4.11. The *Exxon Valdez* ran aground on Bligh Reef on March 24, 1989. Photo was taken three days after the vessel grounded, just before a storm arrived. *Photo credit*: National Oceanic and Atmospheric Administration.

charge at the helm. Many other factors contributed to the wreck and the resulting environmental disaster. The tanker had a state-of-the-science radar system, but it was not functioning when the wreck occurred. The company was aware of the broken radar system, but had not replaced it for a year. Had the third mate been able to use this radar system, he should have been able to avoid the reef.

After the ship ran aground, there was little and late response to the resulting oil spill; this was a direct result of the lack of preparedness. The *Exxon Valdez* had loaded its oil at the port in Prince William Sound. According to a letter written to the Exxon executives meeting several months before the spill, there was not enough oil containment equipment, which was required by law, to control a spill that had occurred in the middle of Prince William Sound. Instead, the Exxon executives hoped to disperse the oil and then let the rest drift away. Finally, early warning signs were ignored. Several smaller oil spills occurred in Prince William Sound prior to the *Exxon Valdez* that were not disclosed or were concealed by Exxon employees. There may have even been fraudulent testing, for example, using clean water instead of testing the water from the sound, so that no notice needed to be made to governmental authorities about the lack of containment equipment at the Alaskan port. The tragedy is highlighted in using a decision tree (see Figure 4.12), which illustrates that had the two factors not co-occurred—poor maintenance and an impaired captain—the likelihood of a bad consequence, a massive oil spill, could have been avoided.

The *Exxon Valdez* spill was the sixth largest in terms of volume of contaminants released, but the toll on wildlife from the spill was quite possibly the worst of all oil spills. An estimated 250,000 seabirds, 2,800 sea otters, 300 harbor seals, 250 bald eagles, and as many as 22 killer whales were killed as a direct result of the spill. Large but unknown numbers of important fish species, especially salmon and herring, were lost. And the persistence of the problem is also unprecedented. Ten years after the spill, only two animal species, bald eagles and sea otters, had completely recovered from the spill.

In some ways, the *Valdez* incident showed environmental success, particularly demonstrating that by the end of the 1980s a number of emergency response procedures had advanced, circumventing even worse and more widespread damage. As an example, after the ship ran aground and the leakage of large volumes of oil was apparent, oil was transferred to another tanker, the *Exxon Baton Rouge* (see Figure 4.13). This kept most of the oil originally carried in the Valdez from spilling into Prince William Sound. About 20% of oil carried by the *Valdez* was spilled, but over 160 million liters was transferred to the *Baton Rouge*. Other protective measures included shielding sensitive habitats, such as fish spawning areas, from the encroaching oil slick (see Figure 4.14).

Cleanup operations included skimming oil from the water surface using boats tied to tow booms (see Figure 4.15). Oil is collected within the

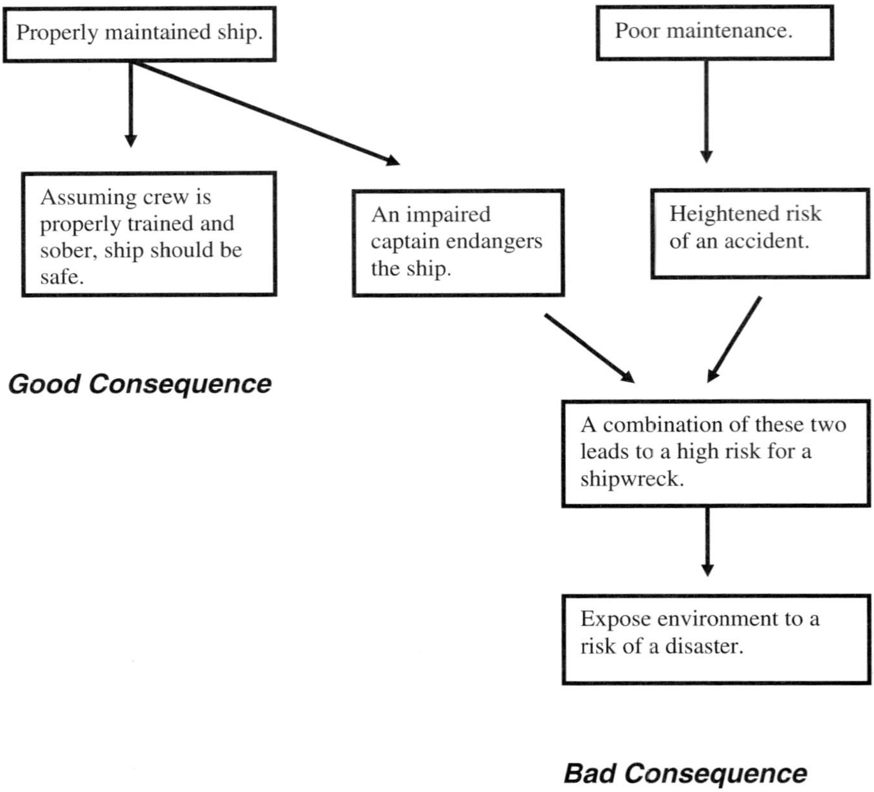

FIGURE 4.12. Decision tree.

boom, while a small skimmer at the apex of the boom is removing the oil from the water surface. The skimmed oil is pumped to the barge behind the skimmer.

Prestige Oil Spill[14]

In 2002, the *Prestige* was a 26-year-old tanker two years from retirement. The ship had a single hull and was such a large risk that no major oil companies would hire her, and she was not allowed near ports in the United States. In 2000, she was dry-docked in China where workers reinforced her hull. Following this work, engineers from the American Bureau of Shipping certified that the vessel was seaworthy.

The *Prestige* left St. Petersburg, Russia, on October 30, 2002, fully loaded with fuel oil, proceeding southerly along the coast toward Spain. Off the Spanish coast, the tanker encountered a storm with large waves. Because such large ships are built to withstand dangerous storms, Captain

FIGURE 4.13. Oil was transferred from the *Exxon Valdez* to the *Exxon Baton Rouge* as one of the emergency response measures during the spill.
Photo credit: National Oceanic and Atmospheric Administration.

FIGURE 4.14. Emergency response following the *Valdez* spill included protection of sensitive habitats using oil barriers, such as the pictured boom floats around a salmon hatchery in the eastern part of Prince William Sound.
Photo credit: National Oceanic and Atmospheric Administration.

FIGURE 4.15. *Valdez* cleanup included skimming the water surface by using boats towing a collection boom. Oil that has been concentrated within the boom is gathered by the skimmer, that is, the vessel at the apex of the boom.
Source and photo credit: National Oceanic and Atmospheric Administration.

Apostolos Mangouras saw no need to take precautions. On the fourth day of the storm, the size of the waves continued to grow, and Captain Mangouras decided to slow the ship and ride out the storm. The waves approached the ship from the right-hand side of the vessel and broke over the deck. One particularly large wave approached the vessel and as it broke over the deck the officers on board heard a large explosion. After the water cleared the deck, the captain noticed that some of the cargo lids had broken loose and some oil was flowing from the tanks into the ocean. In addition, there was a large hole in the starboard side of the ship. Eventually the ship began to list, and the Spanish Coast Guard helped to evacuate all but Captain Mangouras and the chief officer and chief engineer. At this point, the captain flooded the ballast tanks on the side opposite the list in order to level the tanker.

The engines shut off and a Spanish tugboat attempted to fasten a line in order to keep the oil tanker from drifting onto the Spanish coast, but all attempts were unsuccessful. It was rumored that the tugboat took so long to get to the oil tanker because the company that owned the oil tanker was haggling over the price of the assistance from the Spanish tugboat. Therefore, Seragin Diaz Requiero, the head port authority of the Spanish town La Coruna, was airlifted onto the tanker in order to get the engines started. Requiero tried to restart the engines, but according to Requiero most of the problems that prevented the engine from starting were the result of poor maintenance and possibly even sabotage. From his time on the boat, Requiero claims that Captain Mangouras was trying to prevent the engines from being restarted because he wanted to beach the boat on the coast of Spain. If the boat were beached, there would be a large oil spill, but the company that owned the boat would collect insurance money. Eventually Requiero restarted the engines and directed the boat away from the coast. Neither the Spanish nor the Portuguese government would allow the tanker refuge in its ports for fear of a major environmental disaster. Unfortunately the boat broke up and sank several days later. Captain Mangouras was extradited to Spain, charged with sabotage, and placed under a $3.2 million bail.

Lessons Learned: Two-Edged Swords

The lessons from the spills and sudden releases of contaminants to waterways are legion. But two are particularly relevant: economies that depend on crude oil are particularly vulnerable to extensive environmental problems, and environmental policies must be proactive and highly adaptive to looming threats.

Dependence on fossil fuels is the underlying problem. Until we become significantly less dependent, however, aggressive risk reduction and ecological protection steps will need to be pursued.

Pfiesteria piscicida: Nature Out of Sync

A different type of water pollution case occurs when contaminant releases lead to changes in the population dynamics of an ecological system. This occurred recently in North Carolina where fresh and saline waters meet, in the estuaries and sounds between rivers and the Atlantic Ocean.

Pfiesteria piscicida is a toxic dinoflagellate, often classified as algae, which recently has been linked to fish lesions and fish kills in coastal waters from Delaware to North Carolina (see Figure 4.16). Most dinoflagellates, however, are not toxic.

Although many dinoflagellates act like plants, for example, obtaining energy via photosynthesis and processing inorganic nutrients, *Pfiesteria* act more like animals, obtaining some or all of their energy by consuming other organisms. *Pfiesteria piscicida* is now known to have a highly complex life cycle with 24 reported forms, some that can release toxins. Although *Pfiesteria* normally exists in nontoxic forms (see Figure 4.17), feeding on algae and bacteria in the water and in sediments of tidal rivers and estuaries, it can become toxic within fish, especially schooling fish like the Atlantic menhaden (*Brevoortia* tyrannus). The toxicity is initiated by the fish's secretions or excrement in the water. When these substances are sensed by *Pfiesteria*, its cells shift forms to emit a powerful toxin that stuns and makes the fish lethargic. Other toxins may even break down fish skin

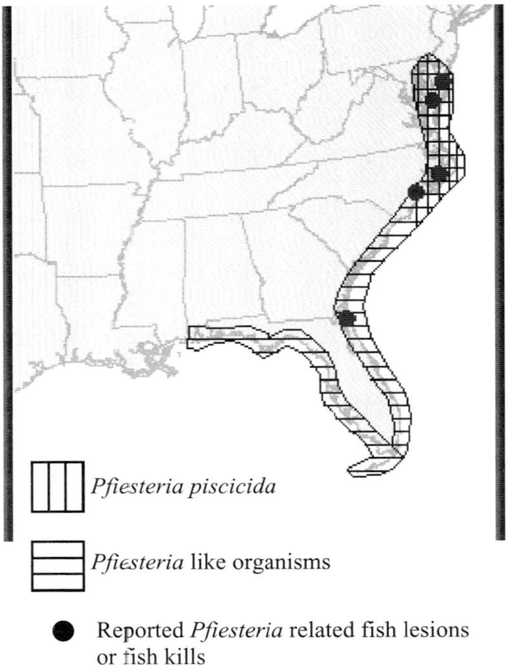

FIGURE 4.16. Range of *Pfiesteria* in the Eastern United States.
Source: U.S. Environmental Protection Agency.

tissue, opening bleeding sores or lesions (Figure 4.18). These chemical assaults are often fatal to the fish. After the fish is stunned, the *Pfiesteria* cells feed on their tissues and blood. Thus, *Pfiesteria* is not an infectious agent, *per se,* so the fish are not "infected." Rather, the fish die from toxins either directly or by secondary infections that attack the fish once the toxins have caused lesions to develop.

Pfiesteria piscicida has been implicated in a number of major fish kills, amounting to millions of fish, along the North Carolina coast, notably the New River and the Albemarle-Pamlico estuarine system, which includes the Neuse and Tar-Pamlico Rivers. In a particularly severe fish kill in 1997, *Pfiesteria* was implicated in killing thousands of fish in several Eastern Shore tributaries of the Chesapeake Bay, including the Chicamacomico and Manokin Rivers and King's Creek in Maryland, and the lower Pocomoke River in Maryland and Virginia. *Pfiesteria piscicida* is also suspected as the cause for a 1987 fish kill in Delaware's Indian River. Fish with lesions similar to those from *Pfiesteria* toxicity have been found in Maryland and Virginia tributaries of the Chesapeake Bay, in many coastal areas of North Carolina, and in the St. John's River in Florida.[15]

Amoeboid form

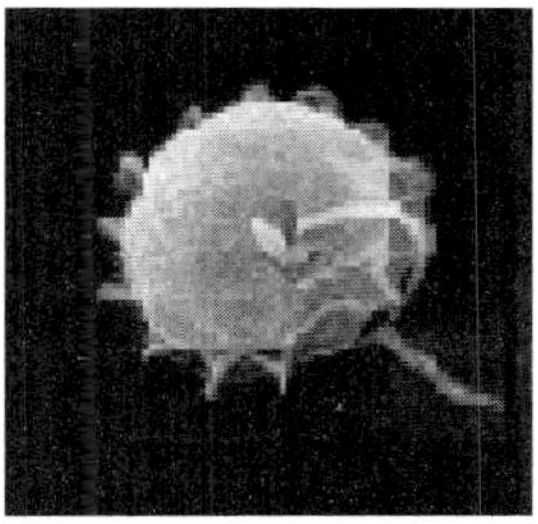

Flagellated form

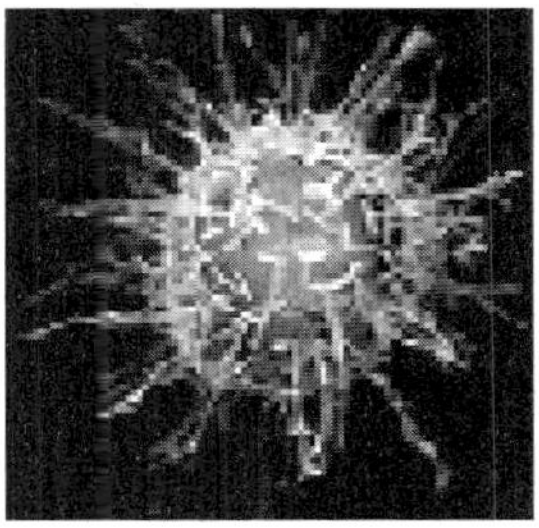

Encysted form

FIGURE 4.17. Three forms of the dinoflagellate *Pfiesteria*.
Source: U.S. Environmental Protection Agency.
Photo credit: North Carolina State University, Aquatic Botany Laboratory.

Lesson Being Learned

The lesson from *Pfiesteria* is a recurring one; that is, there is a delicate balance among physical, chemical, and biological components of an ecosystem. Often, overloading a system, for example, with too much nitrogen and/or phosphorous, can allow opportunistic organisms to grow at the expense of other organisms. This may be the case for *Pfiesteria*, since excess nutrients are common pollutants in the Atlantic coastal waters. Major sources of nutrient pollution include sewage treatment plants, septic tanks,

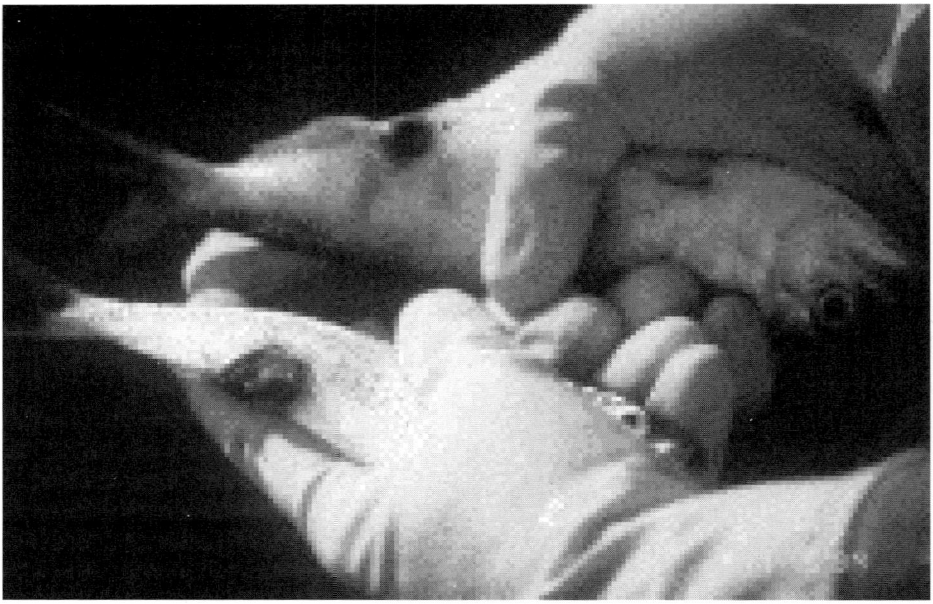

FIGURE 4.18. Fish with lesicns caused by exposure to *Pfiesteria* toxins.
Source: U.S. Environmental Protection Agency.
Photo credit: North Carolina State University, Aquatic Botany Laboratory.

polluted runoff from suburban landscape practices and agricultural operations, combined animal feeding operations (swine and poultry), and air pollutants that settle on the land and water. A systematic approach is needed to study and control the release of these pollutants, a lesson that this dinoflagellate is teaching us the hard way.

Another similar lesscn is that pollutants come in many forms. In this case, the original pollution is chemical (nutrient loads), but the actual agent is biological. So, stressors may be chemical, biological, or physical (e.g., noise, electromagnetic radiation—including ultraviolet light and nuclear radiation—and habitat destruction—the "bulldozer").

Notes and Commentary

1. The principal sources for this discussion are American Planning Association, 1998. "Regional Planning." Chapter 6 in *The Growing Smart*[SM] *Legislative Guidebook, Phases I & II Interim Edition,* APA Press, Chicago, IL. B. Canada, 2003. Congressional Research Service, Library of Congress, Federal Grants to State and Local Governments: A Brief History, CRS Report No. RL30705 (updated February 19, 2003).

2. Water Resources Planning Act of 1965 (as amended), Pub. L. 89–80, 79 Stat. 244, 42 U.S.C. 1962c; Federal Grant and Cooperative Agreement Act of 1977, Pub. L. 95–224, 92 Stat. 3, 41 U.S.C. 501 *et seq.*; E.O. 12044, 43 FR 12660.
3. A. Bettman, 1925. *Planning Problems of Town, City, and Region: Papers and Discussion,* Remington, Baltimore, MD. J. Friedmann, "The Concept of a Planning Region The Evolution of an Idea in the United States," in John Friedmann and William Alonso, eds., 1964. *Regional Development and Planning: A Reader,* MIT Press, Cambridge, MA.
4. D.H. Burnham and E.H. Bennett, 1909. *Plan of Chicago,* DaCapo Press, New York, NY.
5. B.D. McDowell, 1986. "The Evolution of American Planning," *The Practice of State and Regional Panning,* Frank So, Irving Hand, and Bruce D. McDowell, eds., American Planning Association and International City Managers Association, Washington, D.C.
6. "Demonstration Cities and Metropolitan Development Act of 1966," P.L. 89–754, sec. 101; 80 Stat. 1255 and 80 Stat. 1261. "Intergovernmental Cooperation Act of 1968," P.L. 90–577; 96 Stat. 1103.
7. M. Mogulof, 1972. "Metropolitan Councils of Government and the Federal Government," *Urban Affairs Quarterly,* 492.
8. B.D. McDowell, 1995. "Regionalism: What It Is, Where We Are, and Where It May Be Headed," 1995 Annual Conference of the Virginia and National Capital Area Chapters of the American Planning Association, Falls Church, VA, December 4, 1995.
9. Source of this section is the U.S. Environmental Protection Agency's Web site: http://www.epa.gov/glnpo/aoc/index.html.
10. Maritime and Coast Guard Agency (UK), 2005. "Safer Lives, Safer Ships, Cleaner Seas," http://www.mcga.gov.uk/c4mca/mcga-environmental/mcga-dops_cp_environmental-counter-pollution.htm; accessed April 16, 2005.
11. The potential for oil to pollute the marine environment was recognized by the International Convention for the Prevention of Pollution of the Sea by Oil, 1954 (OILPOL 1954). The Conference adopting the Convention was organized by the U.K. government, and the Convention provided for certain functions to be undertaken by IMO when it came into being. In fact, the Convention establishing IMO entered into force in 1958 just a few months before the OILPOL convention entered into force, so IMO effectively managed OILPOL from the start, initially through its Maritime Safety Committee.

    The OILPOL Convention recognized that most oil pollution resulted from routine shipboard operations such as the cleaning of cargo tanks. In the 1950s, the normal practice was simply to wash the tanks out with water and then pump the resulting mixture of oil and water into the sea.
12. U.S. Environmental Protection Agency, 1993. National oil and hazardous substances pollution contingency plan; Final Rule 40 CFR Part 300, Federal Register 59(178):47384–47495.
13. Santa Barbara Wildlife Care Network, 2005. http://www.sbwcn.org/spill.shtml; accessed April 15, 2005.

14. The research for this case was conducted in April of 2005 by Chris Sundberg, a Duke University student, as part of his requirements for EGR 108S, Ethics in Professions, an undergraduate engineering course.
15. The principal source of this discussion is U.S. Environmental Protection Agency, 2005, "Pfiesteria piscicida, Fact Sheet: What You Should Know about *Pfiesteria piscicida*," http://www.epa.gov/owow/estuaries/pfiesteria/fact.html; accessed April 20, 2005.

CHAPTER 5

Landmark Cases

We did not inherit the land from our fathers. We are borrowing it from our children.

Amish Proverb

Significant progress was being made in the fight against air and water pollution up to the mid-1970s. Key pieces of legislation were passed. In the United States and Western Europe, increasingly stringent environmental rules and regulations were put into place. Favorable court rulings supported these protection regulations. Most of these actions were aimed at addressing what we now call traditional pollutants. But waiting to thrust itself into the public arena were the so-called toxic pollutants and hazardous substances. The pollutants so familiar to sanitary engineers and environmental chemists would soon be joined by a litany of problems associated with compounds that were previously unknown or that had not been associated by the scientific community with environmental quality. At the same time, epidemiological studies were improving, especially those linking lifestyles and environmental factors to cancer. Theo Colburn, who is probably best known for her publications on the increasing exposure and effects of environmental endocrine disruptors, sums up the problem of synthetic chemicals that only began to be appreciated in the late 1970s:

> Every one of you sitting here today is carrying at least 500 measurable chemicals in your body that were never in anybody's body before the 1920s . . . We have dusted the globe with man-made chemicals that can undermine the development of the brain and behavior, and the endocrine, immune and reproductive systems, vital systems that assure perpetuity . . . Everyone is exposed. You are not exposed to one chemical at a time, but a complex mixture of chemicals that changes day by day, hour by hour, depending on where you are and the environment you are in . . . In the United States alone it is estimated that over 72,000 different chemicals are used regularly. Two thousand five hundred new chemicals are introduced annually—and of these, only 15 are partially tested

for their safety. Not one of the chemicals in use today has been adequately tested for these intergenerational effects that are initiated in the womb.[1]

A handful of cases exemplified this new era of hazardous wastes. Outrage and frustration over the toxic substances found at Love Canal in New York, Times Beach, Missouri, and the Valley of the Drums in Kentucky led to the passage of numerous environmental laws, especially the Comprehensive Environmental Response, Compensation, and Liability Act, better known as the Superfund, in 1980. These cases also ushered in a litigious approach to solving environmental problems, being among the first hazardous waste cases. Prior to this, much of the legal precedence for environmental jurisprudence had been more on the order of nuisance laws. With the greater recognition of public health risks associated with toxic substances like those found in these hazardous waste cases, the public and environmental professionals called for a more aggressive and scientifically based approach.

Love Canal, New York

The seminal and arguably most infamous case is the contamination in and around Love Canal, New York. The beneficent beginnings of the case belie its infamy. In the nineteenth century, William T. Love saw an opportunity for electricity generation from Niagara Falls and the potential for industrial development. To achieve this, Love planned to build a canal that would also allow ships to pass around the Niagara falls and travel between the two great lakes, Erie and Ontario. The project started in the 1890s, but soon floundered due to inadequate financing and also due to the development of alternating current, which made it unnecessary for industries to locate near a source of power production. Hooker Chemical Company purchased the land adjacent to the canal in the early 1900s and constructed a production facility. In 1942, Hooker Chemical began disposal of its industrial waste in the canal. This was wartime in the United States, and there was little concern for possible environmental consequences. Hooker Chemical (which later became Occidental Chemical Corporation) disposed of over 21,000 tons of chemical wastes including halogenated pesticides, chlorobenzenes, and other hazardous materials into the old Love Canal. The disposal continued until 1952, at which time the company covered the site with soil and deeded it to the City of Niagara Falls, which wanted to use it for a public park. In the transfer of the deed, Hooker specifically stated that the site was used for the burial of hazardous materials, and warned the city that that fact should govern future decisions on the use of the land. Everything Hooker Chemical did during those years was apparently legal and aboveboard.

About this time, the Niagara Falls Board of Education was looking around for a place to construct a new elementary school, and the old Love Canal seemed like a perfect spot. This area was a growing suburb with densely packed single-family residences on streets paralleling the old canal. A school on this site seemed like a perfect solution, and so it was built.

In the 1960s, the first complaints began, and intensified during the early 1970s. The groundwater table rose during those years and brought to the surface some of the buried chemicals. Children in the school playground were seen playing with strange 55-gallon drums that popped out of the ground. The contaminated liquids started to ooze into the basements of the nearby residents, causing odor and health problems. More importantly perhaps, the contaminated liquid was found to have entered the storm sewers and was being discharged upstream of the water intake for the Niagara Falls water treatment plant.

The situation reached a crisis point and President Jimmy Carter declared an environmental emergency in 1978, resulting in the evacuation of 950 families in an area of 10 square blocks around the canal. But the solution presented a difficult engineering problem. Excavating the waste would have been dangerous work and would probably have caused the death of some of the workers. Digging up the waste would also have exposed it to the atmosphere, resulting in uncontrolled toxic air emissions. Finally, there was the question as to what would be done with the waste. Since it was a diverse mixture of contaminants, no single solution such as incineration would have been appropriate. The U.S. EPA finally decided that the only thing to do with this dump was to isolate it and continue to monitor and to treat the groundwater. The contaminated soil on the school site was excavated, detoxified, and stabilized and the building itself was razed. All the sewers were cleaned, removing 62,000 tons of sediment that had to be treated and removed to a remote site. At the present time, the groundwater is still being pumped and treated, thus preventing further contamination.

The cost is staggering, and a final accounting is still not available. Occidental Chemical paid $129 million and continues to pay for oversight and monitoring. The rest of the funds are from the Federal Emergency Management Agency and from the U.S. Army, which was found to have contributed waste to the canal.

The Love Canal story had the effect of galvanizing the American public into understanding the problems of hazardous waste and was the impetus for the passage of several significant pieces of legislation such as the Resource Conservation and Recovery Act; the Comprehensive Environmental Response, Compensation, and Liability Act; and the Toxic Substances Control Act. In particular, a new approach to assessing and addressing these problems has evolved (see the discussion box, "Hazardous Waste Cleanup").

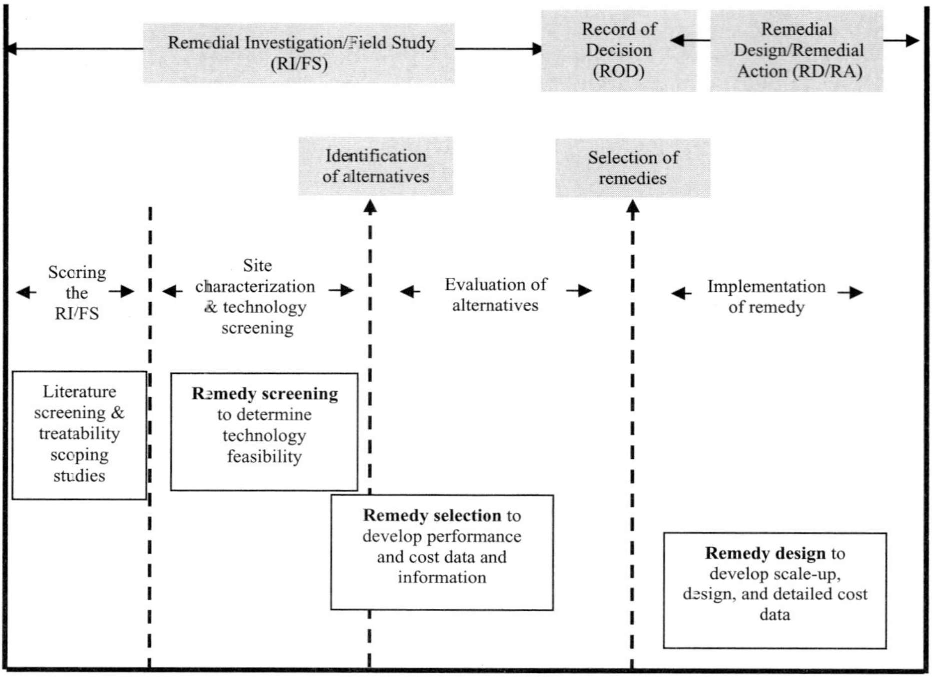

FIGURE 5.1. Steps in a contaminated site cleanup, as mandated by the Superfund. *Source*: U.S. Environmental Protection Agency, 1992. Guide for Conducting Treatability Studies under CERCLA: Thermal Desorption, EPA/540/R-92/074 B.

Hazardous Waste Cleanup

International and domestic agencies have established sets of steps to determine the potential for a release of contaminants and the means for cleaning up a hazardous waste site. In the United States, the steps shown in Figure 5.1 comprise the Superfund Cleanup Process, because they have been developed as regulations under the Comprehensive Environmental Response, Compensation and Liability Act, more popularly known as the Superfund. The first step in this cleanup process is a Preliminary Assessment/Site Inspection (PA/SI), from which the site is ranked in the Agency's Hazard Ranking System (HRS). The HRS is a process that screens the threats of each site to determine if the site should be listed on the National Priority Listing (NPL), which is the list of most serious sites identified for possible long-term cleanup, and what the rank of a listed site should be.

Following the initial investigation, a formal Remedial Investigation/ Feasibility Study (RI/FS) is conducted to assess the nature and the extent of contamination. The next formal step is the Record of Decision (ROD), which describes the various possible alternatives for cleanup to be used at an NPL site. Next, a Remedial Design/Remedial Action (RD/RA) plan is prepared and implemented. The RD/RA specifies which remedies will be undertaken at the site and lays out all plans for meeting cleanup standards for all environmental media. The Construction Completion step identifies the activities that were completed to achieve cleanup. After completion of all actions identified in the RD/RA, a program for Operation and Maintenance (O&M) is carried out to ensure that all actions are as effective as expected and that the measures are operating properly and according to the plan. Finally, after cleanup and demonstrated success, the site may be deleted from the NPL.

Although all sites are unique, a number of steps must be taken for any hazardous waste facility. First, the location of the site and boundaries should be clearly specified, including the formal address and geodetic coordinates. The history of the site, including present and all past owners and operators, should be documented. The search for this background information should include both formal (e.g., public records) and informal documentation (e.g., newspapers and discussions with neighborhood groups[2]).The main or most recent businesses that have operated on the site, as well as any ancillary or previous interests, should be documented and investigated. For example, in the infamous Times Beach, Missouri dioxin contamination incident, the operator's main business was an oiling operation to control dust and to pave roads. Unfortunately, the operator also ran an ancillary waste oil hauling and disposal business. The operator creatively merged these two businesses; that is, spraying waste oil that had been contaminated with dioxins, which led to the widespread problem and numerous Superfund sites in Missouri, including the relocation of the entire town of Times Beach.

The investigation at this point should include *all* past and present owners and operators. Any decisions regarding *de minimus* interests will be made at a later time (by the government agencies and attorneys). At this point, one should be searching for every potentially responsible party (PRP). A particularly important part of this review is to document all sales of the property or any parts of the property. Also, all commercial, manufacturing, and transportation concerns should be known, as these may indicate the types of wastes that have been generated or handled at the site. Even an interest of short duration can be very important, if this interest produced highly persistent and toxic substances that may still be on-site, or that may have migrated off-site. The investigation should also determine whether any attempts were made to dispose of wastes from operations, either on-site or through manifest reports, whether any wastes were shipped off-site. A detailed account should be given of all waste reporting, including air emission and water discharge permits, voluntary audits that include tests like the Toxicity Characteristic Leaching Procedure (TCLP), and these results should be compared to benchmark levels, especially to determine if any of the concentrations of contaminants exceed the U.S. EPA hazardous waste limit (40 CFR 261). For example, the TLCP limit for lead (Pb) is $5\,mg\,L^{-1}$. Any exceedences of this federal limit in the soil or sand on the site must be reported.

Initial monitoring and chemical testing should be conducted to target those contaminants that may have resulted from past activity at the site. A more general surveillance is also needed to identify a broader suite of contaminants. This is particularly important in soil and groundwater, since their rates of migration (Q) are quite slow compared to the rates usually found in air and surface water transport. Thus, the likelihood of finding remnant compounds is greater in soil and groundwater. Also, in addition to parent chemical compounds, chemical degradation products should also be targeted, since decades may have passed since the waste was buried, spilled, or released into the environment.

An important part of the preliminary investigation is the identification of possible exposures, both human and environmental. For example, the investigation should document the proximity of the site to schools, parks, water supplies, residential neighborhoods, shopping areas, and businesses.

One means of efficiently implementing a hazardous waste remedial plan is for the present owners (and past owners, for that matter) to work voluntarily with government health and environmental agencies. States often have voluntary action programs that can be an effective means of expediting the process, which allows companies to participate in, and even lead, the Remedial Investigation and Feasibility Study (RI/FS) consistent with a state-approved work plan (which can be drafted by their consulting engineer).

The feasibility study (FS) delineates potential remedial alternatives, comparing the cost-effectiveness to assess each alternative approach's ability to mitigate potential risks associated with the contamination. The FS also includes a field study to retrieve and chemically analyze (at a state-approved laboratory) water and soil samples from all environmental media on the site. Soil and vadose zone contamination will likely require that test pits be excavated to determine the type and extent of contamination. Samples from the pit are collected for laboratory analysis to determine general chemical composition (e.g., a so-called total analyte list) and TCLP levels (that indicate leaching, i.e., the rate of movement of the contaminants).

An iterative approach may be appropriate as the data are derived. For example, if the results from the screening (e.g., total analytical tests) and the leaching tests indicate the site's main problem is with one or just a few contaminants, then a more focused approach to cleanup may be in order. For example, if the preliminary investigation shows that for most of the site's history a metal foundry was in operation, then the first focus should be on metals. If no other contaminants are identified in the subsequent investigation, a remedial action that best contains metals may be in order. If a clay layer is identified at the site from test pit activities and extends laterally beneath the foundry's more porous overburdened material, the clay layer should be sampled to see if any screening levels have been exceeded. If ground water contamination has not been found beneath the metal-laden material, an appropriate interim action removal may be appropriate, followed by a metal treatment process for any soil or environmental media laden with metal wastes. For example, metal-laden waste has recently been treated by applying a buffered phosphate and stabilizing chemicals to inhibit read (Pb) and other metal leaching and migration.

During and after remediation, water and soil environmental performance standards must be met and confirmed by sampling and analysis; that is, post-stabilization sampling and TCLP analytical methods to assess contaminant leaching (e.g., to ensure that concentrations of heavy metals and organics do not violate the federal standards, that is, Pb concentrations $<5\,\mathrm{mg\,L^{-1}}$). Confirmation samples must be analyzed to verify complete removal of contaminated soil and media in the lateral and vertical extent within the site.

The remediation steps should be clearly delineated in the final plan for remedial action, such as the total surface area of the site to be cleaned up and the total volume of waste to be decontaminated. At a minimum, a Remedial Action is evaluated on the basis of the current and proposed land use around the site; applicable local, state, and federal laws and regulations; and a risk assessment specifically addressing the hazards and possible exposures at or near the site. Any proposed plan should summarize the environmental assessment and the potential risks to public health and the environment posed by the site. The plan should clearly delineate all remedial alternatives that have been considered. It should also include data and information on the background and history of the property, the results of the previous investigations, and the objectives of the remedial actions. Since this is an official document, the state environmental agency must abide by federal and state requirements for public notice, as well as to provide a sufficient public comment period (about 20 days).

The final plan must address all comments. The Final Plan of Remedial Action must clearly designate the selected remedial action, which will include the target cleanup values for the contaminants, as well as all monitoring that will be undertaken during and after the remediation It must include both quantitative (e.g., to mitigate risks posed by metal-laden material with total [Pb] >1,000 $mg\,kg^{-1}$ and TCLP [Pb] ≥5.0 $mg\,L^{-1}$), and qualitative objectives (e.g., control measures and management to ensure limited exposures during cleanup). The plan should also include a discussion on planned and potential uses of the site following remediation (e.g., will it be zoned for industrial use or changed to another land use?). The plan should distinguish between interim and final actions, as well as interim and final cleanup standards. The Proposed Plan and the Final Plan then constitute the Remedial Decision Record.

The ultimate goal of the remediation is to ensure that all hazardous material on the site has either been removed or rendered nonhazardous through treatment and stabilization. The nonhazardous, stabilized material can then be disposed of properly, for example, in a nonhazardous waste landfill.

A Fire That Sparked Controversy: Chester, Pennsylvania

In 1978, a fire in Chester, Pennsylvania, ushered in a new era of emergency response in the United States. Firefighters and other first responders[3] were not ready for what they found—large quantities of illegally stored chemicals, among them highly flammable, extremely toxic, and highly volatile materials. This is a potentially deadly combination of physical and chemical factors. Generally, the higher the substance's vapor pressure at a given temperature, the lower the boiling point. So, compounds with high vapor pressures are classified as *volatile*, meaning they form higher concentrations of vapor above the liquid.

Volatility is an important aspect of measuring toxic compounds as well. Depending on whether they are inorganic (e.g., metals and ions) or organic compounds, various analytical techniques must be used to identify contaminants (see Table 5.1). For example, the presence of organic compounds in water and soil are determined by chromatography; that is, the separation of complex mixtures of compounds that is based upon the compounds' differential affinities for a gas or liquid mobile medium.

TABLE 5.1
Analytical methods used to determine the presence and to quantify concentrations of substances in the environment.

Method	*Description*	*Applications*
A: INORGANIC SPECIES		
1. Atomic Absorption Spectroscopy (AAS)	A solution containing the element to be analyzed is drawn into a flame or placed in a pyrolytic carbon furnace. At about 1,000°C, energy is absorbed at characteristic frequencies. Comparison with standard solutions gives the concentration of the species.	A whole variety of elements can be analyzed; e.g., arsenic, cadmium, chromium, lead, mercury, selenium, vanadium, and zinc. The correct preparation of the sample and correction for background interference is important. The technique is widely used in South Africa and is relatively cheap.
2. Atomic Emission Spectroscopy (AES)	The technique uses similar principles to AAS except a plasma is used to heat the sample to about 2,500°C, and the emission of energy is monitored.	Similar elements can be analyzed as with AAS but usually with greater sensitivity and less interferences. It can also be more readily adapted for sequential analysis of many species in the same sample. The instrument can cost up to 10 times those used for AAS.
3. X-Ray Fluorescence Spectroscopy (XRF)	Measurement of the secondary X-rays or fluorescent radiation emitted after bombarding a target material with primary X-rays gives quantitative information on the composition of the material.	The rapid analysis of elements from sodium to uranium can be done routinely from 0,1% and above. Lighter elements down to boron can be determined with the appropriate attachments. It is widely used for the rapid element analysis of solid wastes; e.g., ash.
4. Ion Chromatography	Separation of positive ions (cations) or negative ions (anions) is achieved by passing a solution through special cation or anion exchange resins. The ions can be detected by conductivity after chemical suppression or by using special eluents and with UV/visible and electrochemical detectors.	The rapid multicomponent analysis of anions, e.g., fluoride, chloride, nitrate, nitrite, and sulphate; or cations, e.g., sodium, potassium, calcium, and magnesium can be achieved. Heavy metal ions such as iron, cobalt, and nickel and their complex ions plus species such as arsenate and fatty acid anions can also be analyzed.

TABLE 5.1
Continued

Method	*Description*	*Applications*
5. Wet Methods	Traditional analytical techniques such as titrimetric, gravimetric, and colorimetric analysis.	Methods exist for the analysis of many elements and their compounds. These methods can be accomplished by relatively unskilled personnel and are well suited to ad-hoc analyses without the need for expensive equipment.
B: ORGANIC SPECIES		
6. Organic Indicator Analysis	**Chemical Oxygen Demand—COD:** total organic carbon is measured by oxidation with chromic acid. **Biological Oxygen Demand—BOD:** total biodegradable carbon and sometimes the oxidizable nitrogen. **Total Organic Carbon—TOC:** total carbon including inorganic carbon dioxide, bicarbonate, and carbonate measured as carbon dioxide. **Dissolved Organic Carbon—DOC:** similar to TOC.	All methods are widely used and give a gross measure of the organic content. The results must be interpreted with caution. The tests should be compared to each other for a better understanding of the nature of the organic content.

	<u>Total Organic Halogen</u>—TOX: all organic halogen* compounds are converted to chloride, bromide, and iodide and analyzed by conventional methods. **<u>Permanganate Value</u>—PV:** similar to COD except that permanganate is used under less rigorous conditions.	
7. Gas Chromatography (GC)	The organic components of a waste are split into their individual components by vaporizing and passing the resulting gas through a column of material which has a different affinity for each compound. They are detected as they come off the column and identified by comparison with a standard or by a mass spectrometer or mass selective detector.	The procedure can be applied only to those species that are volatile. Thousands of compounds can be analyzed including aromatics, solvents, halogenated compounds including PCBs and dioxins, organic acids and bases, and aliphatic compounds.
8. High Performance Liquid Chromatography (HPLC)	The principles are similar to IC and GC with the organic being in the liquid phase and being a neutral species. Detection of the individual components by UV/visible spectroscopy, fluorescence, and electrochemical means.	A technique that lacks the general versatility of GC but is finding increasing application in the analysis of many organic compounds including polycyclic aromatic hydrocarbons and large molecules.

Source: South African Department of Water Affairs and Forestry, 2005. Waste Management Series: http://www.dwaf.gov.za/dir_wqm/docs/Pol_Hazardous.pdf; accessed April 17, 2005.
*Note: "×" is a common symbol for halogens, i.e. TOX = total organic halogens.

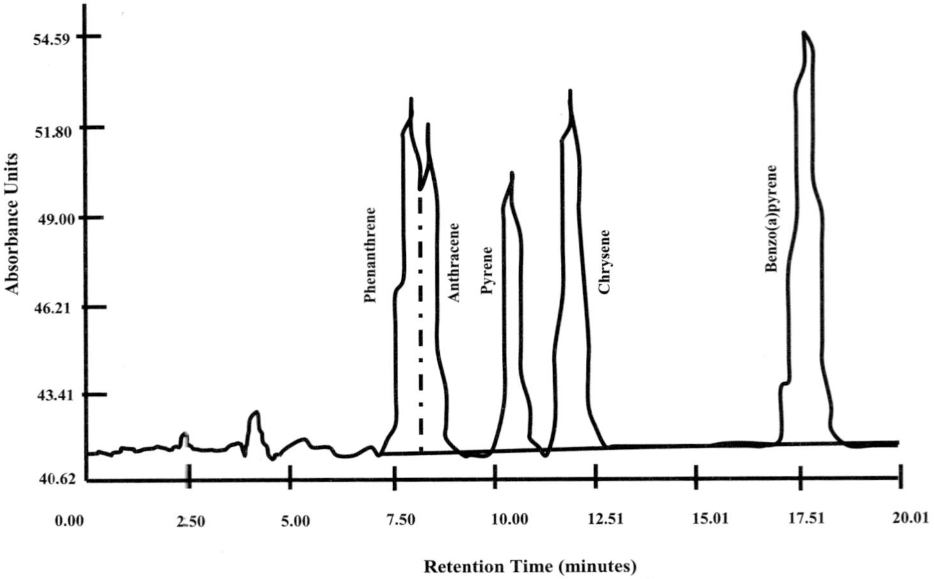

FIGURE 5.2. Chromatogram of five polycyclic aromatic hydrocarbons (PAHs), determined using reverse phase high performance liquid chromatography, a 25-cm × 4.6-mm inside diameter stainless steel column packed with 6-μm DuPont Zorbax ODS sorbant. The mobile phase was 85% acetonitrile and 15% water mixture (by volume). The detection wavelength was in the ultraviolet range (254 nanometers). Analyte injection volume = 10 μL, with a retention time (RT) range of 7 to 18 minutes.
Source: U.S. Occupational Safety and Health Administration, 2005. Sampling and Analytical Methods, Coal Tar Pitch Volatiles (CTPV), Coke Oven Emissions (COE), and Selected Polynuclear Aromatic Hydrocarbons (PAHs), http://www.osha-slc.gov/dts/sltc/methods/index.html; accessed April 17, 2005.

The different times that it takes the compound to separate, the retention time (RT), helps identify the contaminant, which is detected by various methods (see Figure 5.2), such as mass spectroscopy (MS) or flame ionization detection (FID). The lower boiling point compounds, the VOCs, usually come off the column first. That is, as the gas chromatograph's oven increases the temperature of the column, the more volatile compounds leave the column first, and they hit the detector first, so their peaks show up before the less volatile compounds. Other chemical factors such as halogenation and sorption affect this relationship, which are characteristics of the polychlorinated biphenyls (PCBs) that were stored at Chester. So this is not always the case, but as a general rule, a compound's boiling point is a good first indicator of residence time on a column. This means that they are potential air pollutants from storage tanks and such, and it also means that they can present problems to first responders. For example, if a volatile compound is also flammable, as was the case in several of the compounds stored at Chester, the fire and explosion hazard is greater than if the substance were less volatile.

The intense fire destroyed one building and caused extensive damage to two others used for stockpiling drummed wastes. Forty-seven firefighters were hospitalized. In the fire's aftermath, a controversial three-acre site, located on the west bank of the Delaware River in a light industrial area was found. Homes are within 300 meters from the site. From the 1950s to the 1970s the site was used as a rubber recycling facility, and then it was converted to an illegal industrial waste storage and disposal facility until 1978. Numerous 55-gallon drums were stored on the site or their contents were dumped either directly onto the ground or into trenches, severely contaminating soil and groundwater. Wastes included toxic chemicals and PCBs, as well as acids and cyanide salts. Burned building debris, exploded drums, tires, shredded rubber, and contaminated earth littered the property. About 600,000 liters of waste materials remained on-site after the fire. Most of the wastes were in 55-gallon drums stored in the fire-damaged buildings. Because of the dumping of contaminants and the fire, the groundwater and soil were contaminated with heavy metals including arsenic, chromium, mercury, and lead; PCBs; plastic resins; and volatile organic compounds (VOCs) from past disposal activities. In addition to the public health menace, the fire and latent effects of the stored contaminants took place in an ecologically sensitive area, including nearby wetlands and other habitat for wildlife and marine animals. Several cleanup actions were conducted until the site was ultimately removed from the Superfund list of most hazardous sites. Currently, with federal and state approval, the site is undergoing construction to provide parking for the City of Chester's adjacent Barry Bridge Park redevelopment.

Chester demonstrated the importance of physical and chemical properties in environmental problems. For example, the manner in which substances form during combustion is dependent on a number of factors, including temperature, presence of precursor compounds, and the available sites for sorption. For example, complete or efficient combustion (thermal oxidation) of hydrocarbon compounds yields carbon dioxide and water:

$$(CH)_x + O_2 \rightarrow CO_2 + H_2O \tag{5.1}$$

Combustion is actually the combination of O_2 in the presence of heat (as in burning fuel), such as the combustion of octane:

$$C_8H_{18}\ (l) + 17\ O_2\ (g) \rightarrow 8\ CO_2\ (g) + 9\ H_2O\ (g) \tag{5.2}$$

Complete combustion may also result in the production of molecular nitrogen (N_2) when nitrogen-containing organics are burned, such as in the combustion of methylamine:

$$4CH_3NH_2\ (l) + 9\ O_2\ (g) \rightarrow 4\ CO_2\ (g) + 10\ H_2O\ (g) + 2\ N_2\ (g) \qquad (5.3)$$

In addition, fires change the chemical species of stored substances, often making them more hazardous products of incomplete combustion, such as when hydrocarbons are oxidized to form polycyclic aromatic hydrocarbons (PAHs), dioxins, furans, and carbon monoxide (CO).

A particularly toxic chemical class formed from incomplete combustion are the halogenated dioxins. Chlorinated dioxins have 75 different forms and there are 135 different chlorinated furans, depending on the number and arrangement of chlorine atoms on the molecules. The compounds can be separated into groups that have the same number of chlorine atoms attached to the furan or dioxin ring. Each form varies in its chemical, physical, and toxicological characteristics (see Figure 5.3).

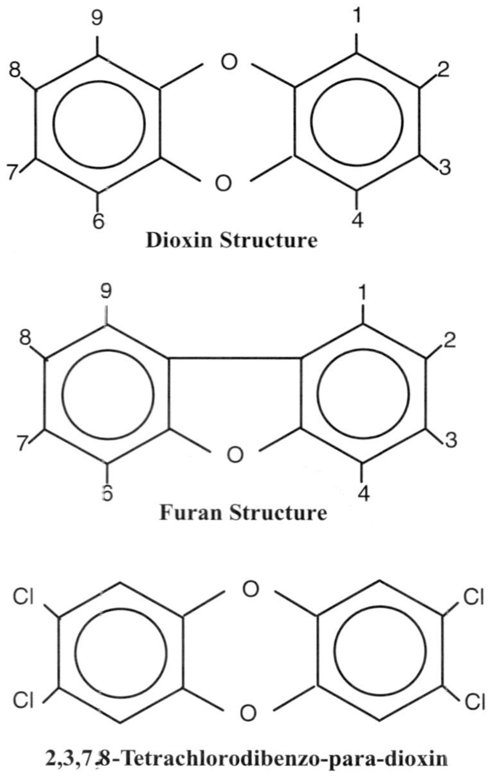

FIGURE 5.3. Molecular structures of dioxins and furans. Bottom structure is of the most toxic dioxin congener, tetrachlorodibenzo-*para*-dioxin (TCDD), formed by the substitution of chlorine for hydrogen atoms at positions 2, 3, 7, and 8 on the molecule.

Dioxins are only created unintentionally during combustion processes; that is, they have never been synthesized for any other reason than for scientific investigation, for example, to make analytical standards for testing. The most toxic form is the 2,3,7,8-tetrachlorodibenzo-p-dioxin (TCDD) isomer. Other isomers with the 2,3,7,8 configuration are also considered to have higher toxicity than the dioxins and furans with different chlorine atom arrangements. Incinerators of chlorinated wastes represent the most common environmental sources of dioxins, accounting for about 95% of the volume. The emission of dioxins and furans from combustion processes may follow three general formation pathways. The first occurs when the fuel material already contains dioxins and/or furans and a fraction of these compounds survives thermal breakdown mechanisms and volatilizes. This is not considered to account for a large volume of dioxin released to the environment, but it may account for the production of dioxin-like, coplanar PCBs.

The second process is the formation of dioxins and furan from the thermal breakdown and molecular rearrangement of precursor compounds, such as the chlorinated benzenes, chlorinated phenols (such as pentachlorophenol, PCP), and PCBs, which are chlorinated aromatic compounds with structural resemblances to the chlorinated dioxin and furan molecules. Dioxins appear to form after the precursor has condensed and adsorbed onto the surface of particles, such as fly ash. This is a heterogeneous process; that is, the reaction occurs in more than one phase (in this case, in the solid and gas phases). The active sorption sites on the particles allow for the chemical reactions, which are catalyzed by the presence of inorganic chloride compounds and ions sorbed to the particle surface. The process occurs within the temperature range of 250°C to 450°C, so most of the dioxin formation under the precursor mechanism occurs away from the high temperature regions of the fire, where the gases and smoke derived from combustion of the organic materials have cooled.

The third way that dioxins and furans form is *de novo*, wherein dioxins are formed from moieties different from those of the molecular structure of dioxins, furans, or precursor compounds. Generally, these can include a wide range of both halogenated compounds like polyvinylchloride (PVC), and nonhalogenated organic compounds like petroleum products, nonchlorinated plastics (polystyrene), cellulose, and lignin (wood), which are common building materials and which were likely present in the Chester fire. Other substances, such as coal and inorganic compounds like particulate carbon and hydrogen chloride gas can provide the necessary chemical compositions for dioxin formation under the right conditions. Whatever *de novo* compounds are involved, however, the process needs a chlorine donor (a molecule that "donates" a chlorine atom to the precursor molecule). This leads to the formation and chlorination of a chemical intermediate that is a precursor. The reaction steps after this precursor is formed can be identical to the precursor mechanism discussed in the previous paragraph.

TABLE 5.2
De novo formation of chlorinated dioxins and furans after heating Mg-Al silicate, 4% Charcoal, 7% Cl, 1% $CuCl_2 \cdot H_2O$ at 300°C.

	Concentrations ($ng g^{-1}$)				
	Reaction Time (hours)				
Compound	*0.25*	*0.5*	*1*	*2*	*4*
Tetrachlorodioxin	2	4	14	30	100
Pentachlorodioxin	110	120	250	490	820
Hexachlorodioxin	730	780	1,600	2,200	3,800
Heptachlorodioxin	1,700	1,840	3,500	4,100	6,300
Octachlorodioxin	800	1,000	2,000	2,250	6,000
Total Chlorinated Dioxins	**3,342**	**3,744**	**7,364**	**9,070**	**17,020**
Tetrachlorofuran	240	280	670	1,170	1,960
Pentachlorofuran	1,360	1,670	3,720	5,550	8,300
Hexachlorofuran	2,500	3,350	6,240	8,900	14,000
Heptachlorofuran	3,000	3,600	5,500	6,700	9,800
Octachlorofuran	1,260	1,450	1,840	1,840	4,330
Total Chlorinated Furans	**8,360**	**10,350**	**17,970**	**24,160**	**38,390**

Source: L. Stieglitz, G. Zwick, J. Beck, H. Bautz, and W. Roth, 1989. *Chemosphere* 19:283.

De novo formation of dioxins and furans may involve even more fundamental substances than those moieties mentioned earlier. For example, dioxins may be generated[4] by heating of carbon particles absorbed with mixtures of magnesium-aluminum silicate complexes when the catalyst copper chloride ($CuCl_2$) is present (see Table 5.2 and Figure 5.4). The *de novo* formation of chlorinated dioxins and furans from the oxidation of carbonaceous particles seems to occur at around 300°C. Other chlorinated benzenes, chlorinated biphenyls, and chlorinated naphthalene compounds are also generated by this type of mechanism.

The presence of dioxins must always be considered during and after a fire. Also, since dioxin and dioxin-like compounds are lipophilic (fat soluble) and persistent, they accumulate in soils, sediments, and organic matter and can persist in solid and hazardous waste disposal sites.[5] These compounds are semi-volatile, so they may migrate away from these sites and be transported in the atmosphere either as aerosols (solid and liquid phase) or as gases (the portion of the compound that volatilizes). Therefore, the environmental professional must take great care in removal and remediation efforts not to unwittingly cause releases from soil and sediments via volatilization or via perturbations, such as an environmental cleanup after a fire, a leaking tank, or drum removal operations.

It is worth noting that 75% of Chester's 36,000 residents are African American, and 95% of residents in neighborhoods closest to the facilities

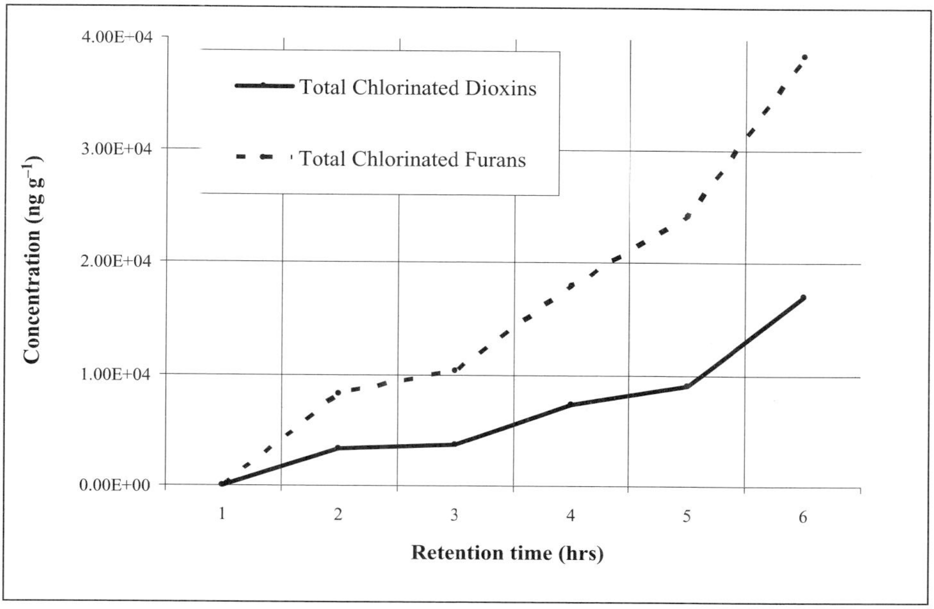

FIGURE 5.4. *De novo* formation of chlorinated dioxins and furans after heating Mg-Al silicate, 4% Charcoal, 7% Cl, 1% $CuCl_2 \cdot H_2O$ at 300°C.
Source: L. Stieglitz, G. Zwick, J. Beck, H. Bautz, and W. Roth, 1989. *Chemosphere* 19:283.

are African American. The poverty rate is 27.2%, three times the national average.[6] These characteristics cause the Chester situation to be one of the first environmental justice cases (see Chapter 11).

Dioxin Contamination of Times Beach

Times Beach was a popular resort community along the Meramec River, about 17 miles west of St. Louis. With few publicly available resources, the roads in the town were not paved and dust on the roads was controlled by spraying oil. For two years, 1972 and 1973, the contract for the road spraying went to a waste oil hauler named Russell Bliss. The roads were paved in 1973 and the spraying ceased.

Bliss obtained his waste oil from the Northeastern Pharmaceutical and Chemical Company in Verona, Missouri, which manufactured hexachlorophene, a bactericidal chemical. In the production of hexachlorophene, considerable quantities of dioxin-laden waste had to be removed and disposed of. A significant amount of the dioxin was contained in the "still bottoms" of chemical reactors, and the company found that having it burned in a chemical incinerator was expensive. The company was taken over by Syntex Agribusiness in 1972, and the new company decided to contract with Russell Bliss to haul away the still bottom waste without telling Bliss what was in the oily substance. Bliss mixed it with other waste oils and this is what he used to oil the roads in Times Beach, unaware that the oil contained a high concentration of dioxin (greater than 2,000 ppm), including the most toxic congener, 2,3,7,8-dibenzo-*para*-dioxin (TCDD).

Bliss also had an oil spraying business where he oiled roads and sprayed oil to control dust, especially in horse arenas. He used the dioxin-laden oil to spray the roads and horse runs in nearby farms. In fact, it was the death of horses at these farms that first alerted the Center for Disease Control to sample the soil at the farms. They found the dioxin, but did not make the connection with Bliss. Finally in 1979, the U.S. EPA became aware of the problem when a former employee of the company told them about the sloppy practices in handling the dioxin-laden waste. The EPA converged on Times Beach in "moonsuits" and panic set in among the populace. The situation was not helped by the message from the EPA to the residents of the town. "If you are in town it is advisable for you to leave and if you are out of town do not go back." In February 1983, on the basis of an advisory from the Centers for Disease Control, the EPA permanently relocated all the residents and businesses at a cost of $33 million. Times Beach was by no means the only problem stemming from the contaminated waste oil. Twenty-seven other sites in Missouri were also contaminated with dioxins.

The concern with dioxin, however, may have been overstated. As a previous accident in Sevesco, Italy, had shown, dioxin is not nearly as acutely toxic to humans as originally feared, causing some to conclude that it is unlikely that the damage to human health in Times Beach was anywhere near the catastrophe originally anticipated. Even some EPA officials later admitted that the evacuation and bulldozing of the community was probably unnecessary. But given the knowledge of dioxin toxicity in 1979, the decision to detoxify the site was not unreasonable. In addition, the carcinogenicity of TCDD was later better established and found to be very high (slope factors $>10^5$ for inhalation, ingestion, and dermal routes).

After everyone had moved out of Times Beach, the houses were razed. The Superfund site eventually was decontaminated at a cost of over $200 million. Cleaning the Times Beach Superfund site was the result of an enormous effort, including the installation of a temporary incinerator to burn the contaminated soil, and the erection of a 15-foot-high barrier around the incinerator to protect it from regular flooding by the Meramec River. A consent decree between the United States, the State of Missouri, and Syntex Agribusiness, the company that assumed responsibility of the site's cleanup, required the implementation of the EPA Record of Decision, which was issued September 29, 1988. This decision called for incineration at Times Beach of dioxin-contaminated soils from 28 sites, including Times Beach, in eastern Missouri. Times Beach was a ghost town since 1983, when it was purchased by the State of Missouri, using Superfund monies. By the end of 1997, cleanup of the site was completed by the EPA and Syntex Agribusiness. More than 265,000 tons of dioxin-contaminated soil from the site and 27 nearby areas had been cleaned. The EPA and the State of Missouri worked closely with Syntex during cleanup to ensure that the restoration made the site suitable for productive use. In 1999, a new 500-acre state park commemorating the famous Route 66 opened on what was once one of the most recognized sites in the country. Thousands of visitors now enjoy the scenic riverside area in Missouri once known as Times Beach.

A Terrifying Discovery: Valley of the Drums

Two of the most important and policy-changing events occurred in the year 1967: the *Torrey Canyon* supertanker oil spill (see Chapter 4) and the identification of an uncontrolled hazardous waste dump by the Kentucky Department of Natural Resources and Environmental Protection (KDNREP)

FIGURE 5.5. The Superfund site known as the Valley of the Drums was among one of the earliest and most severe hazardous waste sites in terms of sheer quantity of illegally disposed contaminants. Discovery of this site helped to motivate the U.S. Congress to pass the Superfund law.
Photo credit: U.S. Environmental Protection Agency, http://www.epa.gov/superfund/programs/er/resource/d1_06.htm; accessed April 5, 2005.

that came to be known as the infamous Valley of the Drums site. The 13-acre property, located in Northern Bullitt County, Kentucky, had over 100,000 drums of waste illegally delivered to the site. About a fourth of the drums were buried and the rest were directly discharging their hazardous contents into pits and trenches. The hydrology of the site allowed for the wastes to move into a nearby creek by storm water runoff. This situation led to a large slug of contaminants reaching the creek in 1979 during a snow melt, precipitating an emergency response action by the U.S. EPA. The subsequent EPA sampling and analysis of the soil and water indicated the presence of elevated concentrations of heavy metals, polychlorinated biphenyls (PCBs), and 140 other chemical contaminants. The EPA required remediation of the site in 1986 and 1987 to reduce exposures and to stem subsequent pollution to the creek and the surrounding environment.

Stringfellow Acid Pits

In southern California, near Glen Avon, five miles northwest of Riverside, the Stringfellow Quarry Company managed a state approved hazardous waste disposal facility during 1956 and 1972 (see Figure 5.6). The Stringfellow Quarry Company disposed of 120 million liters of industrial wastes into an unlined evaporation pond. The contaminants came from production of metal finishing, electroplating, and formulation of the pesticide DDT (1-chloro-4-[2,2,2-trichloro-1-(4-chlorophenyl)-ethyl]-benzene). Due to the pond being unlined, the waste leached into the underlying groundwater table and developed a contaminated plume two miles downstream. The Stringfellow Quarry Company voluntarily closed the site, and the California Regional Water Quality Control Board declared the property a problem area. A policy was adopted to contain waste and keep any further migration of waste to a minimum. Between 1975 and 1980, 2 million liters of liquid waste and DDT contaminants were recovered. In 1980, the EPA recovered another 40 million liters of hazardous waste from the groundwater. The disposal facility was claimed as California's worst environmental hazard in 1983. Since 1983, the EPA required further concentrated effort to be taken at the site on four different occasions. About three quarters of a billion dollars has been spent for remedial action to date.

The groundwater at the site was found to contain various VOCs and heavy metals, including cadmium, nickel, chromium, and manganese. The soil had been contaminated with pesticides, PCBs, sulfates, and heavy metals, all putting the nearby populations at risk, so steps were immediately taken to reduce exposure. For example, the original disposal area is covered by a clay cap, fenced, and guarded by security services. No one currently is drinking water affected by the contaminant plume.

The earliest response actions at the site, taken between 1980 and 1984, included the installation of three groundwater extraction wells; a subsur-

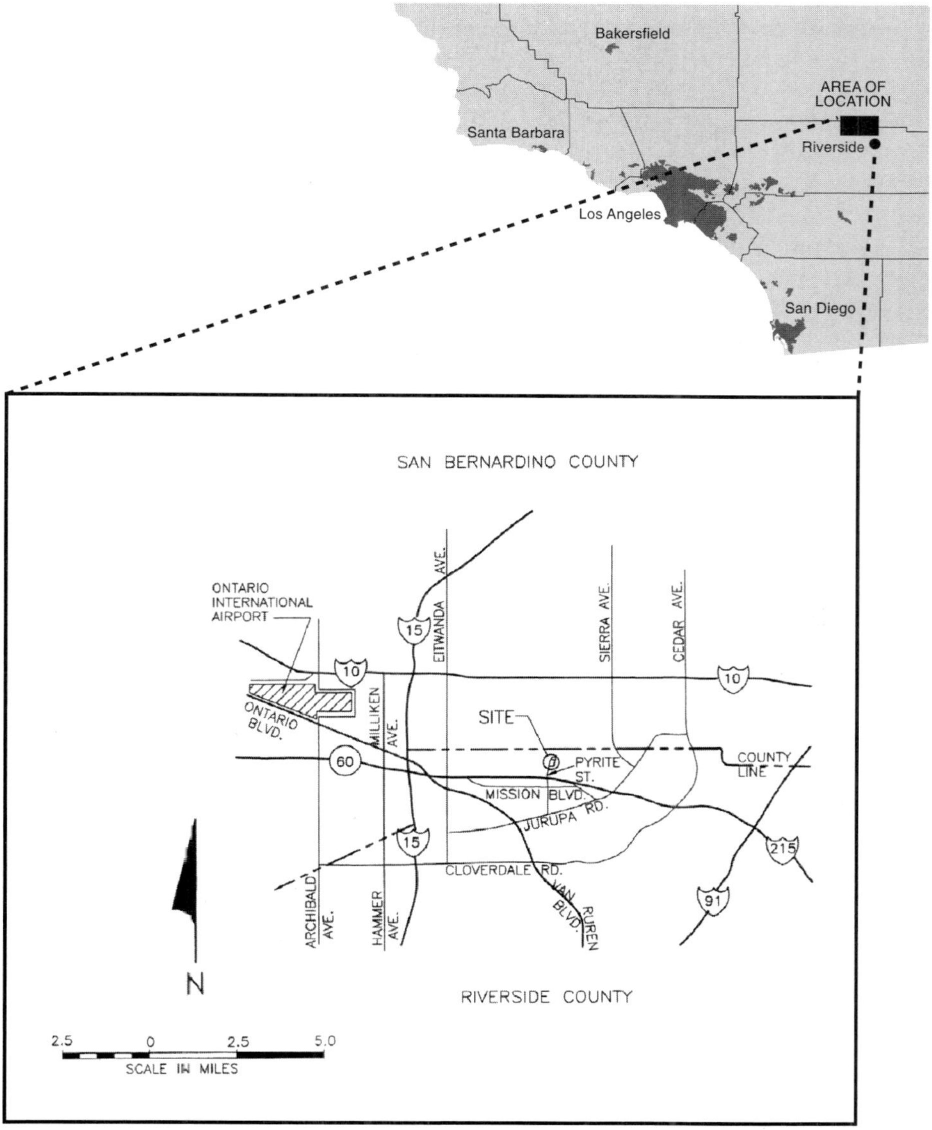

Source: Modified from Environ, 2000

FIGURE 5.6. Site map of the Stringfellow site, Glen Avon, California.
Source: U.S. Environmental Protection Agency, 2001. Five-Year Review for Stringfellow Hazardous Waste Site, Glen Avon, September 27, 2001. Washington, D.C.

FIGURE 5.7. Pretreatment system at Stringfellow site.
Photo credit: U.S. Environmental Protection Agency.

face barrier structure and an on-site surface water drainage system with gunite[7] channels were built. All liquid wastes at the surface of the site were removed to a federally approved hazardous waste disposal facility. With the exception of 1,000 cubic meters of DDT-contaminated soil, which were taken to a federally approved facility, contaminated soils from the site were used to fill waste ponds. The surface was graded, covered with clean soil, and seeded. In 1984, the State of California completed initial cleanup measures including fencing the site, maintaining the existing soil cap, controlling erosion, and disposing of the leachate extracted above and below the on-site clay barrier dam. In 1989, residences that had been receiving bottled water from the state were connected to the Jurupa Community Services District.

Numerous actions have been put in place to remedy the site. In 1984, the EPA selected a remedy for interim treatment of contaminated groundwater. The remedy featured installing a pretreatment system (see Figure 5.7) consisting of lime ($CaCO_3$) precipitation for removing heavy metals and granular activated carbon treatment for removing VOCs. The treated groundwater is discharged to an industrial sewer line, which ultimately discharges to a publicly owned treatment works system. Additional interceptor and monitoring wells were installed to extract contaminated groundwater downgradient of the site. The state completed installation of the pretreatment plant in 1985. As of March 1996, nearly 485 million liters of groundwater had been extracted from the aquifer and treated (known as pump-and-treat). This treatment system will operate until established cleanup levels have been met.

In 1987, the EPA selected a remedy to 1) capture and treat groundwater in the lower canyon area of the site with a groundwater barrier system, 2) install a peripheral channel to divert clean surface water runoff from upgradient areas, 3) extend the existing gunite channels southward to discharge surface water into Pyrite Creek, and 4) reconstruct the Pyrite

Creek Channel. The potentially responsible parties (PRPs) installed the groundwater barrier system and reconstructed the Pyrite Creek Channel. The State of California designed the system and completed construction of the northern channels in 1990. A groundwater extraction system was installed in the community to treat contaminated groundwater that migrates downgradient to the area, possibly followed by reinjection of the treated water. The PRPs have installed an initial community wells extraction system, in an attempt to control the hydraulic conditions of the plume of contaminated groundwater. Further work was begun in September 1997 to install an additional extraction well in order to put the remaining portions of the plume under hydraulic control. In addition, remediation included dewatering of the on-site groundwater—a more aggressive effort to remove water from the water table as an interim source control measure.

Overall, the liquid waste removal, the connection of affected residences to an alternate water supply, and the installation of a ground water capture and treatment system have reduced the potential for exposure to contaminated materials at the Stringfellow site, and the remaining cleanup activities continue to this day.

The March Continues

There are hundreds of hazardous waste sites that could be described here. It suffices to say that the aggregate effect of the sites being found and needing cleanup throughout the world brought about a new paradigm in environmental protection. A few lessons are worth discussing.

Lessons Learned

Failure to Grasp the Land Ethic

In addition to the public health and environmental lessons discussed in Chapter 3, the collapse of the World Trade Center towers illustrates that failure is often a complex mixture of physical principles and human actions or inactions. For example, it is tempting to think that land is a "blank slate" or that buildings are simply three-dimensional structures ready to be built, changed, or demolished as a means to engineering ends. In fact, land and structures are human enterprises that will affect people's lives directly. Another notorious example of an engineering failure as a result of the designers' lack of skills in social sciences is the Pruitt-Igoe public housing project.

The reality of life is that we all fail. What we hope and prepare ourselves for is to design in enough factors of safety to compensate for uncertainty and even remotely possible events. We also hope to learn from failures and successes; those of others (textbook cases and those shared with us by our mentors) as well as our own. Minoru Yamasaki, by most accounts, was a highly successful designer and a prominent figure in the modernist architectural movement of the mid-twentieth century. Tragically and ironically, Yamasaki may be best remembered for two of his projects that failed. Yamasaki and Antonio Brittiochi designed the World Trade Center towers that were to become emblems of Western capitalism. Could it be that Yamasaki's bold modern gothic architecture was an unforeseen invitation to terrorists who despise all things modern?

There is a difference between failure and blame. Failure is simply a statement that the outcome has not, in some manner, met our expectations. We expect buildings to stand. Certainly, Yamasaki cannot be blamed, but the towers failed. In fact, the failure of architects for buildings is seldom structural and often aesthetic or operational (e.g., ugly or an inefficient flow of people). Yamasaki strived to present an aesthetically pleasing structure. One may argue that his architectural success in creating a structure so representative of contemporary America was a factor in its failure, making it a prime target of terrorists. By extension, are we asking architects and engineers to make their works unappealing in appeasement to these radicals? Engineers are blamed when a building fails for structural, environmental, or operational reasons. The lead structural engineer for the WTC towers, Leslie Robertson, has written a gut-wrenching account of what went into the design and how well the buildings met design criteria. He attempted to describe what was known and should have been known and whether this knowledge was incorporated into the design. The best way to express Robertson's anguish in trying to uncover what, if anything could have been foreseen by the engineers to lessen the impact of 9/11 is to quote the last paragraph of his article:

> . . . the events of September 11 have profoundly affected the lives of countless millions of people. To the extent that the structural design of the World Trade Center contributed to the loss of life, the responsibility must surely rest with me. At the same time, the fact that the structures stood long enough for tens of thousands to escape is a tribute to the many talented men and women who spent endless hours toiling over the design and construction of the project . . . making us very proud of our profession. Surely, we have all learned the most important lesson—that the sanctity of human life rises far above all other values.[8]

The difference between now and September 10, 2001, is that we now know better what we are up against. Most assessments have agreed that the structural integrity of the towers was sufficient well beyond the expected contingencies. However, if engineers and planners do not learn the lessons from our failure, they can rightfully be blamed. And the failure will be less a failure of applying physical sciences (withstanding unforeseen stresses and strains) than a failure of imagination. Engineers have been trained to use imagination to envision a better way. Unfortunately, now we must imagine things that were unthinkable before September 11, 2001. Success depends on engaging the social sciences in the planning, design, construction, and maintenance of our projects. This will help to inform us of contingencies not apparent when exclusively applying the physical and natural sciences.

Another of Yamasaki's projects failed in a very different way. The Pruitt Igoe housing development in St. Louis, Missouri, was another modernist monument, but rather than a monument to capitalism, it was supposed to be emblematic of advances in fair housing and progress in the war on poverty. Regrettably, Pruitt Igoe has become an icon of failure of imagination, especially imagination that properly accounts for the human condition. Although we think of public housing projects in terms of housing, they often represent elements of environmental justice. Contemporary understanding of environmental quality is often associated with physical, chemical, and biological contaminants, but in the formative years of the environmental movement, aesthetics and other quality of life considerations were essential parts of environmental quality. Most environmental impact statements addressed cultural and social factors in determining whether a federal project would have a significant effect on the environment. These included historic preservation, economics, psychology (e.g., open space, green areas, and crowding), aesthetics, urban renewal, and the so-called land ethic. Aldo Leopold, in his famous essays, posthumously published as *A Sand County Almanac,* argued for a holistic approach:

> A thing is right when it tends to preserve the integrity, stability and beauty of the biotic community. It is wrong when it tends otherwise.[9]

The land ethic was articulated about a decade after the Pruitt Igoe project was built, so the designers did not benefit from the insights of Leopold and his contemporaries. However, the problems that led to the premature demolition of this costly housing experiment may have been anticipated intuitively if the designers had taken the time to understand what people expected. Then we must ask, who was to blame? There is plenty of culpability to go around. Some blame the inability of the modern architectural style to create livable environments for people living in poverty, largely because as Elizabeth Birmingham has quipped, they "are not the nuanced and sophisticated 'readers' of architectural space the educated architects were."[10] This is a telling observation and an important lesson for engineers. We need to make sure that the use and operation of whatever is designed is sufficiently understood by those living with it.

Other sources of failure have been proposed. Design incompatibility was almost inevitable for high-rise buildings and families with children. However, most large cities have large populations of families with children living in such environments. In fact, St. Louis had successful luxury townhomes not too far from Pruitt Igoe. Another identified culprit was the generalized discrimination and segregation of the era. Actually, when originally inhabited, the Pruitt section was for blacks and Igoe was for whites.

Costs always become a factor. The building contractors' bids were increased to a level where the project construction costs in St. Louis exceeded the national average by 60%. The response to the local housing authority's refusal to raise unit cost ceilings to accommodate the elevated bids was to reduce room sizes, eliminate amenities, and raise densities.[11] As originally designed, the buildings were to become "vertical neighborhoods" with nearby playgrounds, open-air hallways, porches, laundries, and storage areas. The compromises eliminated these features. And some of the removal of amenities led to dangerous situations. Elevators were undersized and stopped only every third floor and lighting was inadequate in the stairwells. So, another lesson must be to know the difference between desirable and essential design elements. No self-respecting structural engineer involved in the building design would have short-cut the factors of safety built into load bearing. Conversely, human elements essential to a vibrant community were eliminated without much if any accommodation.[12]

Finally, the project was mismatched to the people who would live there. Many came from single-family residences. They were moved to a very large, imposing project with 2,800 units and almost 11,000 people living there. This was quadruple the size of the next largest project of the time.

When the failure of the project became overwhelmingly clear, the only reasonable decision was to demolish it, and this spectacular implosion became a lesson in failure for planners, architects, and engineers. In Yamasaki's own words,

> I never thought people were that destructive. As an architect, I doubt if I would think about it now. I suppose we should have quit the job. It's a job I wish I hadn't done.[13]

Engineering is not only applied natural sciences, but many engineers, especially when they advance to leadership positions in engineering, find themselves in professional situations where the social sciences, particularly ethics, would be the more valuable set of skills that would dictate their success as engineers. Teaching our students first to recognize and then to think through ethical problems is like providing a laboratory to see if the ethical mechanism is properly "designed" into our planning. We often overlook "teachable moments." For example, we repeatedly miss opportunities to relate engineering and social science lessons from even the most life- and society-changing events like the fall of the World Trade Center towers.[14]

Thinking of engineering as "applied social science" redefines engineering from a profession that builds things to a profession that helps people. The extension of this conclusion should encourage educators to reevaluate what it is we teach our engineering students. We believe that all engineers should include in their educational quiver at least some arrows that will help them make the difficult ethical and social decisions faced by all professional engineers.

More than any other environmental problems, hazardous waste cases forced engineers and scientists to consider the physicochemical characteristics of the pollutants and match these with the biogeochemical characteristics of the media where these pollutants are found. We had to increase our understanding of the myriad ways that these characteristics would influence the time that these chemicals would remain in the environment, their likelihood to be accumulated in the food chain, and how toxic they would be to humans and other organisms. Those contaminants that have all three of these characteristics worry us the most. In fact, such contaminants have come to be known as PBTs—persistent, bioaccumulating toxicants.

Love Canal, Times Beach, Valley of the Drums, and the many hazardous waste sites that followed them pushed regulators to approach pollutants from the perspective of risk. The principal value added by environmental engineers and other environmental professionals is what we are able to do to improve the quality of human health and ecosystems. The amount of risk is one of the best ways to measure the success of engineers whose projects address environmental injustices. By extension, reliability lets us know how well we are preventing pollution, reducing exposures to pollutants, protecting ecosystems, and even protecting the public welfare (e.g., buildings exposed to low pH precipitation).

Risk, as it is generally understood, is the chance that some unwelcome event will occur. The operation of an automobile, for example, introduces the driver and passengers to the risk of a crash that can cause damage, injuries, and even death. The hazardous waste cases emphasized the need to somehow quantify and manage risks. The understanding of the factors that lead to a risk is known as *risk analysis*. The reduction of this risk (for example, by wearing seat belts in the driving example) is *risk management*. Risk management is often differentiated from *risk assessment*, which is comprised of the scientific considerations of a risk. Risk management includes the policies, laws, and other societal aspects of risk.

Environmental practitioners must consider the interrelationships among factors that put people at risk, suggesting that we are risk analysts. Environmental practitioners provide decision makers with thoughtful studies based upon the sound application of the physical sciences and, therefore, are risk assessors by nature. Engineers control things and, as such, are risk managers. Engineers are held responsible for designing safe products and processes, and the public holds us accountable for its health, safety, and welfare. The public expects environmental practitioners to "give results, not excuses,"[15] and risk and reliability are accountability measures of their success. Engineers design systems to reduce risk and look for ways to enhance the reliability of these systems. Thus, every environmental practitioner deals directly or indirectly with risk and reliability.

Both risk and reliability are probabilities (see discussion of probability in Appendix 3). People living near hazardous waste sites, at least intuitively, assessed the risks and, when presented with solutions by engineers, made decisions about the reliability of the designs. They, for good reason, wanted to be assured that they would be "safe." But safety is a relative term. Calling something safe integrates a value judgment that is invariably accompanied by uncertainties. The safety of a product or process can be described in objective and quantitative terms. Factors of safety are a part of every design. Most of the time, environmental safety is expressed by its opposite term, risk.

Success or failure as environmental practitioners is in large measure determined by what we do compared to what our profession expects us to do. Safety is a fundamental facet of our duties. Thus, we need a set of criteria that tells us when our designs and projects are sufficiently safe. Four safety criteria are applied to test engineering safety:[16]

1. The design must comply with applicable laws.
2. The design must adhere to "acceptable engineering practice."
3. Alternative designs must be sought to see if there are safer practices.
4. Possible misuse of the product or process must be foreseen.

All four of these provisions have been dramatically changed as efforts have been made to deal with hazardous waste incidents. The first two criteria are easier to follow than the third and forth. The well-trained environmental practitioner can look up the physical, chemical, and biological factors to calculate tolerances and factors of safety for specific designs. Laws and regulations are promulgated to protect the public. Love Canal was the key that led to the passage of groundbreaking environmental legislation. Congress passed the Resource Conservation and Recovery Act of 1976 to ensure that wastes are tracked from "cradle to grave" for active hazardous waste facilities; that is, those that are still operational and where an owner/operator is identified.[17] The Comprehensive Environmental Response, Compensation and Liability Act of 1980, better known as the Superfund, was passed to address and clean up abandoned waste sites.[18] These laws authorized the thousands of pages of regulations and guidance that declared that when pollution thresholds are crossed, the design has failed to provide adequate protection. Engineering standards of practice go a step further. Failure here is difficult to recognize. Only other engineers with specific hazardous waste expertise can judge whether the ample margin of safety as dictated by sound engineering principles and practice has been provided in the design. However, finding alternatives and predicting misuse requires quite a bit of creativity and imagination.

But can risks really be quantified? To those of us in the business, the answer is clearly, yes. However, the general public perception is that one person's risk is different from another's and that risk is in the "eye of the beholder." Some of the rationale appears to be rooted in the controversial risks of tobacco use and daily decisions, such as choice of modes of transportation. What most people perceive as risks and how they prioritize those risks is only partly driven by the actual objective assessment of risk; that is, the severity of the hazard combined with the magnitude, duration, and frequency of the exposure to the hazard. For example, the young student smoker may be aware that cigarette smoke contains some nasty compounds, but is not directly aware of what these are (e.g., polycyclic aromatic hydrocarbons and carcinogenic metal compounds). They have probably read the conspicuous warning labels many times as they held the pack in their hands, but these really have not "rung true" to them. They may have never met anyone with emphysema or lung cancer, or they may not be concerned (yet) with the effects on the unborn (i.e., *in utero* exposure). Psychologists also tell us that many in this age group have a feeling of invulnerability. Those who think about it may also think that they will have plenty of time to end the habit before it does any long-term damage. Thus, we should be aware that what we are saying to people, no matter how technically sound and convincing to us as engineers and scientists, may be simply a din to our targeted audience.

The converse is also true. We may be completely persuaded based upon data, facts, and models that something clearly does not cause significant damage, but those we are trying to convince of this finding may not buy it. They may think we have some vested interest, or that they find us guilty by association with a group they do not trust, or that we are simply "guns for hire." (We often are!) They may not understand us because we are using jargon and are not clear in how we communicate the risks. So, do not be surprised if the perception of risk does not match the risk you have quantified.

Risk is a crucial part of the message when environmental professionals communciate with the public, particularly the risk to those that have been or appear to have been exposed to elevated concentrations of contaminants in their food, air, water, or homes. With such communications, people can begin to appreciate that risks can be quantified. For example, when the numbers of tobacco users *and* the incidence of cancer (or other health endpoints) are shown side by side for a population, the strength of association pushes the group to accept that risk is quantitative. People see that there is some numeric link between outcomes and behaviors, but they still rely on intuition and other thought processes that are not always amenable to factual information (see the discussion box "Disasters: Real and Perceived"). People can "do the math" but the math does not hold primacy over what they perceive to be risks. The weight of evidence always includes some nonquantifiable factors.

Disasters: Real and Perceived

What is a disaster? The term gets used frequently, but in myriad ways. Many of the cases in this book have been labelled as environmental disasters. Few would disagree that the occurrences of the December 26, 2004, tsunami in the Indian Ocean and the devestation of New Orleans from Hurricane Katrina and their aftermath were disasters. But a big part of characterizing something as a disaster as opposed to the ubiquitous "failures" is how the public, or at least a substantial part of it, such as the media, perceive it. Failure occurs all the time. In fact, failure is inevitable. Failure becomes a disaster when events in time and space lead us to conclude that the effects were so severe this had to be a disaster. It could also be classified as a disaster when engineers made such a miscalculation or left out some key information. Such mistakes may lead to the public perception that one failure was disastrous, compared to another even more severe outcome that was perceived as less preventable, or even inevitable.

Sometimes, we do not even recognize something as a disaster until long after it occurs. Environmental disasters, for example, may not be noticed for decades. Chronic diseases like cancer have long periods of separation from first exposure to the causative agent and symptoms of disease, the so-called latency period. For example, asbestos workers may be exposed for decades before signs of mesothelioma or lung cancer can be diagnosed. Understudying and underreporting of the exposures and diseases can also obscure linkages between cause and effect, such as the relatively recent linkages between childhood exposures to the metal lead and neurological and developmental diseases. The two problems, latency period and underreporting, can occur together. For example, certain workers may not want to jeopardize their livelihoods and are reluctant to report early symptoms of chronic diseases. Scientists historically have been more likely to study certain demographic groups (e.g., healthy workers) and have avoided others (children, women, and minorities). But when the results do flood in, such as the lead studies in the latter part of the twentieth century or the ongoing arsenic exposures in Bangladesh due to errors in choosing drinking water sources, they are perceived to be "public health disasters."

So, risk perception is a crucial component of risk management. The same facts will be perceived differently by different groups. One group may see the facts as representing a problem that can easily be fixed, while another may perceive the same facts as representing an engineering or public health disaster. Engineers at the State University of New York, Stony Brook,[19] for example, recently compared U.S.

transportation fatalities in 1992 and found that the modes of transportation had similar numbers of fatalities from accidents involving airplanes (775), trains (755), and bicycles (722). To the public, however, air travel has often been considered to have much higher risk associated with it than that for trains and certainly for bicycles. The researchers concluded that two driving factors may lead to these perceptions: (1) a single event in air crashes leads to large loss of life, with much media attention; and (2) people aboard a large aircraft have virtually no control over their situation.

The increased anxiety resulting from highly visible failures and lack of control over outcomes leads to the greater perceived risk. These factors occur in environmental and public health risks. Certain terms are terrifying, like cancer, central nervous system dysfunction, toxics, and ominous-sounding chemical names, like dioxin, PCBs, vinyl chloride, and methyl mercury. In fact, these chemicals *are* ominous, but many chemicals that are less harmful can also elicit anxieties and associated increased perceived risk. For example, I have asked my students at Duke for some years now, as part of a pretest to a professional ethics course, to answer a number of questions. The first two questions on the exam are as follows.

1. The compound dihydrogenmonoxide has several manufacturing and industrial uses. However, it has been associated with acute health effects and death in humans, as a result of displacement of oxygen from vital organs. The compound has been found to form chemical solutions and suspensions with other substances, crossing cellular membranes, and leading to cancer and other chronic diseases in humans. In addition, the compound has been associated with fish kills when supersaturated with molecular oxygen, destruction of wetlands and other habitats, and billions of dollars of material damage each year. A prudent course of action dealing with dihydrogenmonoxide is to:
   a. Ban the substance outright.
   b. Conduct a thorough risk assessment, then take regulatory actions.
   c. Work with industries using the compound to find suitable substitutes.
   d. Restrict the uses of the substance to those of strategic importance to the United States.
   e. Take no action, except to warn the public about the risks.
2. The class of compounds, polychlorinated biphenyls, had several manufacturing and industrial uses during the twentieth

century. However, PCBs were associated with acute health effects and death in humans. The compound has been found to form chemical solutions and suspensions with other substances, crossing cellular membranes, and leading to cancer and other chronic diseases in humans. In addition, the compound has been associated with contaminated sediments, as well as wetlands and other habitats, and billions of dollars of material damage each year. A prudent course of action dealing with PCBs is to:

a. Ban the substances outright.
b. Conduct a thorough risk assessment, then take regulatory actions.
c. Work with industries using the compound to find suitable substitutes.
d. Restrict the uses of the substances to those of strategic importance to the United States.
e. Take no action, except to warn the public about the risks.

The two questions were intentionally worded similarly and the answers worded identically. Everything in the question is factually correct. Most of the students taking the exam have earned A's in high school or college chemistry, physics, and biology, and several are well on their way to completing engineering and other technical degrees. Interestingly, the answers to the two questions have differed very little. Most students seem to be struck by the many negative effects to health and safety. The most frequent answer is "b." conduct a risk assessment. This may, in part, be due to the fact that we as educators have been relentless in reminding students to get their facts straight before deciding. That is the good news. The bad news is that many saw no difference between the two questions and several chose "a." outright bans on both chemicals, the first of which is water!

Actually, the answers to the two questions should have been very different. I would recommend "e" for water and "a" for the polychlorinated biphenyls (simply because they have already been banned since the 1970s following the passage of the Toxic Substances Control Act). Of course, water is not risk free. In fact, it is a contributing factor in many deaths (drowning, electrocution, auto accidents, falls) especially in its solid phase (ice, and workplace incidents, such as steam-related accidents). Water contributed to the worst chemical plant disaster in history in Bhopal, India. We blame methylisocyanate (MIC), but the explosion would not have occurred had MIC not violently reacted with water. In a nuclear meltdown, the ensuing explosion would be caused by contact between the hot core and groundwater below. However, none of us could survive if we banned or placed major restrictions on its general use!

Perceived risks may be much greater than actual risks, or they may be much less. So then, how can technical facts be squared with pubic fears? Like so many engineering concepts, timing and scenarios are crucial. What may be the right manner of saying or writing something in one situation may be very inappropriate in another. Approaches will differ according to whether we need to motivate people to take action, alleviate undue fears, or simply share our findings clearly, no matter whether they are good news or bad. For example, some have accused certain businesses of using pubic relations and advertising tools to lower the perceived risks of their products. The companies may argue that they are simply presenting a counterbalance against unrealistic perceptions.

Environmental practitioners must take care not be used by parties with vested but hidden interests. An emerging risk management technique is the so-called "outrage management," coined by Peter Sandman, a consultant to businesses and governmental agencies.[20] A first step is to present a positive public image, such as a "romantic hero," pointing out all the good things the company or agency provides, such as jobs, modern conveniences, and medical breakthroughs. Although these facts may be accurate, they often have little to do with the decisions at hand, such as the type of pollution controls to be installed on a specific power plant near a particular neighborhood. Another way that public image can be enhanced is to argue that the company is a victim itself, suffering the brunt of unfair media coverage or targeted by politicians. If these do not work, some companies have confessed to being "reformed sinners," who are changing their ways. One of the more interesting strategies put forth by Sandman is that companies can portray themselves as "caged beasts." This approach is used to convince the public that, even though in the past they have engaged in unethical pollution and unfair practices, the industry is so heavily regulated and litigated against that they are no longer able to engage in these acts. So, the public should trust that this new project is different from the company's track record. There is obviously some truth to this tactic, as regulations and court precedents have curtailed a lot of pollution. But the environmental practitioner must be careful to discern the difference between actual improvement and mere spin tactics to eliminate public outrage.

Holding paramount the health, safety, and welfare of the public gives the engineer no room for spin. However, the public often exaggerates risks. So, abating risks that are in fact quite low could mean

TABLE 5.3
Differences between risk assessment and risk perception processes.

Analytical Phase	*Risk Assessment Processes*	*Risk Perception Processes*
Identifying risk	Physical, chemical, and biological monitoring and measuring the event	Personal awareness
	Deductive reasoning	Intuition
	Statistical inference	
Estimating risk	Magnitude, frequency, and duration calculations	Personal experience
	Cost estimation and damage assessment	Intangible losses and nonmonetized valuation
	Economic costs	
Evaluating risk	Cost/benefit analysis	Personality factors
	Community policy analysis	Individual action

Source: Adapted from K. Smith, 1992. *Environmental Hazards: Assessing Risk and Reducing Disaster*, Routledge, London, UK.

unnecessarily complicated and costly measures. It may also mean choosing the less acceptable alternative, one that in the long run may be more costly and deleterious to the environment or public health. The risk assessment and risk perception processes differ markedly, as shown in Table 5.3. Assessment relies on problem identification, data analysis, and risk characterization, including cost-benefit ratios. Perception relies on thought processes, including intuition, personal experiences, and personal preferences. Engineers tend to be more comfortable operating in the middle column (using risk assessment processes), whereas the general public often uses the processes in the far right column. We can liken this to the left-brained engineer trying to communicate with a right-brained audience (or "Engineers Are from Mars but Clients Are from Venus").[21] It can be done, so long as preconceived and conventional approaches do not get in the way.

This was experienced recently in dealing with a predominantly African American community in North Carolina, where an environmental assessment was being proposed. During one of the early scoping meetings, the early stages of the study were discussed. The engineers and scientists were explaining the need to be scientifically objective, to provide adequate quality assurance of the measurements, and to have a sound approach for testing hypotheses and handling data. I must admit that we thought going into the meeting that the subject matter was pretty "dry" and expected little concern or feedback. We expected the neighborhood interest to pique when the quality assured and validated data was shared. However, during the scoping meetings, people expressed concern about what we would do if we "found something." They wanted to know if we would begin interventions then and there. We were not prepared for these questions, because we knew that the data were not truly acceptable until it had been validated and interpreted. So, we recommended patience until the data met the scientists' requirements for rigor. The neighborhood representatives did not see it that way. At best, they thought we were naïve, and at worst disingenuous. It seems that they had been "studied" before, with little action to follow these studies. They had been told previously some of the same things they were being told at our meeting: "Trust us!" We were applying rigorous scientific processes (middle column of Table 5.3), which they had endured previously. Their concerns are explained by their experience and awareness (right-hand column). As a result, our flowcharts were changed to reflect the need to consider actions and interventions before project completion. This compromise was acceptable to all parties.

So, both "lay" groups and our highly motivated and intelligent engineers and scientists of tomorrow have difficulty in parsing perceived and real risks. We can expect the balance between risk assessment and risk perception to be a major challenge in all of our projects.

Sometimes, perception *is* reality.

To scientists and engineers at least, risk is a very straightforward and quantifiable concept: risk equals the probability of some adverse outcome. Risks are thus a function of probability and consequence.[22] The consequence can take many forms. In the medical and environmental sciences, it is called a hazard. Risk, then, is a function of the particular hazard and the chances of a person (or neighborhood or workplace or population) being exposed to the hazard. In the environmental business, this hazard often takes the form of toxicity, although other public health and environmental hazards abound.

Certain minority subpopulations, for example, have higher body burdens of persistent toxicants than what is found in the general population. Subsistent fishing and hunting is more common in Inuit populations in the Arctic regions of North America. Tissue concentrations of PCBs and toxic compounds in fish and top predators (e.g., polar bears) has increased dramatically in the past five decades.[23] Thus, the PCB body burden of the Inuit has also increased. So, any decisions about acceptable levels of exposures to PCBs for Inuit people must take into account the already elevated levels.

Cancer risk is a very important type of risk with a great deal of public concern. It is particularly important to be aware of the possible disparities in decisions that lead to greater cancer risks (see the case study, "'Cancer Alley' and Vinyl Chloride").

"Cancer Alley" and Vinyl Chloride[24]

In southern Louisiana, the lower Mississippi River industrial corridor is home to a predominantly low income, minority (African American and Latino) community who are being exposed to many pollutants.[25] This 80-mile long region, known as "Cancer Alley," between Baton Rouge and New Orleans, has experienced releases of carcinogens, mutagens, teratogens (birth defect agents), and endocrine disruptors in the atmosphere, soil, ground water, and surface water. More than 100 oil refineries and petrochemical facilities are located in this region. It has been reported that per capita releases of toxic air pollutants is about 27 kilograms (kg), nine times greater than the U.S. average of only 3 kg.[26] The U.S. average of 260 kg of toxic air pollutants per square mile is dwarfed by the more than 7,700 kg per square mile in the industrial corridor.

One particular carcinogen of concern is vinyl chloride. In the 1970s, cases of liver cancer (hepatic angiosarcoma) began to be reported in workers at polymer production facilities and other industries where vinyl chloride was present. Since then, the compound has been designated as a potent human carcinogen (inhalation slope factor $= 0.3\,\text{kg} \cdot \text{day}\,\text{mg}^{-1}$).

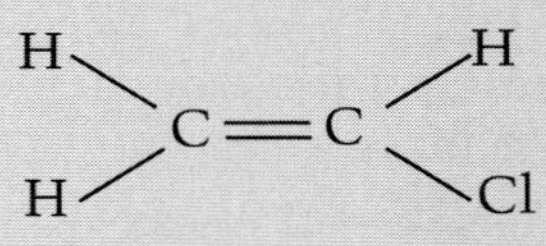

vinyl chloride

Vinyl chloride at first glance may appear to be readily broken down by numerous natural processes, including abiotic chemical and microbial degradation (see Figure 5.8), but numerous studies have shown that vinyl chloride concentrations can remain elevated over long periods of time. In fact, under environmental conditions, vinyl

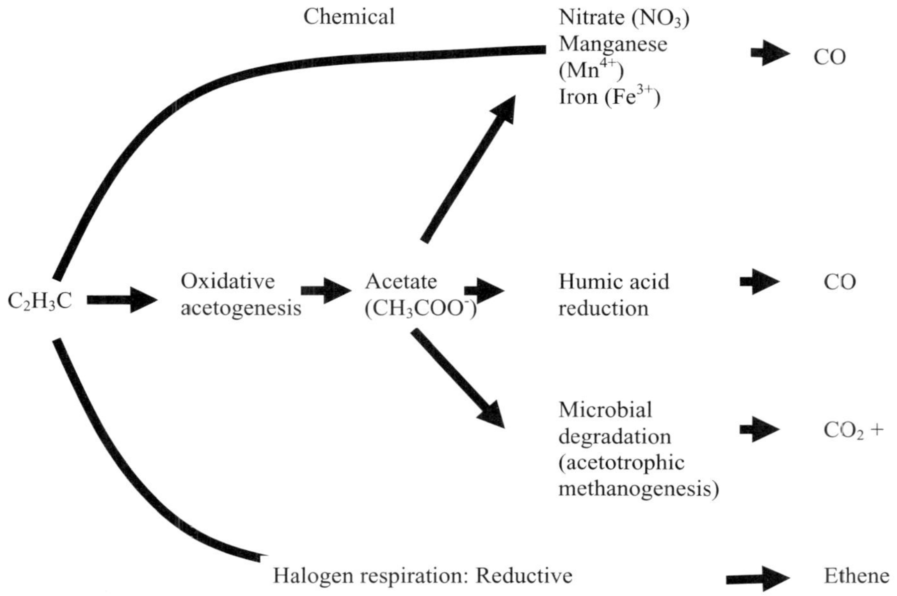

FIGURE 5.8. Biodegradation pathways for vinyl chloride.
Source: U.S. Geological Survey, 2004. Microbial Degradation of Chloroethenes in Ground Water Systems, Toxic Substances Hydrology Program: Investigations, http://toxics.usgs.gov/sites/solvents/chloroethene.html; accessed November 29, 2004.

chloride can be extremely persistent (see Appendix 4), with an anaerobic half-life ($T_{1/2}$) in soil greater than two years. It can also be difficult to treat with conventional engineering methods. For example, aerobic degradation in sewage treatment plants and surface water in an isolated bacteria culture with vinyl chloride concentrations of 20–120 $mg\,L^{-1}$ needs a minimum of 35 days to completely degrade the compound. Nontraditional treatment methods, such as attack by hydroxyl radicals, can significantly reduce the $T_{1/2}$.[27] In heavily polluted areas like Cancer Alley, vinyl chloride repositories can remain intact for decades, serving as a continuous potential source. These repositories can actually be compounds other than vinyl chloride, but which break down to form the compound; for example, chloroethylene solvents degrade to vinyl chloride. With its high vapor pressure (2,300 mm Hg at 20°C) and high aqueous solubility (1,100 $mg\,L^{-1}$), the chances of people being exposed via the air or drinking water once vinyl chloride is formed can be considerable.

Local groups have begun arming themselves with environmental data, such as the emissions and other release information in the Toxic Release Inventory (TRI), which show the inordinately high toxic chemical release rates near their communities. Local communities have challenged nearby industries with possible health effects linked to chemical exposures. For example, residents in Mossville, Louisiana, argued that several health problems in their community could be linked to chemical releases by 17 industrial facilities located within one kilometer of the community. These confrontations led to a number of advocates writing the 2000 report, *Breathing Poison: The Toxic Costs of Industries in Calcasieu Parish, Louisiana*, which called for "pollution reduction, environmental health services, and a fair and just relocation for consenting residents."[28] These efforts have gained the attention of national media and regulatory agencies and have been emblematic of the environmental justice movement.

Bioaccumulation and Its Influence on Risk

As mentioned in Chapter 2, the likelihood that a substance will find its way into the food web is an important aspect of its hazard. Toxicokinetic models predict the dynamics of uptake, distribution, depuration, and elimination of contaminants within organisms. Persistence and bioaccumulation are interdependent. If the substance is likely to be sorbed to organic matter (i.e., high K_{oc} value), it will have an affinity for tissues. If a substance partitions from the aqueous phase to the organic phase (i.e., high K_{ow} value) it is likely to be stored in fats of higher trophic level organisms (e.g., carnivores and omnivores). The bioconcentration factor (BCF) is the ratio of the concentration of the substance in a specific genus to the exposure concentration, at equilibrium. The exposure concentration is the concentration in the environmental compartment (almost always surface water). The BCF is similar to the bioaccumulation factor (BAF), but the BAF is based on the uptake of the organism from both the water and the food. The BCF is based on the direct uptake from the water only. A BCF of 500 means that the organism takes up and sequesters a contaminant to concentrations 500 times greater than the exposure concentraton. Generally, any substance that has a BAF or BCF >5,000 is considered to be highly bioaccumulative, although the cutoff point can differ depending on the chemicals of concern, the regulatory requirements, and the type of ecosystem in need of protection.

It is important to note that genera will vary considerably in reported BCF values and that the same species will bioaccumulate different compounds at various rates. The amount of bioaccumulated contaminant generally increases with the size, age, and fat content of the organism and decreases with increasing growth rate and efficiency. Bioaccumulation also is often higher for males than females and in organisms that are proficient in storing water. Top predators often have elevated concentrations of persistent, bioaccumulating toxic substances (known as PBTs).

The propensity of a substance to bioaccumulate is usually inversely proportional to its aqueous solubility, since hydrophilic compounds are usually more easily eliminated by metabolic processes. In fact, the first stages of metabolism often involve adding or removing functional groups to make it more water soluble. Generally, compounds with $\log K_{ow} > 4$ can be expected to bioaccumulate. However, this is not always the case; for example, very large molecules (e.g., cross-sectional dimensions >9.5 Angstroms (Å) and molecular weights >600) are often too large to pass through organic membranes, which is known as steric hindrance. Since, in general, the larger the molecule, the more lipophilic it becomes, some very lipophilic compounds (i.e., $\log K_{ow} > 7$) will actually have surprisingly low rates of bioaccumulation due to steric hindrance.

Bioaccumulation not only makes it difficult to find and measure toxic compounds, but it complicates how people and ecosystems can become exposed. For example, a release of a persistent, bioaccumulating substance can interfere with treatment plant efficiencies and greatly increase human exposures (see the case study, "The Kepone Tragedy").

The Kepone Tragedy

The Allied Chemical plant in Hopewell, Virginia, has been in operation since 1928 and had produced many different chemicals during its lifetime. In the 1940s the plant started to manufacture organic insecticides, which had recently been invented, DDT being the first and most widely used. In 1949 it started to manufacture chlordecone (trade name Kepone), a particularly potent herbicide that was so highly toxic and carcinogenic (see Table 5.4) that Allied withdrew its application to the Department of Agriculture to sell this chemical to American farmers. It was, however, very effective and cheap to make, and so Allied started to market it overseas.

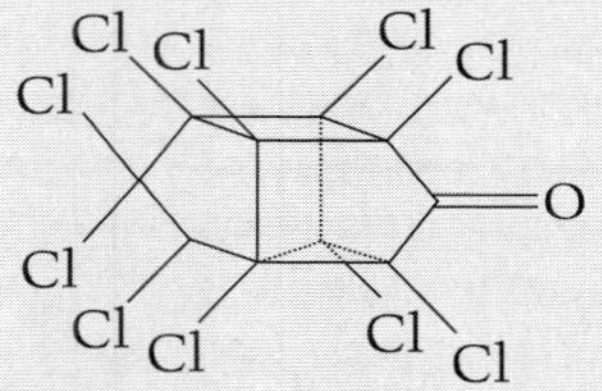

chlordecone
(each intersection is a carbon atom)

TABLE 5.4
Properties of chlordecone (Kepone).

Formula	*Physico-chemical properties*	*Environmental Persistence and Exposure*	*Toxicity*
1,2,3,4,5,5,6,7,9,10,10-dodecachlorooctahydro-1,3,4-metheno-2H-cyclobuta (cd) pentalen-2-one ($C_{10}Cl_{10}O$).	Solubility in water: 7.6 mg L^{-1} at 25°C; vapor pressure: less than 3 × 10^{-5} mmHg at 25°C; log K_{ow}: 4.50	Estimated half-life ($T_{1/2}$) in soils between 1 and 2 years, whereas in air is much higher, up to 50 years. Not expected to hydrolyze or biodegrade in the environment. Also, direct photodegradation and vaporization from water and soil is not significant. General population exposure to chlordecone is mainly through the consumption of contaminated fish and seafood.	Workers exposed to high levels of chlordecone over a long period (more than one year) have displayed harmful effects on the nervous system, skin, liver, and male reproductive system (likely through dermal exposure to chlordecone, although they may have inhaled or ingested some as well). Animal studies with chlordecone have shown effects similar to those seen in people, as well as harmful kidney effects, developmental effects, and effects on the ability of females to reproduce. There are no studies available on whether chlordecone is carcinogenic in people. However, studies in mice and rats have shown that ingesting chlordecone can cause liver, adrenal gland, and kidney tumors. Very highly toxic for some, species such as Atlantic menhaden sheepshead minnow, or Donaldson trout with LC_{50} between 21.4 and 56.9 mg·L^{-1}

Source: United Nations Environmental Programme, 2002. "Chemicals: North American Regional Report," Regionally Based Assessment of Persistent Toxic Substances, Global Environment Facility.

In the 1970s the national pollutant discharge elimination permit system under the Clean Water Act went into effect and Allied was to list all the chemicals it was discharging into the James River. Recognizing the problem with Kepone, Allied decided not to list it as part of their discharge, and a few years later "tolled" the manufacture of Kepone to a small company called Life Science Products Co., set up by two former Allied employees, William Moore and Virgil Hundtofte. The practice of tolling, long-standing in chemical manufacture, involves giving all the technical information to another company as well as an exclusive right to manufacture a certain chemical—for the payment of certain fees, of course. Life Sciences Products set up a small plant in Hopewell and started to manufacture Kepone, discharging all its wastes into the sewerage system.

The operator of the Hopewell wastewater treatment plant soon found that he had a dead anaerobic digester. He had no idea what killed the microbes in his digester, and tried vainly to restart it by lowering the acidity. (Methane-producing organisms in anaerobic digesters are quite sensitive to chemical changes, especially pH).

In 1975, one of the workers at the Life Sciences Products plant visited his physician, complaining of tremors, shakes, and weight loss. The physician took a sample of blood and sent it to the Center for Disease Control in Atlanta for analysis. What they discovered was that the worker had an alarmingly high $8\,mg\,L^{-1}$ of Kepone in his blood. The State of Virginia immediately closed down the plant and took everyone into a health program. Over 75 people were found to have Kepone poisoning. It is unknown how many of these people eventually developed cancer. The Kepone that killed the digester in the wastewater treatment plant flowed into the James River, and over 100 miles of the river was closed to fishing due to the Kepone contamination. The sewers through which the waste from Life Science flowed was so contaminated that it was abandoned and new sewers built. These sealed sewers are still under the streets of Hopewell, and serve as a reminder of corporate decision making based on the wrong priorities.

Biological Response

Even if a substance persists and is taken up by an organism, its hazards are still dependent upon the response of the organism after it comes into contact with the substance. This is the essence of the hazard; that is, does the chemical, physical, or biological agent elicit an adverse response?

This response is measurable. When a contaminant interacts with an organism, substances like enzymes are generated as a response. Thus, measuring such substances in fluids and tissues can provide an indication or "marker" of contaminant exposure and biological effects resulting from the exposure. The term *biomarker* includes any such measurement that indicates an interaction between an environmental hazard and a biological system.[29] In fact, biomarkers may indicate any type of hazard—chemical, physical, and biological. An exposure biomarker is often an actual measurement of the contaminant itself or any chemical substance resulting from the metabolism and detoxification processes that take place in an organism. For example, measuring total lead (Pb) in the blood may be an acceptable exposure biomarker for people's exposures to Pb. However, other contaminants are better reflected by measuring chemical byproducts, such as compounds that are rapidly metabolized upon entering an organism. Nicotine, for example, is not a very good indicator of smoking, but the metabolite, cotinine, can be a reliable indicator of nicotine exposure. Likewise, when breath is analyzed to see if someone has been drinking alcohol, the alcohol itself (i.e., ethanol) is not usually a good indicator, but various metabolites, such as acetaldehyde, that have been formed as the body metabolizes the ethanol are excellent markers.

Exposure to ethanol by the oral pathway (i.e., drinking alcoholic beverages) illustrates the continuum of steps between exposure and response (see Figure 5.9). Table 5.5 gives examples of the types of biomarkers for a specific type of exposure, maternal alcohol consumption. Interestingly, the

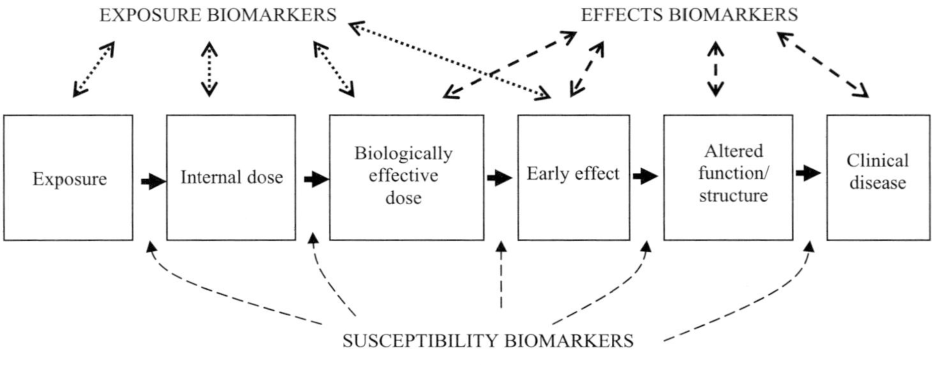

FIGURE 5.9. Continuum from exposure to a toxic substance to clinically diagnosed disease. The continuum is a time sequence, but the chemical to which the organism is exposed is not necessarily the same chemical in subsequent stages; that is, metabolites are formed, which can serve as the biomarker. Enzymes produced to enhance metabolism or detoxification can also serve as biomarkers. Susceptibility biomarkers indicate increased vulnerability between the steps.
Source: Adapted from C.F. Bearer, 2001. "Markers to detect drinking during pregnancy," *Alcohol Research and Health*, 25 (3), 210–218.

TABLE 5.5
Examples of biomarkers following an oral exposure to a toxic substance, ethanol, in pregnant women.

Exposure/Effect Step	*Biomarker Type*	*Example Biomarkers*
Internal dose	Alcohol ingestion	Blood ethanol concentration
Biologically effective dose	Ethanol metabolites	Acetaldehyde Ethyl glucuronide Fatty acid ethyl esters (FAEEs) Cocaethylene
Early effects	Enzymes in ethanol metabolic reactions	Cytochrome P450 2E1 Catalase FAEE synthase
Alter function or structure	Target protein alteration	Carbonhydrate-deficient transferring Serum proteins Urinary dolichols Sialic acid
	Early target organ damage	Gamma glutamyltransferase Aspartate aminotransferase/ alanine aminotransferase Mean corpuscular volume B-hexosaminidase
Clinical disease	Physiological response, including neurological damage and low birth weight, in newborn baby	Fetal alcohol syndrome

Adapted from C.F. Bearer, 2001. "Markers to detect drinking during pregnancy, *Alcohol Research and Health*," 25 (3), 210–218.

response and biomarkers for alcohol consumption are similar to those for some environmental contaminants, such as Pb, mercury (Hg), and PCBs.

Exposure biomarkers are also useful as an indication of the contamination of fish and wildlife in ecosystems. For example, measuring the activity of certain enzymes, such as ethoxyresorufin-*O*-deethylase (EROD), in aquatic fauna *in vivo* biomarker, indicates that the organism has been exposed to planar halogenated hydrocarbons (e.g., certain dioxins and PCBs), PAHs, or other similar contaminants. The mechanism for EROD activity in the aquatic fauna is the receptor-mediated induction of cytochrome P450-dependent mono-oxygenases when exposed to these contaminants.[30] The biological response does not necessarily have to respond to chemical stress.

Stresses to environmental quality also can come about from ecosystem stress, such as loss of important habitats and decreases in the size of the population of sensitive species.

A substance may also be a "public welfare hazard" that damages property values or physical materials, expressed, for example, as its corrosiveness or acidity. The hazard may be inherent to the substance, but like toxicity, a welfare hazard usually depends on the situation and conditions where the exposure may occur.

Situations are most hazardous when a number of conditions exist simultaneously; witness the hazard to firefighters using water in the presence of oxidizers. The challenge to the environmental practitioner is how to remove or modify the characteristics of a substance that render it hazardous, or to relocate the substance to a situation where it has value.

Organic versus Inorganic Toxicants

We have been talking about a number of different pollutants, so we should try to distinguish some of their more important characteristics. Environmental contaminants fall into two major categories, organic and inorganic. Organic compounds are those that have at least one covalent bond between two carbon atoms or between a carbon and a hydrogen atom. Thus, the simplest hydrocarbon, methane (CH_4), has a bond between carbon and four hydrogen atoms. Organic compounds are subdivided between aliphatic (chains) and aromatic (rings) compounds. A common group of aliphatic compounds are the chain structures known as alkanes, which are hydrocarbons with the generic formula C_nH_{2n+2}. If these compounds have all the carbon atoms in a straight line, they are considered normal, and are known as *n*-alkanes. The simplest aromatic, benzene (C_6H_6), has bonds between carbon atoms and between carbon and hydrogen atoms (see Figure 5.10).

The structure of the compound determines its persistence, toxicity, and ability to accumulate in living tissue. Subtle structural differences can lead to very different environmental behaviors. Even arrangements with identical chemical formulae, that is, isomers, can exhibit very different chemical characteristics. For example, the boiling points at 1 atm for *n*-pentane, isopentane, and neopentane (all C_5H_{12}) are 36.1°C, 27.8°C, and 9.5°C, respectively. Among the most important factors are the length of the chains in aliphatic compounds and the number and configurations of the rings in aromatics. Arguably, substitutions are even more critical. For example, methane is a gas under environmental conditions, but it becomes a very toxic and bioaccumulating liquid (carbon tetrachloride or tetrachloromethane) when the hydrogen atoms are substituted with chlorine atoms (CCl_4). Naphthalene, the simplest polycyclic aromatic hydrocarbon ($C_{10}H_8$), is considered to be a possible human carcinogen, but the data are not sufficient to calculate a slope factor. However, when an amine group (NH_2) substitutes for a hydrogen atom to form 2-naphthylamine ($C_{10}H_9N$),

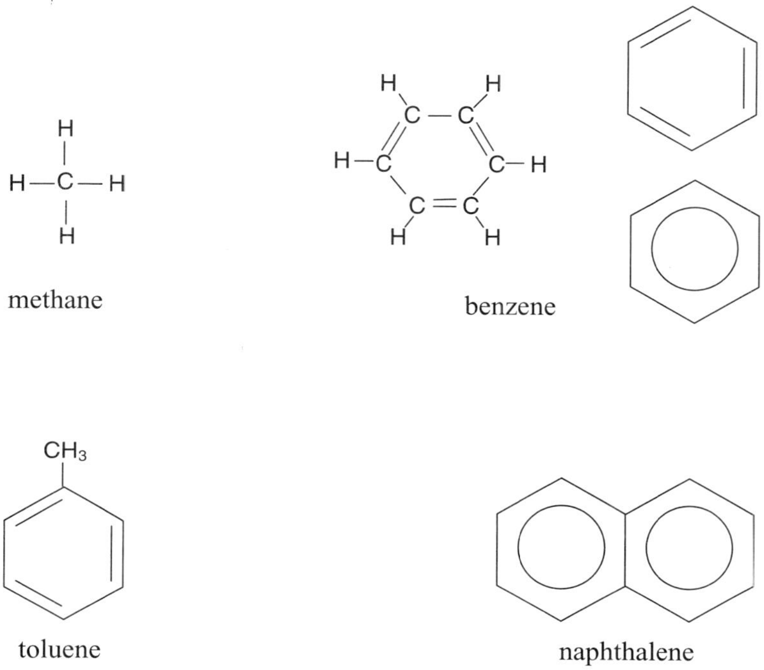

FIGURE 5.10. Organic compound structures. Methane is the simplest aliphatic structure and benzene is the simplest aromatic structure. Note that the benzene molecule has alternating double and single bonds between the carbon atoms. The double and single bonds flip, or resonate. This is why the benzene ring is also shown as the two structures on the right, which are the commonly used condensed form in aromatic compounds, such as the solvent toluene and the polycyclic aromatic hydrocarbon, naphthalene.

the inhalation cancer slope factor is $1.8\,kg \cdot day\,mg^{-1}$; which is quite steep. The formulation of pesticides takes advantage of the dramatic increases in toxicity by substitution reactions (see the case study, "Pesticides and Sterility").

Pesticides and Sterility

For many years both Shell Oil and Dow Chemical supplied a pesticide containing dibromochloropropane (DBCP) to Standard Fruit Company for use on its banana plantations, in spite of evidence since the 1950s that DBCP causes sterility in laboratory animals. Even after it was shown that DBCP also causes sterility in humans and it was banned in the United States, Shell continued to market the pesticide in Central America.

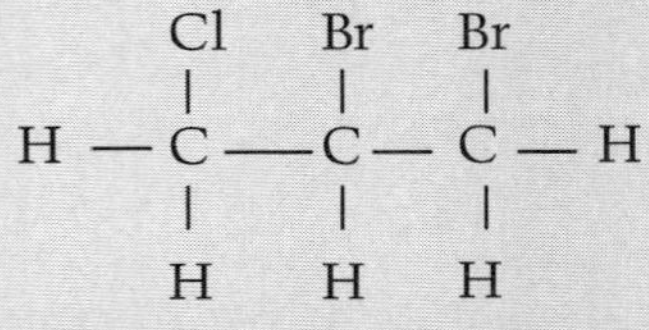

dibromochloropropane (DBCP)

In 1984, banana plantation workers from several Central American countries filed a class action suit against Shell, claiming that they became sterile and faced a high risk of cancer. In response, Shell claimed that it was inconvenient to continue the case because the workers were in Costa Rica, a claim that was quickly thrown out of court. Shell finally settled out of court with the Costa Rican workers and paid $20 million in damages to the 16,000 claimants. A particularly insensitive scientist from Shell is quoted as saying: "Anyway, from what I hear they could use a little birth control down there."[31]

Congeners are configurations of a common chemical structure. For example, all polychlorinated biphenyls (PCBs) have two benzene rings bonded together at two carbon atoms. They also have at least one chlorine substitution around the rings, so that there are 209 possible configurations, or 209 PCB congeners. Since the two benzene rings can rotate freely on the connecting bond, for any PCB congener (except decachlorobiphenyl, in which every hydrogen has been substituted by a chlorine), the location of chlorines can differ; for example, 2,3,4-trichlorobiphenyl is the same as 2′,3,4-trichlorobiphenyl and the same as 2,4′,6′-trichlorobiphenyl. The location of the chlorine atoms can lead to different physical, chemical, and biological characteristics of molecules, including their toxicity, persistence, and bioaccumulation potential.

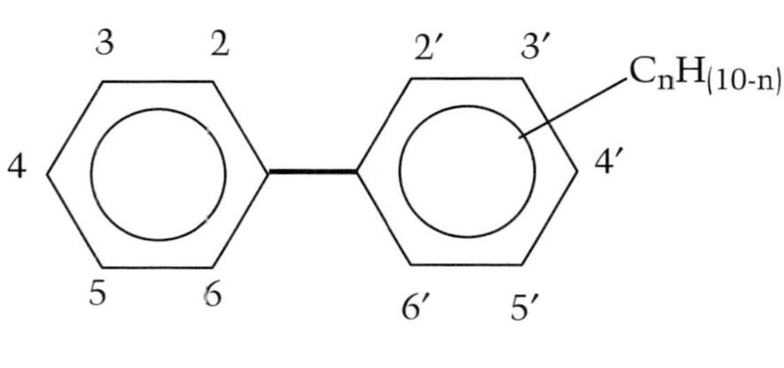

Polychlorinated Biphenyl Structure

Numerous acids are organic, because they contain the C—C and C—H bonds. For example, acetic acid ($HC_2H_3O_2$), benzoic acid ($HC_7H_5O_2$), and cyanoacetic acid ($C_3H_3NO_2$) are organic acids. Like other compounds, organic acids can have substitutions that change their hazard, such as when acetic acid's hydrogen atoms are substituted with chlorines to form trichloroacetic acid ($C_2HCl_3O_2$).

Inorganic compounds are those that do not contain carbon-to-carbon or carbon-to-hydrogen covalent bonds. Thus, even carbon-containing compounds can be inorganic. For example, the pesticides sodium cyanide (NaCN) and potassium cyanide (KCN) are inorganic compounds, as are the gases carbon monoxide (CO) and carbon dioxide (CO_2), compounds that contain the anions carbonate (CO_3^{2-}) and bicarbonate (HCO_3^-), and inorganic acids, such as carbonic acid (H_2CO_3) and cyanic acid (HCNO).

Metals are particularly important in environmental situations. Like other elements, the compounds formed by metals vary in their toxicity and how rapidly they move and change in the environment. However, certain metals, no matter what their form, are hazardous. Unlike carbon, hydrogen, oxygen, and many other elements, which in certain configurations are essential and in others are toxic, heavy metals and metalloids are considered hazardous, no matter what the chemical species. For example, any amount of lead or mercury in any form is considered toxic, although some forms are much more toxic than others. And, since metals and metalloids are elements, we are not going to be able to "destroy" them as we do organic compounds by using chemical, thermal, and biological processes. Destruction simply means that we are changing compounds into simpler compounds (e.g., hydrocarbons are broken down to CO_2 and H_2O). But metals are already in elemental form, so the engineer must attempt to change the metal or metalloid to make it less toxic and less mobile, and once that is done, to take measures to keep the metal wastes away from people, wildlife, and other receptors.

The oxidation state or *valence* of metals and metalloids is the most important factor in their toxicity and mobility. The outermost electrons determine how readily an element will enter into a chemical reaction and what type of reaction will occur. This is the oxidation number of the element. Most metals contain more than one oxidation state, each with its own toxicity and mobility characteristics. However, in most cleanup situations, all forms of the metal, even those with low toxicity and mobility, must be removed since when environmental conditions change, the metals may change to more toxic and mobile forms (see the case study, "Jersey City Chromium").

Jersey City Chromium

Jersey City, in Hudson County, New Jersey, was once the chromium processing capital of America and over the years, 20 million tons of chromate ore processing residue was sold or given away as fill. There are at least 120 contaminated sites, which include ball fields and basements underlying homes and businesses. It is not uncommon for brightly colored chromium compounds to crystallize on damp basement walls and to "bloom" on soil surfaces where soil moisture evaporates, creating something like an orange hoar frost of hexavalent chromium, Cr^{6+}. A broken water main in the wintertime resulted in the formation of bright green ice due to the presence of trivalent chromium, Cr^{3+}.

The companies that created the chromium waste problem no longer exist, but liability was inherited by three conglomerates through a series of takeovers. In 1991, Florence Trum, a local resident, successfully sued Maxus Energy, a subsidiary of one of the conglomerates, for the death of her husband, who loaded trucks in a warehouse built directly over a chromium waste disposal site. He developed a hole in the roof of his mouth and cancer of the thorax, and it was determined by autopsy that his death was caused by chromium poisoning. Even though the subsidiary company did not produce the chromium contamination, the judge ruled that company managers knew about the hazards of chromium, making the company culpable.

The State of New Jersey initially spent $30 million to locate, excavate, and remove some of the contaminated soil. But the extent of the problem was overwhelming and they stopped these efforts. The director of toxic waste cleanup for New Jersey admitted that even if the risks of living or working near chromium were known, the state does not have the money to remove it. Initial estimates for site remediation are well over a billion dollars.[15] Citizens of Hudson County are angry and afraid. Those sick with cancer wonder if it could have been prevented. Mrs. Trum perceived the perpetrators as well-dressed businesspeople who were willing to take chances with other peoples' lives. "Big business can do this to the little man. . . ," she said.

The contamination in Jersey City is from industries that used chromium in their processes, including metal plating, leather tanning, and textile manufacturing. The deposition of this chromium in dumps has resulted in chromium contaminated water, soils, and sludge. Chromium is particularly difficult to regulate because of the complexity of its chemical behavior and toxicity, which translates into scientific uncertainty. Uncertainty exacerbates the tendency of regulatory agencies to make conservative and protective assumptions, the tendency of the regulated to question the scientific basis for regulations, and the tendency of potentially exposed citizens to fear potential risk.

Chromium exists in nature primarily in one of two oxidation states—Cr^{3+} and Cr^{6+}. In the reduced form of chromium, Cr^{3+}, there is a tendency to form hydroxides that are relatively insoluble in water at neutral pH values. Cr^{3+} does not appear to be carcinogenic in animal and bioassays. In fact, organically complexed Cr^{3+} has recently become one of the more popular dietary supplements in the United States and can be purchased commercially as chromium picolinate ($C_{18}H_{12}CrN_3O_6$) or with trade names like Chromalene to help with proper glucose metabolism, control of blood fat concentrations, an aid to weight loss and muscle tone, and essential to gene expression.

When Cr^{3+} oxidizes as Cr^{6+}, however, chromium is highly toxic. It is implicated in the development of lung cancer and skin lesions in industrial workers. In contrast to Cr^{3+}, nearly all Cr^{6+} compounds have been shown to be potent mutagens. The U.S. EPA has classified chromium as a human carcinogen by inhalation based on evidence that Cr^{6+} causes lung cancer. However, by ingestion, chromium has not been shown to be carcinogenic.

What confounds the understanding of chromium chemistry is that under certain environmental conditions, Cr^{3+} and Cr^{6+} can interconvert. In soils containing manganese, Cr^{3+} can be oxidized to Cr^{6+}. Given the heterogeneous nature of soils, these redox reactions can occur simultaneously. Although organic matter may serve to reduce Cr^{6+}, it may also complex Cr^{3+} and may make it more soluble—facilitating its transport in groundwater and increasing the likelihood of encountering oxidized manganese present in the soil.

Cleanup limits for chromium are still undecided but through the controversy, there have evolved some useful technologies to aid in resolution of the disputes. For example, analytical tests to measure and distinguish between Cr^{3+} and Cr^{6+} in soils have been developed. Earlier in the history of New Jersey's chromium problem, these assays were not reliable and would have necessitated remediating to soil concentrations based on total chromium. Other technical/scientific advances include remediation strategies designed to reduce Cr^{6+} to Cr^{3+} in order to decrease risk without excavation and removal of soil designated as hazardous waste. The establishment of clean-up standards is anticipated but the proposed endpoint based on contact dermatitis is controversial. While some perceive contact dermatitis as a legitimate claim to harm, others have jokingly suggested regulatory limits for poison ivy, which also causes contact dermatitis. The methodology by which dermatitis-based soil limits were determined has come under attack by those who question the validity of skin patch tests and the inferences by which patch test results translate into soil Cr^{6+} levels. The value of dermatitis-based limits is that they provide a modicum of safety, i.e., an early warning to prevent more serious problems.

The frustration with slow cleanup and what the citizens perceive as double-talk by scientists finally culminated in the unusual step of amending the state constitution so as to provide funds for hazardous waste cleanups. State environmentalists depicted the constitutional amendment as a referendum on Gov. Christine Todd Whitman's environmental record, which relaxed enforcement and reduced cleanups. (Whitman was the first administrator of the U.S. Environmental Protection Agency named by President George W. Bush.)

Radioisotopes

Different atomic weights of the same element are the result of different numbers of neutrons. The number of electrons and protons of stable atoms must be the same.

Elements with differing atomic weights are known as *isotopes*. An element may have numerous isotopes. Stable isotopes do not undergo natural radioactive decay, whereas radioactive isotopes involve spontaneous radioactive decay, as their nuclei disintegrate. Thus, these are known as radioisotopes. This decay leads to the formation of new isotopes or new elements. The stable product of an element's radioactive decay is known as a radiogenic isotope. For example, lead (Pb; atomic number = 82) has four naturally occurring isotopes of different masses (^{204}Pb, ^{206}Pb, ^{207}Pb, ^{208}Pb). Only ^{204}Pb is stable. The isotopes ^{206}Pb and ^{207}Pb are daughter (or progeny) products from the radioactive decay of uranium (U); ^{208}Pb is a product from thorium (Th) decay. Owing to the radioactive decay, the heavier isotopes of lead will increase in abundance compared to ^{204}Pb. The toxicity of a radioisotope can be twofold—chemical toxicity and radioactive toxicity (see the case study, "Radiation Poisoning in Goiania, Brazil"). For example, Pb is neurotoxic no matter what the atomic weight, but if people are exposed to its unstable isotopes, they also are threatened by radiation emitted from the nucleus' decay. The energy of the radioactive decay can alter genetic material and lead to mutations, including cancer.

Radiation Poisoning in Goiania, Brazil[32]

In the early 1980s, a small cancer clinic was opened in Goiana, but business was not good, and the clinic closed five years later. Left behind in the abandoned building were a radiation therapy machine and some canisters containing waste radioactive material—1,400 curies of Cesium 137, which has a half-life of 30 years. In 1987 the container of Cesium 137 was discovered by local residents and was opened, revealing a luminous blue powder. The material was a local curiosity and children even used it to paint their bodies, which caused them to sparkle. One of the little girls went home for lunch and ate a sandwich without first washing her hands. Six days later she was diagnosed with radiation illness, having received an estimated five to six times the lethal radiation exposure for adults. The ensuing investigation identified the true content of the curious barrel. In all, over 200 persons had been contaminated and 54 were serious enough to be hospitalized, with four people dying from the exposure (including the little girl with the sandwich). Treatment of radiation disease is challenging. The International Atomic Energy Commission characterized the treatment of the Goianian patients as:

> . . . the first task was to attempt to rid their bodies of cesium. For this, they administered Prussian blue, an iron compound that bonds with cesium, aiding its excretion. The problem in this case was the substantial delay—at least a week—from initial exposure to treatment. By that time much of the cesium had moved from the bloodstream into the tissues, where it is far more difficult to remove . . . the patients were also treated with antibiotics as needed to combat infections and with cell infusions to prevent bleeding. . . .[33]

By the time the government mobilized to respond to the disaster, the damage was done. A large fraction of the population had received excessive radiation, and the export of produce from Goiania dropped to zero, creating a severe economic crisis. The disaster is now recognized as the second worst radiation accident in the world, second only to the explosion of the nuclear power plant in Chernobyl.

Source: www.nbc-med.org/sitecontent/medref/onlineref/casestudies/csgiania.html.

Factors of Safety

Not everything causes disease. In fact, with the myriad of chemicals in the environment, workplace, and home, relatively few have been associated with chronic diseases like cancer. However, for those that do, risk seldom is zero. Simple mathematics tells us that if the hazard is zero, then risk must be zero. So, only a carcinogen can cause cancer. No matter what the dose, the cancer risk from a noncarcinogen is zero. A prominent hypothesis in carcinogenesis is the two-hit theory, suggested by A.G. Knudson[34] in 1971. The theory argues that cancer develops after genetic material (i.e., usually deoxyribonucleic acid, DNA) is damaged. The first damage is known as *initiation*. This step may, but does not necessarily, lead to cancer. The next step, *promotion*, changes the cell's makeup and nature, such as the loss of normal homeostasis (cellular self-regulation), and the rapid division of clonal tumor cells. Promoters may or may not be carcinogens. So, when we say that a noncarcinogen dose cannot lead to cancer, we are talking specifically of compounds that initiate cancer, since exposure to noncarcinogenic promoters, such as excessive dietary fats, can hasten the onset of cancer cells.

Health researchers use the reference dose (RfD)[35] to assign a level of exposure that is "safe" in terms of health hazard for all diseases except cancer. The RfD represents the highest allowable daily exposure associated with a noncancerous disease. It is calculated from the threshold value below which no adverse effects are observed (the so-called *no observable adverse effect level*, or NOAEL), along with uncertainty and modifying factors based upon the quality of data and the reliability and representativeness of the studies that produced the dose-response curve:

$$\text{RfD} = \frac{(\text{NOAEL})}{(\text{UF}_{1\ldots n}) \times (\text{MF}_{1\ldots n})} \tag{5.4}$$

where

RfD = Reference dose ($\text{mg kg}^{-1}\,\text{d}^{-1}$)
$\text{UF}_{1\ldots n}$ = Uncertainty factors related to the exposed population and chemical characteristics (dimensionless, usually factors of 10)
$\text{MF}_{1\ldots n}$ = Modifying factors that reflect the results of qualitative assessments of the studies used to determine the threshold values (dimensionless, usually factors of 10)

The uncertainty factors address the robustness and quality of data used to derive the RfD, especially to be protective of sensitive populations (e.g., children and the elderly). It also addresses extrapolation of animal data from comparative biological studies to humans, accounting for differences in dose-response among different species. An uncertainty factor can also be applied when the studies upon which the RfD is based are conducted with various study designs; for example, if an acute or subchronic exposure is administered to determine the NOAEL, but the RfD is addressing a chronic disease, or if a fundamental study used a lowest observed adverse effect level (LOEAL) as the threshold value, requiring that the NOAEL be extrapolated from the LOAEL. The modifying factors address the uncertainties associated with the quality of data used to derive the threshold values, mainly from qualitative, scientific assessments of the data. For airborne contaminants, a *reference concentration* (RfC) is used in the same way as the RfD. That is, the RfC is an estimate of the daily inhalation exposure that is likely to be without appreciable risk of adverse effects during a lifetime. The chronic RfD is used with administered oral doses under long-term exposures (i.e., exposure duration >7 years), and the oral subchronic RfD is applied for shorter exposures of two weeks to seven years.

The slope factor (SF) is the principal hazard characteristic for carcinogens (Appendix 5 provides SF values for a number of compounds). Both the RfD and the SF are developed from a mix of mutagenicity studies, animal testing, and epidemiology. Unlike the RfD, which provides a safe level of exposure, cancer risk assessments generally assume there is no threshold. Thus, the NOAEL and LOAEL are meaningless for cancer risk. Instead, cancer slope factors are used to calculate the estimated probability of increased cancer incidence over a person's lifetime (the so-called "excess lifetime cancer risk" or ELCR). Slope factors are expressed in inverse exposure units since the slope of the dose-response curve is an indication of risk per exposure. Thus, the units are the inverse of mass per mass per time, usually $(\mathrm{mg\,kg^{-1}\,day^{-1}})^{-1} = \mathrm{kg\,day\,mg^{-1}}$. This means that the product of the cancer slope factor and exposure (risk) is dimensionless. This should make sense because risk is a unitless probability of adverse outcomes. The SF values are contaminant-specific and route-specific. Thus, we must not only know the contaminant, but how a person is exposed (e.g., via inhalation, via ingestion, or through the skin).

The more potent the carcinogen, the larger the slope factor will be (i.e., the steeper the slope of the dose-response curve). Note, for example, that when inhaled, ingested, or dermally exposed, the slope for the most carcinogenic dioxin, tetrachlorodibenzo-*p*-dioxin, is eight orders of magnitude steeper than the slope for aniline. Keep in mind that this is the linear part of the curve. The curve is actually sigmoidal because at higher doses the effect is dampened; that is, the response continues to increase, but at a decreasing rate. This process is sometimes called the saturation effect. One way to think about this is to consider that if the dose-response curve comes from animal tests of various doses there is a point at which increasing the dose of a chemical adds little to the onset of tumors. The dosage approaches an effective limit and becomes asymptotic. So, if chemical A is given to 1,000 rats at increasing dosages, an incremental increase in rats with tumors is seen. This is the linear range. Doubling the dose doubles the effect. But at some inflection point the increased dosage, say after 50 rats with tumors, if the dose is doubled, half as many additional rats with tumors are seen. The rate continues to decrease up to a point where even very large doses do not produce many additional tumors. This is one of the challenges of animal experiments and models. We are trading dose for time; the assumed lifetime of humans is about 70 years and the doses to carcinogens are usually very small (e.g., parts per billion or trillion). Animal doses may last only a few months and use relatively high doses. We have to extrapolate long-term effects from limited data from short-term studies. The same is somewhat true for human studies, where we try to extrapolate effects from a small number of cases to a much larger population; (e.g., a small study comparing cases to controls in one hospital or a retrospective view of risk factors that may have led to a cluster of cases of cancer).

It can be argued that addressing rare and chronic diseases like cancer, endocrine dysfunction, reproductive disorders, and neurological diseases is an effort in controlling the variables to reduce the possibility of an improbable (thankfully!) event. In fact, new statistical devices are being developed to deal with rare events (see the discussion box, "Small Numbers and Rare Events").

Small Numbers and Rare Events

In statistics, there is an interesting but logical observation that when we deal with rare events, small changes can be very profound. If you think about it, when you start with very small numbers, a slight change can make a difference. Stockbrokers and retailers use this phenomenon often. For example, a company may be the fastest growing company in its field this year. Upon investigation, its sales may have been only \$5.00 last year, but they grew to \$5,000.00 this year. This is a thousand-fold increase; real estate investors might say that sales grew 100,000% this year! Engineers and scientists often prefer absolute terms and might say that the growth rate was $\$4.995 \times 10^3\,yr^{-1}$. These are all correct statements. But would you rather invest in a company that had \$10 million in sales last year and grew to \$20 million this year? That is only a two-fold increase and only 100% growth. But, the absolute growth is $\$1 \times 10^6\,yr^{-1}$, or three orders of magnitude greater than the small firm. What does this tell us about rare outcomes, like cancer?

First, we must be certain that we understand what the numbers mean. In reviewing epidemiological information, is the data given as an incidence of disease or prevalence? Disease incidence is the number of new cases diagnosed each year; prevalence is the number of cases at any given time. Next, we must be careful to ascertain whether the values are absolute or relative. For example, are the values given as a year-over-year change or are they simply a one-time event? In environmental and public health reports, especially risk assessments, the values are often presented as probabilities in engineering notation; for example, a common target of cleanup of hazardous waste sites is that no more than one additional case of cancer per million population should result from the clean site, that is, the added risk is less than or equal to 10^{-6}. Like all probabilities, this is simply a fraction and a decimal. However, if the environmental practitioner uses it in a public forum, it can be very disarming and not clearly understood. In fact, the whole concept of population risk is foreign to most people. The point is that when the environmental practitioner goes about explaining rare events like cancer, great care must be taken.

Toxicology deals with even smaller values and often very limited data. In fact, one of the raging toxicological debates is that of cancer dose-response and where to literally draw the line. As a matter of scientific policy, in what is known as the precautionary principle, many health agencies around the world assume that a single molecule of a carcinogen *can* cause cancer. In other words, there is no threshold under which a dose, no matter how small, would be safe; one hit potentially leads to a tumor. This approach is commonly known as the one-hit model. Most other diseases have such a threshold dose, known as the no observed adverse effect level or the NOAEL (as shown in Figure 5.11). The precautionary principle is in large part due to our lack of understanding of how things work at the molecular level. Toxicological models work better when they use observed data, but at a level below this, we are guessing (albeit a very educated guess) as to what is happening (see Figure 5.12). Since risk at very low doses is not directly measurable using animal experiments or from epidemiology, mathematical models are used to extrapolate from high to low doses.

Various extrapolation models or procedures may reasonably fit the observed data, however, extremely large differences of risk at low doses can be calculated. Scientists must use different models depending on the particular chemical compound, as well as use information about how cancer seems to be occurring (i.e., the biological "mechanism of action" at work in the cell).[36] When such biological information is limited, the default is to assume linearity and since there is no threshold, the curve intersects the *y*-axis at 0. For example, the U.S. Environmental Protection Agency usually recommends a linearized multistage procedure as the default model unless sufficient information to the contrary exists. The linearized multistage procedure calls

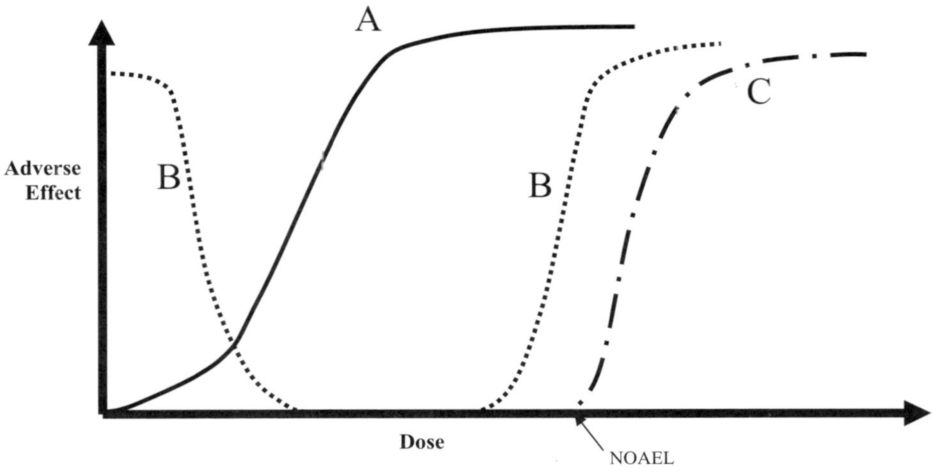

FIGURE 5.11. Three prototypical dose-response curves. Curve A represents the no-threshold curve, which expects a response (e.g., cancer) even if exposed to a single molecule (this is the most conservative curve). Curve B represents the essential nutrient dose-response relationship, and includes essential metals, such as trivalent chromium or selenium, where an organism is harmed at the low dose due to a deficiency (left side) and at the high dose due to toxicity (right side). Curve C represents toxicity above a certain threshold (noncancer). This threshold curve expects a dose at the low end where no disease is present. Just below this threshold is the NOEAL. *Sources*: U.S. Environmental Protection Agency; and D.A. Vallero, 2004. *Environmental Contaminants: Assessment and Control*, Elsevier Academic Press, Burlington, MA.

for the fitting of a multistage model to the data. Multistage models are exponential models approaching 100% risk at high doses, with a shape at low doses given by a polynomial function. If this is first degree, the model is equivalent to a so-called one-hit model, yielding almost a linear relationship between low dose and cancer risk. An upper bound risk is estimated by applying an appropriate linear term into the statistical bound for the polynomial. At sufficiently small exposures, any higher-order terms in the polynomial are assumed to be negligible, and the graph of the upper bound will appear to be a straight line. The slope of this line is called the slope factor, which is a measure of the cancer potency of the compound; the steeper the slope the more potent the carcinogen.[37]

A key engineering lesson from the hazardous waste cases is the need for understandable information and effective risk communica-

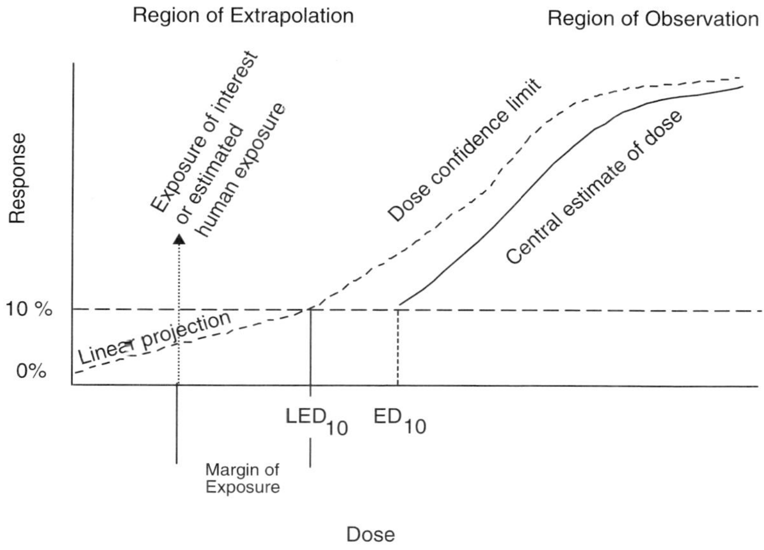

FIGURE 5.12. Linearized multistage dose-response curve showing the two major regions of data availability. LED_{10} = lower 95% confidence limit on a dose associated with 10% extra risk; ED_{10} = estimate of the dose that would lead to 10% increase in the response (in this case, cancer).
Sources: U.S. Environmental Protection Agency; and D.A. Vallero, 2004. *Environmental Contaminants: Assessment and Control*, Elsevier Academic Press, Burlington, MA.

tion in environmental projects. The term treatment is expressed as pollutant removal efficiency, such as percent removal. For example, assessing how good an incinerator is in destroying a hazardous substance, engineers measure and report the removal efficiency for that compound. Environmental and chemical engineers use the "rule of six nines" for extremely hazardous compounds; that is, the quantity (mass) of a compound in a waste stream must be reduced by 99.9999%. For instance, if the most toxic form of dioxin, tetrachlorodibenzo-*para*-dioxin (TCDD), is in a waste stream, the incinerator must destroy 99.9999% (six nines) of the TCDD. If the incinerator is destroying 99.9998% then it theoretically is out of compliance (assuming the means to quantify the pollutant removal are within the range of six significant figures). Often, however, the removal is reported in units of mass or concentration. If a waste contains a total of 100 mg (mass), or 100 mg L^{-1} (concentration) of TCDD, after treatment in a properly operating incinerator, we are left with 0.0001 mg if we started with 100 mg (100 mg – 0.999999 × 100 mg). If the incinerator increases its efficiency by seven nines (99.99999% removal), we would have 0.00001 mg TCDD left. That is, the improvement allowed us to remove only 0.00009 mg of TCDD. This leaves the engineer open to "spin." For example, the incinerator improvement may look better if the removal rate is reported as nanograms (ng) removed (additional 0.09 ng removed). To make the difference look insignificant, you could report the removal as grams removed (only 0.00000009 removed with new expensive equipment). But, both removal efficiencies are the same, only the units differ! Units can be challenging. For example, in hazardous waste engineering, we often use parts per billion (ppb). That is a small concentration. For example, in the language of bartending, one ppb is equivalent to a gin and tonic, where the shot of gin is added to a volume of tonic carried by a train with six miles of tanker cars![38]

A further problem is that removal efficiency is a relative measure of success. If a waste has a large amount of a contaminant, even relatively inefficient operations look good. Taking the TCDD example, if waste A has 100 grams of TCDD (scary thought!) and waste B has 100 ng of TCDD and they both comply with the rules of six nines, the waste A incinerator is releasing 0.0001 grams or 100 ng of the contaminant to the atmosphere, but the waste B incinerator is emitting only 0.0001 ng. That is why environmental laws also set limits on the maximum mass or concentration of a contaminant leaving the stack (or pipe for water discharges). In addition, the laws require that for some pollutants that ambient concentration not be exceeded. However, for many very toxic compounds that require elaborate and expensive monitoring devices, such ambient monitoring is infrequent and highly localized (e.g., near a known polluter). Regulators often depend on self-reporting by the facilities, with an occasional audit (analogous to the IRS accepting a taxpayers self-reporting, which is verified to some extend by audits of a certain sample of taxpayers).

Statistics and probabilities for extreme and rare events can be perplexing. People want to know about trends and differences in exposures and diseases between their town or neighborhood and those of others. Normal statistical information about central tendencies, like the mean, median, and mode or ranges and deviations, fail us when we analyze rare events. Normal statistics allows us to characterize the typical behaviors in our data in terms of differences between groups and trends, focusing on the center of the data. Extreme value theory (EVT), conversely, lets us focus on the points far out on the tail of our data, with the intent of characterizing a rare event. For example, perhaps we have been collecting health data for 10 years for thousands of workers exposed to a contaminant. What is special about those who have been most highly exposed (e.g., those at the 99th percentile)? What can we expect as the highest exposures over the next 50 years? EVT is one means of answering these questions. The first question can be handled with traditional statistics, but the second is an extrapolation (50 years hence) beyond our data set.

Such extrapolations in EVT are justified by a combination of mathematics and statistics—probability theory and inference and prediction, respectively. This can be a very powerful analytical tool. However, the challenge may come after the engineer has completed the analysis. The engineer may be confident that the neighborhood is not much of an additional risk based upon EVT and traditional methods. But how does the engineer explain how such a conclusion was derived? Many in the audience have not taken a formal course in basic statistics, let alone a course that deviates from the foundations of statistics, such as EVT! Senol Utku, a former colleague at Duke, was fond of saying: "To understand a non-banana, one must first understand a banana." This was in the context of discussing the value of linear relationships in engineering. Everyone recognizes that many engineering and scientific processes and relationships are nonlinear in their behavior, but the students must first learn to apply linear mathematics. My advice is to use the best science possible, but be ready to retrace your approaches. Otherwise, it comes off as "smoke and mirrors!"

Exposure Estimation

Risk is a function of both hazard and exposure. An exposure is any contact with an agent. For chemical and biological agents this contact can come about from a number of exposure pathways; that is, routes taken by a substance, beginning with its source to its endpoint (i.e., a target organ, like the liver, or a location short of that, such as in fat tissues). The substances often change to other chemical species as a result of the body's metabolic and detoxification processes. These new substances are known as degradation products or metabolites.

Physical agents, such as electromagnetic radiation, ultraviolet (UV) light, and noise, do not follow this pathway exactly. The contact with these sources of energy can elicit a physiological response that may generate endogenous chemical changes that behave somewhat like the metabolites. For example, UV light may infiltrate and damage skin cells. The UV light helps to promote skin-tumor promotion by activating the transcription factor complex activator protein-1 (AP-1) and enhancing the expression of the gene that produces the enzyme cyclooxygenase-2 (*COX2*). Noise (acoustical energy) can also elicit physiological responses that affect an organism's chemical messaging systems, that is, endocrine, immune, and neural.

The exposure pathway also includes the manner in which people can come into contact with (i.e., be exposed to) the agent. The pathway has five parts:

1. The source of contamination (e.g., a leaking landfill).
2. An environmental medium and transport mechanism (e.g., soil with water moving through it).
3. A point of exposure (such as a well used for drinking water).
4. A route of exposure (e.g., inhalation, dietary ingestion, nondietary ingestion, dermal contact, and nasal).
5. A receptor population (those who are actually exposed or who are where there is a potential for exposure).

If all five parts are present, the exposure pathway is known as a completed exposure pathway. In addition, the exposure may be short-term, intermediate, or long-term. Short-term contact is known as an acute exposure, that is, occurring as a single event or for only a short period of time (up to 14 days). An intermediate exposure is one that lasts from 14 days to less than one year. Long-term or chronic exposures are greater than one year in duration.

Determining the exposure for a neighborhood can be complicated. For example, even if we do a good job identifying all the contaminants of concern and their possible sources (no small task), we may have little idea of the extent to which the receptor population has come into contact with these contaminants (steps 2 through 4). Thus, assessing exposure involves not only the physical sciences, but the social sciences, especially psychology and behavioral. People's activities greatly affect the amount and type of exposures. That is why exposure scientists use a number of techniques to establish activity patterns, such as asking potentially exposed individuals to keep diaries, videotaping, and using telemetry to monitor vital information, such as heart and ventilation rates.

General ambient measurements, such as air pollution monitoring equipment located throughout cities, are often not good indicators of actual population exposures. As indicated in Figure 5.13, lead (Pb) and mercury (Hg) compounds comprise the greatest mass of toxic substances released into the U.S. environment. This is largely due to the large volume and surface areas involved in metal extraction and refining operations. However, this does not necessarily mean that more people will be exposed at higher concentrations or more frequently to these compounds than to others. The mere fact that a substance is released, or even that it is found in the ambient environment, is not tantamount to its coming in contact with people. Conversely, even a small amount of a substance under the right circumstances can lead to very high levels of exposure (e.g., in an occupational setting, in certain indoor environments, and through certain pathways, such as nondietary ingestion of paint chips by children).

A recent study by the Lawrence Berkley Laboratory demonstrates the importance of not simply assuming that the released or even background concentrations are a good indicator of actual exposure.[39] The researchers

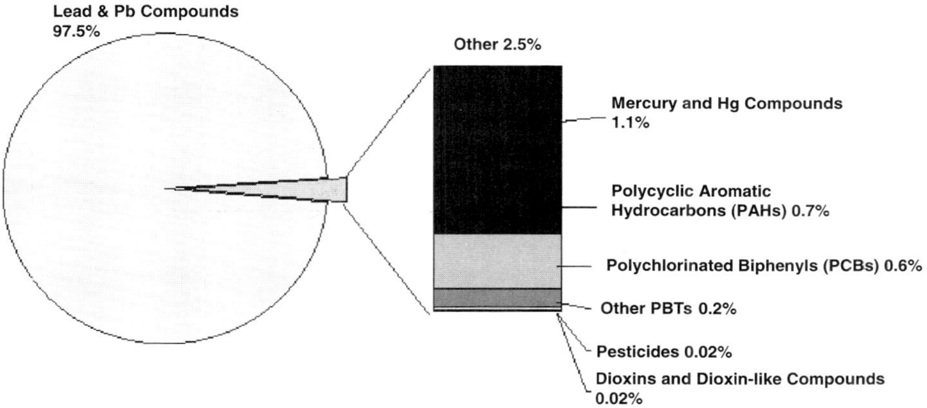

FIGURE 5.13. Total U.S. releases of contaminants in 2001, as reported to the Toxic Release Inventory (TRI). Total releases = 2.8 billion kg. Note: Off-site releases include metals and metal compounds transferred off-site for solidification/stabilization and for wastewater treatment, including to publicly owned treatment works. Off-site releases do not include transfers to disposal sent to other TRI facilities that reported the amount as an on-site release.
Source: U.S. Environmental Protection Agency.

were interested in how sorption may affect indoor environments, so they set up a room (chamber) made up of typical building materials and furnished with actual furniture like that found in most residential settings. A number of air pollutants were released into the room and monitored. Figure 5.14 shows the organic solvent, xylene, exhibiting the effects of sorption.

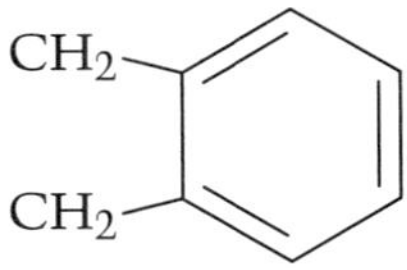

xylene

With the room initially sealed, the observed decay in vapor phase concentrations indicates that the compound is adsorbing onto surfaces (walls, furniture, etc.). The adsorption continues for hours, with xylene concentrations reaching a quasi-steady state. At this point the room is flushed with clean air to free all vapor phase xylene. The xylene concentrations shortly after the flush began to rise again until reaching a new steady-state. This rise must be the result of desorption of the previously sorbed xylene, since the initial source is gone. Sorption is one of the processes that must be considered to account for differences in the temporal pattern of indoor versus outdoor concentrations.

Figure 5.15 shows a number of the ways that contaminants can enter and leave an indoor environment. People's activities as they move from one

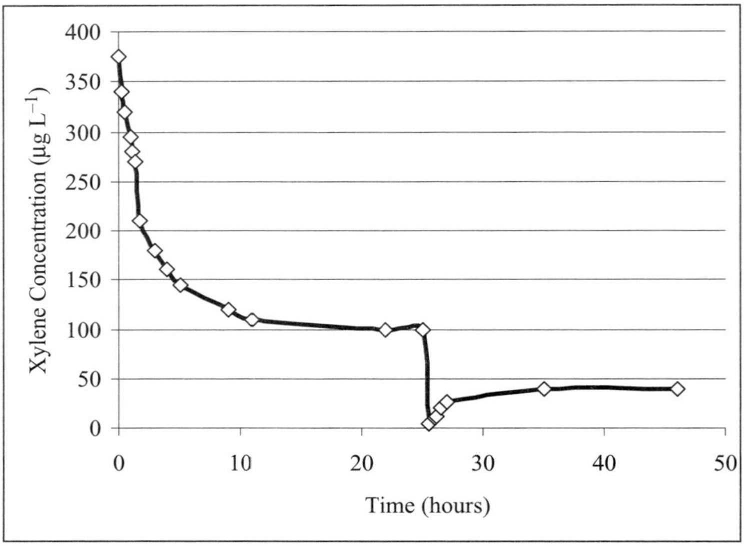

FIGURE 5.14. Vapor phase concentrations of xylene measured in a chamber sealed during adsorption and desorption periods.
Source: Adapted from B. Singer, 2003. "A Tool to Predict Exposure to Hazardous Air Pollutants," *Environmental Energy Technologies Division News,* 4(4), 5.

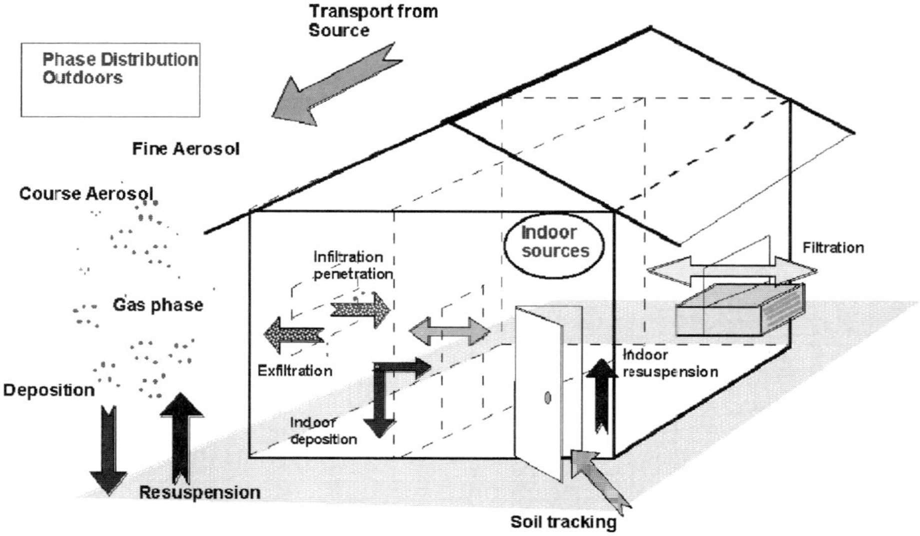

FIGURE 5.15. Movement of an agent into and out of a home. Accounting for the movement and change of a chemical compound (the mass balance) is a key component of an exposure assessment.
Source: U.S. Department of Energy, Lawrence Berkley Laboratory, 2003. http://eetd.lbl.gov/ied/ERA/CalEx/partmatter.html.

location to another make for unique exposures. For example, people generally spend much more time indoors than outdoors. The simplest quantitative expression of exposure is:

$$E = D/t \tag{5.5}$$

where

E = human exposure during the time period, t (units of concentration [mass per volume] per time) (mg kg^{-1} day^{-1})
D = mass of pollutant per body mass (mg kg^{-1})
t = time (day)

Usually, to obtain D, the chemical concentration of a pollutant is measured near the interface of the person and the environment during a specified time period. This measurement is sometimes referred to as the potential dose (i.e., the chemical has not yet crossed the boundary into the body, but is present where it may enter the person, such as on the skin, at the mouth, or at the nose).

Expressed quantitatively, exposure is a function of the concentration of the agent and time. It is an expression of the magnitude and duration of the contact. That is, exposure to a contaminant is the concentration of that contact in a medium integrated over the time of contact:

$$E = \int_{t=t_1}^{t=t_2} C(t)dt \tag{5.6}$$

where

E = exposure during the time period from t_1 to t_2
$C(t)$ = concentration at the interface between the organism and the environment, at time t

Equation 5.6 is interesting for reasons beyond the physical sciences and engineering. It shows that exposure is a function of the physicochemical characteristics of the pollution scenario (i.e., the toxicant *and* the substrate determine the concentration). But it also shows that the social sciences and humanities come into play in determining possible exposures, since the exposure is also a function of time (i.e., the dt term). Whether people are indoors or outdoors, how they get to work, what they do at home, how long they sleep, and myriad other sociometric factors, known as activity patterns, are needed to estimate the time of exposure.

The concentration at the interface is the potential dose (i.e., the chemical might enter the person). Since the amount of a chemical agent that penetrates from the ambient atmosphere into a building affects the concentration term of the exposure equation, a complete mass balance of the contaminant must be understood and accounted for; otherwise, exposure estimates will be incorrect. The mass balance consists of all inputs and outputs, as well as chemical changes to the contaminant:

$$\begin{gathered}\text{Accumulation or loss of contaminant A} = \\ \text{Mass of A transported in} - \text{Mass of A transported out} \pm \text{Reactions}\end{gathered} \tag{5.7}$$

The reactions may be either those that generate chemical A (i.e., *sources*), or those that destroy chemical A (i.e., *sinks*). Thus, the amount of mass transported in is the inflow to the system that includes pollutant discharges, transfer from other control volumes and other media (for example, if the control volume is soil, the water and air may contribute mass of chemical A), and formation of chemical A by abiotic chemistry and biological transformation. Conversely, the outflow is the mass transported out of the control volume, which includes uptake by biota, transfer to other compartments (e.g., volatilization to the atmosphere), and abiotic and biological degradation of chemical A. This means the rate of change of mass in a control volume is equal to the rate of chemical A transported in less the rate of chemical A transported out, plus the rate of production from sources, and minus the rate of elimination by sinks. Stated as a differential equation, the rate of change of chemical A is:

$$\frac{d[A]}{dt} = -v \cdot \frac{d[A]}{dx} + \frac{d}{dx}\left(\Gamma \cdot \frac{d[A]}{dx}\right) + r \tag{5.8}$$

where

v = fluid velocity
Γ = a rate constant specific to the environmental medium
$\frac{d[A]}{dx}$ = concentration gradient of chemical A
r = internal sinks and sources within the control volume

Reactive compounds can be particularly difficult to measure. For example, many volatile organic compounds in the air can be measured by first collecting them in stainless steel canisters and analyzing by chromatography in the lab. However, some of these compounds, like the carbonyls (notably aldehydes like formaldehyde and acetaldehyde), are prone to react inside the canister, meaning that by the time the sample is analyzed, a portion of the carbonyls are degraded (under-reported). Therefore, other methods must be used, such as trapping the compounds with dinitrophenyl hydrazine (DNPH)-treated silica gel tubes that are frozen

TABLE 5.6
Preservation and holding times for anion sampling and analysis.

PART A: Common Anions

Analyte	Preservation	Holding Time
Bromide	None required	28 days
Chloride	None required	28 days
Fluoride	None required	28 days
Nitrate-N	Cool to 4°C	48 hours
Nitrite-N	Cool to 4°C	48 hours
Ortho-Phosphate-P	Cool to 4°C	48 hours
Sulfate	Cool to 4°C	28 days

PART B: Inorganic Disinfection Byproducts

Analyte	Preservation	Holding Time
Bromate	$50\,mg\,L^{-1}$ EDA	28 days
Bromide	None required	28 days
Chlorate	$50\,mg\,L^{-1}$ EDA	28 days
Chlorite	$50\,mg\,L^{-1}$ EDA, cool to 4°C	14 days

Source: U.S. Environmental Protection Agency, 1997. EPA Method 300.1: Determination of Inorganic Anions in Drinking Water by Ion Chromatography, Revision 1.0.

until being extracted for chromatographic analysis. The purpose of the measurement is to see what is in the air, water, soil, sediment, or biota at the time of sampling, so any reactions before the analysis give measurement error.

It is important to keep in mind that the chemical that is released or to which one is exposed is not necessarily what needs to be measured. For example, if the released chemical is reactive, some or all of it may have changed into another form (i.e., *speciated*) by the time it is measured. Even relatively nonreactive compounds may speciate between when the sample is collected (e.g., in a water sample, an air canister, a soil core, or a bag) and when the sample is analyzed. In fact, each contaminant has unique characteristics that vary according to the type of media in which it exists and extrinsic conditions like temperature and pressure. Sample preservation and holding times for the anions according to EPA Method 300.1, Determination of Inorganic Anions in Drinking Water by Ion Chromatography, are shown in Table 5.6. These methods vary according to the contaminant of concern and the environmental medium from which it is collected, so the environmental practitioner needs to find and follow the correct methods.

The general exposure equation (5.6) is rewritten to address each route of exposure, accounting for chemical concentration and the activities that affect the time of contact. The exposure calculated from these equations is

TABLE 5.7
Commonly used human exposure factors.

Exposure Factor	*Adult Male*	*Adult Female*	*Child (3–12 years of age)*[41]
Body weight (kg)	70	60	15–40
Total fluids ingested (L d^{-1})	2	1.4	1.0
Surface area of skin, without clothing (m^2)	1.8	1.6	0.9
Surface area of skin, wearing clothes (m^2)	0.1–0.3	0.1–0.3	0.05–0.15
Respiration/ventilation rate (L min^{-1})—Resting	7.5	6.0	5.0
Respiration/ventilation rate (L min^{-1})—Light activity	20	19	13
Volume of air breathed (m^3 d^{-1})	23	21	15
Typical lifetime (years)	70	70	NA
National upper-bound time (90th percentile) at one residence (years)	30	30	NA
National median time (50th percentile) at one residence (years)	9	9	NA

Sources: U.S. Environmental Protection Agency, 2003. *Exposure Factor Handbook*; and Agency for Toxic Substances and Disease Registry, 2003, ATSDR Public Health Assessment Guidance Manual.[40]

actually the chemical intake (I) in units of concentration (mass per volume or mass per mass) per time, such as $mg\,kg^{-1}\,day^{-1}$:

$$I = \frac{C \cdot CR \cdot EF \cdot ED \cdot AF}{BW \cdot AT} \tag{5.9}$$

where

C = chemical concentration of contaminant (mass per volume)
CR = contact rate (mass per time)
EF = exposure frequency (number of events, dimensionless)
ED = exposure duration (time)
AF = absorption factor (equals 1 if C is completely absorbed)
BW = body weight (mass)
AT = averaging time (if lifetime exposure = 70 yrs)

These factors are further specified for each route of exposure, such as the lifetime average daily dose (LADD) as shown in Appendix 6. Some of the default values often used in exposure assessments are given in Table 5.7. The LADD is based on a chronic, long-term exposure. Acute and subchronic exposures require different equations, since the exposure duration (ED) is much shorter. For example, instead of LADD, acute exposures to noncarcingens may use maximum daily dose (MDD) to calculate exposure. However, even these exposures follow the general model given in Equation 5.9.

The hazard and exposure information must be combined to determine the risk. Two methods of doing so are given in Appendix 7.

Risk-Based Cleanup Standards

The hazardous waste cases as well as those affecting the air and water have changed the way pollutants are regulated. For most of the second half of the twentieth century, environmental protection was based on two types of controls, technology-based and quality-based. Technology-based controls are set according to what is achievable from the current state of the science and engineering. These are feasibility-based standards. The Clean Air Act has called for best achievable control technologies (BACT), and more recently for maximally achievable control technologies (MACT). Both standards reflect the reality that even though from an air quality standpoint it would be best to have extremely low levels of pollutants, technologies are not available or are not sufficiently reliable to reach these levels. Requiring unproven or unreliable technologies can even exacerbate the pollution, such as in the early days of wet scrubbers on coal-fired power plants. Theoretically, the removal of sulfur dioxide could be accomplished by venting the power plant flue through a slurry of carbonate, but the technology at the time was unproven and unreliable, allowing all-too-frequent releases of untreated emissions while the slurry systems were being repaired. Selecting a new technology over older proven techniques is unwise if the trade-off of the benefit of improved treatment over older methods is outweighed by the numerous failures (i.e., no treatment).

Technology-based standards are a part of most environmental programs. Wastewater treatment, ground water remediation, soil cleaning, sediment reclamation, drinking water supply, air emission controls, and hazardous waste site cleanup all, in part, are determined by availability and feasibility of control technologies.

Quality-based controls are those that are required to ensure that an environmental resource is in good enough condition to support a particular use. For example, a stream may need to be improved so that people can swim in it and so that it can be a source of water supply. Certain streams may need higher levels of protection than others, such as the so-called "wild and scenic rivers." The parameters will vary, but usually include minimum levels of dissolved oxygen and maximum levels of contaminants. The same goes for air quality, where ambient air quality must be achieved so that concentrations of contaminants listed as National Ambient Air Quality Standards, as well as certain toxic pollutants, are below levels established to protect health and welfare.

Recently, environmental protection has become increasingly risk-based. Risk-based approaches to environmental protection, especially contaminant target concentrations, are designed to require engineering controls and preventive measures to ensure that risks are not exceeded. The risk-based approach actually embodies elements of both technology-based and quality-based standards. The technology assessment helps determine how realistic it will be to meet certain contaminant concentrations, and the quality of the environment sets the goals and means to achieve cleanup. Environmental practitioners are often asked, "How clean is clean?" When do we know that we have done a sufficient job of cleaning up a spill or hazardous waste site? It is often not possible to have nondetectable concentrations of a pollutant. Commonly, the threshold for cancer risk to a population is one in a million excess cancers. However, we may find that the contaminant is so difficult to remove that we almost give up on dealing with the contamination and put in measures to prevent exposures, such as fencing the area in and prohibiting access. This is often done as a first step in remediation, but is unsatisfying and controversial (and usually politically and legally unacceptable). Thus, even if costs are high and technology unreliable, the environmental practitioner must find suitable and creative ways to clean up the mess and meet risk-based standards.

Risk-based target concentrations can be calculated by solving for the target contaminant concentration in the exposure and risk equations. Since risk is the hazard (e.g., slope factor) times the exposure (e.g., LADD), a cancer risk-based cleanup standard can be found by enumerating the exposure equation (5.9) within the risk equation. For example, the exposure (LADD) equation for drinking water is:

$$\mathrm{LADD} = \frac{\mathrm{C} \cdot \mathrm{CR} \cdot \mathrm{ED} \cdot \mathrm{AF}}{\mathrm{BW} \cdot \mathrm{AT}} \tag{5.10}$$

where CR = water consumption rate (L day^{-1}). Thus, since risk is the product of exposure (LADD) and hazard (slope factor for cancer) the cancer drinking water risk equation is

$$\mathrm{Risk} = \frac{\mathrm{C} \cdot \mathrm{CR} \cdot \mathrm{EF} \cdot \mathrm{ED} \cdot \mathrm{AF} \cdot \mathrm{SF}}{\mathrm{BW} \cdot \mathrm{AT}} \tag{5.11}$$

and solving for C:

$$\mathrm{C} = \frac{\mathrm{Risk} \cdot \mathrm{BW} \cdot \mathrm{AT}}{\mathrm{CR} \cdot \mathrm{EF} \cdot \mathrm{ED} \cdot \mathrm{AF} \cdot \mathrm{SF}} \tag{5.12}$$

This is the target concentration for each contaminant needed to protect the population from the specified risk, for example, 10^{-6} would be inserted for the risk term in equation 5.12. In other words, C is the concentration that must not be exceeded in order to protect a population having an average body weight and over a specified averaging time from an exposure of certain duration and frequency that leads to a risk of one in a million.

Although one-in-a-million added risk is a commonly used benchmark, cleanup may not always be required to achieve this level. For example, if a site is considered to be a removal action—that is, the principal objective is to get rid of a sufficient amount of contaminated soil to reduce possible exposures—the risk reduction target may be as high as one additional cancer per 10,000 (i.e., 10^{-4}). An example of a risk-based cleanup calculation is given in Appendix 8.

The Drake Chemical Company Superfund Site: A Risk-Based Case[42]

The Drake Chemical Company of Lock Haven, PA, was a major producer of chemicals during the Second World War and continued to provide employment opportunities to the economically depressed town after the war. One of its waste chemicals that the company disposed of in an open pit was beta-naphthylamine (also known as 2-naphthylamine), a compound used as dye.

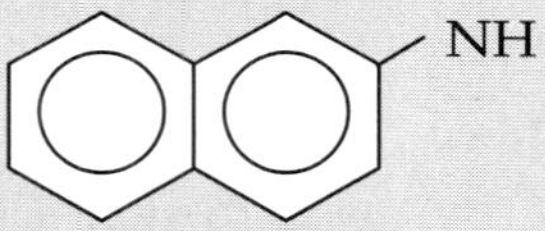

beta-naphthylamine

Unfortunately, beta-naphthylamine is also a potent carcinogen (inhalation and oral cancer slope factor = 1.8),[43] having been found to be a known human carcinogen based on sufficient evidence of carcinogenicity in humans. Epidemiological studies have shown that occupational exposure to beta-naphthylamine alone or when present as an impurity in other compounds is causally associated with bladder cancer in workers.[44]

In 1962 the State of Pennsylvania banned the production of this chemical, but the damage to the ground water had already been done with the disposal of beta-naphthylamine into the uncontrolled pit. The order from the state caused Drake to stop manufacturing beta-naphthylamine, but it continued to produce other chemicals, seemingly without much concern for the environment or the health of the people in Lock Haven. Finally in 1981, the U.S. EPA closed them down and took control of the property. They discovered several unlined lagoons and hundreds of often unmarked barrels of chemicals stored in makeshift buildings. After removing the drums and draining the lagoons, inspectors then found that the beta-naphthylamine had seeped into nearby property and into creeks, creating a serious health hazard. The EPA's attempts to clean the soil and the water was, however, met with public opposition. Much of the public blamed the EPA for forcing Drake Chemical, a major local employer, to close the plant. In addition, the best way to treat the contaminated soil was to burn it in an incinerator, and the EPA made plans to bring in a portable unit. Now the public, not at all happy with the EPA being there in the first place, became concerned with the emissions from the incinerator. After many studies and the involvement of the U.S. Army Corps of Engineers, the incinerator was finally allowed to burn the soil, which was then spread out and covered with 3.5 feet of topsoil. The groundwater was pumped and treated, and this continued until the levels of beta-naphthylamine reached background concentrations. The project was not completed until 1999, with the EPA paying for the legal fees of the lawyers who argued against the cleanup.

Some general principles have been almost universally adopted by regulatory agencies in determining risks, especially those concerned with cancer risks from environmental exposures (see Table 5.8).

Zero risk can occur only when either the hazard (e.g., toxicity) does not exist or the exposure to that hazard is zero. A substance found to be associated with cancers based upon animal testing or observations of human populations can be further characterized. Association of two factors, such as the level of exposure to a compound and the occurrence of a disease, does not necessarily mean that one causes the other. Often, after study, a third variable explains the relationship. However, it is important for science to do what it can to link causes with effects. Otherwise, corrective and preventive actions cannot be identified. So, strength of association is a beginning step toward cause and effect. A major consideration in strength of association is the application of sound technical judgment of the weight of evidence. For example, characterizing the weight of evidence for carcinogenicity in humans consists of three major steps:[45]

1. Characterization of the evidence from human studies and from animal studies individually.
2. Combination of the characterizations of these two types of data to show the overall weight of evidence for human carcinogenicity.
3. Evaluation of all supporting information to determine if the overall weight of evidence should be changed.

Note that none of these steps is absolutely certain.

TABLE 5.8
General principles applied to health and environmental risk assessments in the United States.

Principle	*Explanation*
Human data are preferable to animal data.	For purposes of hazard identification and dose-response evaluation, epidemiological and other human data better predict health effects than animal models.
Animal data can be used in lieu of sufficient, meaningful human data.	Although epidemiological data are preferred, agencies are allowed to extrapolate hazards and to generate dose-response curves from animal models.
Animal studies can be used as a basis for risk assessment.	Risk assessments can be based upon data from the most highly sensitive animal studies.
Route of exposure in animal study should be analogous to human routes.	Animal studies are best if from the same route of exposure as those in humans, e.g., inhalation, dermal, or ingestion routes. For example, if an air pollutant is being studied in rats, inhalation is a better indicator of effect than if the rats are dosed on the skin or if the exposure is dietary.
Threshold is assumed for noncarcinogens.	For noncancer effects, e.g., neurotoxicity, endocrine dysfunction, and immunosuppression, there is assumed to be a safe level under which no effect would occur (e.g., NOAEL, which is preferred, but also LOAEL).
Threshold is calculated as a reference dose or reference concentration (air).	Reference dose (RfD) or concentration (RfC) is the quotient of the threshold (NOAEL) divided by factors of safety (uncertainty factors and modifying factors; each usually multiples of 10): $\mathrm{RfD} = \dfrac{\mathrm{NOAEL}}{\mathrm{UF} \times \mathrm{MF}}$
Sources of uncertainty must be identified.	Uncertainty factors (UFs) address: • Interindividual variability in testing • Interspecies extrapolation • LOAEL-to-NOAEL extrapolation • Subchronic-to-chronic extrapolation • Route-to-route extrapolation • Data quality (precision, accuracy, completeness, and representativeness) Modifying factors (MFs) address uncertainties that are less explicit than the UFs.
Factors of safety can be generalized.	The uncertainty and modifying factors should follow certain protocols, e.g., 10 = for extrapolation from a sensitive individual to a

TABLE 5.8
Continued

Principle	*Explanation*
	population; 10 = rat-to-human extrapolation, 10 = subchronic-to-chronic data extrapolation, and 10 = LOAEL used instead of NOAEL.
No threshold is assumed for carcinogens.	No safe level of exposure is assumed for cancer-causing agents.
Precautionary principle is applied to cancer model.	A linear, no-threshold dose-response model is used to estimate cancer effects at low doses; i.e., to draw the unknown part of the dose-response curve from the region of observation (where data are available) to the region of extrapolation.
Precautionary principle is applied to cancer exposure assessment.	The most highly exposed individual generally is used in the risk assessment (upper-bound exposure assumptions). Agencies are reconsidering this worst-case policy, and considering more realistic exposure scenarios.

Source: U.S. Environmental Protection Agency, 2001. General Principles for Performing Aggregate Exposure and Risk Assessment, Office of Pesticides Programs, Washington, D.C.

Environmental risk by nature addresses probable impossibilities. From a statistical perspective, it is extremely likely that cancer will not be eliminated during our lifetimes. But the efforts to date have shown great progress toward reducing risks from several forms of cancer. This risk reduction can be attributed to a number of factors, including changes in behavior (smoking cessation, dietary changes, and improved lifestyles), source controls (less environmental releases of cancer-causing agents), and the reformulation of products (substitution of chemicals in manufacturing processes).

Risk Assessment: The First Step

Risk characterization is the stage where the environmental practitioner summarizes the necessary assumptions, describes the scientific uncertainties, and determines the strengths and limitations of the analyses. The risks are articulated by integrating the analytical results, interpreting adverse outcomes, and describing the uncertainties and weights of evidence. The emphasis varies, for example, much of their culture and livelihood is directly linked to ecosystems, such as Native American subsistence agriculture, silviculture, and fishing, and African American communities in or near riparian and littoral habitats.

A reliable risk assessment is the groundwork for determining whether risks are disproportionate in a given neighborhood or region. Exposures to hazards can be disproportionate, which leads to disproportionate risk. There are also situations where certain groups of people are more sensitive to the effects of pollutants. Such things are hard to quantify, but need to be addressed.

Risk assessment is a process distinct from risk management, where actions are taken to address and reduce the risks. But the two are deeply interrelated and require continuous feedback with each other. Engineers are key players in both efforts. In addition, risk communication between the engineer and the client further complicate the implementation of the risk assessment and management processes. What really sets risk assessment apart from the actual management and policy decisions is that the risk assessment must follow the prototypical rigors of scientific investigation and interpretation outlined in this chapter. Risk management draws upon technical risk assessment, but must also factor in other social considerations.

Notes and Commentary

1. T. Colburn, 1996. Speech at *The State of the Word Forum*, San Francisco, CA.
2. Many community resources are available, from formal public meetings held by governmental authorities to informal groups, such as homeowner association meetings and neighborhood "watch" and crime prevention group meetings. Any research-related activities should adhere to federal and other governmental regulations regarding privacy, intrusion, and human subject considerations. Privacy rules have been written according to the Privacy Act and the Paperwork Reduction Act (e.g., the Office of Management and Budget limits the type and amount of information that U.S. agencies may collect in what is referred to as an Information Collection Budget).

   Any research that affects human subjects, at a minimum, should have prior approval for informed consent of participants and thoughtful consideration of the need for an institutional review board (IRB) approval.
3. As the name implies, first responders are the teams who first arrive on the scene of an emergency. They include firefighters, HAZMAT teams, police, and medical personnel. These people are particularly vulnerable to exposures. Often, the contents of items and areas needing response are not well known, so the wrong treatment or response can be dangerous, such as spraying water on low-density or water-reactive substances. Other vulnerabilities include the frenetic nature of an emergency response. For example, the first responders to the World Trade Center attacks on September 11, 2001, had incompatible radios and, since cell phone networks had collapsed, they were not able to communicate well with each other. This undoubtedly contributed to a number of deaths. The vulnerability has been articulated well by Captain Larry R. Collins,

a 24-year member of the Los Angeles County Fire Department *(Frontline First Responder*, April 5, 2003):

> A truly accurate assessment of the stability of damaged structures often requires the skill, experience, training, and knowledge of a certified structural engineer who is prepared to perform a risk analysis and make certain calculations about the weight of the material, the status of key structural members, how the loads have been redistributed after the event, and the need for stabilization or evacuation. Unfortunately, first responders typically don't have those capabilities, and when lives are hanging in the balance, they don't have the luxury of time to wait for a structural engineer. Someone needs to make immediate decisions about firefighting, search and rescue, and other emergency operations.

4. L. Stieglitz, G. Zwick, J. Beck, H. Bautz, and W. Roth, 1989. *Chemosphere*, 19:283.
5. For a discussion of the transport of dioxins, see C. Koester and R. Hites, 1992. "Wet and dry deposition of chlorinated dioxins and furans," *Environmental Science and Technology*, 26:1375–1382. R. Hites, 1991. Atmospheric transport and deposition of polychlorinated dibenzo-p-dioxins and dibenzofurans, Research Triangle Park, NC. EPA/600/3-91/002.
6. U.S. Census Bureau, 2003; reported in U.S. Commission on Civil Rights, 2003. Not in My Backyard: Executive Order 12,898 and Title VI as Tools for Achieving Environmental Justice, Washington, D.C.
7. Gunite is a patented construction material composed of cement, sand, or crushed slag and water mixed pneumatically. Often used in the construction of swimming pools, it provides a waterproof lining.
8. L. Robertson, 2002. Reflections on the World Trade Center, *The Bridge*, 32.
9. A. Leopold, 1949. *A Sand County Almanac*, Oxford University Press (1987), New York, NY.
10. E. Birmingham, 1998. Position Paper: "Reframing the Ruins: Pruitt-Igoe, Structural Racism, and African American Rhetoric as a Space for Cultural Critique," Brandenburgische Technische Universität, Cottbus, Germany. See also: C. Jencks, 1987. *The Language of Post-Modern Architecture*, 5e, Rizzoli, New York, NY.
11. A. von Hoffman, 2002. "Why They Built Pruitt-Igoe." Taubman Center Publications, A. Alfred Taubman Center for State and Local Government, Harvard, University, Cambridge, MA.
12. J. Bailey, 1965. "A Case History of Failure," *Architectural Forum*, 122 (9).
13. *Ibid.*
14. See, for example, D.A. Vallero, 2002. "Teachable Moments and the Tyranny of the Syllabus: September 11 Case, *Journal of Professional Issues in Engineering Education and Practice*, 129 (2), 100–105.
15. C. Mitcham and R.S. Duval, 2000. *Engineering Ethics*, Chapter 8, "Responsibility in Engineering," Prentice-Hall, Upper Saddle River, NJ.

16. C.B. Fleddermann, 1999. *Engineering Ethics*, Chapter 5, "Safety and Risk," Prentice-Hall, Upper Saddle River, NJ.
17. Resource Conservation and Recovery Act of 1976 (42 U.S.C. s/s 321 et seq.).
18. Comprehensive Environmental Response, Compensation and Liability Act of 1980 (42 U.S.C. 9601-9675). December 11, 1980. In 1986, CERCLA was updated and improved under the Superfund Amendments and Reauthorization Act (42 U.S.C. 9601 et seq.), October 17, 1986.
19. State University of New York, Stony Brook, 2004. http://www.matscieng.sunysb.edu/disaster/; accessed November 6, 2004.
20. P. Sandman's advice is found in S. Rampton and J. Stauber, 2001. *Trust Us, We're Experts: How Industry Manipulates Science and Gambles with Your Future,* Jeffrey B. Tarcher/Putnam, New York, NY.
21. Akin to John Gray's bestselling book *Men Are From Mars, Women Are from Venus: A Practical Guide for Improving Communications and Getting What You Want in Your Relationships,* 1ed Harper Collins, 1992, New York, NY. Another analogy is that of the popular personality tests, such as the Myers-Briggs typologies. Often, engineers and scientists direct their intellectual energies toward the inner world, at least while they are on the job. They attempt to be clear about data and information in order to understand what it is they are studying. They trust experience (i.e., they adhere to experimental findings). Conversely, many of their clients direct their energies outwardly, speaking before they have completely formulated an idea. This is not necessarily "sloppiness," but scientists tend to perceive it to be. It is often an attempt to explore possible alternatives to address a problem. In other words, when it comes to science, the client is often more comfortable with ambiguity than is the engineer. Interestingly, some of the great scientists, like Einstein, Bohr, and the contempories like Gould and Hawking evidenced a great deal of comfortable ambiguous and yet-to-be-explored paradigms.
22. H.W. Lewis, 1990. *Technological Risk*, Chapter 5, "The Assessment of Risk," W.W. Norton & Company, Inc., New York, NY.
23. C. Tesar, 2000. "POPs: What They Are; How They Are Used; How They Are Transported," *Northern Perspectives,* 26 (1), 2–5.
24. The source of this discussion is the U.S. Commission on Civil Rights report, "Not in My Backyard."
25. Chatham College, "Leaders of Cancer Alley," http://www.chatham.edu/rci/well/women21-30/canceralley.html; accessed April 10, 2003.
26. Elizabeth Teel, deputy director, Environmental Law Clinic, Tulane Law School, testimony before the U.S. Commission on Civil Rights, hearing, Washington, D.C., Jan. 11, 2002, official transcript, p. 117.
27. German Federal Ministry for Economic Cooperation and Development, 2004. Environmental Handbook: Documentation on monitoring and evaluating environmental impacts, Volume III, Compendium of Environmental Standards: http://www.gtz.de/uvp/publika/English/vol369.htm; accessed November 29, 2004.

28. Mossville Environmental Action Network, 2000. "Breathing Poison: The Toxic Costs of Industries in Calcasieu Parish, Louisiana": http://www.mapCruzin.com/mossville/reportondioxin.htm.
29. State of Georgia, 2003. *Watershed Protection Plan Development Guidebook.*
30. National Research Council, 1989. *Biologic Markers in Reproductive Toxicology,* National Academy Press, Washington, D.C.
31. David Weir and Constance Matthiessen, "Will the Circle Be Unbroken?" *Mother Jones,* June 1989.
32. General source of information for this case is NBC-Med: http://www.nbc-med.org/SiteContent/MedRef/OnlineRef/CaseStudies/csGoiania.html; accessed December 3, 2004.
33. M. Sun, 1987. "Radiation Accident Grips Goiania," *Science,* 238, 1028–1031.
34. A.G. Knudson, 1985. "Hereditary Cancer, Oncogenes, and Antioncogenes," *Cancer Research,* 45 (4), 1437–1443.
35. For air pollutants, the reference concentration (RfC) is used. It is applied in exactly the same manner as the RfD.
36. E.E. McConnell, H.A. Solleveld, J.A. Swenberg, and G.A. Boorman, 1986. "Guidelines for Combining Neoplasms for Evaluation of Rodent Carcinogenesis Studies," *Journal of the National Cancer Institute,* 76(2): 283–289.
37. U.S. Environmental Protection Agency, 1992. Background Document 2, EPA Approach for Assessing the Risks Associated with Chronic Exposures to Carcinogens, Integrated Risk Information System.
38. T. Colburn, D. Dumanoski, and J.P. Myers, 1996. *Our Stolen Future: Are We Threatening Our Fertility, Intelligence, and Survival? A Scientific Detective Story,* Dutton, New York, NY.
39. B. Singer, 2003. "A Tool to Predict Exposure to Hazardous Air Pollutants," *Environmental Energy Technologies Division News,* 4(4), 5.
40. These factors are updated periodically by the U.S. EPA in the *Exposure Factor Handbook* at www.epa.gov/ncea/exposfac.htm.
41. There is no consensus on the definiton of "child" in risk assessment. The *Exposure Factor Handbook* uses these values for children between the ages of 3 and 12 years.
42. L.D. Budnick, D.C. Sokal, H. Falk, J.N. Logue, and J.M. Fox, 1984. "Cancer and Birth Defects Near the Drake Superfund Site," *Pennsylvania Archives of Environmental Health,* 39, 409–413.
43. California Office of Environmental Health Hazard Assessment, 2002. California Cancer Potency Values: http://www.oehha.ca.gov/risk/chemicalDB/index.asp; accessed November 23, 2004.
44. International Agency for Research on Cancer, 1974. IARC Monographs on the Evaluation of the Carcinogenic Risk of Chemicals to Man. Some Aromatic Amines, Hydrazine and Related Substances, N-Nitroso Compounds and Miscellaneous Alkylating Agents. Vol. 4, Lyon, France. International Agency for Research on Cancer, 1979. IARC Monographs on the Evaluation of the Carcinogenic Risk of Chemicals to Humans. Chemicals and Industrial Processes Associated with Cancer in Humans, Supplement 1, Lyon, France. International

Agency for Research on Cancer, 1982. IARC Monographs on the Evaluation of the Carcinogenic Risk of Chemicals to Humans. Chemicals, Industrial Processes and Industries Associated with Cancer in Humans. Supplement 4, Lyon, France. International Agency for Research on Cancer, 1987. IARC Monographs on the Evaluation of Carcinogenic Risks to Humans: Overall Evaluations of Carcinogenicity. Supplement 7, Lyon, France.

45. U.S. Environmental Protection Agency, 1986. Guidelines for Carcinogen Risk Assessment, Report No. EPA/630/R-00/004, *Federal Register* 51(185):33992–34003, Washington, D.C.

CHAPTER 6

By Way of Introduction

Only with absolute fearlessness can we slay the dragons of mediocrity that invade our gardens.

George Lois, Twentieth Century U.S. Advertising Designer

Although Lois was probably talking about society's tendencies away from excellence; many of the "dragons" in our environment have come at our own invitation, as intentionally introduced species. Unlike the chemical stressors we discussed in Chapters 3 through 5, these are living organisms. They are usually introduced into ecosystems where no natural predators can check the newly arrived species' numbers and geographic range.

Introduced species cause us to rethink our concept of pollution. Like so many other issues in this book, the type of pollution caused when an opportunistic, invasive organism colonizes and outlives its welcome is a systematic one. The threat is not usually due to a single species, although this could be very important if a species were already threatened or endangered prior to the invasion. It is usually a problem of the whole ecosystem. Something is out of balance.

The concept of invasion is not one that is consistently applied. For example, the very diligent Invasive Species Specialist Group (ISSG) has reluctantly listed the 100 "worst" invasive species in the world (see Table 6.1). In ISSG's own words, the task is difficult:

> Species and their interactions with ecosystems are very complex. Some species may have invaded only a restricted region, but have a huge probability of expanding, and causing further great damage (for example, see *Boiga irregularis*: the brown tree snake). Other species may already be globally widespread, and causing cumulative but less visible damage. Many biological families or genera contain large numbers of invasive species, often with similar impacts; in these cases one representative species was chosen. The one hundred species aim to collectively illustrate the range of impacts caused by biological invasion.[1]

TABLE 6.1
Worst invasive species listed in the Global Invasive Species Database.

Genus and species	*Common Names*
1. *Acacia mearnsii* (shrub, tree)	acácia-negra, Australian acacia, Australische akazie, black wattle, swartwattel, uwatela
2. *Achatina fulica* (mollusc)	Afrikanische Riesenschnecke, giant African land snail, giant African snail
3. *Acridotheres tristis* (bird)	common myna, Hirtenmaina, Indian myna, Indian mynah, mynah
4. *Aedes albopictus* (insect)	Asian tiger mosquito, forest day mosquito, zanzare tigre
5. *Anopheles quadrimaculatus* (insect)	common malaria mosquito, Gabelmücke
6. *Anoplolepis gracilipes* (insect)	ashinaga-ki-ari, crazy ant, Gelbe Spinnerameise, gramang ant, long-legged ant, Maldive ant, yellow crazy ant
7. *Anoplophora glabripennis* (insect)	Asian longhorned beetle, Asiatischer Laubholzkäfer, longicorne Asiatique, starry sky beetle
8. *Aphanomyces astaci* (fungus)	crayfish plague, Wasserschimmel
9. *Ardisia elliptica* (tree)	ati popa'a, shoebutton ardisia
10. *Arundo donax* (grass)	arundo grass, bamboo reed, cana, cane, canne de Provence, carrizo grande, cow cane, donax cane, giant cane, giant reed, narkhat, ngasau ni vavalangi, Pfahlrohr, reedgrass, river cane, Spanisches Rohr, Spanish cane, Spanish reed
11. *Asterias amurensis* (starfish)	Flatbottom seastar, Japanese seastar, Japanese starfish, Nordpazifischer Seestern, North Pacific seastar, northern Pacific seastar, Purple-orange seastar
12. *Banana Bunchy Top Virus (BBTV)* (micro-organism)	BTV, Bunchy top virus
13. *Batrachochytrium dendrobatidis* (fungus)	Chytrid-Pilz, chytridiomycosis, frog chytrid fungus
14. *Bemisia tabaci* (insect)	mosca Blanca, sweet potato whitefly, Weisse Fliege
15. *Boiga irregularis* (reptile)	Braune Nachtbaumnatter, brown tree snake, brown treesnake, culepla

TABLE 6.1
Continued

Genus and species	*Common Names*
16. *Bufo marinus* (amphibian)	Aga-Kröte, bufo toad, bullfrog, cane toad, crapaud, giant American toad, giant toad, kwapp, macao, maco pempen, Maco toro, marine Toad, Suriname toad
17. *Capra hircus* (mammal)	goat, Hausziege
18. *Carcinus maenas* (crustacean)	European shore crab, green crab, strandkrabbe
19. *Caulerpa taxifolia* (alga)	caulerpa, killer alga, lukay-lukay, Schlauchalge, sea weed
20. *Cecropia peltata* (tree)	Ameisenbaum, faux-ricin, parasolier, pisse-roux, pumpwood, trumpet tree, yagrumo hembra
21. *Cercopagis pengoi* (crustacean)	fishhook waterflea, Kaspischer Wasserfloh
22. *Cervus elaphus* (mammal)	cerf elaphe, Ciervo colorado, deer, Edelhirsch, elk, European red deer, red deer, Rothirsch, Rotwild, Rothirsch, wapiti
23. *Chromolaena odorata* (herb)	agonoi, bitter bush, chromolaena, hagonoy, herbe du Laos, huluhagonoi, jack in the bush, kesengesil, mahsrihsrihk, masigsig, ngesngesil, otuot, rumput belalang, rumput golkar, rumput putih, Siam weed, Siam-Kraut, triffid weed, wisolmatenrehwei
24. *Cinara cupressi* (insect)	cypress aphid, cypress aphid, Zypressen Blattlaus
25. *Cinchona pubescens* (tree)	cascarilla, chinarindenbaum, hoja ahumada, hoja de zambo, quinine, quinoa, quinquinia rouge, red cinchona, roja, rosada, Roter Chinarindenbaum
26. *Clarias batrachus* (fish)	alimudan, cá trê tráng, cá trèn trang, clarias catfish, climbing perch, freshwater catfish, Froschwels, hito, htong batukan, ikan keling, ikan lele, Ito, kawatsi, keli, klarievyi som, koi, konnamonni, kug-ga, leleh, magur, mah-gur, mangri, marpoo, masarai, mungri, nga-khoo, pa douk, paltat, pantat, pla duk, pla duk dam, pla duk dan, pla duk nam jued, pla duk nam juend, Thai hito, Thailand catfish, trey andaing roueng, trey andeng, walking catfish, wanderwels, Yerivahlay

TABLE 6.1
Continued

Genus and species	*Common Names*
27. *Clidemia hirta* (shrub)	Hirten-Schwarzmundgewaechs, kaurasiga, Koster's curse, kui, mbona na mbulamakau, roinisinga, soap bush, soapbush
28. *Coptotermes formosanus* (insect)	Formosa Termite, formosan subterranean termite
29. *Cryphonectria parasitica* (fungus)	chestnut blight, Edelkastanienkrebs
30. *Cyprinus carpio* (fish)	carp, carpa, carpat, carpe, carpe, carpe commune, carpeau, carpo, cerpyn, ciortan, ciortanica, ciortocrap, ciuciulean, common carp, crap, crapcean, cyprinos, escarpo, Europäischer Karpfen, European carp, German carp, grass carp, grivadi, ikan mas, kapoor-e-maamoli, kapor, kapr obecný, karp, karp, karp, karp, karp, karp dziki a. sazan, karpa, karpar, karpe, Karpe, karpen, karper, karpfen, karpion, karppi, kerpaille, koi, koi carp, korop, krap, krapi, kyprinos, læderkarpe, lauk mas, leather carp, leekoh, lei ue, mas massan, mirror carp, olocari, pa nai, pba ni, pla nai, ponty, punjabe gad, rata pethiya, saran, Saran, sarmão, sazan, sazan baligi, scale carp, sharan, skælkarpe, soneri masha, spejlkarpe, sulari, suloi, tikure, trey carp samahn, trey kap, ulucari, weißfische, wild carp, wildkarpfen
31. *Dreissena polymorpha* (mollusc)	moule zebra, racicznica zmienna, zebra mussel, Zebra-Muschel
32. *Eichhornia crassipes* (aquatic plant)	aguapé, bung el ralm, jacinthe d'eau, jacinto de agua, jacinto-aquatico, jal kumbhi, lechuguilla, lila de agua, mbekambekairanga, wasserhyazinthe, water hyacinth
33. *Eleutherodactylus coqui* (amphibian)	Caribbean tree frog, common coqui, Coqui, Puerto Rican treefrog
34. *Eriocheir sineusis* (crustacean)	Chinese freshwater edible crab, Chinese mitten crab, chinesische wolhandkrab, chinesische wollhandkrabbe, crabe chinois, kinesisk ullhandskrabba, kinesiske uldhandskrabbe, kinijos krabas, kitajskij mokhnatorukij krab, krab welnistoreki, kraba welnistoreki, villasaksirapu

TABLE 6.1
Continued

Genus and species	Common Names
35. *Euglandina rosea* (mollusc)	cannibal snail, Rosige Wolfsschnecke, rosy wolf snail
36. *Euphorbia esula* (herb)	Esels-Wolfsmilch, leafy spurge, spurge, wolf's milk
37. *Fallopia japonica* (herb, shrub)	crimson beauty, donkey rhubarb, German sausage, huzhang, itadori, Japanese bamboo, Japanese fleece flower, Japanese knotweed, Japanese polygonum, kontiki bamboo, Mexican-bamboo, peashooter plant, renouée du Japon, reynoutria fleece flower, sally rhubarb
38. *Felis catus* (mammal)	cat, domestic cat, feral cat, Hauskatze, house cat, moggy, poti, pusiniveikau
39. *Gambusia affinis* (fish)	Barkaleci, Dai to ue, Gambusia, Gambusie, Gambusino, Gambuzia, Gambuzia pospolita, Gambuzija, guayacon mosquito, Isdang canal, Kadayashi, Koboldkärpfling, Kounoupopsaro, Live-bearing tooth-carp, Mosquito fish, Obyknovennaya gambuziya, pez mosquito, San hang ue, Silberkärpfling, tes, Texaskärpfling, Topminnow, western mosquitofish, Western mosquitofish
40. *Hedychium gardnerianum* (herb)	awapuhi kahili, cevuga dromodromo, conteira, Girlandenblume, kahila garland-lily, kahili, kahili ginger, kopi, sinter weitahta, wild ginger
41. *Herpestes javanicus* (mammal)	beji, Kleiner Mungo, mangouste, mangus, mweyba, newla, small Indian mongoose
42. *Hiptage benghalensis* (shrub, vine, climber)	adimurtte, adirganti, atimukta, benghalen-Liane, chandravalli, haldavel, hiptage, kampti, kamuka, liane de cerf, madhalata, madhavi, Madhavi, Madhavi, madhumalati, madmalati, ragotpiti, vasantduti
43. *Imperata cylindrica* (grass)	alang-alang, blady grass, Blutgras, carrizo, cogon grass, gi, impérata cylindrique, japgrass, kunai, lalang, ngi, paille de dys, paillotte, satintail, speargrass

TABLE 6.1
Continued

Genus and species	*Common Names*
44. *Lantana camara* (shrub)	ach man, angel lips, ayam, big sage, blacksage, bunga tayi, cambara de espinto, cuasquito, flowered sage, lantana, lantana wildtype, largeleaf lantana, latora moa, pha-ka-krong, prickly lantana, shrub verbean, supirrosa, Wandelroeschen, white sage, wild sage
45. *Lates niloticus* (fish)	chengu, mbuta, nijlbaars, nilabborre, Nilbarsch, nile perch, perca di nilo, perche du nil, persico del nilo, sangara, Victoria perch, victoriabaars, victoriabarsch
46. *Leucaena leucocephala* (tree)	acacia palida, aroma blanca, balori, bo chet, cassis, false koa, faux mimosa, faux-acacia, fua pepe, ganitnityuwan tangantan, graines de lin, guaje, guaslim, guaxin, huaxin, horse/wild tamarind, huaxin, ipil-ipil, jumbie bean, kan thin, kanthum thect, koa haole, koa-haole, kra thin, kratin, lamtoro, lead tree, Leucaena, leucaena, liliak, lino criollo, lopa samoa, lusina, nito, pepe, rohbohtin, schemu, siale mohemohe, subabul, tamarindo silvestre, tangan-tangan, tangantangan, te kaitetua, telentund, tuhngantuhngan, uaxim, vaivai, vaivai dina, vaivai ni vavalangi, wild mimosa, wild tamarind, zarcilla
47. *Ligustrum robustum* (shrub, tree)	bora-bora, Ceylon Privét, Sri Lankan privet, tree privet, troene
48. *Linepithema humile* (insect)	Argentine ant, Argentinische Ameise, formiga-argentina
49. *Lymantria dispar* (insect)	Asian gypsy moth, erdei gyapjaslepke, gubar, gypsy moth, lagarta peluda, limantria, løVstraesnonne, maimai-ga, mniska vel'kohlava, Schwammspinner, spongieuse
50. *Lythrum salicaria* (aquatic plant, herb)	Blutweiderich, purple loosestrife, rainbow weed, salicaire, spiked loosestrife
51. *Macaca fascicularis* (mammal)	crab-eating macaque, long-tailed macaque
52. *Melaleuca quinquenervia* (tree)	cajeput, Mao-Holzrose, melaleuca, niaouli, paper bark tree, punk tree

TABLE 6.1
Continued

Genus and species	Common Names
53. *Miconia calvescens* (tree)	bush currant, cancer vert, miconia, purple plague, velvet tree
54. *Micropterus salmoides* (fish)	achigã, achigan, achigan à grande bouche, American black bass, bas dehanbozorg, bas wielkogeby, bass, bass wielkgebowy, biban cu gura mare, black bass, bol'sherotyi chernyi okun', bolsherotnyi amerikanskii tscherny okun, buraku basu, fekete sügér, forelbaars, forellenbarsch, green bass, green trout, großmäuliger Schwarzbarsch, huro, isobassi, khorshid Mahi Baleh Kuchak, lakseabbor, largemouth bass, largemouth black bass, lobina negra, lobina-truche, northern largemouth bass, okounek pstruhový, okuchibasu, Öringsaborre, Ørredaborre, ostracka, ostracka lososovitá, perca americana, perche d'Amérique, perche noire, perche truite, persico trota, stormundet black bass, stormundet ørredaborre, tam suy lo ue, zwarte baars
55. *Mikania micrantha* (vine, climber)	American rope, Chinese creeper, Chinesischer Sommerefeu, fue saina, liane americaine, mile-a-minute weed, ovaova, usuvanua, wa bosucu, wa mbosuthu, wa mbosuvu, wa mbutako, wa ndamele
56. *Mimosa pigra* (shrub)	bashful plant, catclaw, catclaw mimosa, chi yop, columbi-da-lagoa, eomrmidera, espino, giant sensitive plant, giant sensitive tree, giant trembling plant, juquiri, juquiri grand, kembang gajah, mai yah raap yak, maiyarap ton, malicia-de-boi, mimosa, mimosa, mimose, putri malu, semalu gajah, sensitiva, trinh nu nhon, una de gato, xao ho
57. *Mnemiopsis leidyi* (comb jelly)	American comb jelly, comb jelly, comb jellyfish, Rippenqualle, sea gooseberry, sea walnut, Venus' girdle, warty comb jelly
58. *Mus musculus* (mammal)	biganuelo, field mouse, Hausmaus, house mouse, kiore-iti, raton casero, souris commune, wood mouse

TABLE 6.1
Continued

Genus and species	*Common Names*
59. *Mustela erminea* (mammal)	ermine, ermine, Grosswiesel, Hermelin, hermine, short-tailed weasel, short-tailed weasel, stoat
60. *Myocastor coypus* (mammal)	Biberratte, coipù, coypu, nutria, ragondin, ratão-do-banhado, Sumpfbiber
61. *Morella faya* (tree)	Feuerbaum, fire tree
62. *Mytilus galloprovincialis* (mollusc)	Mediterranean mussel, Mittelmeer-Miesmuschel
63. *Oncorhynchus mykiss* (fish)	pstrag teczowy, rainbow trout, redband trout, Regenbogenforelle, steelhead trout, trucha arco iris, truite arc-en-ciel
64. *Ophiostoma ulmi sensu lato* (fungus)	Dutch elm disease, Schlauchpilz
65. *Opuntia stricta* (shrub)	Araluen pear, Australian pest pear, chumbera, common pest pear, common prickly pear, erect prickly pear, Feigenkaktus, gayndah pear, nopal estricto, pest pear of Australia, sour prickly pear, spiny pest pear, suurturksvy
66. *Oreochromis mossambicus* (fish)	blou kurper, common tilapia, fai chau chak ue, Java tilapia, kawasuzume, kurper bream, malea, mojarra, mosambik-maulbrüter, Mozambikskaya tilapiya, Mozambique cichlid, Mozambique mouth-breeder, Mozambique mouthbrooder, Mozambique tilapia, mphende, mujair, nkobue, tilapia, tilapia del Mozambique, tilapia du Mozambique, tilapia mossambica, tilapia mozámbica, trey tilapia khmao, weißkehlbarsch, wu-kuo yu
67. *Oryctolagus cuniculus* (mammal)	Europäisches Wildkaninchen, kaninchen, lapin, rabbit
68. *Pheidole megacephala* (insect)	big-headed ant, brown house-ant, coastal brown-ant, Grosskopfameise, lion ant
69. *Phytophthora cinnamomi* (fungus)	Phytophthora Faeule der Scheinzypresse, phytophthora root rot
70. *Pinus pinaster* (tree)	cluster pine, maritime Pine
71. *Plasmodium relictum* (micro-organism)	avian malaria, paludisme des oiseaux, Vogelmalaria

TABLE 6.1
Continued

Genus and species	*Common Names*
72. *Platydemus manokwari* (flatworm)	Flachwurm, flatworm
73. *Pomacea canaliculata* (mollusc)	apple snail, channeled apple snail, Gelbe Apfelschnecke, golden apple snail, golden kuhol, miracle snail
74. *Potamocorbula amurensis*	Amur river clam, Amur river corbula, Asian bivalve, Asian clam, brackish-water corbula, Chinese clam, marine clam, Nordpazifik-Venusmuschel, Numakodaki
75. *Prosopis glandulosa* (tree)	honey mesquite, mesquite, Mesquite-Busch, Texas mesquite
76. *Psidium cattleianum* (shrub, tree)	cattley guava, cherry guava, Chinese guava, Erdbeer-Guave, goyave de Chine, kuahpa, ngguava, purple strawberry guava, strawberry guava, tuava tinito, waiawi
77. *Pueraria montana var. lobata* (vine, climber)	kudzu, kudzu vine, Kudzu-Kletterwein
78. *Pycnonotus cafer* (bird)	red-vented bulbul, Rußbülbül
79. *Rana catesbeiana* (amphibian)	bullfrog, North American bullfrog, Ochsenfrosch, rana toro
80. *Rattus rattus* (mammal)	black rat, blue rat, bush rat, European house rat, Hausratte, roof rat, ship rat
81. *Rinderpest virus* (micro-organism)	cattle plague
82. *Rubus ellipticus* (shrub)	Asian wild raspberry, broadleafed bramble, Ceylon blackberry, eelkek, Himalaya-Wildhimbeere, kohkihl, Molucca berry, Molucca bramble, Molucca raspberry, piquant lou-lou, robust blackberry, soni, wa ngandrongandro, wa sori, wa votovotoa, wild blackberry, wild raspberry, yellow Himalayan raspberry
83. *Salmo trutta* (fish)	an breac geal, aure, bachforelle, blacktail, breac geal, brook trout, brown trout, denizalabaligi, denizalasi, Europäische Forelle, finnock, forelle, galway sea trout, gillaroo, gwyniedyn, havørred, havsöring, herling, hirling, k'wsech, kumzha, lachförch, lachsforelle, lassföhren,

TABLE 6.1
Continued

Genus and species	*Common Names*
	losos taimen, losos' taimen, mahiazad-e-daryaye khazar, meerforelle, meritaimen, morska postrv, morskaya forel', orange fin, öring, orkney sea trout, ørred, ørret, pastrav de mare, peal, pstruh morsky, pstruh obecný, pstruh obecný severomorský, pstruh obycajný, salmo trota, salmon trout, sea trout, sewin, siwin, sjøaure, sjøørret, sjourrioi, taimen, thalasopestrofa, troc, troc wedrowna, trota fario, trout, trucha, trucha común, trucha marina, truita, truite brune, truite brune de mer, truite d'europe, truite de mer, truta marisca, truta-de-lago, truta-fário, truta-marisca, urriði, whiting, whitling, zeeforel
84. *Schinus terebinthifolius* (tree)	Brazilian holly, Brazilian pepper, Brazilian pepper tree, Christmas berry, faux poivrier, Florida holly, Mexican pepper, pimienta de Brasil, poivre rose, Rosapfeffer, warui
85. *Sciurus carolinensis* (mammal)	Grauhoernchen, gray squirrel, grey squirrel, scoiattolo grigio
86. *Solenopsis invicta* (insect)	red imported fire ant (RIFA), rote importierte Feuerameise
87. *Spartina anglica* (grass)	common cord grass, Englisches Schlickgras, rice grass, townsends grass
88. *Spathodea campanulata* (tree)	African tulip tree, Afrikanischer Tulpenbaum, amapola, apär, baton du sorcier, fa'apasi, fireball, flame of the forest, fountain tree, Indian Cedar, ko'i'i, mata ko'i'I, mimi, orsachel kui, patiti vai, pisse-pisse, pititi vai, rarningobchey, Santo Domingo Mahogany, taga mimi, tiulipe, tuhke dulip, tulipan africano, tulipier du Gabon
89. *Sturnus vulgaris* (bird)	blackbird, common starling, English starling, estornino pinto, etourneau sansonnet, étourneau sansonnet, Europäischer Star, European starling
90. *Sus scrofa* (mammal)	kuhukuhu, kune-kune, petapeta, pig, poretere, razorback, te poaka, Wildschwein

TABLE 6.1
Continued

Genus and species	Common Names
91. *Tamarix ramosissima* (shrub, tree)	salt cedar, Sommertamariske, tamarisk, tamarix
92. *Trachemys scripta elegans* (reptile)	Gelbwangen-Schmuckschildkroete, red-eared slider, red-eared slider terrapin
93. *Trichosurus vulpecula* (mammal)	brushtail possum, Fuchskusu
94. *Trogoderma granarium* (insect)	escarabajo khapra, khapra beetle, khaprakäfer, trogoderma (dermeste) du grain
95. *Ulex europaeus* (shrub)	gorse, kolcolist zachodni, Stechginster
96. *Undaria pinnatifida* (alga)	apron-ribbon vegetable, Asian kelp, haijiecai, Japanese kelp, miyeuk, qundaicai, wakame
97. *Vespula vulgaris* (insect)	common wasp, common yellowjacket, Gemeine Wespe
98. *Vulpes vulpes* (mammal)	fuchs, lape, lis, raposa, red fox, renard, rev, Rotfuchs, silver, black or cross fox, volpe, vos, zorro
99. *Wasmannia auropunctata* (insect)	albayalde, cocoa tree-ant, formi électrique, formiga pixixica, fourmi rouge, hormiga colorada, hormiga roja, hormiguilla, little fire ant, little introduced fire ant, little red fire ant, pequena hormiga de fuego, petit fourmi de feu, Rote Feuerameise, sangunagenta, satanica, small fire ant, tsangonawenda, West Indian stinging ant, yerba de Guinea
100. *Sphagneticola trilobata* (herb)	ate, atiat, creeping ox-eye, dihpw ongohng, Hasenfuss, ngesil ra ngebard, rosrangrang, Singapore daisy, trailing daisy, tuhke ongohng, ut mõkadkad, ut telia, wedelia

Source: The IUCN/SSC Invasive Species Specialist Group (ISSG) (http://www.issg.org)

One could also add that the list includes pets, like the house cat and ferret and even a college mascot, the terrapin. And, anglers (and diners) will appreciate the rainbow trout. So, there is definitely a sociological and psychological aspect to the ranking.

Invasive species are organisms that are not native to an ecosystem. They are problematic when they cause harm, such as loss of diversity and

FIGURE 6.1. Infestation by kudzu (*Pueraria* spp.) in Southeastern United States. *Photo Credit*: U.S. Department of Agriculture, Forest Service, James H. Miller.

other environmental damage, economic problems, or even human health concerns. These organisms can be any biota, that is, microbes, plants (see Figure 6.1), and animals (see Figure 6.2), but usually, at least their presence and impact are in part due to human activity.

In this chapter, we consider two invasive aquatic species, the Asian shore crab and the zebra mussel. These were chosen for a number of reasons. The crab is at an early stage of invasion, but the potential stress it places on coastal ecosystems could be quite substantial. The mussel is well established in a number of Great Lakes ecosystems. So, the two species provide an opportunity to compare prevention, management, and control approaches to address the problems presented by the invaders.

Asian Shore Crab

The Asian shore crab (*Hemigrapsus sanguineus*) is indigenous to the western Pacific Ocean from Russia, along the Korean and Chinese coasts to Hong Kong, and the Japanese archipelago (see Figure 6.3). It is a highly adaptive and very opportunistic omnivore, feeding on algae, salt marsh grass, larval and juvenile fish, and small invertebrates such as amphipods, gastropods, bivalves, barnacles, and polychaetes.

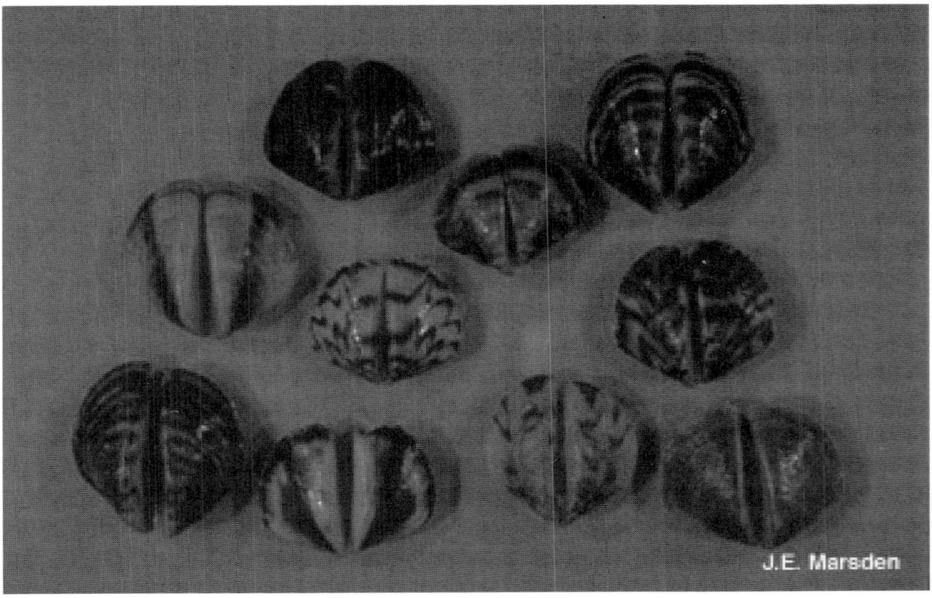

FIGURE 6.2. Zebra mussels (*Dreissena polymorpha*), an invasive species in the Great Lakes.
Photo credit: Lake Michigan Biological Station, Zion, Illinois, J. E. Marsden.

The Asian shore crab is a prolific breeder, with a reproductive season (May to September) twice the length of native crabs. Female Asian shore crabs can lay 50,000 eggs per clutch with up to four clutches each breeding season. Since the larvae are suspended in the water for approximately one month before developing into juvenile crabs, the larvae can move great distances, which makes them very invasive and allows them to be introduced into new habitats. This versatile creature lives in a hard-bottom intertidal or sometimes subtidal habitat. It can live on artificial structures and on mussel beds, on oyster reefs, and under rocks where its habitat overlaps that of native crab species.

Hemigrapsus was first recorded in the United States at Townsend Inlet, Cape May County, New Jersey, in 1988. This species is now well established and exceptionally abundant along the Atlantic intertidal coastline of the United States from Maine to North Carolina (see Figure 6.4). Since it withstands a wide range of environmental conditions, the Asian shore crab's invasion will likely continue along the U.S. eastern coastline.

The manner in which this crab species was introduced to the United States Atlantic coast is not known, although scientists speculate that adults or larvae were brought by incoming ships of global trade via ballast water

FIGURE 6.3. Asian shore crab (*Hemigrapsus sanguineus*).
Photo credit: U.S. Geological Survey, Center for Aquatic Resource Studies.

discharge. The Asian shore crab has a diverse choice of food, so its potential effect on populations of native aquatic species such as crabs, fish, and shellfish could be profound, with major disruptions to the food web. It also occupies habitats very similar to our native mud crabs, possibly overwhelming and dominating their habitat. This potential impact on native species populations may result from direct predation or competition for the same food source. For example, *Hemigrapsus* may compete with larger species, such as the blue crab, rock crab, lobster, and the nonnative green crab. Recent trends show numbers of shore crabs are steadily increasing with a corresponding decline in native crab populations. Thus, *Hemigrapsus* may also pose real threats to coastline ecosystems and aquaculture operations. Early findings from scientific investigations show that rockfish and

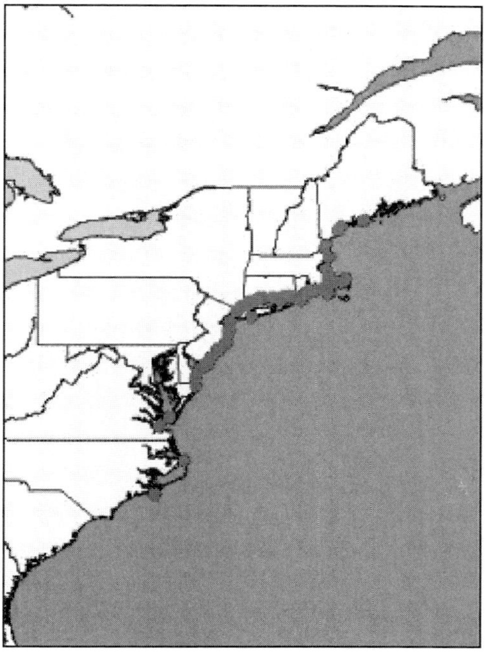

FIGURE 6.4. *Hemigrapsus sanguineus* locations in the United States.
Source: U.S. Geological Survey, Center for Aquatic Resource Studies.

seagulls may be predators of *Hemigrapsus*. However, the normal controls, such as parasites that help keep populations of *Hemigrapsus* in check in its native range, are not present along the U.S. Atlantic coast. Therefore, there is a distinct possibility that the Asian shore crab will continue to enlarge its range along the Atlantic coastline until it reaches tolerance levels, especially for salinity and temperature. Scientists are tracking changes in native species, studying the shore crab's spread along the coastline, and conducting field and laboratory experiments to understand the biological and ecological characteristics of the shore crab in various aquatic habitats of this species.

At a minimum, the continued invasion needs to be managed. For example, ballast water needs to be managed to reduce the entry of *Hemigrapsus* to new habitats. There may never be a "silver bullet" (e.g., a native predator that will eat *Hemigrapsus* yet not become an invader itself). Ecological problems such as this are complex and often never completely "solved." The best we can hope for in many cases is to catch the problem sufficiently early and set in place a system of prevention and control of activities that encourage expansion of the invader's range.

Zebra Mussel Invasion of the Great Lakes

Clear water can be deceiving. We often equate clarity with quality. Indeed, turbidity is an indication of the presence of high concentrations of certain contaminants, such as suspended and dissolved solids. However, sometimes the water is so clear that it is a cause for concern. Could there be an invisible toxicant that is killing everything and in its wake, we have a very clear, but highly polluted water body? Sometimes water clarity is the result of a lack of diversity, that is, one or a few species are destroying the habitat of many other species. This is what can happen when certain opportunistic aquatic organisms invade a water ecosystem. With few or no predators, these invaders consume the food sources much faster than their competitors. Well on their way to recovering from the problems of eutrophication in the 1960s and 1970s (see Chapter 5), the Great Lakes are once again threatened. But, instead of the problem of chemical pollutant loads, this time it is a biological stressor.

The zebra mussel (*Dreissena polymorpha*), native to the Caspian Sea region of East-Central Asia, is one such invader to the Great Lakes of North America. The zebra mussel can grow to an adult size of approximately 5 cm long, weighing less than two grams. They have a D-shaped shell that has alternating light and dark stripes, which explains the name. The shell is thin and fragile. These mollusks prefer subdued light and flowing water, and feed by filtering phytoplankton at a rate of up to 2 L water day^{-1}. The female mussel lays 30,000 to 40,000 eggs per spawning, and spawning several times each year. Larvae, known as veligers, are able to swim within eight hours of fertilization. Veligers can stay suspended in the water for several weeks until settling onto a hospitable substrate. *Dreissena* produce byssal threads to adhere to rocks, structures, pipes, and other hard substrate, including other mussel shells. Growing in clusters at shallow depth (2 to 5 meters), mussel colonies can move via ships and other vessels and, when detached, can colonize new aquatic habitats. They are also quite durable, being able to survive for two weeks outside of water in cool, moist environments.

Increasing numbers of zebra mussels directly impact plankton populations due to the mussels' highly efficient filtering capacity when they form large colonies, potentially shifting system energetics and reducing available food resources for higher organisms. For example, populations of native clams are threatened due to the zebra mussels' colonization of their shells (Figure 6.5).[2] Recent data indicate that snails are also being used as substrate for mussel attachment. Snails play a key role as grazers in the benthic (bottom sediment) community of the Great Lakes and as food for fish, such as perch, sunfish, and whitefish.[3]

One of the most obvious demonstrations of the rapid increase in zebra mussel densities recently seen in the open waters of the Great Lakes is the species' colonization of municipal and industrial water-intake pipes. In 1991 and 1992, facilities drawing raw water from Lake Michigan began

FIGURE 6.5. Zebra mussel (*Dreissena polymorpha*) colonizing on papershell mussel (*Leptodea fragilis*).
Photo credit: National Biological Survey, D.W. Schlorsser.

treatment programs to reduce infestation of intake pipes.[4] This is leading to both ecological and economic costs. Retrofitting plants in Chicago and northern Illinois shoreline communities, which alone totaled $1,778,000 by 1992, not including indirect costs like greater human resources needed for maintenance and additional chemicals required for cleanup.[5] Also, retrofitting and chemical treatments increase the risks associated with accidents, spills, and leaks.

These creatures are also having substantial effects on recreational and aesthetic values in the Great Lakes. The hulls of vessels are being fouled, engine cooling systems clogged, and broken mussel shells littering beaches. They are even affecting the international price of pearls, since native clams from the Illinois River are shipped to Japan for use in the cultured pearl industry; with a value of about $1.4 million annually. The infestation of zebra mussels on these clams, however, is increasing clam mortality considerably, with a concomitant loss of revenues. Chemical and mechanical controls for zebra mussels are only useful in localized areas such as intake pipes and other artificial structures, but not in the open waters of the lake. There is hope that native predators, such as freshwater drum (*Aplodinotus grunniens*), diving ducks, and crayfish, may also keep mussel populations in check in some lake systems, but there is a very strong likelihood that much more damage will occur in the years ahead.

Lesson Learned: Need for Meaningful Ecological Risk Assessments

A good way to consider invasive species is to assess the actual or potential damage they do to an ecosystem. Ecological risk assessment[6] is a process employed to determine the likelihood that adverse outcomes may occur in an ecosystem as a result of exposure to one or more stressors. The process systematically reduces and organizes data, gathers information, forms assumptions, and identifies areas of uncertainty to characterize the relationships between stressors and effects. As is the case for human health risk assessments, the stressors may be chemicals. However, ecological risk assessments must also address physical and biological stressors. For example, the placement of a roadway or the changes brought about by bulldozers and earthmovers are considered to be physical stressors to habitats. The accidental or intentional introduction of invasive biota, such as grass carp (fauna) and kudzu (flora) in the Southern United States, are examples of biological stressors.

The identification of possible adverse outcomes is crucial. These outcomes alter essential structures or functions of an ecosystem. The severity of outcomes is characterized as to their type, intensity, and scale of the effect and the likelihood of an ecosystem to recover from the damage imposed by a single or multiple stressors. The characterization of adverse ecological outcomes can range from qualitative, expert judgments to statistical probabilities. The emergent fields of eco-toxicology and eco-risk have several things in common with human toxicology and risk assessment, such as concern about ambient concentration of contaminants and uptake in water, air, and soil. In some ways, however, ecological dose-response and exposure research differs from that in human systems. First, ecologists deal with many different species, some more sensitive than others to the effects of contaminants. Second, the means of calculating exposure are different, especially if we are concerned about the exposure of an entire ecosystem.

Ecosystems are complex. Ecologists characterize them by evaluating their composition, structure, and functions. Ecosystem composition is a listing, a taxonomy if you will, of every living and nonliving part of the ecosystem. Ecological structure, as the term implies, is how all the parts of the system are linked to form physical patterns of life forms from single forest stands to biological associations and plant communities. A single wetland or prairie, for example, has a much simpler structure than does a multilayered forest, which consists of plant and microbial life in the detritus, herbs, saplings, newer trees, and canopy trees. Ecosystem functions include cycles of nitrogen, carbon, and phosphorous that lead to biotic processes such as production, consumption, and decomposition.

Indicators of an ecosystem's condition include:

- Diversity—"Biodiversity" has been defined as the ". . . composition, structure, and function (that) determine, and in fact constitute, the biodiversity of an area. Composition has to do with the identity and variety of elements in a collection, and includes species lists and measures of species diversity and genetic diversity. Structure is the physical organization or pattern of a system, from habitat complexity as measured within communities to the pattern of patches and other elements at a landscape scale. Function involves ecological and evolutionary processes, including gene flow, disturbances, and nutrient cycling."[7] (See Appendix 9 for a way to estimate diversity).
- Productivity—This is an expression of how economical a system is with its energy. It tells how much biomass is produced from abiotic (e.g., nutrients and minerals) and biotic resources (from microbial populations to canopy plant species to top predator fauna). One common measure is net primary productivity which is the difference between two energy rates:

$$P_1 = k_p - k_e \tag{6.1}$$

where P_1 = Net primary productivity
k_p = Rate of chemical energy storage by primary producers
k_e = Rate at which the producers use energy (via respiration)

- Sustainability—How likely is it that the diversity and productivity will hold up? Even though an ecosystem appears to be diverse and highly productive, is there something looming that threatens the continuation of these conditions? For example, is an essential nutrient being leached out of the soil, or are atmospheric conditions changing that may threaten a key species of animal, plant, or microbe? Sustainability is difficult to quantify precisely.

Ecological risk assessments may be prospective or retrospective, but often are both. The Florida Everglades provides an example of an integrated risk approach. In the 1990s, the population of panthers, a top terrestrial carnivore in Southern Florida, was found to contain elevated concentrations of mercury (Hg). This was observed through retrospective eco-epidemiological studies. The findings were also used as scientists recommended possible measures to reduce Hg concentrations in sediment and water in Florida. Prospective risk assessments can help to estimate expected declines in Hg in panthers and other organisms in the food chain from a mass balance perspective. That is, as the Hg mass entering the environment through the air, water, and soil is decreased, how has the risk to sensitive species concomitantly been reduced? Integrated retrospective and prospective risk assessments are employed where ecosystems have a history of previous impacts and the potential for future effects from a wide range of stressors. This may be the case for hazardous waste sites.

The ecological risk assessment process embodies two elements, characterizing the adverse outcomes and characterizing the exposures. From these elements, three steps are undertaken:

1. Problem Formulation
2. Analysis
3. Risk Characterization

In problem formulation, the rationale for conducting the assessment is fully described, the specific problem or problems are defined, and the plan for analysis and risk characterization is laid out. Tasks include integrating available information about the potential sources; the description of all stressors and effects; and the characterization of the ecosystem and the receptors. Two basic products result from this stage of eco-risk assessment: assessment endpoints and conceptual models.

The analysis phase consists of evaluating the available data to conduct an exposure assessment, when exposure to stressors is likely to occur or to have occurred. From these exposure assessments, the next step is to determine the possible effects and how widespread and severe these outcomes will be. During analysis, the environmental practitioner should investigate the strengths and limitations of data on exposure, effects, and ecosystem and receptor characteristics. Using these data, the nature of potential or actual exposure and the ecological changes under the circumstances defined in the conceptual model can be determined. The analysis phase provides an exposure profile and stressor-response profile, which together form the basis for risk characterization.

Thus, the ecological risk assessment provides valuable information by:

- Providing information to complement human health information, thereby improving environmental decision making
- Expressing changes in ecological effects as a function of changes in exposure to stressors, which is particularly useful to the decision maker who must evaluate trade-offs, examine different options, and determine the extent to which stressors must be reduced to achieve a given outcome
- Characterizing uncertainty as a degree of confidence in the assessment, which aids the focus on those areas that will lead to the greatest reductions in uncertainty
- Providing a basis for comparing, ranking, and prioritizing risks, as well as information to conduct cost-benefit and cost-effectiveness analyses of various remedial options

- Considering management needs, goals, and objectives, in combination with engineering and scientific principles, to develop assessment endpoints and conceptual models during problem formulation.

Notes and Commentary

1. IUCN/SSC Invasive Species Specialist Group (ISSG): http://www.issg.org/database/species/search.asp?st=100ss&fr=1&sts=; Accessed on April 20, 2005.
2. G.L. Mackie, 1991, "Biology of the Exotic Zebra Mussel," *Dreissena polymorpha*, in Relation to Native Bivalves and Its Potential Impact on Lake St. Clair." *Hydrobiologia*, 219:251–268.
3. W.B. Scott and E.J. Crossman, 1973. Bulletin 184: Fisheries Research Board of Canada, Ottawa, Ontario.
4. U.S. Department of the Interior, 2005. National Biological Service: A Report to the Nation on the Distribution, Abundance, and Health of U.S. Plants, Animals, and Ecosystems, Washington, D.C., http://biology.usgs.gov/s+t/index.htm; accessed April 18, 2005.
5. S. Nelson, 1992, "A Pound of Cure for a Ton of Mussels," *Aquaticus: Journal of the Shedd Aquarium*, 23:28–29.
6. See G. Suter, 1993. *Ecological Risk Assessment*, Boca Raton, FL: Lewis Publishers; U.S. Environmental Protection Agency, 1992, Framework for Ecological Risk Assessment. Washington, D.C., EPA/630/R-92/001; and Federal Register 63(93):26846–26924.
7. R. Noss, 1990, "Indicators for Monitoring Biodiversity: A Hierarchical Approach," *Conservation Biology*, 4(4), pp.355–364.

CHAPTER 7

Environmental Swords of Damocles

Yes, I know there is a sword above your head, and that it may fall at any moment. But why should that trouble you? I have a sword over my head all the time. I am every moment in dread lest something may cause me to lose my life.

King Dionysius in *The Sword of Damocles*[1]

According to ancient Greek mythology, during a banquet Damocles was required by King Dionysius to sit under a sword that was suspended from the ceiling by a single horse hair. Thus, in modern parlance, an ever-present peril is known as a "sword of Damocles." Certain contemporary situations seem to leave us in similar peril, where a slight change or a continuation in what we are doing can lead to disaster. In certain matters of environmental consequence, even a small miscalculation or misstep can lead to large-scale environmental damage and may irreversibly imperil public health. Present debates regarding global climate change, suspected carcinogens, chemicals that can alter hormones, and genetic engineering are laced with concerns that they may in fact be swords of Damocles waiting to fall soon or in the distant future.

Whether our actions are deemed paranoid or are simply judicious is often not known until it is too late. In such situations where the stakes are very high and uncertainties are large, the prudent course of action may be to take preventative and precautionary measures. The so-called "precautionary principle" is called for when an activity threatens harm to human health or the environment, so that precautionary measures are taken even if some cause and effect relationships are not established scientifically. It was first articulated in 1992 as an outcome of the Earth Summit in Rio de Janeiro, Brazil.[2] The principle states that the proponent of the activity (such as a pharmaceutical company's development of a new chemical or a biotechnology company's research in genetic engineering), rather than the public, bears the burden of proof in these cases.[3] The precautionary principle provides a margin of safety beyond what may exist directly from science. It changes the onus of proof that harm exists to proving that the harm does *not* exist at the outset. Some have argued that if the principle is carried to an extreme, however, it could severely reduce technological advancement because it could severely limit risk-taking that has led to many scientific and medical breakthroughs.[4] Perhaps one way to balance risks is to consider any harm that can result from even very positive outcomes (see the discussion box, "The Tragedy of the Commons").

The Tragedy of the Commons

In his classic work "Tragedy of the Commons," Garrett Hardin gives an example of the individual herder and the utility of a single cow and what is best for the pasture.[5] If everyone takes the egocentric view, the pasture will surely be overgrazed. So, the farmer who will gain immediate financial gain by adding a cow to the herd must decide the utility of the cow *versus* the collective utility of the pasture. The utilities, for the herder, are not equal. The individual utility is one, but the collective utility is less than one. In other words, the farmer may be aware that the collective cost to each herder on the pasture adding a cow is that overgrazing will cause the pasture to be unproductive for all herders at some threshold. So, the utility becomes inelastic at some point. The damage may even be permanent, or at least it may take a very long time to recover to the point where it may sustain any cows, including those of the individual herder.

Hardin's parable demonstrates that even though the individual sees the utility of preservation (no new cows) in a collective sense, the ethical egoistic view may well push the decision toward the immediate gratification of the individual at the expense of the collective good.

Libertarians argue that the overall collective good will come as a result of the social contract. John Stuart Mill is recognized as the principal author of utilitarianism (i.e., the outcome determines whether something is morally acceptable). Utilitarianism determines that a moral act should produce the greatest amount of good consequences for the greatest number of beings. Even Mill, however, saw the need for the "harm principle" to counterbalance that temptation to use good ends to rationalize immoral methods, that is, the "ends justifying the means." The harm principle states:

> . . . the sole end for which mankind is warranted, individually or collectively, in interfering with the liberty of action of any of their number, is self-protection. That the only purpose for

which power can be rightfully exercised over any member of a civilized community, against his will, is to prevent harm to others. His own good, either physical or moral, is not sufficient warrant. He cannot rightfully be compelled to do or forbear because it will be better for him to do so, because it will make him happier, because, in the opinion of others, to do so would be wise, or even right . . . The only part of the conduct of anyone, for which he is amenable to society, is that which concerns others. In the part which merely concerns himself, his independence is, of right, absolute. Over himself, over his own body and mind, the individual is sovereign.[6]

More recently, John Rawls conceptualized a "veil of ignorance." Rawls argues that rational people will adopt principles of justice when they reason from general considerations, rather than from their own personal situation.[7] Reasoning without personal perspective and considering how to protect the weakest members comprise Rawls' veil of ignorance. Both the harm principle and the veil of ignorance are buffers against pure ethical egoism. That is, the utilitarian view requires that one not be so self-centered that a personal decision causes harm to another, and the Rawlsian view requires that the weakest members of society be protected from the expressions of free will of others. So, the need to "protect the pasture" must be balanced against decisions based on individual utilities.

When something threatens our very survival or may lead to irreversible changes in our species, precaution is in order.

That said, the cases in this chapter have little in common with each other, except that they are swords of Damocles. They are looming threats to the environment. They add risk presently. They also increase future risks. Thus, the actual risks are difficult to predict.

Global Climate Change

The science of global warming is considered in detail in Chapter 9, but for now, it worth discussing as a sword of Damocles. The earth's atmosphere serves as a blanket that modulates swings in temperatures. Basically, the chemical composition of the atmosphere allows the penetration of certain wavelengths of electromagnetic radiation (e.g., visible light and ultraviolet (UV) radiation) more readily than other wavelengths (e.g., infrared radiation (heat)). This process of differential electromagnetic penetration, absorption by surfaces, reradiation, and trapping of heat is known as the greenhouse effect (see the discussion box, "The Greenhouse Effect"). Carbon dioxide (CO_2) is a major greenhouse gas. Other gases, like methane, are even more potent greenhouse gases, but their concentrations are much lower than that of CO_2. Also, atmospheric concentrations of CO_2 have been increasing and are expected to nearly double in the next few decades from the preindustrial levels (about 280 ppm). In addition, numerous models predict that global atmospheric temperatures are likely to rise concomitantly with increased CO_2 concentrations.

The more CO_2 in the atmosphere, the more worried many of us become that the climate will change. Pollution prevention, changes in lifestyle, and other personal choices can reduce our reliance on the need to burn fossil fuels in the first place. Another approach is to add other energy choices to the mix, like solar, wind, and even nuclear energy (although this has a mixed history and is viewed by many to be unacceptable).

The Greenhouse Effect

The fact that we can see through the glass walls and roof of a greenhouse explains differential penetration of electromagnetic radiation. Visible light must easily traverse the glass, since we can see the plants inside. If the walls and roof were opaque this means that part of the light spectrum was absorbed by walls and roof surfaces. The surfaces inside the greenhouse convert the shortwave light to longer wavelengths in the infrared (IR) range, a process known as *black body radiation*. We cannot see, but we can feel, this change. The darker surfaces inside make for a more efficient black body radiator than lighter surfaces. Thus, the interior of a car with black seats, even when the temperature is below freezing, will be quite warm on a sunny day. At night, when the shortwave (UV and visible) light is no longer available, the interior temperature falls because there is no longer any incoming shortwave radiation to convert and eventually the heat dissipates. The thickness of the glass is directly proportional to the trapping of heat.

This same process works in the earth's atmosphere, but instead of glass it is the chemical composition of the atmosphere that determines just how much energy penetrates, is converted, and is trapped. A very thin or nonexistent atmosphere, like that of the moon, cannot trap IR, so the temperature swings are dramatic each lunar day. Conversely, the planet Venus has such an efficient greenhouse system (its atmosphere is about 96.5% carbon dioxide and 3.5% nitrogen), that its temperatures stay very high even at night. The earth's atmosphere is midway between these two extremes. The major gases that trap the reradiated infrared waves (heat) are the same as those that sustain life on earth, including water vapor and CO_2. Other greenhouse gases are released from both natural and human activities, such as methane (CH_4), which is released from reduced conditions like those in bogs and forest understories, as well as from industrial operations. Some, like the chorofluorocarbons (CFCs), are entirely synthetic.

So, increasing the atmosphere's concentration of gases that selectively allow more short-wave radiation to pass through and that absorb relatively more longer wave radiation is analogous to thickening the glass on the greenhouse. That is, we can expect the temperature in the house to increase.

Many believe the United States can do more to prevent global warming. Editorials are commonplace, arguing that the United States should have adopted and ratified the Kyoto Protocol, a legally binding agreement under which industrialized countries will reduce their collective emissions of greenhouse gases, especially CO_2 and ozone. Other gases also contribute, but to much less an extent, to the greenhouse effect, such as methane, nitrous oxide, sulfur hexafluoride (SF_6), and halocarbons. The Kyoto agreement calls for a reduction of 5.2% of emissions of these gases from what was emitted in the year 1990.

On the one hand, reductions make sense since the industrialized nations are the largest sources of greenhouse gas emissions. So, the 5.2% reduction could be supported as a measure of effectiveness in addressing the buildup of CO_2. On the other hand, the reason that the industrialized nations produce so much CO_2 is that they have developed and improved their processes, including combustion, so that much of what is manufactured is done in these countries. Could the protocol push many of these more efficient processes into countries with less effective environmental controls so that not only would the global levels of CO_2 continue to increase and possibly increase at a greater rate, but dirtier technologies and less controls would mean that other pollutants would be generated from these less efficient processes? CO_2 is a measure of complete (efficient) combustion. What happens if we export manufacturing to places where we have less efficient processes that produce very toxic products of incomplete combustion, such as the carcinogenic polycyclic aromatic hydrocarbons (PAHs), dioxins, and furans? Can we expect more dioxin, furans, and PAHs from these other countries? What are the trade-offs?

Vehicles seem to be a different story. If we adopted the Kyoto Protocol or something like it, it might very well cause us to reduce greenhouse gases. We could do this by requiring better fuel economy from cars. If we burn less fossil fuel to get the same amount of horsepower, we have less reactant (hydrocarbons) so we can generate a smaller mass of product (CO_2). We can do this by increasing the number of fuel cell, electric, hydrogen, and hybrid cars. And we can do it by improving the design and connecting where we live, work, learn, and play. In a word, we need better infrastructure and transportation planning. That is, we need environmental planners and engineers to step up to the challenge.

Persistent, Bioaccumulating Toxicants

One of the principal reasons for the concern about the plethora of organic chemicals and heavy metals in the environment has been the connection between exposures to these substances and cancer and other chronic diseases. Intrinsic properties of compounds render them more or less toxic. In addition, physical and chemical properties determine whether the compounds will resist degradation and persist for long time periods and build up in organisms. Such compounds are known as persistent, bioaccumulating toxicants (PBTs). Polycyclic aromatic hydrocarbons (PAHs), a family of large, flat compounds with repeating benzene structures, represent a class of PBTs. The chemical structure (stereochemistry) renders most PAHs to be highly hydrophobic (i.e., fat soluble) and difficult for an organism to eliminate (since most blood and cellular fluids are mainly water). This property also enhances the PAHs' ability to insert themselves into the deoxyribonucleic acid (DNA) molecule, interfering with transcription and replication. This is why some large organic molecules can be mutagenic and carcinogenic. One of the most toxic PAHs is benzo(a)pyrene, which is found in cigarette smoke, combustion of coal, coke oven emissions, and numerous other processes that use combustion. The compound can become even more toxic when it is metabolized, a process known as *activation* (see Figure 7.1).

The Inuit and Persistent Organic Pollutants

Persistent Organic Pollutants (POPs) include a wide range of substances: industrial chemicals (e.g., PCBs) and byproducts of industrial processes (e.g., hexachlorobenzene—HCB, and chlorinated dioxins) which are unintentionally toxic. Other POPs have characteristics that are intentionally toxic, such as insecticides (e.g., DDT) and herbicides (e.g., 2,4-dichlorophenoxyacetic acid, better known as 2,4-D), or fungicides (e.g., vinclozolin). Those POPs with substituted chlorines are referred to as organochlorines.

Interest in the presence of POPs in the arctic environment arises in particular because of the concern that indigenous peoples and other northern residents subsist on traditional food for all or part of their diet. Studies have shown that even very remote arctic regions have been chronically

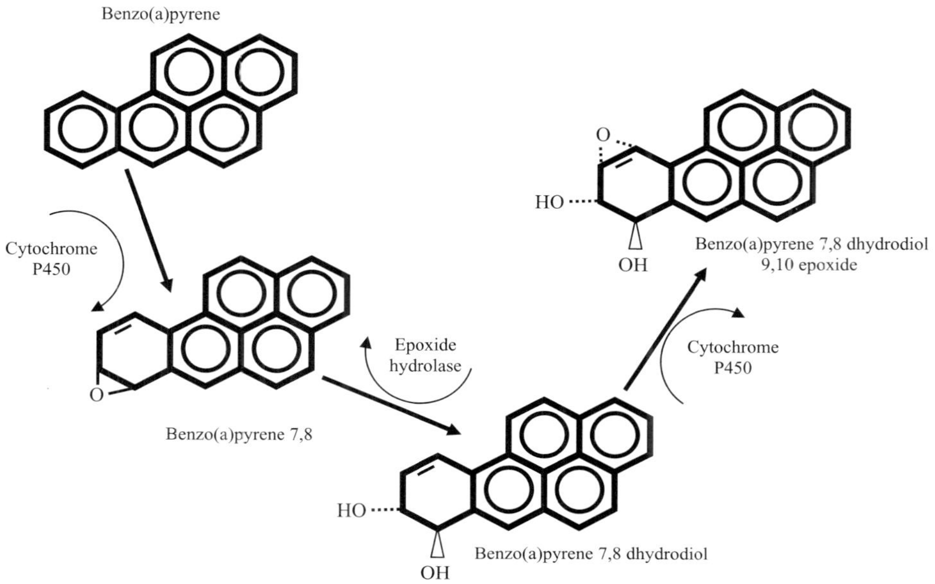

FIGURE 7.1. Biological activation of benzo(a)pyrene to form the carcinogenic active metabolite, Benzo(a)pyrene 7,8 dihydrodiol 9,10 epoxide. During metabolism, the biological catalysts (enzymes), cytochrome P-450 and epoxide hydrolase, are employed to make the molecule more polar, and in the process form diols and epoxides. These metabolites are more toxic than the parent compound.

exposed to POPs, so these subpopulations are vulnerable to be adversely affected. POPs are of particular concern because:

1. They persist in the environment for long periods of time, which allows them to be transported large distances from their sources; are often toxic; and have a tendency to bioaccumulate. Many POPs biomagnify in food chains.
2. Many indigenous people in the arctic depend on traditional diets that are both an important part of their cultural identity and a vital source of nourishment. Alternative sources of food often do not exist; however, traditional diets are often high in fat and POPs tend to accumulate in fatty tissue of the animals that are eaten.
3. Most northern residents have not used or directly benefited from the activities associated with the production and use of these chemicals; however, indigenous peoples in the arctic have some of the highest known exposures to these chemicals.

Due to these physicochemical properties, POPs can move many hundreds of kilometers away from their sources, either in the gas phase or

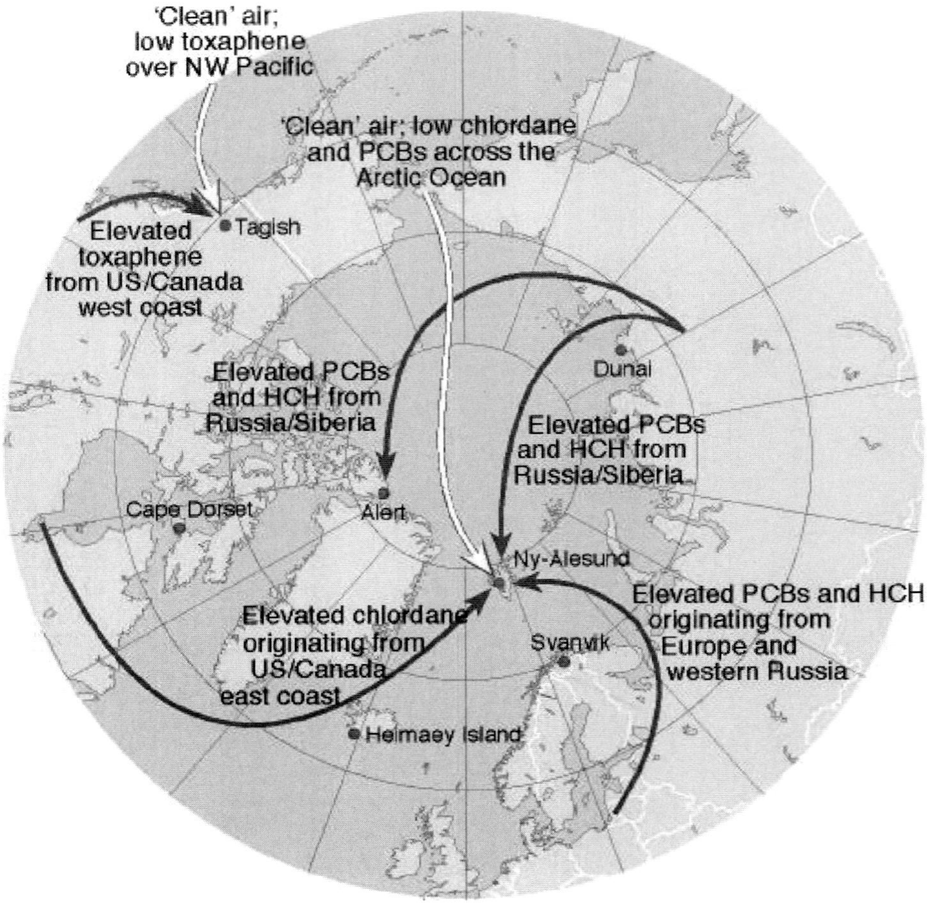

FIGURE 7.2. Long-range transport of persistent organic pollutants in the arctic regions.
Source: Russian Chairmanship of the Arctic Council, 2005. Draft Fact Sheet.

attached to particles. They are generally moved by advection, along with the movement of air masses. Some of the routes of long-range transport of POPs are shown in Figure 7.2.

A particularly vulnerable group is the Inuit. Lactating Inuit mothers' breast milk, for example, contains elevated levels of PCBs, DDT, and its metabolites, chlorinated dioxins and furans, and brominated organics, such as residues from fire retardants (polybrominated diphenyl ethers (PBDEs)), and heavy metals.[8] These compounds are encountered to varying extents among women in industrially developed as well as in developing nations. Some of the highest levels of contaminants have been detected in the

Canadian Inuit, whose diet consists of seal, whale, and other species high on the marine food chain. In the process the Inuit body burden of POPs is quite high.[9] These elevated exposures have led to adverse health effects. A study of Inuit women from Hudson Bay[10] indicated very high levels of PCBs and the DDT breakdown product dichlorodiphenylethene (DDE) in breast milk; these results prompted an examination of the health status of Inuit newborns.[11] Correlation analysis revealed a statistically significant negative association between male birth length and levels of hexachlorobenzene, mirex, PCBs, and chlorinated dibenzodioxins in the fat of mothers' milk. No significant differences were observed between male and female newborns for birth weight, head circumference, or thyroid-stimulating hormone. Immune system effects have also been detected in Inuit infants suspected of receiving elevated levels of PCBs and dioxins during lactation. These babies had a drop in the ratio of the CD4+ (helper) to CD8+ (cytotoxic) T-cells at ages 6 and 12 months (but not at 3 months).[12]

The Inuit situation demonstrates the critical ties between humans and their environment and the importance of physical properties of contaminants (e.g., persistence, bioaccumulation, and toxicity potentials), the conditions of the environment (e.g., the lower arctic temperatures increase the persistence of many POPs), and the complexities of human activities (e.g., diet and lifestyle) in order to assess risks and, ultimately, to take actions to reduce exposures. The combination of these factors leave the Inuit in a tragic dilemma. Since they are subsistence anglers and hunters, they depend almost entirely on a tightly defined portion of the earth for food. Their lifestyle and diet dictate dependence on food sources high in POPs.

The lesson extends even further, since exposures also include mother's milk. Pediatricians rightly encourage breast feeding for its many attributes, including enhancing the infant's immune system in the critical first weeks after birth. So, in terms of risk trade-offs, it is dangerous to discourage breast feeding. This lesson applies not only to the Inuit, or even just subsistence farmers, hunters, and anglers, but to all of us. We need to be finding ways to ensure that breast milk everywhere does not contain hazardous levels of PBTs and other contaminants. The only way to do this is to consider the entire life cycle of the pollutants and find ways to prevent their entry into the environment in the first place.

Extrinsic Factors

The greater persistence of POPs in the arctic regions compared to temperate and tropical regions is a direct result of temperature. Toxicity properties of environmental contaminants are also affected by extrinsic conditions, such as whether the substances are found in the air, water, sediment, or soil, along with the conditions of these media (e.g., oxidation-reduction, pH, and grain size). For example, the metal mercury is usually more toxic in reduced and anaerobic conditions because it is more likely to

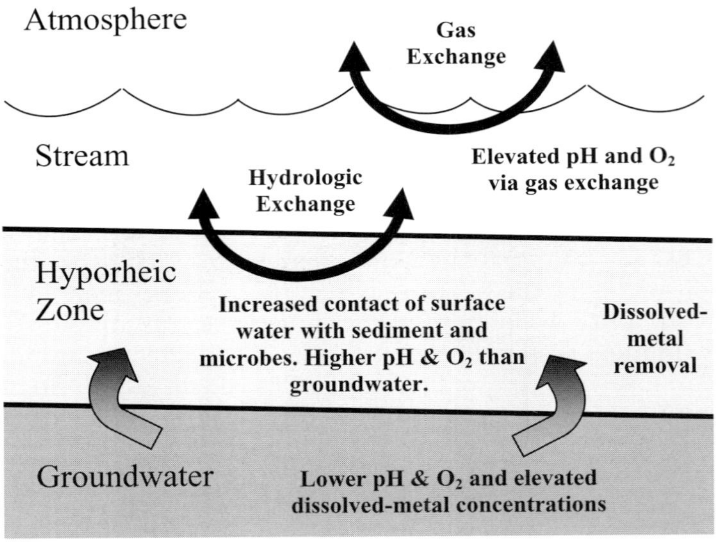

FIGURE 7.3. Exchanges and reactions that can occur in groundwater, sediment, and surface water. Some of the stream water moves into and out of the sediment and in shallow groundwater (i.e., the hyporheic zone). The process can increase the mobility of dissolved metallic compounds.
Source: Adapted from U.S. Geological Survey and D.A. Vallero, 2004. *Environmental Contaminants: Assessment and Control,* Academic Press, Elsevier Sciences, Burlington, MA.

form alkylated organometallic compounds, like monomethyl mercury and the extremely toxic dimethyl mercury. These reduced chemical species are likely to form when buried under layers of sediment where dissolved oxygen levels approach zero. Ironically, engineers have unwittingly participated in increasing potential exposures to these toxic compounds. With the good intention of attempting to clean up contaminated lakes in the 1970s, engineers recommended and implemented dredging programs. In the process of removing the sediment, however, the metals and other toxic chemicals that had been relatively inert and encapsulated in buried sediment were released to the lake waters. In turn, the compounds were also more likely to find their way to the atmosphere (see Figure 7.3). This is a lesson to engineers to take care in considering the many physical, chemical, and biological characteristics of the compound and the environment where it exists.

Some of these important physical and chemical characteristics that account for persistence, bioaccumulation, and toxicity of substances in the environment are shown in Table 7.1.

TABLE 7.1
Physicochemical properties important in hazard identification of chemical compounds.

Property of substance or environment	*Chemical importance*	*Physical importance*
Molecular Weight (MW)	Contaminants with MW >600 may not be bioavailable because they are too large to pass through membranes (known as steric hindrance). Larger molecules tend to be initially attacked and degraded at more vulnerable functional groups (e.g., microbial degradation often first removes certain functional groups).	The heavier the molecule, the lower the vapor pressures. For example, the more carbon atoms in an organic compound the less likely it will exist in gas phase under common environmental conditions. Heavier molecules are more likely to remain sorbed to soil and sediment particles.
Chemical Bonding	Chemical bonds determine the resistance to degradation. Ring structures are generally more stable than chains. Double and triple bonds add persistence to molecules compared to single-bonded molecules.	Large, aromatic compounds have affinity for lipids in soil and sediment. Solubility in water is enhanced by the presence of polar groups in structure. Sorption is affected by presence of functional groups and ionization potential.
Stereochemistry	Stereochemistry is the spatial configuration or shape of a molecule. Neutral molecules with cross-sectional dimensions >9.5 Angstroms (Å) have been considered to be sterically hindered in their ability to penetrate the polar surfaces of the cell membranes. A number of persistence, bioaccumulation, and toxicity properties of chemicals are determined, at least in part by a molecule's stereochemistry.	Lipophilicity (i.e., solubility in fats) of neutral molecules generally increases with molecular mass, volume, or surface area. Solubility and transport across biological membranes are affected by a molecule's size and shape. Molecules that are planar, such as polycyclic aromatic hydrocarbons, dioxins, or certain forms of polychlorinated biphenyls, are generally more lipophilic than are globular molecules of similar molecular weight.

TABLE 7.1
Continued

Property of substance or environment	*Chemical importance*	*Physical importance*
		However, the restricted rate of bioaccumulation of octachlorodibenzo-p-dioxin (9.8 Å) and decabromobiphenyl (9.6 Å) has been associated with these compounds' steric hindrance.
Solubility	Lipophilic compounds can be very difficult to remove from particles and may require highly destructive (e.g., combustion) remediation techniques. Insoluble forms (e.g., valence states) may precipitate out of water column or be sorbed to . particles.	Hydrophilic compounds are more likely to exist in surface water and in solution in interstices of pore water of soil, vadose zone, and aquifers underground. Lipophilic compounds are more likely to exist in organic matter of soil and sediment.
Co-Solvation	If a compound is hydrophobic and nonpolar, but is easily dissolved in acetone or methanol, it can still be found in water because these organic solvents are highly miscible in water. The organic solvent and water mix easily, and a hydrophobic compound will remain in the water column because it is dissolved in the organic solvent, which in turn has mixed with the water.	An important mechanism for getting a highly lipophilic and hydrophobic compound into water, where the compound can then move by advection, dispersion, and diffusion. PBTs like PCBs and dioxins may be transported as co-solutes in water by this means.
Vapor Pressure or Volatility	Volatile organic compounds (VOCs) exist almost entirely in the gas phase since their vapor pressures in the environment are usually greater than 10^{-2} kilopascals; semivolatile organic compounds (SVOCs) have vapor pressures between 10^{-2} and 10^{-5} kilopascals, and	Volatility is a major factor in where a compound is likely to be found in the environment. Higher vapor pressures mean larger fluxes from the soil and water to the atmosphere. Lower vapor pressures, conversely, cause

TABLE 7.1
Continued

Property of substance or environment	*Chemical importance*	*Physical importance*
	nonvolatile organic compounds (NVOCs) have vapor pressures $<10^{-5}$ kilopascals.	chemicals to have a greater affinity for the aerosol phase.
Fugacity	Often expressed as Henry's Law Constant (K_H)—the vapor pressure of the chemical divided by its solubility of water. Thus, high fugacity compounds are likely candidates for remediation using the air (e.g., pump-and-treat and air-stripping).	Compounds with high fugacity have a greater affinity for the gas phase and are more likely to be transported in the atmosphere than those with low fugacity. Care must be taken not to allow these compounds to escape prior to treatment.
Octanol-Water Coefficient (K_{ow})	Substances with high K_{ow} values are more likely to be found in the organic phase of soil and sediment complexes than in the aqueous phase. They may also be more likely to accumulate in organic tissue.	Transport of substances with higher K_{ow} values is more likely to be on particles (aerosols in the atmosphere and sorbed to fugitive soil and sediment particles in water), rather than in water solutions.
Sorption	Adsorption (onto surfaces) dominates in soils and sediments low in organic carbon (solutes precipitate onto soil surface). Absorption (three-dimensional sorption) is important in soils and sediments high in organic carbon (partitioning into organic phase/aqueous phase matrix surrounding mineral particles), so the organic partitioning coefficient (K_{oc}) is often a good indicator of the sorption potential of a PBT.	Partitioning determines which environmental media will dominate. Strong sorption constants indicate that soil and sediment may need to be treated in place. Phase distributions favoring the gas phase indicate that contaminants may be off-gassed and treated in their vapor phase. This is particularly important for semivolatile PBTs that under typical environmental conditions exist in both the gas and solid phase.

TABLE 7.1
Continued

Property of substance or environment	*Chemical importance*	*Physical importance*
Substitution, Addition, and Elimination	These processes are important for treatment and remediation of PBT-contamination. For example, dehalogenation (e.g., removal of chlorine atoms) of organic compounds by anaerobic treatment processes often renders them much less toxic. Adding or substituting a functional group can make the compound more or less toxic. Hydrolysis is an important substitution mechanism where a water molecule or hydroxide ion substitutes for an atom or group of molecules. Phase 1 metabolism by organisms also uses hydrolysis and redox reactions (discussed later) to break down complex molecules at the cellular level.	These processes can change the physical phase of a compound (e.g., dechlorination can change an organic compound from a liquid to a gas) and can change their affinity to or from one medium (e.g., air, soil, and water) to another. That is, properties such as fugacity, solubility, and sorption will change and may allow for more efficient treatment and disposal. New species produced by hydrolysis are more polar and, thus, more hydrophilic than their parent compounds, so they are more likely to be found in the water column.
Dissociation	Molecules break down by a number of types of dissociation, including hydrolysis, acid-base reactions, photolysis, dissociation of complexes, and nucleophilic substitution; i.e., a nucleophile ("nucleus lover") is attracted to a positive charge in a chemical reaction, and donates electrons to the other compound (i.e., an electrophile) to form a chemical bond.	Hydrolysis involves the dissociation of compounds via acid-base equilibria among hydroxyl ions and protons and weak and strong acids and bases. Dissociation may also occur by photolysis directly by the molecules absorbing light energy, and indirectly by energy or electrons transferred from another molecule that has been broken down photolytically.
Reduction-Oxidation	Reduction is the chemical process where at least one electron is transferred to another compound. Oxidation is the companion reaction	Reductions and oxidations are paired into so-called redox reactions. Such reactions occur in the environment, leading to

TABLE 7.1
Continued

Property of substance or environment	*Chemical importance*	*Physical importance*
	where an electron is transferred from a molecule. These reactions are important in hazardous waste remediation. Often, toxic organic compounds can be broken down ultimately to CO_2 and H_2O by oxidation processes, including the reagents ozone, hydrogen peroxide, and molecular oxygen (i.e., aeration). Reduction is also used in treatment processes. For example, hexavalent chromium is reduced to the less toxic trivalent form in the presence of ferrous sulfate: $2CrO_3 + 6FeSO_4 \rightarrow 3Fe_2(SO_4)_3 + Cr_2(SO_4)_3 + 6H_2O$ The trivalent form is removed by the addition of lime, where it precipitates as $Cr(OH)_3$.	chemical speciation of parent compounds into more or less mobile species. For example elemental or divalent mercury is reduced to the toxic species, mono- and dimethyl mercury in sediment and soil low in free oxygen. The methylated metal species have greater affinity than the inorganic species for animal tissue.
Diffusion	Diffusion is the mass flux of a chemical species across a unit surface area. It is a function of the concentration gradient of the substance. A compound may move by diffusion from one compartment to another (e.g., from the water to the soil particle).	The concentration gradients within soil, underground water, and air determine to some degree the direction and rate that the contaminant will move. This is a very slow process in most environmental systems. However, in rather quiescent systems[13] ($<2.5 \times 10^{-4}\,cm\,s^{-1}$), such as aquifers and deep sediments, the process can be very important.
Isomerization	A congener is any a chemical compound that is a member of a chemical family, the	The fate and transport of chemicals can vary significantly depending

TABLE 7.1
Continued

Property of substance or environment	*Chemical importance*	*Physical importance*
	members of which have different molecular weights and various substitutions; e.g., there are 75 congeners of chlorinated dibenzo-*p*-dioxins. Isomers are chemical species with identical molecular formulae, but that differ in atomic connectivity (including bond multiplicity) or spatial arrangement. An enantiomer is one of a pair of molecular species that are nonsuperimposable mirror images of each other.	upon the isomeric form. For example, the rates of degradation of left-handed chiral compounds (mirror images) are often more rapid than for right-handed compounds (possibly because left-handed chirals are more commonly found in nature, and microbes have acclimated their metabolic processes to break them down). Isomeric forms also vary in their fate.
Biotransformation	Many of the processes discussed in this table can occur in or be catalyzed by microbes. These are biologically mediated processes. Reactions that may require long periods of time to occur can be sped up by biological catalysts, i.e., enzymes. Many fungi and bacteria reduce compounds to simpler species to obtain energy. Biodegradation is possible for almost any organic compound, although it is more difficult in very large molecules, insoluble species, and completely halogenated compounds.	Microbial processes will transform parent compounds into species that have their own transport properties. Under aerobic conditions, the compounds can become more water soluble and are transported more readily than their parent compounds in surface and groundwater. The fungicide example given later in this chapter is an example of how biological processes change the flux from soil to the atmosphere.
Availability of Free Oxygen	Complete microbial processing degrades hydrocarbons by oxidation to CO_2 and H_2O when free O_2 is available (aerobic digestion). In absence of free O_2, microbes	Aerobic and anaerobic processes may need to be used in a series for persistent compounds. For example, aerobic processes can cleave the aromatic

TABLE 7.1
Continued

Property of substance or environment	*Chemical importance*	*Physical importance*
	completely degrade organic compounds to CH_4 and H_2O (anaerobic digestion).	ring of polychlorinated biphenyls (PCBs) that have up to three chlorines. PCBs with four or more chlorines may first need to be treated anaerobically to remove the excess chlorines, before the rings can be cleaved.
Potential to Bioaccumulate	Bioaccumulation is the process by which an organism takes up and stores chemicals from its environment through all environmental media. This includes bioconcentration, i.e., the direct uptake of chemicals from an environmental medium alone, and is distinguished from biomagnification, i.e., the increase in chemical residues in organisms that have been taken up through two or more levels of a food chain.	Numerous physical, biological, and chemical factors affect the rates of bioaccumulation needed to conduct environmental risk assessment. For chemicals to bioaccumulate, they must be sufficiently stable, conservative, and resistant to chemical degradation. Elements, especially metals, are inherently conservative, and are taken up by organisms either as ions in solution or via organometallic complexes, such as chelates. Complexation of metals may facilitate bioaccumulation by taking forms of higher bioavailability, such as methylated forms. The organisms will metabolize by hydrolysis that allows the free metal ion to bond ionically or covalently with functional groups in the cell, like sulfhydryl, amino, purine, and other reactive groups. Organic compounds with structures that shield them from enzymatic actions or from

TABLE 7.1
Continued

Property of substance or environment	*Chemical importance*	*Physical importance*
		nonenzymatic hydrolysis have a propensity to bioaccumulate. However, readily hydrolyzed and eliminated compounds are less likely to bioaccumulate (e.g., phosphate ester pesticides like parathion and malathion). Substitution of hydrogen atoms by electron-withdrawing groups tends to stabilize organic compounds like the polycyclic aromatic hydrocarbons (PAHs). For example, the chlorine atoms are large and highly electronegative, so chlorine substitution shields the PAH molecule against chemical attack. Highly chlorinated organic compounds, such as the PCBs, bioaccumulate to high levels since they possess properties that allow them to be easily taken up, but do not allow easy metabolic breakdown and elimination.

Adapted from D.A. Vallero, 2004. *Environmental Contaminants: Assessment and Control,* Academic Press, Elsevier Sciences, Burlington, MA.

Persistence

As demonstrated in the Inuit exposure case, substances that remain in the environment long after their release are more likely to continue to cause problems or to be a threat to environmental quality. Persistence is commonly expressed as the chemical half-life ($T_{1/2}$) of a substance; that is, the time it takes to degrade one-half of the mass. The U.S. Environmental

Protection Agency considers a compound to be persistent if it has a $T_{1/2}$ in water, soil, or sediment that is greater than 60 days and very persistent if the $T_{1/2}$ is greater than 180 days. In air, the compound is considered persistent if its $T_{1/2}$ is greater than two days. Some of the most notoriously toxic chemicals are also very persistent.

The concept of persistence elucidates the notion of tradeoffs that are frequently needed as part of many responses to environmental insults. It also underlines the importance that good science is necessary but never sufficient to provide an acceptable response to environmental challenges. Let us consider the pesticide DDT (1,1,1-trichloro-2,2-bis-(4-chlorophenyl)-ethane ($C_{14}H_9Cl_5$)). DDT is relatively insoluble in water (1.2–5.5 mgL^{-1} at 25°C) and is not very volatile (vapor pressure: 0.02×10^{-5} mmHg at 25°C).[14] Looking at the water solubility and vapor pressures alone may lead us to believe that people and wildlife are not likely to be exposed in the air or water. However, the compound is highly persistent in soils, with a $T_{1/2}$ of about 1.1 to 3.4 years, so it may still end up in drinking water in the form of suspended particles or in the air sorbed to fine particles. DDT also exhibits high bioconcentration factors (in the order of 50,000 for fish and 500,000 for bivalves), so once organisms are exposed, they tend to increase body burdens of DDT over their lifetimes. In the environment, the parent DDT is metabolized mainly to 1,1-dichloro-2,2-bis(*p*-chlorophenyl)ethane (DDD) and DDE.[15]

The physicochemical properties of a substance determine how readily it will move among the environmental compartments, to and from sediment, surface water, soil, groundwater, air, and in the food web, including humans. So, if a substance is likely to leave the water, it is not persistent in water. However, if the compound moves from the water to the sediment, where it persists for long periods of time, it must be considered environmentally persistent. This is an example of how terminology can differ between chemists and engineers. Chemists often define persistence as an intrinsic chemical property of a compound, whereas engineers see it as both intrinsic and extrinsic (i.e., a function of the media, energy and mass balances, and equilibria). So, engineers usually want to know not only about the molecular weight, functional groups, and ionic form of the compound, but also whether it is found in the air or water, and what is the condition of the media (e.g., pH, soil moisture, sorption potential, and microbial populations). The movement among phases and environmental compartments is known as partitioning (see Chapter 2).

It is important to keep in mind the difference between chemical persistence and environmental persistence. For example, we can look at Henry's Law, solubility, vapor pressure, and sorption coefficients for a compound and determine that the compound is not persistent. However, in real-life scenarios, this may not be the case. For example, there may be a repository of a source of a nonpersistent compound that leads to a continuous, persistent exposure of a neighborhood population (see the case study,

"'Cancer Alley' and Vinyl Chloride" in Chapter 5). Or, a compound that is ordinarily not very persistent may become persistent under the right circumstances; for example, a reactive pesticide that is tracked into a home and becomes entrapped in carpet fibers. The lower rate of photolysis (degradation by light energy) indoors and the sorptive characteristics of the carpet twill can lead to dramatically increased environmental half-lives of certain substances.

Endocrine Disrupting Compounds

A potentially important threat looming at present with long-term implications is presented by a suite of chemicals that appear to alter hormonal functions in animals, including mammals. Such chemicals, known as hormonally active agents or endocrine disrupting compounds (or simply endocrine disruptors), have been associated with abnormal spermatogenesis; feminization of males; masculinization of females; dysfunction of adrenal, pineal, and thyroid glands; auto-regulatory problems; and other hormonally related problems. They are diverse in molecular structure (see Table 7.2), come from a myriad of sources, and have been detected throughout the environment (i.e., in food, water, air, soil, and plant and animal tissues).

Lake Apopka: A Natural Experiment

Recent research has shown that many endocrine disruptors are present in the environment at levels capable of negatively affecting wildlife. Our understanding of endocrine disruption has been enhanced by scientific research in three major areas: laboratory studies (including animal testing and cellular receptor binding studies); epidemiology (the study of the incidence and distribution of diseases in human populations and ecosystems); and natural experiments (unplanned events, uncontrolled *per se* by scientists, but that allow for "before and after" comparisons).

Among the first heavily researched hormonal agents was DDT.[16] Throughout the 1980s, exposure to this pesticide was associated with abnormal sexual differentiation in seagulls, as well as thinning and cracking of bald eagle eggs[17] and the feminization and loss of fertility found in reptiles.[18] A profound endocrine disruption natural experiment was that of a large spill of a pesticide mixture into Lake Apopka in central Florida. Studies following the spill have indicated that male alligators, the lake's top predators, showed marked reductions in

TABLE 7.2
Selected compounds found in the environment suspected of adversely affecting hormonal function, based on *in vitro*, *in vivo*, cell proliferation, or receptor-binding studies. (For full list, study references, study types, and cellular mechanisms of action, see Chapter 2 of National Research Council, *Hormonally Active Agents in the Environment*, National Academy Press, Washington, D.C., 2000. Source for asterisked (*) compounds is Colburn, *et al.*, http://www.ourstolenfuture.org/Basics/chemlist.htm.)

Compound[1]	*Endocrine Effect*[2]	*Potential Source*
2,2′,3,4′,5,5′-Hexachloro-4-biphenylol and other chlorinated biphenylols	Anti-estrogenic	Degradation of PCBs released into the environment
4′,7-Dihydroxy daidzein and other isoflavones, flavones, and flavonals	Estrogenic	Natural flora
Aldrin*	Estrogenic	Insecticide
Alkylphenols	Estrogenic	Industrial uses, surfactants
Bisphenol A and phenolics	Estrogenic	Plastics manufacturing
DDE (1,1-dichoro-2,2-bis(p-chlorophenyl)ethylene)	Anti-androgenic	DDT metabolite
DDT and metabolites	Estrogenic	Insecticide
Dichlorobromopropane and other halogenated hydrocarbons	Estrogenic	Nematocide
Dicofol	Estrogenic or anti-androgenic in top predator wildlife	Insecticide
Dieldrin	Estrogenic	Insecticide
Diethylstilbestrol (DES)	Estrogenic	Pharmaceutical
Endosulfan	Estrogenic	Insecticide
Hydroxy-PCB congeners	Anti-estrogenic (competitive binding at estrogen receptor)	Dielectric fluids
Kepone (Chlorodecone)	Estrogenic	Insecticide
Lindane (-hexachlorocyclohexane) and other HCH isomers	Estrogenic and thyroid agonistic	Miticide, insecticide
Lutolin, quercetin, and naringen	Anti-estrogenic (e.g., uterine hyperplasia)	Natural dietary compounds

TABLE 7.2
Continued

Compound[1]	*Endocrine Effect*[2]	*Potential Source*
Malathion*	Thyroid antagonist	Insecticide
Methoxychlor	Estrogenic	Insecticide
Octachlorostyrene*	Thyroid agonist	Electrolyte production
Pentachloronitrobenzene*	Thyroid antagonist	Fungicide, herbicide
Pentachlorophenol	Anti-estrogenic (competitive binding at estrogen receptor)	Preservative
Phthalates and their ester compounds	Estrogenic	Plasticizers, emulsifiers
Polychlorinated biphenyls (PCBs)	Estrogenic	Dielectric fluid
Polybrominated diphenyl ethers (PDBEs)*	Estrogenic	Fire retardants, including *in utero* exposures
Polycyclic aromatic hydrocarbons (PAHs)	Anti-androgenic (Aryl hydrocarbon-receptor agonist)	Combustion byproducts
Tetrachlorodibenzo-*para*-dioxin and other halogenated dioxins and furans*	Anti-androgenic (Aryl hydrocarbon-receptor agonist)	Combustion and manufacturing (e.g., halogenation) byproduct
Toxaphene	Estrogenic	Animal pesticide dip
Tributyl tin and tin organometallic compounds*	Sexual development of gastropods and other aquatic species	Paints and coatings
Vinclozolin and metabolites	Anti-androgenic	Fungicide
Zineb*	Thyroid antagonist	Fungicide, insecticide
Ziram*	Thyroid antagonist	Fungicide, insecticide

[1] Not every isomer or congener included in a listed chemical group (e.g., PAHs, PCBs, phenolics, phthalates, and flavinoids) has been shown to have endocrine effects. However, since more than one compound has been associated with hormonal activity, the whole chemical group is listed here.
[2] Note that the antagonists' mechanisms result in an opposite net effect. In other words an anti-androgen feminizes and an anti-estrogen masculinizes an organism.

gonad size. This well publicized spill was among the first to highlight the suspected ecological link between chemical exposure and endocrine disruption. These alligators exhibited reproductive and hormonal abnormalities, including elevated levels of estrogen, abnormal seminal vesicles, and smaller than normal penises, after an exposure due to a large spill of dicofol (containing 15% DDT).

Since the spill, other pesticides and chemicals have been associated with endocrine-related abnormalities in wildlife, including the inducement of feminine traits, such as secretion of the egg-laying hormone, vitellogenin, in males of numerous fish and other aquatic species downstream from wastewater treatment plants.[19] Birds and terrestrial animals are also affected by endocrine disrupting compounds (EDCs).[20] Recently, these problems have found their way to humans exposed to halogenated compounds and pesticides.[21] A nationwide survey of pharmaceuticals in U.S. surface water found EDCs at $ng\,L^{-1}$ levels in 139 stream sites. Several of these EDCs were found at even $\mu g\,L^{-1}$ levels, including nonylphenol ($40\,\mu g\,L^{-1}$), bisphenol A ($12\,\mu g\,L^{-1}$), and ethinyl estradiol ($0.831\,\mu g\,L^{-1}$).[22] Many of these compounds are extremely persistent in the environment, so their removal before entering environmental media is paramount to reducing exposures.

As in many other cases where DDT is involved, there is no unanimity within the scientific community on its link to the problems in the wildlife in Lake Apopka. For example, it has been argued that sexual problems of alligators in Lake Apopka, Florida, may be a manifestation of exposures to already existing endocrine disruptors prior to the spill, and the physiological and hormonal problems are simply the result of these remnant exposures.[23] Another potent substance, dibromochloropropane, well known to sterilize factory workers in California, was formulated, and wastes with high concentrations of its residues were stored in unlined ponds near the Florida lake's shores. Thus, the dibromochloropropane may well have leached into the lake, giving rise to the endocrine disruptive effects in the alligators.

Although many pesticides and some industrial chemicals have undergone toxicity testing (i.e. potential for adverse health effects), these tests are in many ways inadequate to ascertain the degree to which a substance will interact with the endocrine system. Scientific knowledge related to endocrine disruptors is evolving, but there is general scientific agreement that better endocrine screening and testing of existing and new chemicals are needed.

Genetic Engineering

If done wrong, nanotechnology in combination with genetic manipulation can wreak havoc to ecosystems and potentially threaten the public health. Ironically, the major justification for many of these emergent technologies is improved health. The challenge is to avoid opening Pandora's boxes while opening the knowledge treasure troves that these technologies hold.

There is clearly an anti-genetic-engineering voice, or often a shout, out there. Some opponents are clearly neo-Luddites, but others are simply asking for protections and commitments to precaution before going into an unbridled mode of genetic engineering research and applications. Some would also say that that day has already passed, and trying to control or even moderate the direction of genetic engineering is akin to changing a tire on a bus as it moves down the highway. It is physically possible, but certainly not very probable. So, should we resolve to ready our efforts to address the more than likely problems, or even disasters, which will result? The prudent approach is to embrace the best by using the technologies for good, but prepare for the worst.

It is not the purpose of this book to address the specifics of genetic engineering, but it would be negligent not to mention this as an emerging paradigm that will affect the environment in many ways. Some are already apparent, such as the enhanced treatment of hazardous wastes with genetically engineered microbes and the development of biologically altered organisms to obviate the need for chemical pesticides. However, any downsides to even these beneficial applications may not be known for some time.

Nuclear Fission

Since humans were first able to split the atom, there has been great concern about living in the nuclear age. Nuclear energy hinges on a rather simple reaction, the nuclear reaction, which occurs between the nuclei of atoms. When the so-called strong forces holding the nucleus are released during this reaction, extremely large amounts of energy are produced. A nuclear reaction contrasts with a chemical reaction wherein the reaction occurs exclusively between an atom's exterior electrons. The release of energy in a chemical reaction is that of weak forces. The nuclear reaction results from the bombardment of a nucleus with atomic or subatomic particles or very high energy radiation. Possible reactions are emission of other particles or the splitting of the nucleus, known as fission. During nuclear fission, a very heavy nucleus is split into two roughly equal fragments. This process, known as a chain reaction, releases several neutrons, which in turn split other nuclei in neighboring atoms. If the chain reaction is not stopped, a nuclear explosion can occur.

So, the energy production itself can be a problem. That is, the release of energy results in the conversion to large amounts of heat. That is a good thing if you are trying to make steam to turn a turbine, which converts mechanical energy into electricity. It is a bad thing if the heat finds its way to places where it can cause harm, such as surface water where aquatic life is often very sensitive to even small changes; for example, an increase in temperature of two or three degrees Celsius can completely alter the aquatic biota. However, the heat is not the biggest perceived problem with nuclear energy production.

The most intractable and potentially damaging aspects of nuclear power is its generation of very persistent and toxic radioactive wastes and the potential for accidents. The worst nuclear power disaster took place in the Ukraine.

Meltdown at Chernobyl[24]

On April 26, 1986, the world's most disastrous nuclear power accident occurred in Chernobyl. Located in northern Ukraine 100 km north of the capital Kiev, the Chernobyl power plant is 7 km from the border of Belarus. The reactor is on the river Pripyat, which joins the Dnieper 12 km away in the town of Chernobyl. The reactor has been inactive since December 12, 2000; however, people are still affected by the spread of radiation. The contaminated territories lie to the north of Ukraine, the south and east of Belarus, and in the western border area between Russia and Belarus. An estimated 125,000 to 146,000 square kilometers in Ukraine, Belarus, and Russia are contaminated with high levels of cesium 137. At the time of the explosion, approximately 7 million people lived in the area of contamination; 5.5 million people continue to live in these territories (see Figure 7.4).

The incident was a combined result of a flawed reactor operated by poorly trained staff without proper regard for safety measures. There was also a lack of communication between the personnel operating the facility and the team in charge. First of all, although plant operators and the USSR Committee of State Security were aware of the reactor's design weaknesses (as stated in a memorandum that outlined the plant's inadequate monitoring of safety equipment), they did not make changes nor take precautions. Chernobyl's RBMK reactor type was known to suffer from instability at low power (a positive void coefficient), which could result in uncontrollable increases in power. Other plants developed preventative designs, but the Chernobyl power plant did not.

The initiating events were the demands for increased power generation. These led to the formation of heat and steam pockets, resulting in less neutron absorption. Liquid water is used to absorb escaping neutrons, but steam does not have the ability to slow down the reaction.

During a routine maintenance check, the station technicians neglected to follow proper safety procedures. As Reactor 4 was to be shut off and

FIGURE 7.4. Map of region surrounding the Chernobyl nuclear power facility.[25]

checked, it was decided to test whether, in an event of a shutdown, enough energy could be generated to maintain the plant until the diesel power supply came into effect. The standard 30 control rods was neglected, and instead the test only used six to eight rods. Many were withdrawn due to the buildup of xenon that absorbed neutrons and decreased power. As the flow of coolant water fell, the poor design of the plant caused a large power surge that was 100 times the nominal power output. The slowing turbines and decreased cooling caused a positive void coefficient in the cooling channels. At 1:23 A.M., the rise in heat caused the fuel casing to rupture and resulted in a steam explosion destroying the reactor core, leading to a second explosion that cast fragments of burning fuel (see Figures 7.5 and 7.6).

Firefighters were called in to put out the fire in what remained of the Unit 4 building. A group of 14 firemen first arrived minutes after the explosion, and hundreds more arrived soon after. By 2:30 A.M., the largest fires on the roof of the reactor were under control and were finally put out by 5:00 A.M. "However, a graphite fire had started, so the firefighters who had already been exposed to the highest levels of radiation stayed on to try to extinguish it. The graphite fire posed a much bigger problem than the conventional fires because little experience at the local or even the international level was available on how to control this type of fire. In fact, 31 firefighters died at the scene. Yet, the greatest fear was the possibility of further spreading of the plume of radionuclides."

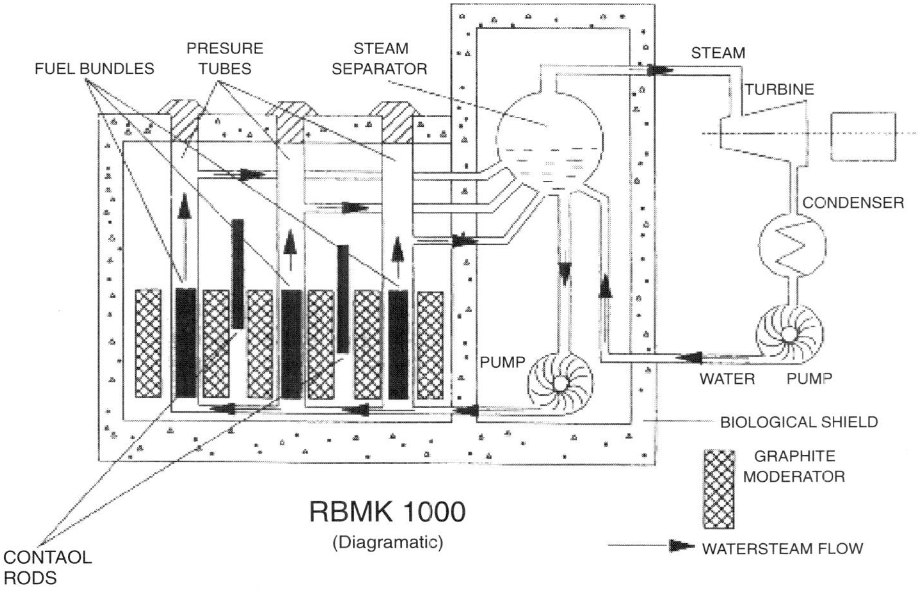

FIGURE 7.5. Reactor core in Chernobyl nuclear power plant.
Source and schematic credit: Nuclear Energy Agency, Organization for Economic Co-operation and Development, 2002, *Chernobyl: Assessment of Radiological and Health Impacts, 2002 Update of Chernobyl: Ten Years On,* Paris, France, page 25.

It was decided to attack the fire by dumping neutron-absorbing materials onto the site from helicopters. Many of the compounds, however, were not dropped directly onto the target and instead acted as an insulator, raising temperatures, and causing a greater dispersion of radionuclides within the next week. The graphite fire was finally extinguished on May 9. For the next three years, about 800,000 people who assisted in the cleanup of Chernobyl still suffer from the effects of radiation today, if indeed they are still alive.

The explosion was kept secret from the local population and the world for several days. Life went on as normal in the nearby towns and villages, and it was not until days later, upon the arrival of military tanks, that the explosion and its effects became known. One hundred and sixteen thousands people were ordered to evacuate their homes immediately with minimal possessions. The evacuation zone was fairly densely populated. Between 1990 and 1995, 210,000 people resettled in a new town, Slavutich, that had been built for the personnel of the power plant.

Radioactive particles were carried by wind toward the west and north to prime farm land. Radioactivity were highest in places receiving the most rain within the few days after the explosion, and even today particles con-

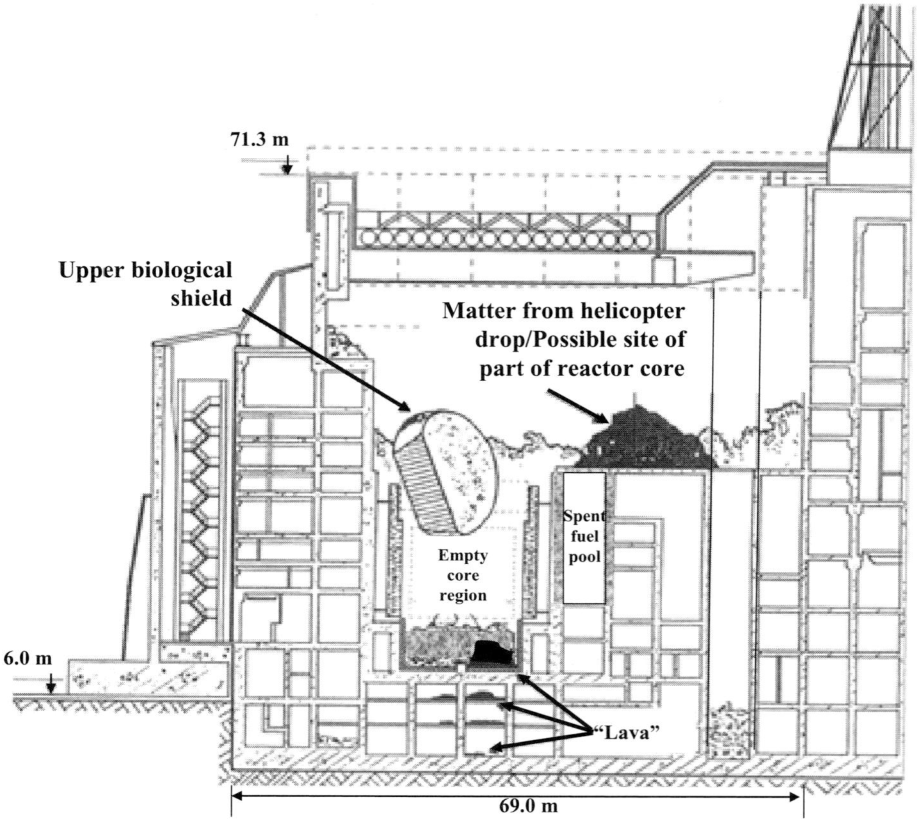

FIGURE 7.6. Schematic of reactor at Chernobyl. Adapted from: World Nuclear Association, 2005, http://www.world-nuclear.org/info/chernobyl/chernowreck2.gif; accessed on April 21, 2005.

tinue to seep into the ground. People were exposed to both internal and external radiation. "The major routes of human exposure to radiation were from ingestion of cow's milk contaminated with iodine[131] (resulting in internal exposure), contact with gamma/beta radiation from the radioactive cloud, and contact with cesium-137 deposited on the ground (resulting in external exposure)."[26] The water supply, plants, and animals in the area are radioactive. On average, 70% of radiation poisons the body through food and drink, and 30% is breathed into the body as air particles which has been linked to thyroid cancer, lymphatic cancer, heart conditions, and poor eyesight. The explosion has also resulted in a decrease in IQ in children, and genetic mutations are twice as likely in families exposed to radiation, causing permanent damage to DNA passed down to their children.

Preceding the explosion, the average rate of thyroid cancer in Ukrainian children was four to six per million (see Figure 7.7), which rose to an

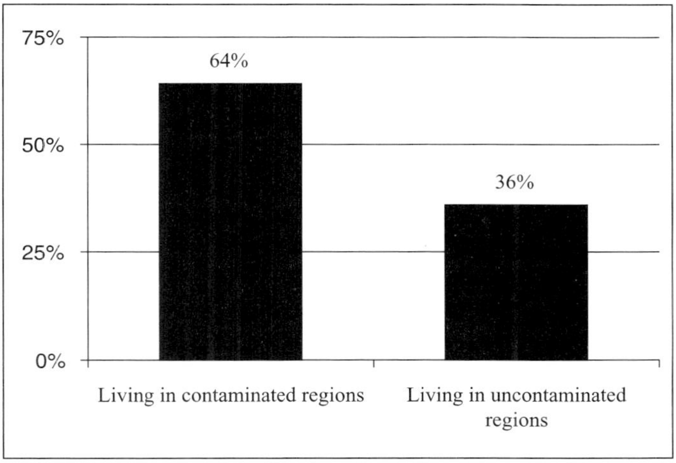

FIGURE 7.7. Precentage of Ukranian children with thyroid cancer, by residence.

astonishing 45 per million from 1986 to 1997. Sixty-four percent of these children lived in the most contaminated regions (Kiev, Chernigov, Zhitomir, Cherkassy, and Rovno).[27]

Besides the massive physical destruction and health effects, the explosion in Chernobyl's power plant also had serious psychological consequences, along with social, economic, and political effects. The evacuation of the contaminated areas caused a shortage of labor in the area, resulting in serious economic difficulties. People have lived in a constant fear of radiation and employers often discriminate against accident survivors.

Terrorism

As discussed in Chapter 3, the recent concern with terrorism, particularly following the attacks on the World Trade Center and the Pentagon on September 11, 2001, has changed the role of engineers and environmental scientists in numerous ways. For example, we are charged with providing safe water to drink and cleaning up polluted water before it enters streams and lakes, but both of these efforts usually require a centralized facility, such as a drinking water plant or a wastewater treatment plant. The challenge is to ensure that these facilities are free from terrorist acts. We are also charged with greater roles in emergency response. Many professional disciplines need to receive ample training in emergency response. Providing high-quality air, water, and food, handling chemical and radioactive wastes, and protecting indoor and occupational settings all have obvious needs to design facilities that can resist failures, that have a need to respond to such failures, and in the event that these measures do not work, that have explicit plans for recovery. Terrorism is truly a sword of Damocles.

Ecosystem Habitat Destruction

Environmental science benefited from the ecological view that emerged from the 1960s. That is, sanitary engineering projects were perceived to be within a larger, overall framework than simply free-standing structures. But other things were beginning to change. In fact, a whole new field of ecology was becoming legitimized in the 1970s. It is hard to believe that only a generation ago, few people had any idea of the meaning of the term ecosystem. Ecology is the study of relationships among organisms and their surroundings (i.e., their environment). These relationships are actually complicated fault trees, if you will, that make up an ecosystem. The operative word here is "system." So it stands to reason that engineers, being the systematic types that we are, were ready to incorporate this ecological perspective into environmental engineering.

Another important collective observation among engineers and scientists in the middle part of the twentieth century was that problems were really multifaceted and called for comprehensive solutions. Early in the environmental awareness movement, the news carried stories of heavily contaminated rivers and lakes, where one or a few sources were readily linked to the problems. For example, if the pH of a river seemed to drop suddenly after passing by a pickling operation of a steel mill, it was pretty obvious that the effluent from that facility was the major problem. However, as biological and chemical feedback systems became better understood, the importance of subtle and incremental contributions of even small sources became more apparent. For example, a large lake may have received significant loadings of suspended solids and oxygen-depleting pollutants from adjacent municipal waste effluents, but the lake was also receiving heavy nutrient (e.g., phosphorus and nitrogen compounds) loadings from individual homes' septic tanks and overland flow (what came to be called nonpoint sources) from farms and highways. This meant that even if we did a great job at designing, engineering, and operating conventional structures (e.g., wastewater treatment plants), only part of the job was accomplished. Thus, the structural solutions (i.e., sanitary engineering) began to be viewed within the framework of an overarching ecological (i.e., environmental science and engineering) approach.

The ecological view called for multimedia solutions. In other words, we could no longer be entirely comfortable that our designs and solutions were unique and independent. This is still a problem (as discussed in Chapter 1). In the 1980s, the term "intermedia transfer" gained currency within the environmental sciences. This meant that you may have solved a problem within one media, only to have shifted it to and maybe even worsened it in another media. For example, sludge is the byproduct of a well-designed wastewater treatment facility. If the facility is not working well, no sludge is generated. But when it is, it turns a potential water pollution problem (discharge to a lake or stream) into a solid waste problem (sludge management). Thus, the original problem (nutrients and contaminants in effluent discharge) has become concentrated and transferred to another form (biosolids). What we do with this "new" form of pollution determines our real success in dealing with the original source. If we burn it (biosolids have a caloric value), we may be sending toxic heavy metals and chlorine compounds out into the atmosphere where they may be inhaled or deposited onto soil, biota, or surface waters. In fact, some of those chlorine compounds did not even exist until we decided to burn the sludge, but now some very nasty substances are out there (e.g., dioxins, hexachlorobenzene, and polycyclic aromatic hydrocarbons). The heavy metals were there, but at low concentrations compared to what we are now releasing. If we decide to apply the sludge to land, then we may be creating soil or ground water contamination. So, the new challenge for the modern engineers of the 1980s and beyond was to see their projects within a larger scope. To borrow economics terms, we had to consider the externalities and contingencies of our structures well beyond their specific design and operation and maintenance (O&M).

Harkening back to the old name for environmental engineers, the term "sanitary" is tied directly to public health. Although many, arguably most, environmental engineering projects still target public health, many strive for ecological endpoints, such as improved biodiversity and productivity. Wetlands protection became a national priority in the United States in the last quarter of the twentieth century, for good reason. We were losing them at an alarming pace (see Figure 7.8, for example). For some time, many environmental experts had maintained that once a wetland was lost, it could not be restored (we heard similar arguments about contaminated sediments and ground water, as well). However, being the technological optimists that engineers often are, they found ways to restore, design, and construct new wetlands. But they also needed to learn ways to preserve existing wetlands. So, the "sanitary" (health) focus continued to grow, but the newer emphasis on ecological resources called for an environmental (health *and* ecological) perspective.

Having made these distinctions between the traditional sanitary *versus* environmental engineering, I must add that many of the great sanitary engineers, in fact, were behaving like environmental engineers. The likes of McKinney, Eckenfelder, and Vesilind met many of the criteria of the environmental engineer, including an interdisciplinary approach, an ecological view, and a systematic process for solving environmental problems. In a way, they "spawned" the environmental engineers of the next generation.

Lessons Learned

Predicting the future is often difficult and always wrong. But this does not mean that environmental practitioners and researchers should not try. To

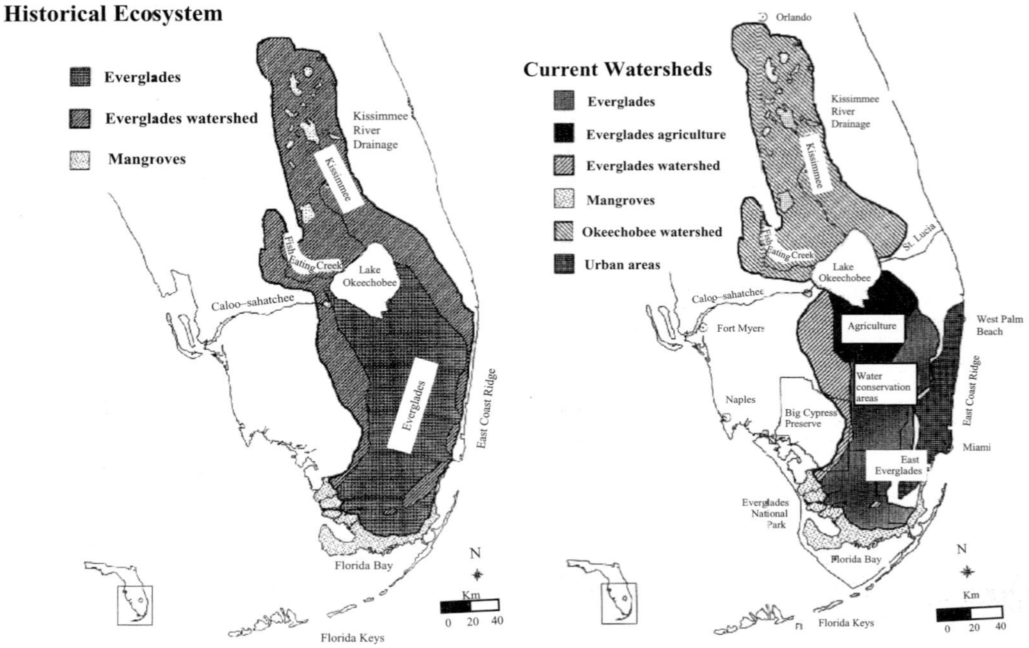

FIGURE 7.8. The extent of the Everglades ecosystem prior to agricultural and urban development (left map) and after development (right map).
Source: G.T. Bancroft, 1996. "Case Study: United States of America," *Human Population, Biodiversity and Protected Areas: Science and Policy Issues, Report of a Workshop, April 20–21, 1995,* Washington, D.C., V. Dompka, ed., American Association for the Advancement of Science, 1996.

the contrary, we must use the best available scientific information and build better models to project the future. Our view biases any such project. As evidence, let us consider the varying approaches of two engineering professions in how they address environmental problems.[28] One of the largest engineering societies in the United States is the American Society of Mechanical Engineers (ASME). The ASME has shown its concern for environmental problems by adding a fundamental canon to their code of ethics:

> Engineers shall consider environmental impact in the performance of their professional duties.

It is important to note that in this canon the strong *shall* is nullified by the weak *consider*. Apparently all that an engineer has to do is to *think* about environmental impact and this requirement is fulfilled. Perhaps, at the end of a long day of negotiations, the group may settle on the language stated in their canon, with the hope that the very mention of "environmental impact" will instill a modicum of environmental awareness in the engineer. It can be argued that this is better than nothing. For example, in the event that an engineer is called before a panel of peers to explain why she dumped hazardous waste during the performance of her duties, the fact that environmental impact has been codified will be part of the indictment. This is no "sword of Damocles" hanging over the engineer, but it is a means of making her aware at the outset that environmental considerations are important.

Further, in the ASME professional practice curriculum,[29] under this canon (Canon 8) is listed the following statement:

> Engineers shall concern themselves with the impact of their plans and designs on the environment. When the impact is a clear threat to health or safety of the public, then the guidelines for this Canon revert to those of Canon 1.

Canon 1 reads:

> Engineers shall hold paramount the safety, health and welfare of the public in the performance of their professional duties.

The ASME is saying that, practically speaking, there will never be a situation where the concerns for the nonhuman environment will take precedence over human health, safety, and welfare issues. A strict interpretation of this guidance is that the engineer who wants to ignore effects on the environment and on future people can safely hide behind the ASME Code of Ethics. However, this is inconsistent with another module of AMSE's professional practice curriculum, sustainability, which states that every engineering project must include the following elements:

1. Baseline studies of natural and built environments
2. Analyses of project alternatives
3. Feasibility studies
4. Environmental impact studies
5. Assistance in project planning, approval, and financing
6. Design and development of systems, processes, and products
7. Design and development of construction plans
8. Project management
9. Construction supervision and testing
10. Process design
11. Start-up operations and training
12. Assistance in operations
13. Management consulting
14. Environmental monitoring
15. Decommissioning of facilities

16. Restoration of sites for other uses
17. Resource management
18. Measuring progress for sustainable development

Furthermore, the curriculum admonishes:

> Just as engineers use safety factors due to the overriding need for safety, they should similarly use a sustainability factor because of the overriding need for sustainability. The safety of the human race in the future demands it no less than the safety of the human race in the present demands a safety factor.[30]

The American Society of Civil Engineers (ASCE) has taken a different tack than that of the ASME in addressing the environmental problem. Their code is one of the earliest engineering code of ethics, adopted in 1914. Based in spirit on the original Code of Hammurabi,[31] the 1914 ASCE Code addressed the interactions between engineers and their clients, and between engineers themselves. Only in the 1963 revisions did the ASCE Code include statements about the engineer's responsibility to the general public, stating as a fundamental canon the engineer's responsibility for the health, safety, and welfare of the public.

Responding to the general growth of environmental awareness, and conscious of the popular image of civil engineers as the perpetrators of environmental destruction, the ASCE Code was revised in 1977 to include the following statement:

> Engineers should be committed to improving the environment to enhance the quality of life.

Note first that "Engineers *should* . . ." which is very different from "Engineers *shall*. . . ." The use of "should" in effect precludes the enforcement of this section of the Code. All enforceable sections begin with the statement "Engineers shall . . ." Further, note that environmental effects relate solely to quality of life. Although the Code is vague on the matter, the phrase "quality of life" presumably applies only to *human* life. The Code in no way suggests that nature has intrinsic value beyond its utility, or instrumental value, to humans.

Recognizing this deficiency in the Code, the Environmental Impact Analysis Research Council (EIARC) of the Technical Council on Research (a committee of the ASCE) proposed in 1983 an eighth fundamental canon[32] (ASCE 1984). The proposed canon reads:

> Engineers shall perform service in such a manner as to husband the world's resources and the natural and cultured environment for the benefit of present and future generations.

Listed under the canon are nine guidelines that elaborate on the canon. For example, guideline 8.g reads:

> Engineers, while giving proper attention to the economic well-being of mankind and the need to provide for responsible human activity, *shall* [our emphasis] be concerned with the preservation of high quality, unique and rare natural systems and natural areas and *shall* [our emphasis] oppose or correct proposed actions which they consider, or which are considered by a reasonable consensus of recognized knowledgeable opinion, to be detrimental to those systems or areas.

The proposal struck many people as relatively modest and uncontroversial. It is explicitly anthropocentric: the environment is to be protected "for the benefit of present and future generations"—of humans, obviously. Nonetheless, in their January 1984 meeting, the Professional Activities Committee voted *unanimously against* recommending approval, and the canon died there.

In spite of these setbacks, many of the members and leaders of ASCE recognized the need for some definitive statement about the engineer's responsibilities to future generations. In response to these concerns, ASCE modified the code of ethics in 1997 to include a reference to "sustainable development"—a phrase that seemed to be sufficiently ambiguous to permit approval by the ASCE leadership. Recognizing this as a positive step in defining the responsibilities of engineers toward the environment, the First Fundamental Canon in the 1997 revisions of the ASCE Code of Ethics was changed to read:

> Engineers shall hold paramount the safety, health and welfare of the public and shall strive to comply with the principles of sustainable development in the performance of their professional duties.

The term "sustainable development" was first popularized by the World Commission on Environment and Development (also known as the Brundtland Commission), sponsored by the United Nations. Within this report sustainable development is defined as "development that meets the needs of the present without compromising the ability of future generations to meet their own needs."[33]

Even in light of the problems associated with the actual application of the sustainable development clause to engineering practice,[34] the addition of this requirement to the ASCE Code of Ethics is a laudable development. The major problem is with the wording in the Code. Note that "the engineer *shall* [my emphasis] *strive to* [my emphasis] comply with the principles of sustainable development . . ."

The strong first word is undone by the subsequent word by requiring the engineer only to try his or her best to comply with the principles of sustainable development. In effect this canon is just like the ASME environmental statement in that it does not require the engineer to consider environmental effects if these will interfere with his or her paramount duty to protect the health, safety, and welfare of the public.

Thus, two of the largest and oldest engineering societies have yet to grapple successfully with the addition of a useful statement concerning environmental issues into their codes of ethics. What is needed is a strong, workable, ethical statement that would express the attitudes of many engineers toward their self-perceived role as guardians of the environment for future generations. This is a problem of any emergent need. Recently, for example, ethicists and engineers have begun to draw the distinction between macro-ethics and micro-ethics. A potential macro-ethical issue, is one that the individual researcher or engineer cannot do much about, but that the whole scientific community needs to address. These are the potential Pandora's boxes, where the incremental contribution of hundreds or thousands of researchers is not significant in their threats individually, but *en masse*, the problems that could ensue are monumental. Environmental science and technology are rife with unknowns. Small changes can lead to widespread effects. For example, the introduction of the zebra mussel as a hitchhiker in the hulls of a barge has led to biome-scale reductions in biodiversity in the Great Lakes. Likewise, the widespread use of estrogenic compounds has affected aquatic life and even human health as it reaches the waterways.

Preventing environmental problems is a dynamic process. Due diligence, precaution, and factors of safety are not static, but depend on interdependent relationships of internal factors that vary from situation to situation. Also, there have been enough disasters and near catastrophes that prudence is required. Such calamities do not usually occur in a nice neat linear projection, but may be the consequence of an iteration of unlikely occurrences (see the discussion box, "The Butterfly Effect"). On the other hand, engineers cannot simply add up all the factors of safety to arrive at an overall factor of safety for their project. Unnecessarily cautious approaches are not only expensive in terms of time and resources; it well may not be protective. In fact, this approach could lead to unjust, unsafe, and unacceptable outcomes.

The Butterfly Effect

The butterfly effect characterizes "sensitive dependence upon initial conditions,"[35] as a postulate of chaos theory. A small change for good or bad can reap exponential rewards and costs. Edward Lorenz, at a 1963

New York Academy of Sciences meeting, related the comments of a "meteorologist who had remarked that if the theory were correct, one flap of a seagull's wings would be enough to alter the course of the weather forever." Lorenz later revised the seagull example to be that of a butterfly in his 1972 paper, "Predictability: Does the Flap of a Butterfly's Wings in Brazil set off a Tornado in Texas?" at a meeting of the American Association for the Advancement of Science, Washington, D.C. In both instances, Lorenz argued that future outcomes are determined by seemingly small events cascading through time. The butterfly effect has become well known in society, even to the point that this scenario has been presented in at least one television commercial. Engineers and mathematicians struggle with means to explain (and predict) such outcomes of so-called ill-posed problems. The scientific method is such that orderly systems generally are preferred, especially those with a well-posed problem, that is, one that is uniquely solvable (i.e., a unique solution exists) and one that is dependent upon a continuous application of data. The LaPlace equation is an example. By contrast, an ill-posed problem does not have a unique solution and can be solved only by discontinuous applications of data, meaning that even very small errors or perturbations can lead to large deviations in possible solutions.[36] Finding the appropriate times and places to solve ill-posed problems is a gravid area of mathematical and scientific research. For example, an ill-posed problem may be solved by what are known as inverse methods, such as restricting the class of admissible solutions using *a priori* knowledge. *A priori* methods include variational regularization using a quadratic stabilizer. Usually this requires stochastic approaches, assumptions that the processes and systems will behave in a random fashion.

By extension, small changes for good or bad can produce unexpectedly large effects. Including neighbors and the potentially affected public early in engineering decisions can prevent large problems down the road. An idea from one citizen, if given proper attention, can lead to a successful study and help to avoid costly mistakes. We often hear after the fact how someone had noticed a strange odor and odd behaviors, but whose questions and complaints were ignored until a larger leak of a toxic substance caused unnecessary ill effects. If we go back and review the comments of citizens early in the planning stages of large public projects, there may be evidence that the road, or treatment plant, or landfill that seemed so right to the technical folks was obviously problematic for the neighbors. After all, they live there. They often have intricate and detailed intelligence gathered over decades. Engineers and scientists ignore this information at their peril.

Ignoring small details can lead to big problems!

The role of micro-ethics is to ensure that the individual practitioner behaves in an upright manner. To date, most engineering and professional ethics has been attentive to micro-ethics, but is only beginning to find ways to look at actual and potential problems at the macro scale. The ASME and ASCE, for example, are struggling with balancing environmental quality with the daily demands on the individual engineer. The major lesson is that to be ready for the swords of Damocles environmental science and engineering must find ways to bridge the big issues with the day-to-day decision making of the practitioners; that is, to get out in front of potential environmental problems and prescribe approaches to be taken by environmental professionals. That way, the individual practitioners will be more able to "hold paramount the health, safety and welfare of the public" and "comply with the principles of sustainable development" as we do our jobs.

Notes and Commentary

1. James Baldwin, 1896. *Fifty Famous Stories Retold,* American Book Company, New York, NY.
2. United Nations Conference on Environment and Development, 1992. Principle 15, Declaration on Environment and Development, Rio de Janeiro, Brazil. The principle reads:

   > In order to protect the environment, the precautionary approach shall be widely applied by States according to their capabilities. Where there are threats of serious or irreversible damage, lack of full scientific certainty shall not be used as a reason for postponing cost-effective measures to prevent environmental degradation.

3. I. Goklany, 2001. *The Precautionary Principle,* Cato Institute, Washington, D.C. The Goklany reference is the definition by Carolyn Raffensperger and Joel Tickner paraphrased here.
4. J. Morris, 2000. *Rethinking Risk and the Precautionary Principle,* Butterworth-Heinemann, Burlington, MA.
5. Garrett Hardin, 1968. "Tragedy of the Commons," *Science,* 162, December 13, 1968.
6. J.S. Mill, 1869. *On Liberty,* Longman, Roberts & Green, London, U.K.
7. J. Rawls, 1999. *A Theory of Justice* (1971), Belknap Press Reprint, Cambridge, MA.
8. B.R. Sonawane, 1995. "Chemical contaminants in human milk: An overview," *Environmental Health Perspectives,* 103(suppl 6):197–205. K. Hooper and T.A. McDonald, 2000. "The PBDEs: An emerging environmental challenge and another reason for breast-milk monitoring programs," *Environmental Health Perspectives,* 108:387–392.
9. E. Dewailly, P. Ayotte, S. Bruneau, C. Laliberté, D.C.G. Muir, and R.J. Norstrom, 1993. "Inuit exposure to organochlorines through the aquatic food chain in Arctic Quebec," *Environmental Health Perspectives,* 101:618–620.

10. E. Dewailly, A.J. Nantel, J.P. Weber, and F. Meyer, 1989. "High levels of PCBs in breast milk of Inuit women from arctic Quebec," *Bulletin of Environmental Contamination Toxicology*, 43:641–646.
11. Dewailly et al., 1993.
12. E. Dewailly, S. Bruneau, C. Laliberté, et al., 1993. "Breast milk contamination by PCBs and PCDDs/PCDFs in Arctic Quebec: Preliminary results on the immune status of Inuit infants," *Dioxin* 1993. 13th International Symposium on Chlorinated Dioxins and Related Compounds; Vienna, Austria, p. 403–406.
13. W. Tucker and L. Nelken, 1982. "Diffusion Coefficients in Air and Water," *Handbook of Chemical Property Estimation Methods*. McGraw-Hill, New York, NY.
14. The source for the physicochemical properties of DDT and its metabolites is the United Nations Environmental Programme, 2002. "Chemicals: North American Regional Report," Regionally Based Assessment of Persistent Toxic Substances, Global Environment Facility.
15. The two principal isomers of DDD are *p,p'*-2,2-bis(4-chlorophenyl)-1,1-dichloroethane; and *o,p'*-1-(2-chlorophenyl)-1-(4-chlorophenyl)-2,2-dichloroethane. The principal isomer of DDE is *p,p'*-1,1'-(2,2-dichloroethenylidene)-bis [4-chlorobenzene]
16. D. Fry and C. Toone, 1981, "DDT–induced feminization of gull embryos" *Science* 213(4510), 922.
17. S. Weimeyer et al., 1984, "Organochlorine, pesticide, polychlorobiphenyl, and mercury residues in bald eagle eggs—1969–79—and their relationships to shell thinning and reproduction" *Archives of Environmental Contamination and Toxicology* 13(5), 529.
18. L. Guillette et al., 1994, "Developmental abnormalities of the gonad and abnormal sex-hormone concentrations in juvenile alligators from contaminated and control lakes in Florida," *Environmental Health Perspectives* 102(8), 680.
19. See, for example, C. Purdom, et al., (1994). "Estrogenic effects from sewage treatment works" *Chem. Ecol.* 8, 275; and S. Joblin et al., 1996, "Inhibition of testicular growth in rainbow trout (Oncorhynchus mikiss) exposed to estrogenic alkyphenolic chemicals," *Environmental Toxicology and Chemistry* 15(2), 194.
20. G. Fox, 2001, "Effects of endocrine disrupting chemicals on wildlife in Canada: past, present and future," *Water Quality Research Journal of Canada* 36(2), 233.
21. See Sheiner EK et al., 2003. "Effect of occupational exposures on male fertility: Literature review," *Industrial Health* 41(2), 55; and P. Guzelian, 1982, "Comparative toxicology of chlordecone (kepone) in humans and experimental animals," *Annual Reviews of Pharmacology and Toxicology* 22, 89; and T. Hayes et al., 2002. "Hermaphroditic, demasculinized frogs after exposure to the herbicide atrazine at low ecologically relevant doses," *Proceedings of the National Academy of Sciences of the United States of America*, 99(8), 5476.
22. D. Koplin et al., 2002, "Pharmaceuticals, hormones, and other organic wastewater contaminants in U.S. streams, 1999–2000: A national reconnaissance," *Environmental Science and Technology* 36(11), 1202.

23. R.W. Risebrough, 1999. "Endocrine disruption: Questions for the environmental community," *SETAC News*, 19(July):16–17.
24. The principal source for this case was written by Lisa Gerovich, a Duke University undergraduate student, as part of the requirements for the course, Ethics in Professions, EGR 108S.
25. World Nuclear Association, 2005. http://www.world-nuclear.org/info/chernobyl/ukr_map.gif; accessed April 21, 2005.
26. U.S. Department of Energy, 2005. http://www.eh.doe.gov/health/ihp/chernobyl/chernobyl.html; accessed April 21, 2005.
27. U.S. Department of Energy, 2005. Ukrainian Cancer Study published on July 1, 1999. http://www.eh.doe.gov/health/ihp/chernobyl/chernobyl.html; accessed April 21, 2005.
28. This discussion is taken in part from some discussions and writings with P. Aarne Vesilind, Rooke Professor of Civil Engineering at Bucknell University.
29. American Society of Mechanical Engineers, 2004. Professional Practice Curriculum: Engineering Ethics, http://www.professionalpractice.asme.org/engineering/ethics/8.htm; accessed November 2, 2004.
30. Ibid. http://www.professionalpractice.asme.org/communications/sustainability/8.htm.
31. M.W. Martin and R. Schinzinger, 2005. *Ethics in Engineering*, 4e, McGraw-Hill Higher Education, New York, NY.
32. American Society of Civil Engineers, 1984. The Environmental Impact Analysis Research Council of the Technical Council on Research, "A Proposed Eighth Fundamental Canon for the ASCE Code of Ethics," *Journal of Professional Issues in Engineering*, 110, 3. Also, much of the ethics discussion in this chapter has risen from my ongoing discussions and collaborations with P. Aarne Vesilind.
33. World Commission on Environment, 1987. *Our common future: Report of the World Commission on Environment and Development*, Oxford University Press, Oxford, UK.
34. P.A. Vesilind and A.S. Gunn, 1998. *Engineering, Ethics, and the Environment*, Cambridge University Press, New York, NY.
35. R.C. Hilborn, 1994. *Chaos and Nonlinear Dynamics*, Oxford University Press, UK.
36. J. Hadamard, 1923. *Lectures on the Cauchy Problem in Linear Partial Differential Equations*, Yale University Press, New Haven, CN.

Part III

Other Paradigms

Many environmental issues and problems do not properly fit within a single media. Metals are particularly troublesome, since they can readily change forms and move within and among the water, air, sediment, soil, and biota. Some problems are better defined by scale, such as planetary problems of ozone depletion and climate change. Other problems are better addressed from the standpoint of risk assessment and risk perception, such as product scares. And other problems, though having technical aspects, are best approached as societal and cultural problems, such as those that have come about because of injustices.

This eclectic mix of cases in Part III provides insights into some of the most important emerging environmental problems and opportunities to address them.

CHAPTER 8

Dropping Acid and Heavy Metal Reactions

Men's mighty mine-machines digging in the ground,
Stealing rare minerals where they can be found.
Concrete caves with iron doors bury it again,
While a starving frightened world fills the sea with grain.

Mike Pender, *Why Is It We Are Here?* 1970

Since the Iron Age, humans have depended on metals in their quests to modernize. Metals are needed for industry, farming, defense, housing, transportation, and almost every aspect of contemporary life. But, as Pender points out, finding and extracting metals is a costly enterprise. Pender's song is part of a collection from the Moody Blues' album *A Question of Balance*. Like much of their work, the album poses questions about what is to be valued. Having experienced the tumultuous 1960s, environmental values were among those questioned, along with peace and justice. I could have selected some of the more obvious protest songs, such as those by Joan Baez, Bob Dylan or, arguably the most pessimistic, *Eve of Destruction* by Barry McGuire. But, what makes Pender's observations so telling is the importance of concentration and form in determining whether an element is a resource or a pollutant.

This is the stuff of inorganic chemistry, the subject matter of this chapter, especially those where acid-base relationships and reactions between metals and nonmetals have contributed to the problem. Although this book, like many other environmental texts, devotes significant attention to the organic pollutants, the lion's share of pollutants are actually inorganic (see Figure 5.13).

Metals usually exist in relatively low concentrations in ores, so they must be refined. In addition to completely changing the physical landscape, mining leaves behind tailings consisting of large amounts of slag or dross with elevated amounts of toxic metals and metalloids. Separating the desired metals from the host rock and other minerals in the ore generates large quantities of wastes.[1] Mining minerals other than coal in the United

States totaled 50 billion metric tons in 1985.[2] So, even if the chemical form of these metals remains the same as they were prior to mining, their concentrations have increased exponentially. They are also now more exposed to meteorological events, erosion, and mass wasting, so that they may contaminate ground and surface waters, as well as pollute the air.

By and large, these deposits are dominated by metal carbonates, oxides, and sulfides. Of these, the mining of sulfide minerals like copper, nickel, lead, zinc, and silver sulfide deposits are the most environmentally important because the process of extracting metals allows the sulfur to become oxidized and released to the atmosphere as SO_2, which reacts with water and other atmospheric constituents to produce sulfuric acid (see Chapter 2, "Contaminants of Concern: Sulfur and Nitrogen Compounds"). Remaining ore materials are often mixtures of minerals with very high concentrations of toxic metals and metalloids, for example, arsenic, cadmium, and mercury. Following extraction, the oxidation of metal sulfides remaining in the solid wastes increases the acidity of surrounding waters.

Case of the Negative pH: Iron Mountain, California[3]

The acidity is so strong in some waters receiving runoff from mining activities that the calculated pH values are negative!

The nice thing about software is that it does what you tell it to do. For example, a common data validation check is to instruct validation software to look for physically impossible values, such as negative areas and volumes (e.g., –20 acres or –10 liters). Another instruction is to look for unrealistic values, such as negative pH values or pH values greater than 14. The pH value is one of the few very complicated concepts that has gained currency in most areas of engineering and medicine, and is widely grasped by the general public. They may know that their shampoo is "pH balanced" and even that that means it is chemically neutral, with a pH value near 7. Elementary students are taught that the range of pH is between 0 and 14. However, to a theoretical chemist, the range is not so limited.

Water not only exists as molecular water (H_2O), but also includes hydrogen (H^+) and hydroxide (OH^-) ions:

$$H_2O \leftrightarrow H^+ + OH^- \tag{8.1}$$

The negative logarithm of the molar concentration of hydrogen ions, $[H^+]$ in a solution (usually water in the environmental sciences), is referred to as pH. This convention is used because the actual number of ions is extremely small. Thus pH is defined as:

$$pH = -\log_{10}[H^+] = \log_{10}([H^+]^{-1}) \tag{8.2}$$

The brackets refer to the molar concentrations of chemicals, and in this case it is the ionic concentration in moles of hydrogen ions per liter. The reciprocal relationship of molar concentrations and pH means that the more hydrogen ions you have in solution the lower your pH value will be.

Likewise, the negative logarithm of the molar concentration of hydroxide ions, $[OH^-]$ in a solution, is pOH:

$$pOH = -\log_{10}[OH^-] = \log_{10}([OH^-]^{-1}) \tag{8.3}$$

The relationship between pH and pOH is constant under equilibrium conditions:

$$K = [H^+]\,[OH^-] = 10^{-14} \tag{8.4}$$

When expressed as a negative log, it becomes obvious that the pH and pOH scales are reciprocal to one another and that they both range from 0 to 14. Thus, a pH 7 must be neutral (just as many hydrogen ions as hydroxide ions). The log relationship means that for each factor pH unit change there is a factor of 10 change in the molar concentration of hydrogen ions. Thus, a pH 2 solution has 100,000 times more hydrogen ions than neutral water (pH 7), or $[H^+] = 10^{12}$ versus $[H^+] = 10^7$, respectively.

However, for very strong acids, the molarity of hydrogen ions greater than 1 yields a negative value of pH. For example, a 12 molar (i.e., 12 *M*) HCl solution has a theoretical pH of $-\log(12) = -1.1$. So for highly acidic, extremely hazardous conditions (such as the occasional mine drainage system), we may measure negative pH values in environmental situations.

I have never observed these situations in engineering practice. In fact, I have never seen a pH near zero. Even the strongest acids do not completely dissociate when they are dissolved in very high concentrations. In the 12 *M* HCl solution, a portion of the hydrogen will remain bound to the chlorine, so the pH will be higher than what is predicted from the acid molarity. In addition, the number of water molecules are dwarfed by the number of acid molecules, the influence of the H^+ ions is increased (the effective H^+ concentration, i.e., the activity) and is highly elevated above the actual concentration. So, although we commonly show pH to be the negative log of the molar concentration of hydrogen ions, it is actually the negative log of the hydrogen ion activity (i.e., $pH = -\log a_{H^+}$).

Even if we see a negative pH it is very likely to be a measurement error. At very high pH, using existing equipment (glass pH electrode), we regularly see a large positive measurement bias (measured pH >> true pH), which is very difficult to correct, but is commonly known as the acid error.

By extension, a pH value >14 is theoretically possible for highly caustic solutions (strong bases at very high concentrations). However, such conditions are seldom found in the natural environment. But they do

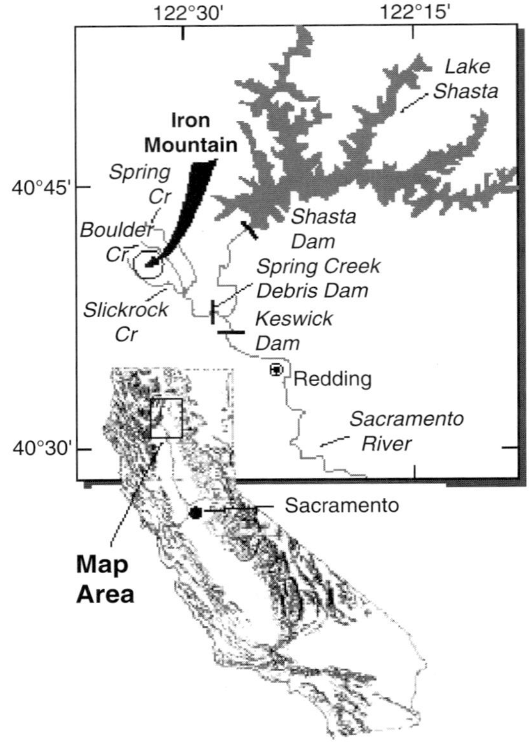

FIGURE 8.1. Site map of Iron Mountain Mine.
Source: U.S. Environmental Protection Agency.

uncommonly occur in highly contaminated sites, such as the Superfund site at Iron Mountain, California.[4] From the 1860s through 1963, the 4,400-acre Iron Mountain Mine (IMM) site (see Figure 8.1) periodically was mined for iron, silver, gold, copper, zinc, and pyrite. Though mining operations were discontinued in 1963, underground mine workings, waste rock dumps, piles of mine tailings, and an open mine pit still remain at the site.[5] Historic mining activity at IMM has fractured the mountain, exposing minerals in the mountain to surface water, rainwater, and oxygen; for example, when pyrite is exposed to moisture and oxygen, sulfuric acid forms. This sulfuric acid runs through the mountain and leaches out copper, cadmium, zinc, and other heavy metals. This acid flows out of the seeps and portals of the mine. Much of the acidic mine drainage ultimately is channeled into the Spring Creek Reservoir by creeks surrounding IMM.

The Bureau of Reclamation periodically releases the stored acid mine drainage into Keswick Reservoir. Planned releases are timed to coincide with the presence of diluting releases of water from Shasta Dam. On occasion, uncontrolled spills and excessive waste releases have occurred when Spring Creek Reservoir reached capacity. Without sufficient dilution, this results in the release of harmful quantities of heavy metals into the Sacramento River. Approximately 70,000 people use surface water within three miles as their source of drinking water. The low pH level and the heavy metal contamination from the mine have caused the virtual elimination of aquatic life in sections of Slickrock Creek, Boulder Creek, and Spring Creek. Since 1940, high levels of contamination in the Sacramento River have caused numerous fish kills. The continuous release of metals from IMM has contributed to a steady decline in the fisheries population in the Sacramento River. In 1989, the National Marine Fisheries Service took emergency action to list the Winter Run Chinook Salmon as threatened under the Endangered Species Act and to designate the Sacramento River from Red Bluff Diversion Dam to Keswick Dam as a critical habitat. In January 1994, the National Marine Fisheries Services issued its final rule reclassifying the Winter Run Chinook Salmon as an endangered species.

Aquatic biota can also readily accumulate toxic metals which is not only a human health threat, but releases if contaminants are acutely toxic to aquatic life. This contributed to the steady decline in fish populations and has contributed to the listing of the Winter Run Chinook Salmon as an endangered species.

High concentrations of heavy metals exist in sediments and pore waters in the Spring Creek Arm of Keswick Reservoir. These sediments, when mixed with water, can produce conditions that are toxic to aquatic life. The sediments are located immediately downgradient from the discharge of the Spring Creek Power Plant operated by the Bureau of Reclamation. Power house operations or large floods spilling from the Spring Creek Debris Dam could remobilize the contaminated sediments. Potential cleanup of these sediments is a high priority.[6]

Since 1983, the U.S. EPA has been conducting a remedial investigation at the site under Superfund to characterize the nature and extent of the acid mine drainage and its effects on receiving waters, including Spring Creek and the Spring Creek Arm of Keswick Reservoir. The EPA signed four records of decisions, selecting remedial actions for parts of the Iron Mountain site in 1986, 1992, 1993, and 1997. Remedial investigations are continuing so that additional remedial approaches can be evaluated by EPA for potential selection. The U.S. EPA filed an action in federal court seeking to recover the costs of its cleanup activities for the Iron Mountain Mine site. Pretrial litigation activities are proceeding at this writing.

In engineering practice, it is usually prudent to be highly suspicious of any reported pH value less than zero or greater than 14. And, it is reasonable to assume that these values are artifacts of either improper data logging or sampling error. But the Iron Mountain case demonstrates that although such extreme pH values are highly improbable, they are certainly not impossible. The protocol for removing the data points and how to treat

FIGURE 8.2. Tailings left behind after lead and zinc mining near Desloge, Missouri. *Source*: U.S. Geological Survey.

the gaps they leave must be detailed in the quality assurance plan for the project.

Dropping erroneous values makes scientific sense, but what is done about the deletion? If we just leave it blank, we may not really be properly representing the water body's quality. If we put a value in (i.e., "impute"), such as a neutral pH 7, we have changed the representativeness of the data. Even a more sophisticated method, like interpolating a data point between the two nearest neighbors' values, is not necessarily good. For example, we might miss an important, but highly localized "hot spot" of pollution. So, even at this very early step of manipulating data, bias (i.e., systematic error) can be introduced.

Mining can devastate the environment (see Figure 8.2). Extraction and removing impurities to refine gold, silver, copper, lead, and other metals from sulfide ores is an old technology, first introduced in the Western Hemisphere as the Europeans colonized and migrated westward in North America. The "gold rushes" in the West during the mid-nineteenth century accelerated the quest for precious metals.

Although extraction efficiency also improved, the nineteenth and twentieth centuries brought tremendous increases in scale to the mining industry through the advent of steam, internal combustion, and electric power. The basic methods for extracting metals, though, continued to rely largely on mechanical separation, gravity, water, and heat. Consequently, the increasingly large quantities of toxic metals and byproducts released into the environment left their marks on U.S. biota. No information on the cumulative effects of metals mining and refining on biota exists, but 557,650 abandoned mines in the United States are estimated to have contaminated 728 square kilometers of lakes and reservoirs and 19,000 kilometers of streams and rivers.[7] Areas where toxic releases to the environment from mining and smelting have caused large-scale effects on biological diversity or have jeopardized particularly rare or valuable living resources are numerous; some examples are listed in Table 8.1.

TABLE 8.1
Examples of pollution resulting from mining and extraction activities.

Location	*Effect*
Copper Basin, Tennessee	SO_2 emissions from copper smelting beginning in 1843 eliminated vegetation over a 130-square-kilometer area and may have contributed to the endangered status of Ruth's golden aster, a federally listed plant endemic to the Ocoee Valley. Metals and sediment have also contaminated Tennessee Valley Authority reservoirs on the Ocoee River.
Palmerton, Pennsylvania	Zinc smelting emissions from 1898 to 1980 completely denuded an 8-square-kilometer area and affected plants and animals for a much greater distance. Stream aquatic communities were not measurably affected.
Tri-State Mining District—Missouri, Kansas, Oklahoma	A century of zinc mining and smelting and attendant acidification and toxic metals releases have left bare areas and eliminated animal life from small streams. Among affected species are the Neosho madtom, a federally listed threatened fish, and the Neosho mucket, a rare mussel that is a candidate for federal listing.
Torch Lake, Michigan	The disposal of tailings and other copper mining wastes from the late 1860s to the 1960s is believed responsible for an outbreak of liver cancer in Sauger.
Leadville, Colorado	Mining in the headwaters of the Arkansas River system since the 1860s has resulted in acidification and toxic metals pollution that continues to affect aquatic communities for 50 kilometers downstream.

TABLE 8.1
Continued

Location	*Effect*
Clark Fork River System, Montana	Some 15 million cubic meters of mine tailings containing elevated metal concentrations, generated since mining began in 1880, have visibly contaminated and affected the aquatic biota in more than 230 kilometers of the Clark Fork mainstem (acidic, metals-laden mine drainage has also affected the benthic and fish communities and reduced the productivity of sport fisheries in the Blackfoot River, a tributary).
Blackbird Mine, Idaho	Mining contaminated 40 kilometers of Panther Creek, a tributary of the Salmon River. Releases of copper, cobalt, and other metals for more than 50 years decimated the resident fishery and the spring-summer run of chinook salmon, a threatened species.
Coeur d'Alene Valley, Idaho	Mining and smelting in and around Kellogg since 1879 have contaminated the South Fork of the Coeur d'Alene River, obliterated area vegetation, and contaminated biota. Mining wastes were responsible for repeated deaths of tundra swans into the late 1980s.
Iron Mountain Mine, Redding, California	Fish kills caused by metals released from mines and mine wastes have occurred in the Sacramento River for more than 100 years. Threatened are populations of steelhead trout (the sea-run form of rainbow trout) and chinook salmon, which have been denied access to all but 32 kilometers of Sacramento River spawning habitat since construction of the Shasta Dam. Metals from Iron Mountain Mine, together with warm summer discharges from Shasta Dam, may be responsible for the imperiled status of the spring-run chinook salmon population.

Source: U.S. Geological Survey.

Acid Mine Drainage

The most direct and immediately obvious toxicological effect of the use of coal has been acid mine drainage (see Figure 8.3). Acid drainage is caused by the oxidation of metallic compounds such as the mineral constituent of rocks and soils that is often present in coal mine slag. Most streams affected by coal mine drainage are acidic (pH 2.5 to 6.0), with high iron and sulfate concentrations. Ferric hydroxide often precipitates as a fine floc that may coat stream bottoms and further harm aquatic life. Acid mine drainage can

FIGURE. 8.3. Iron hydroxide precipitate in a Missouri stream receiving acid drainage from surface coal mining.
Photo credit: U.S. Geological Survey, D. Hardesty.

also leach toxic metals such as copper, aluminum, and zinc from rocks and soils (copper is particularly toxic to fish). The oxidation, or weathering, of pyrite and other metal sulfides is ongoing as minerals are exposed to air, and these reactions may be sped up by bacteria. For example, a number of streams in the Southeastern United States are naturally acidic due to the surrounding pyritic rock formations.

By the mid-1960s, a century of U.S. surface mining had disturbed about 8,000 square kilometers, including 21,000 kilometers of streams (totaling 550 square kilometers), 281 natural lakes (419 square kilometers), and 168 reservoirs (168 square kilometers). Coal mining accounted for 41% of the total disturbed lands, the bulk of pollution being from acid mine drainage in the

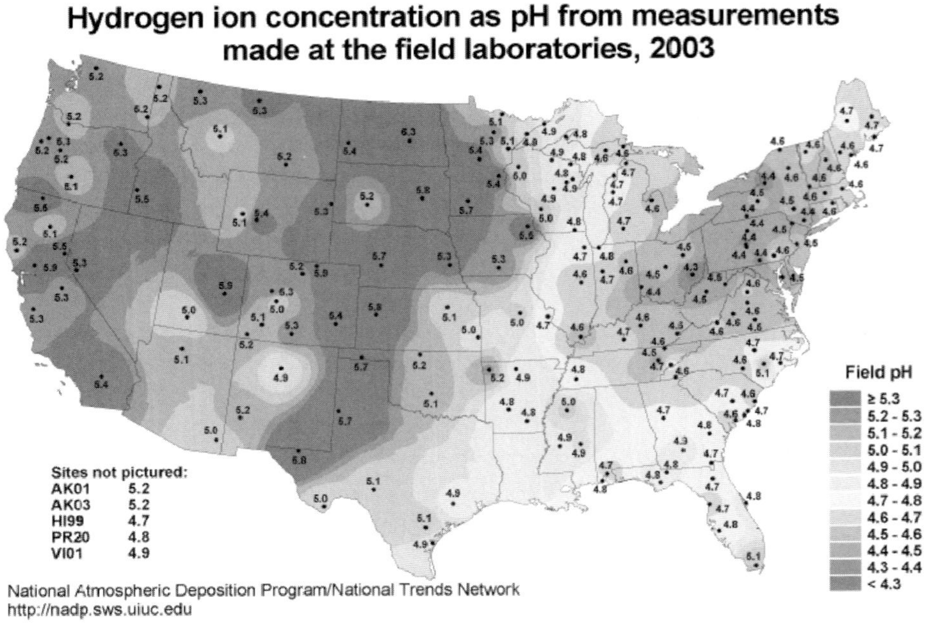

FIGURE 8.4. United States rainfall pH, 2003, as weighted mean values based on measurements at about 200 sites maintained by the National Acid Deposition Program. *Source*: National Acid Deposition Program/National Trends Network; http://nadp.sws.uiuc.edu/isopleths/maps2003/phfield.gif, accessed on August 23, 2005.

East and the Midwest portions of the United States. Current U.S. surface mining regulations mandate that disturbed lands be restored, but much remains to be done to address past and ongoing drainage-related problems.

Acid Precipitation

Eastern and midwestern coals contain significant quantities of sulfur, so burning them releases large quantities of sulfur dioxide (SO_2), the major component of acid precipitation, to the atmosphere. Most of the high-sulfur coal consumed in the United States during this century has been used to make steel and to generate electricity in the East and Midwest. From there, atmospheric pollutants responsible for acid precipitation are transported northward and eastward by prevailing winds and storms. These trends are reflected in the geographic distribution of rainfall pH (see Figure 8.4). Emissions from coal-fired electric generating plants presently constitute the largest source of atmospheric SO_2. Other constituents of acid precipitation, including those from automotive exhausts, are distributed similarly.

Areas with underlying crystalline rock, shale, and sandstone are more prone to acidification than those underlain by rock systems that buffer the acids, such as limestone and other carbonate-rich rock. Potentially sensitive areas are widely distributed in North America and include much of the Appalachian Mountains, where rainfall is most acidic (see Figure 8.4); the Canadian Shield region of the upper Midwest (that is, the northern parts of Michigan and Wisconsin, as well as eastern Minnesota and parts of eastern and central Canada); the higher elevations of the Sierra Nevada, Rocky Mountains, and Cascade Range; and parts of the Ozark and Ouachita uplands, mid-Atlantic Coastal Plain, and Florida. Buffering by ions in ground water and constituents leached from watersheds makes large lakes and rivers at lower elevations less susceptible to acidification than smaller, higher-elevation lakes and streams.

The interactions of ions in precipitation (i.e., H^+, SO_4^{+2}, NO_3^{-2}) with organic and inorganic constituents of soil and water affect toxicity. Particularly important is the leaching of potentially toxic elements, especially aluminum, from rocks and soils by acidic precipitation. Toxicity attributable to pH and aluminum is often episodic, occurring during high surface-water discharge in the spring months. Spring is also the time when spawning and larvae releases occur for many aquatic organisms, making them vulnerable to reduced pH conditions.

By definition, acid rain is rainfall with a pH lower than about 5.0; the pH of distilled water in equilibrium with atmospheric CO_2 is 5.6, but other atmospheric constituents tend to make rainfall more acidic even in areas unaffected by air pollution. In addition to sulfur, the combustion of coal emits other potentially toxic elements, including arsenic, cadmium, lead, mercury, and selenium. Cadmium and selenium are concentrated in coal ash, from which they may be leached into surface waters and accumulated to toxic concentrations by aquatic organisms. Mercury, along with selenium and other elements in coal are released into the atmosphere in stack emissions and can move long distances. Mercury and selenium readily bioaccumulate in birds, mammals, and predatory fishes. Mercury is generally released from point sources (e.g., caustic soda, that is, sodium hydroxide (NaOH) plants and paper mills). Bioaccumulation of mercury in remote lakes in the Northeast seems to indicate that atmospheric transport and natural chemical processes tend to keep mercury available for accumulation by organisms. According to the U.S. EPA, coal-fired electric generating plants are the greatest sources of atmospheric mercury; other important sources include municipal and hospital waste incinerators.

Metals are the elements listed on the left side of the periodic table of elements (see Figure 8.5). They form positive ions (*cations*), are reducing agents, have low electron affinities, and have positive valences (oxidation numbers). Nonmetals, listed on the right side of the periodic table, form negative ions (*anions*), are oxidizing agents, have high electron affinities,

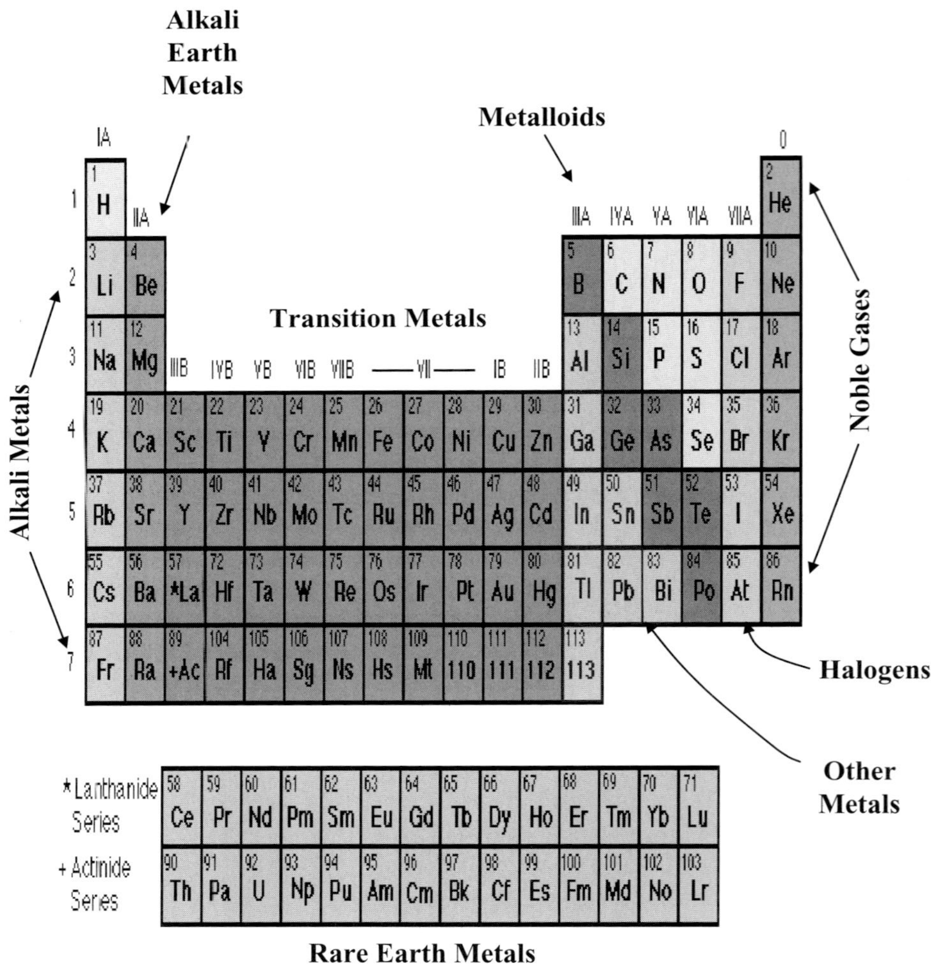

FIGURE 8.5. Periodic table of elements.

and have negative valences. Metalloids have properties of both metals and nonmetals, but two environmentally important metalloids, arsenic (As) and antimony (Sb), behave much like metals in terms of their toxicity and mobility, so they are often grouped with the heavy metals.

For most metals, the chemical form determines just how toxic it is. The form also determines how readily the metal moves in the environment and how rapidly it is taken up and stored by organisms. The chemical form is determined by the oxidation state or *valence* of the metal.

At some concentration, every element except those generated artificially by fission in nuclear reactors are found in nature, especially in soils.

Thus, it would be absurd to address metal contamination problems by trying to "eliminate" them. This is a common misconception, especially with regard to heavy metal and metalloid contamination. For example, mercury (Hg) and lead (Pb) are known to be important contaminants that cause neurotoxic and other human health effects and environmental pollution. The global mass balance of these metals, however, does not change; only the location and form (i.e., chemical species) can be changed. Therefore, protecting health and ecological resources is a matter of reducing and eliminating exposures and changing the form of the compounds of these elements to render them less mobile and less toxic. The first place to start such a strategy is to consider the oxidation states, or valence, of elements (see Chapter 2).

Let us consider two metals, one metalloid, and a mineral fiber known to cause environmental problems.

Lead: The Ubiquitous Element

Since the late 1970s, the U.S. government has taken specific actions to reduce lead exposures in the national population, largely by mandating that manufacturers eliminate lead from their products. This has occurred through a number of actions:

- In 1976, a total of 186.47 million kg of lead was used in gasoline in the United States. By 1983, this amount had dropped to 51.59 million kg; and by 1990, lead used in gasoline had been reduced to 0.47 million kg.
- The amount of lead used in soldered cans decreased markedly throughout the 1980s. In 1980, 47% of food and soft drink cans were lead soldered. By 1985, this figure had dropped to 14%; by 1990, only 0.85% of food and soft drink cans were lead soldered. As of November 1991, lead-soldered food or soft drink cans were no longer manufactured in the United States.
- In 1978, the amount of lead in lead-based paint was limited to less than 0.06% by weight.

These measures have been effective in reducing overall exposures to lead hazards. Still, lead-based paint remains a problem, predominantly in older, deteriorating housing stock. Eliminating the hazards of lead-based paint will require more than just removing lead from manufactured products; instead, it must be addressed as a holistic environmental justice concern, not simply a housing, health, or environmental issue. A new strategy that considers the economic and racial parameters of lead exposures and how to address them is needed to reduce lead hazards for all populations.

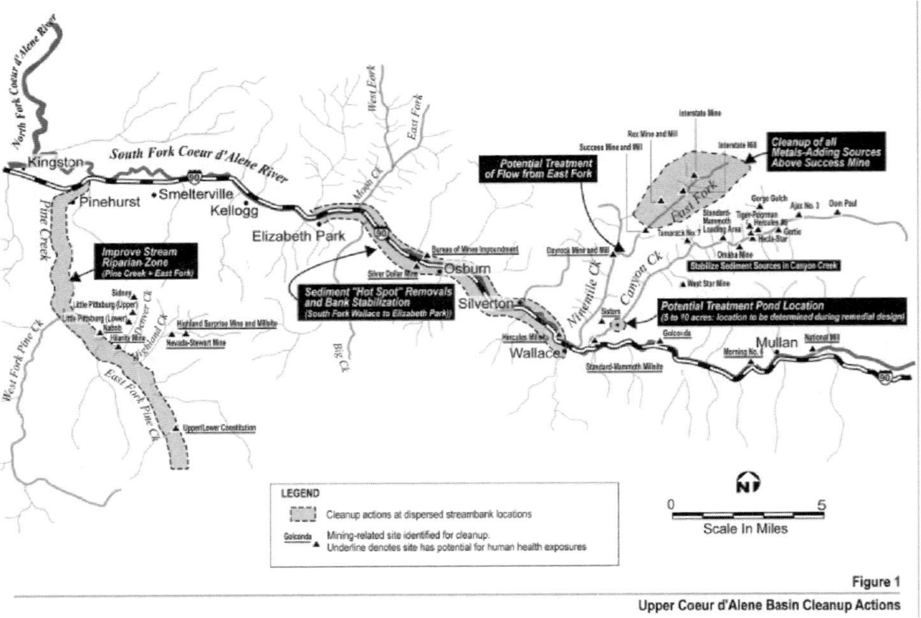

FIGURE 8.6. Site map of Bunker Hill, Idaho.
Source: U.S. Environmental Protection Agency, Region 10 Superfund: Bunker Hill/Coeur d'Alene Basin: http://yosemite.epa.gov/R10/CLEANUP.NSF/fb6a4e3291f5d28388256d140051048b/a2887c971c1dd0f588256cce00070aac!OpenDocument; accessed April 21, 2005.

Coeur d'Alene Valley and the Bunker Hill Lead Smelter[8]

Commercial mining for lead, zinc, silver, and other metals began in the Bunker Hill area in Idaho in 1883, and in the next few years mineral processing and smelting began (in the early 1900s) and continued until 1981. Throughout most of the twentieth century, this area, known as the Silver Valley, became a prominent center for mining and processing metals. Large quantities of tailings were left behind and often disposed in surface waters. A plank and pile dam was built in 1910 along the South Fork of the Coeur d'Alene River at the Pinehurst Narrows to retain the tailings (see the map in Figure 8.6). The dam deposited the tailings throughout the floodplain of the South Fork in an area referred to as Smelterville Flats. The dam failed in 1933, further dispersing tailings downstream. Another repository for tailing, known as the Central Impoundment Area (CIA), was constructed in 1928. This tailing's impoundment was expanded on numerous occasions when tailing quantities dictated. It eventually took up about 80 hectares of surface area.

Surface water, ground water, soil, and sediment contamination occurred throughout the valley as a result of the mining, milling, and smelting processes. Vegetation was either removed for logging or died from acid precipitaton that resulted from the smelter's large emissions of SO_2. The biggest problem, however, was that blood lead levels in children in the valley were very high; far exceeding health standards set by the federal government.

In 1983, the federal government listed the site on its National Priorities List, the listing of the worst Superfund hazardous waste sites. This was followed shortly by notices to the potentially responsible parties (PRPs) that the site needed to be remediated. The PRP investigation and cleanup took about 10 years. Cleanup plans included a Remedial Investigation and Feasibility Study, initial cleanup of the smelter complex, terracing of the denuded hillsides, and some revegetation work.

The U.S. EPA issued a Record of Decision (ROD) in 1992 detailing the required remedy for the nonpopulated part of the site (about $55\,km^2$). Two of the PRPs filed for bankruptcy in 1992 and 1994, so that the U.S. EPA and the State of Idaho had to assume direct responsibility for the cleanup. The remaining PRPs signed consent decrees with the EPA and did commit to a share of the remediation. The remediation steps are summarized in Table 8.2. The Bunker Hill smelter site and the surrounding area are still the subject of big environmental debates, including the lack of consensus on the target cleanup levels needed, and the debate on how to make the polluters pay, instead of passing the costs along to the taxpayers. In a way, this case is a hybrid, since both the PRPs and the governmental agencies shared responsibilities, including costs.

Mercury: Lessons from Minamata

Minamata, a small factory town on Japan's Shiranui Sea, seemed destined for industry. The name itself means "nitrogen," emblematic of the town's production of commercial fertilizer by the Chisso Corporation for decades, beginning in 1907.[9] Beginning in 1932, the company produced pharmaceutical products, perfumes, plastics, and processed petrochemicals. Chisso became highly profitable, notably because it became the only Japanese source of a high-demand primary chemical, D.O.P. (diotyl phthalate), a plasticizing agent. These processes needed the reactive organic compound, acetaldehyde, which is produced using mercury. The residents of Minamata paid a huge price for this industrial heritage. Records indicate that from 1932 to 1968, the company released approximately 27 tons of mercury com-

TABLE 8.2
Summary of remedial actions implented by the U.S. Environmental Protection Agency and the State of Idaho at Bunker Hill Superfund site.

Remedial Action	*General Description of Remedial Action*
Hillsides	Reduce erosion, increase infiltration, and minimize direct contact by contouring, terracing, and revegetating hillside areas that are essentially denuded. Provide surface armor or soil cover on mine waste rock dumps and remove solid waste landfills to on-site consolidation areas.
Gulches (Grouse, Government, Magnet, and Deadwood)	Reduce erosion, minimize direct contact, and minimize migration of contaminants to surface and groundwater by constructing erosion control structures and sediment basins, removing contaminated soils above cleanup levels, relocating the A-1 Gypsum Pond from Magnet Gulch to the CIA, reconstructing Government and Magnet Creeks, and installing surface barriers consistent with future land use.
Smelterville Flats (north and south of Interstate 90)	Minimize direct contact, surface water erosion, and migration of contaminants to surface and groundwater by conducting extensive tailings removals throughout the floodplain, depositing removed tailings on the CIA, reconstructing portions of the SFCDR, providing soil barriers and revegetation as necessary. Construct storm drain/swale conveyance system for surface water generated south of I-90 highway.

Source: U.S. Environmental Protection Agency.

pounds into the adjacent Minamata Bay. This directly affected the dietary intake of the toxic mercury of fisherman, farmers, and their families in Kumamoto, a small village about 900 km from Tokyo. The consumed fish contained extremely elevated concentrations of a number of mercury compounds, including the highly toxic methylated forms (i.e., monomethyl mercury and dimethyl mercury), leading to classic symptoms of methyl mercury poisoning. In fact, the symptoms were so pronounced the syndrome of these effects came to be known as the Minamata Disease.

In the middle of the 1950s, residents began to report what they called the "strange disease," including the classic form of mercury toxicity, which are disorders of the central and peripheral nervous systems (PNS and CNS, respectively). Diagnoses included numbness in lips and limbs, slurred speech, and constricted vision. A number of people engaged in uncontrollable shouting. Pets and domestic animals also demonstrated mercury toxicity, including "cat suicides"[10] and birds dying in flight. These events met with panic by the townspeople.

The physician Hajime Hosokawa from the Chisso Corporation Hospital, reported in 1956 that, "an unclarified disease of the central nervous system has broken out." Hosokawa correctly associated the fish dietary exposure to the health effects. Soon after this initial public health declaration, government investigators linked the dietary exposures to the bay water. Chisso denied the linkages and continued the chemical production, but within two years, they moved their chemical releases upstream from Minamata Bay to the Minamata River, with the intent of reducing the public outcry. The mercury pollution became more widespread. For example, towns along the Minamata River were also contaminated. Hachimon residents also showed symptoms of the "strange disease" within a few months. This led to a partial ban by the Kumamoto Prefecture government, who responded by allowing fisherman to catch, but not to sell, fish from Minamata Bay. The ban did not reduce the local people's primary exposure, since they depended on the bay's fish for sustenance. The ban did acquit the government from further liability, however.

Some three years after the initial public health declaration, in 1959, Kumamoto University researchers determined that the organic forms of mercury were the cause of the Minamata Disease. A number of panels and committees, which included Chisso Corporation membership, studied the problem. They rejected the scientific findings and any direct linkages between the symptoms and the mercury-tainted water. After Dr. Hosokawa performed cat experiments that dramatically demonstrated the effects of mercury poisoning for Chisso managers, he was no longer allowed to conduct such research and his findings were concealed from the public.[11] Realizing the links were true, the Chisso Corporation began to settle with the victims. The desperate and relatively illiterate residents signed agreements with the company for payment, but these agreements also released the company from any responsibility. The agreement included the exclusion: ". . . if Chisso Corporation were later proven guilty, the company would not be liable for further compensation." However, Minamata also represents one of the first cases of environmental activism. Residents began protests in 1959, demanding monetary compensation. These protests led to threats and intimation by Chisso, however, so victims settled for fear of losing even the limited compensation.

Chisso installed a mercury removal device on the outfall, known as a *cyclator*, but they omitted a key production phase, so the removal was not effective. Finally, in 1968, the Chisso Corporation stopped releasing mercury compounds into the Minamata River and Bay. Ironically, the decision was not an environmental one, nor even an engineering solution. The decision was made because the old mercury production method had become antiquated. Subsequently, the courts found the Chisso Corporation repeatedly and persistently contaminated Minamata Bay from 1932 to 1968.

Victim compensation has been slow. About 4,000 people have either been officially recognized as having Minamata Disease or are in the queue

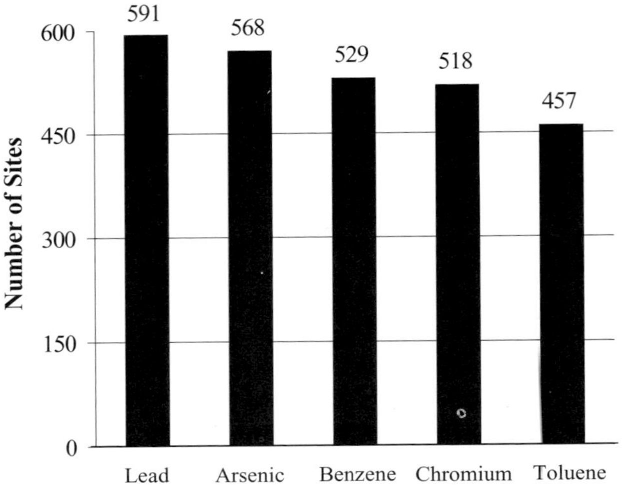

FIGURE 8.7. Five most commonly found contaminants at high-priority waste sites in the United States (National Priority Listing sites).
Source: U.S. Environmental Protection Agency, 2002. Proven Alternatives for Aboveground Treatment of Arsenic in Groundwater, Engineering Forum Issue Paper, EPA-542-S-02-002 (revised), www.epa.gov/tio/tsp.

for verification from the board of physicians in Kumamoto Prefecture. Fish consumption from the bay has never stopped, but mercury levels appear to have dropped, since cases of severe poisoning are no longer reported.

Arsenic Tragedy in Bangledesh

Arsenic is actually a metalloid; it is a lot like a metal but it does have some nonmetallic qualities. It shares, for example, some properties with phosphorus and nitrogen (all Group V-A elements on the periodic table). For general environmental purposes, however, it is usually lumped in with the heavy metals. The principal reasons for this are that it is generally removed from water and soil with technologies that work for metals (such as precipitation/coprecipitation techniques), its toxicity and bioaccumulation behavior is similar to that of metals, and it is often found in nature and in contaminated sites along with metals. In fact, it is the second most commonly found contaminant in hazardous waste sites in the United States (see Figure 8.7).

Arsenic has been used in industrial products and processes, including wood preservatives, paints, dyes, metals, pharmaceuticals, pesticides, herbicides, soaps, and semiconductors, but since it is also a rather commonly occurring element, it is found in natural backgrounds in rocks, soils, and sediment. The range of potential sources makes dealing with arsenic complicated. For example, some water supplies happen to be in areas where arsenic and metals are found in relatively high concentrations because of leaching from surrounding rocks and soils.

This is a real problem. Try to empathize with the municipal engineer attempting to adhere to federal and state drinking water standards (known as maximum contaminant levels, or MCLs), who must also rely on wells that receive water from arsenic-laden rock formations. It is not difficult to remove large concentrations of chemicals from water, but it becomes increasingly difficult and expensive as the required concentrations decrease. For example, it is a general rule that if it costs $1 per gallon to remove 90% of a contaminant, it will require another $1 to remove 99% of it, and another dollar to remove 99.9% of the contaminant. Thus the cost of removal as the required concentration approaches zero is exponential.

For metals and arsenic, the available technologies are more limited than for organic contaminants. For example, many organic contaminants (especially those that are not chlorinated) can be thermally treated, where they are broken down into harmless elemental constituents such as carbon, hydrogen, and oxygen. Since arsenic is an element, this is not possible. All we can design for is moving arsenic from one place to another where people are less likely to be exposed to it. Like heavy metals, arsenic's mobility and toxicity are determined by its oxidation state, or valence. As^{3+}, for example, is up to ten times more water soluble and is more toxic to humans than when it is reduced to As^{5+}. Arsenic in some valence states is much less likely to move in the environment or to cause health problems than in others. However, once a person is exposed to the arsenic, metabolic processes can change these less toxic forms back to highly toxic forms.[12]

Exposure to any form of arsenic is bad. Engineers need to know the forms (valence states) of the arsenic to optimize treatment and removal, but health scientists are often concerned about total arsenic exposures.

The physical and chemical properties of arsenic are complex, but protecting people from the exposures to arsenic is even more complicated. All three branches of the federal government have become involved. Congress has passed numerous laws addressing arsenic exposure, such as the Safe Drinking Water Act, which requires that the executive branch (in this case, the EPA) establish a standard (MCL) for contaminants in drinking water. The actual concentration allowed is based on scientific evidence, professional judgment, and ample margin of safety (commensurate with uncertainties, *and there are always uncertainties!*). The courts become involved when there is disagreement on whether the law is being upheld and whether the standards are sufficient. For local water supplies (e.g., towns), this can translate into hundreds or even thousands of plaintiffs (i.e., people living in the town that is being sued).

Even though everyone agrees that arsenic is toxic, they cannot agree on where to draw the line on allowable exposures. Recently, the MCL was lowered from $50 \mu g\ L^{-1}$ to $10 \mu g\ L^{-1}$. This meant that water supplies just meeting the old standard would have to remove five times more arsenic.

The town engineer may know that the present equipment at the plant would have to be replaced or upgraded, but the way such information is shared can affect what people perceive. For example, the town engineer may quote Robert Goyer, Chair of the National Research Council, Subcommittee to Update the 1999 Arsenic in Drinking Water Report, from his 2001 testimony before Congress:

> . . . . chronic exposure to arsenic is associated with an increased incidence of bladder and lung cancer at arsenic concentrations below the current MCL. This conclusion was strengthened by new epidemiological studies.[13]

However, after delving a bit further, the town engineer may have found that the National Research Council also said in 1999 that:

> No human studies of sufficient statistical power or scope have examined whether consumption of arsenic in drinking water at the current maximum contaminant level . . . results in an increased incidence of cancer or noncancer effects.[14]

Had the science changed that much in the two years between the 1999 report and Goyer's testimony? Had new studies or better interpretations of those studies led to the change? Or is it simply a matter on whose perspective carries the day? The National Research Council is a highly respected science organization. The committee members are at the top of their fields, but they come from different organizations and often differ on how data and information should be interpreted. Although their sources are the same epidemiological studies and models, it is not uncommon for subcommittee members to log minority opinions, based upon differences in professional judgment.

What complicates controversies such as the acceptable level of arsenic in water is that groups with strong and divergent ideologies, such as the Sierra Club *versus* the Heritage Foundation, will buttress their positions based on political differences. Pity the engineer who has to tell the town council at a public meeting that they will have to spend money for improved arsenic removal. The engineer will inevitably be asked to justify the request. Although the correct answer is that the MCL is set by the EPA and is now mandated, politics will influence the perception of the decision makers. Although engineers are prone to emphasize science and professional ethics, they need to listen for the third factor, politics, as well. And the town engineer must listen both to the nonscientific and to the scientific types.

For countries with sufficient financial and technical means and infrastructures, the arsenic debate represents a trade-off of values. It gets into some very complicated and controversial issues, such as the costs of preventing one cancer. Some have argued that if you include all costs of clean-

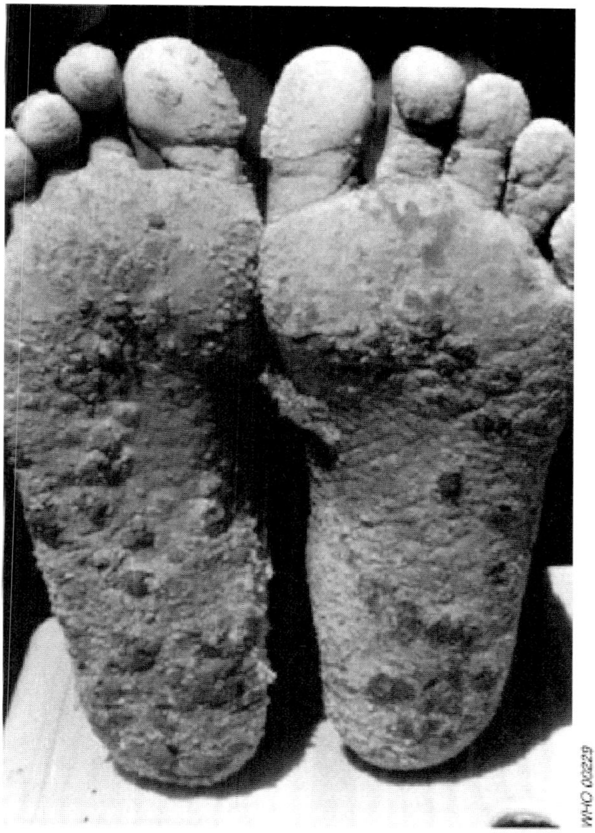

FIGURE 8.8. Skin lesions resulting from arsenic exposure in Bangladesh.
Source: A.H. Smith, E.O. Lingas, and M. Rahman, 2000. "Contamination of drinking-water by arsenic in Bangladesh," *Bulletin of the World Health Organization*, 2000, 78 (9), 1093–1103. *Photo credit*: World Health Organization.

ing up hazardous waste sites, certain sites would amount to billions of dollars to prevent a single cancer. Obviously, that is worth it if the cancer is your own or that of someone you care about, but what if it is some anonymous, statistical person? Is there a threshold when something is just too costly? If so, are we not defining that point as the "value of one human life?" This is an important matter for those writing health and environmental regulations.

In Bangladesh in the 1990s, elevated levels of arsenic in drinking water had become epidemic. As many as 77 million of the 125 million Bangladeshi people are being exposed to elevated concentrations of arsenic in their drinking water, already resulting in about 100,000 related, debilitating skin lesions (see Figure 8.8), with chronic diseases expected to increase with time.[15] Sad to say, an engineering solution to another problem has played a major role in exacerbating the arsenic problem. Surface water sources, especially standing ponds, in Bangladesh historically have contained significant microbial pathogens causing acute gastrointestinal disease in infants and children. To address this problem, the United Nations Children's Fund (UNICEF) in the 1970s began working with Bangladesh's Department of Public Health Engineering to fabricate and install tube-wells in an attempt to give an alternative and safer source of water, groundwater. Tube wells are mechanisms that consist of a series of 5-cm diameter tubes inserted into the ground at depths of usually less than 200 m. Metal hand pumps at the top of each tube are used to extract water.

The engineering solution appeared to be a straightforward application of the physical sciences. In fact, when the tube wells were first installed, the water was not tested for arsenic. This was in spite of the fact that local people had originally protested the use of ground water in some locations as "the devil's water." Was it possible that the indigenous folklore was rooted in information about possible contamination that would have been valuable for the foreign engineers to know? Is it also possible that the educational, cultural, and technical differences contributed to poor listening by the engineers? Either way, the engineers unwittingly worsened the problem by exposing vulnerable people to toxic levels of arsenic.

Asbestos in Australia[16]

Australia's National Occupational Health and Safety Commission is responsible for developing regulations to protect workers from asbestos exposures. In so doing, the commission must consider scientific and economic information. However, the assumptions that are used can greatly affect the expected costs. In other words, the goal is to reduce the number of future deaths and diseases, such as mesothelioma and asbestosis, to a virulent form of asbestos, chrysotile fibers. This exposure occurs when products containing chrysotile are imported, manufactured, and processed. Regulators must choose from several alternatives based on safety, health, and cost/effectiveness. The cost differences can be dramatic. In this instance, the Australian commission chose from three alternatives (see Tables 8.3 and 8.4):

1. Maintaining the status quo (base case)
2. Legislative prohibition/ban
3. Reduction in the national exposure standard

The commission recommended the second option, the legislative ban, because of the lack of sufficient information on the safety of alternative materials, the cost of compliance compared to net benefits and, if and when chrysotile products are prohibited, the expected exemptions needed when

TABLE 8.3
Comparisons of quantifiable cost impacts of proposed phase-out of chrisotile in Australian products, based on the national exposure standard of 1.0 fiber per mL of air, the maximum number of exposed employees, the lower figure used for the value of human life, and a 5% annual cost for mesothelioma.

Item	*Phase-Out Option Assumptions*	*Present Value $ (Over 40 yrs @ 8%)*
Savings in Death and Illness:		
Exposure Standard	1.0 fiber per mL	
Number of Persons Exposed	22,300	
Value of Human Life	$1.5 million	
Cost of Lung Cancer + Mesothelioma	$667,000*1.05	$24,187,596
Savings in Business Compliance Costs:		
Savings in OHS Controls	Waste disposal and medical exams only	$29,511,511
	Present Value Benefits:	$53,699,107
Increase in Costs to Business:		
Increased Cost of Substitutes to Small Business	20% brakes, 17% gaskets	($6,014,403)
Capital and Recurrent Costs to Large Business	$8.3 million Yr 1 $1,098,900 p.a.	($20,789,143)
	Present Value Costs:	($26,803,546)
	Net Result:	$26,895,561

Source: Commonwealth of Australia, National Occupational Health and Safety Commission, 2004. Regulatory Impact Statement of the Proposed Phase-Out of Chrysotile Asbestos, Canberra.

suitable substitute materials are not available or in areas of competing values and national interests, such as defense. This may have been a wise choice. I often say that it seems when everyone is dissatisfied with the environmental decision, it could well be the right one.

Among other matters, the scenario analysis indicates that the net overall benefit of Option 2 diminishes as the phase-out period extends. It would appear that were the phase-out period adopted to approach 10 years, the costs to business would outweigh the offsetting benefits to business and workers.

As complicated as this likely appears to most engineers and other technical professionals, think of how it appears to the general public. They do not have to be paranoid to fear that we might be "pulling one over on them!"

Like many environmental policy decisions, the asbestos regulations must balance credible science with policy and risk management. Politics and science can often be strangers. Science is a quest for truth. Even if politics

TABLE 8.4
Summary of scenarios demonstrating sensitivity of the costs and benefits analysis to key factors, namely: (1) phase-out period; (2) number of workers exposed; (3) savings in compliance costs; and (4) cost convergence of asbestos substitutes. Among other matters, the scenario analysis indicates that the net overall benefit of Option 2 diminishes as the phase-out period extends. It would appear that were the phase-out period adopted to approach ten years, the costs to business would outweigh the offsetting benefits to business and workers.

Key Assumptions	*Net Present Value Over 40 Years at 8% Discount Rate**			*Discussion*
	Scenario 1	*Scenario 2*	*Scenario 3*	
Timeframe for Phase-Out	$26,895,561 3 Years	$17,486,930 5 Years	–$2,327,666 10 Years	Highly sensitive to changes in phase-out period. The shorter the period the higher the NPV. Longer phase-outs continue costs associated with illness and other business costs, which lower the overall NPV (see Note 2).
	Scenario 4	*Scenario 5*		
Workers Exposed	$26,895,561 22,300 workers	$13,880,157 10,300 workers		Highly sensitive to numbers of workers exposed. Halving the estimated number of workers exposed still results in a positive NPV.
	Scenario 6	*Scenario 7*		
Compliance Cost Savings	$26,895,561 Selected cost savings	–$2,615,951 No cost savings		If there are no savings in costs of complying, then the NPV becomes slightly negative.
Substitutes and Cost Convergence	$26,895,561 Costs converge	–$46,507,961 Costs do not converge		Highly sensitive to cost convergence for substitutes. Highly negative NPV if no convergence over next 40 years.

TABLE 8.4
Continued

* Health outcomes have been quantified and expressed as savings in the potential costs of death and illness over 40 years at a discount rate of 8%.
Source: Commonwealth of Australia, National Occupational Health and Safety Commission, 2004. Regulatory Impact Statement of the Proposed Phase-Out of Chrysotile Asbestos, Canberra.
Notes:
1. These findings show the proposal to be highly sensitive to changes in underlying assumptions but it has not been possible to fully quantify the current cost to the community of illnesses such as asbestosis and other malignancies arising from chrysotile exposure. Hence, the Net Present Value is not a complete quantification of all quantitative impacts and should be used as a guide to decision making only.
2. The NPV reflects net benefits derived from the following annual (unless otherwise stated) cash flows:
- Benefits each year from savings in health and illness ($2.194m), from reduction in business costs of waste disposal ($1.179m) and from reduction in business costs of medical exams ($0.64m).
- Benefits every three years from reduction in business costs of medical exams ($2.56m).
- Less costs imposed on small business each year during the phase-out period only ($6.66m).
- Less costs imposed on large business each year ($1.098m) and in Year 1 only from investment in new production equipment ($8.3m).

were not fraught with "spin" and "word-smithing," it would be different from science. For example, the peer review process in science is not democratic or populist. Majority does not rule when it comes to what is acceptable. One finding out of thousands must be heeded. A single study can change a paradigm. A vexing, yet refreshing, characteristic of engineers is their idealism. Engineers are forward thinking and have been selected, usually self-selected, as "can-do" types. They see a problem or an opportunity and think about ways to do it. These attributes can become vexing, however, for the engineer who steps into the policy area and engages in the uncharted waters of politics. Engineers even seem naïve about truth being anything other than what the laws of nature demand. Newton's laws don't lie!

I am reminded of an engineer who was interviewed on September 12, 2001, about the collapse of the World Trade Center towers. After the engineer had shared some of the technical information and gave an excellent explanation of the factors that led to the towers' collapse, the reporter commented that the team of terrorists that planned the attack must have included an engineer. The engineer was visibly shaken by the assertion and made the comment that he hoped that was not the case because "engineering is a helping profession."[17] Yes, indeed it is, but it is folly to assume that the talents and expertise with which we have been blessed will not on rare occasions, be used for malevolence. So, paradigms are shifting not only for policy makers, but for practicing engineers and also for other environmental professionals. Misdeeds are truly evil, but not preparing ourselves for their possibility would be a big mistake, leading to tragic mishaps.

The cases in this chapter cover the gamut, from misdeeds to mistakes to mishaps. The good news is that the professionals, like the physician in the Minamata case, played key roles in identifying and beginning to address the problems. This may be the foremost lesson we can draw from them.

Notes and Commentary

1. The principal source for this discussion is U.S. Geological Survey, 1998. *Status and Trends of the Nation's Biological Resources*, Washington, D.C. Updated online at http://biology.usgs.gov/s+t/SNT/.
2. U.S. Environmental Protection Agency, 1985. Report to Congress. Wastes from the extraction and beneficiation of metallic ores, phosphate rock, asbestos, overburden from uranium mining, and oil shale. Office of Solid Waste, Washington, D.C. EPA/530-SW-85-033.
3. The principal source of this article is D.K. Nordstrom, C.N. Alpers, C.J. Ptacek, and D.W. Blowes, 2000. "Negative pH and Extremely Acidic Mine Waters from Iron Mountain, California," *Environmental Science and Technology*, 34 (2), 254–258.
4. Nordstrom et al.
5. The source for the Iron Mountain discussion is U.S. Environmental Protection Agency, 2004. Site Information: Iron Mountain Mine, EPA ID# CAD980498612. http://yosemite.epa.gov/r9/sfund/overview.nsf/0/7a8166ef298804808825660b007ee658?OpenDocument#descr; updated August 23, 2004.
6. U.S. Geological Survey, 2005. Iron Mountain Superfund Site—EPA Technical Support: http://ca.water.usgs.gov/projects00/ca527.html; accessed April 11, 2005.
7. K. Custer, 2002, Current Status of Hardrock Abandons Mine Land Program Mineral Policy Center, Washington D.C.
8. The principal source for this section is U.S. Environmental Protection Agency, 2000, First 5-Year Review of the Non-Populated Area Operable Unit Bunker Hill Mining and Metallurgical Complex, Shoshone County, Idaho.
9. A principal source for the Minamata case is the Trade & Environment Database, developed by James R. Lee, American University, The School of International Service: http://www.american.edu/TED/; accessed April 19, 2005.
10. Local residents near Minamata reported bizarre feline behavior, such as falling into the sea and drowning. Since cats consume fish, it is possible that they were highly exposed to mercury compounds, which led to such psychotic behaviors. In fact, the cats were omens of what was to happen in the human population later in the 1950s.
11. This is an all too common professional ethics problem—lack of full disclosure. It is often, in retrospect, a very costly decision to withhold information about a product, even if the consequences of releasing the information would adversely affect the "bottom line." Ultimately, as has been seen in numerous ethical case studies, the costs of not disclosing are severe, such as bankruptcy

and massive class action lawsuits, let alone the fact that a company's decision may have led to the death and disease of the very people they claim to be serving, their customers and workers!

12. We see similar metabolic behavior for metals like mercury. In fact, there is debate when a fish is found to have methylated forms of Hg. Was the fish exposed to another form and metabolically reduced the Hg, or was it exposed to methyl mercury, which it simply stored?
13. Testimony of R.A. Goyer before the U.S. House of Representatives, Science Committee, October 4, 2001.
14. National Research Council, 1999. *Arsenic in Drinking Water*, National Academies Press, Washington, D.C.
15. World Health Organization, 2000. Press Release WHO/55, "Researchers Warn of Impending Disaster from Mass Arsenic Poisoning."
16. I could have used any of a large number of cases from most countries, especially the highly developed nations, such as vermiculite mining and products from Libby, Montana, in the United States. In fact, there is a raging legal battle ongoing at this writing. The Australian case, however, allows for discussion of science, economics, and other facets of the problem.
17. The interview occurred on the Cable News Network (CNN), but the names of the engineer or the reporter are not known.

CHAPTER 9

Spaceship Earth

Now there is one outstandingly important fact regarding Spaceship Earth, and that is that no instruction book came with it.

R. Buckminster Fuller, *Operating Manual for Spaceship Earth, 1963*

We live on a planet that is miraculous in its balances. The atmosphere is a delicate blend of all the physical and chemical essentials that allow for life. The feedbacks between biological systems and abiotic components provide an operatic dance allowing for the survival of life forms within a tightly defined range of temperatures, water vapor, essential chemical combinations, and sunlight.

Buckminster Fuller's quote, years before the first Earth Day, is an early recognition that humans could indeed mess things up. Throughout most of our history, our influence was fairly localized. But, with the advent of increasingly large and widespread consequences from industrial, transportation, mining, and other anthropogenic activities, we have become able to change the composition of our atmosphere and its relationship to other geological and hydrological systems. A year after Fuller's quote, in 1964, the Canadian communication expert Marshall McLuhan (1911–1980) said, "There are no passengers on Spaceship Earth. We are all crew." Human beings may not be the first pilot of our "spaceship" but we are doing a lot of harm from the backseat!

The science of the planet at various scales is the province of atmospheric chemistry; that is, the study of the chemical constituents of the earth's atmosphere and the roles these substances play in influencing the atmosphere's temperature, radiation, and dynamics. When the composition is changed even slightly, the earth's fitness to support life is affected. For example the dynamics of climate, protection from harmful radiation, or the quality of air in the troposphere (where we live) is dependent upon a relatively small range of chemical constituents in the atmosphere. This balance is impacted by both natural and human-made (anthropogenic) emissions, trace gas distributions, and the chemical reactions of molecules and atoms in the atmosphere.[1] Of particular concern in recent decades is the emission and changes in the concentrations of chemical species naturally present in the atmosphere, such as the greenhouse gases carbon dioxide (CO_2), methane (CH_4), and nitrous oxide (N_2O), or the addition of anthropogenic chemicals such as chlorofluorocarbons (CFCs), which can affect climate by directly changing the absorption of radiation and hence temperatures and dynamics.

Chemical reactions in the atmosphere can alter the chemical balance in the atmosphere, resulting in potentially devastating global effects. The relationships of the intricately connected atmospheric reactions are highly complex. For example, temperature changes affect the rates of chemical reactions, radiation changes affect the rates of photochemical reactions, and dynamics changes affect the distributions of all chemical species. Thus a systematic approach must be taken to understand the complicated processes of feedback between the atmosphere and the biosphere, including relationships between emissions, transformations, and sequestrations of important chemical species, the transport mechanisms, including atmospheric circulation and the absorption of radiation. Unfortunately, there are large gaps in information and there is a great deal of disagreement about what the scientific data are telling us about the future of the atmosphere. And making scientific progress is not enough; characterizing and predicting global conditions must also account for social, geopolitical, economic, and other social scientific factors. Trying to predict the characteristics of the future atmosphere is difficult. Sometimes, the best we can do is predict the "direction of the arrow"; that is, will a certain constituent or factor improve or degrade?[2]

Changes in the Global Climate[3]

Climate is weather that is averaged over a long period of time:

> Climate in a narrow sense is usually defined as the "average weather," or more rigorously, as the statistical description in terms of the mean and variability of relevant quantities over a period of time ranging from months to thousands or millions of years. The classical period is 30 years, as defined by the World Meteorological Organization (WMO). These quantities are most often surface variables such as temperature, precipitation, and wind. Climate in a wider sense is the state, including a statistical description, of the climate system.[4]

There is little doubt that the earth's troposphere is undergoing change (see Figure 9.1). What is in doubt is to what extent the change is "normal" and the extent to which human activities are causing the change. For example, the global average surface temperature has increased over the twentieth century by about 0.6°C. Sometimes, we scientists fail to make a complete point. You may hear people asking the government "to do something about the greenhouse effect." But this can be likened to doing something about "gravity." Yes, the effects of gravity can be painful and fatal, as evidenced by plane crashes and guillotines, but repealing the law of gravity is not a reasonable approach to these problems (safer aircrafts and avoiding the wrath of French revolutionaries are preferable). By analogy, what is being asked for is not to change the laws of thermodynamics that explain the greenhouse effect, but to manage resources, including controlling emissions and maintaining and increasing plant life to sequester the greenhouse gases.[5]

The Intergovernmental Panel on Climate Change (IPCC), established by the World Meteorological Organization (WMO) and the United Nations Environment Programme (UNEP), has drawn some important links between emissions and other human activities and likely changes in the earth's climate. As evidence, in *Climate Change 2001: Working Group I: The Scientific Basis*, the IPCC reports[6] the following.

The mean surface temperature of the earth is increasing:

- The global average surface temperature (the average of near surface air temperature over land and sea surface temperature) has increased since 1861. Over the twentieth century, the increase has been 0.6 ± 0.2°C (see Figure 9.1a). This value is about 0.15°C larger than that previously estimated for the period up to 1994, owing to the relatively high temperatures of the additional years (1995 to 2000) and improved methods of processing the data. These values take into account various adjustments, including urban heat island effects. The record shows a great deal of variability; for example, most of the warming occurred during the twentieth century, during two periods, 1910 to 1945 and 1976 to 2000.
- Globally, it is very likely that the 1990s was the warmest decade and 1998 the warmest year in the instrumental record since 1861 (see Figure 9.1a).
- New analyses of proxy data for the Northern Hemisphere indicate that the increase in temperature in the twentieth century is likely to have been the largest of any century during the past 1,000 years. It is also likely that, in the Northern Hemisphere, the 1990s was the warmest decade and 1998 the warmest year (see Figure 9.1b). Because less data are available, less is known about annual averages prior to 1,000 years before the present and for conditions prevailing in most of the Southern Hemisphere prior to 1861.
- Between 1950 and 1993, mean nighttime daily minimum air temperatures over land increased by about 0.2°C per decade, doubling the rate of increase in daytime daily maximum air temperatures (0.1°C per decade). The result is a longer freeze-free season in many mid- and high-latitude regions, for example, temperate and arboreal

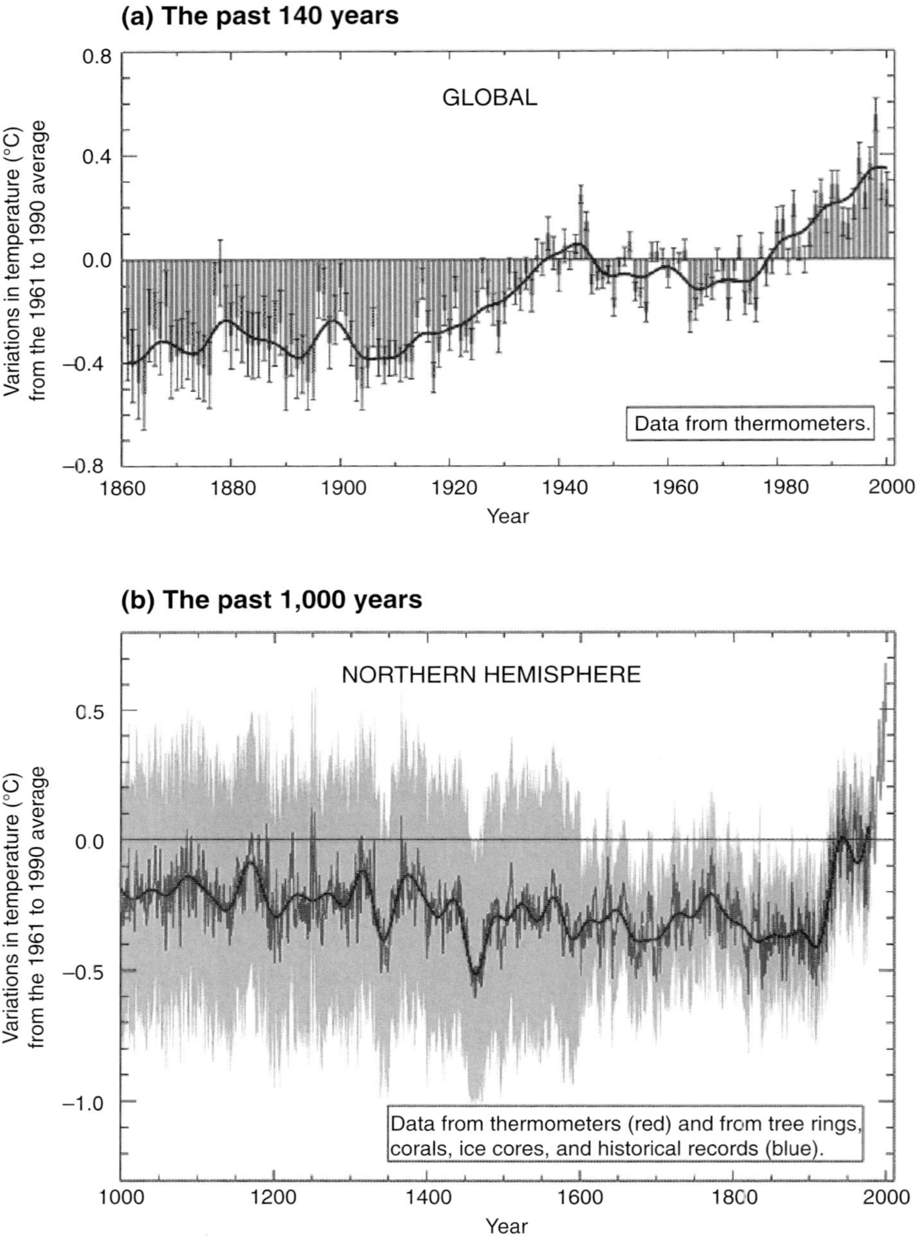
(a) The past 140 years
GLOBAL
Variations in temperature (°C) from the 1961 to 1990 average
0.8
0.4
0.0
−0.4
−0.8
1860
1880
1900
1920
1940
1960
1980
2000
Year
Data from thermometers.
(b) The past 1,000 years
NORTHERN HEMISPHERE
Variations in temperature (°C) from the 1961 to 1990 average
0.5
0.0
−0.5
−1.0
1000
1200
1400
1600
1800
2000
Year
Data from thermometers (red) and from tree rings, corals, ice cores, and historical records (blue).

forests, respectively. The increase in sea surface temperature over this period is about half that of the mean land surface air temperature.

Temperatures have risen during the past four decades in the lowest 8 km of the atmosphere:

- Since the late 1950s (the period of adequate observations from weather balloons), the overall global temperature increases in the lowest 8 km of the atmosphere and in surface temperature have been similar at 0.1°C per decade.
- Since the start of the satellite record in 1979, both satellite and weather balloon measurements show that the global average temperature of the lowest 8 km of the atmosphere has changed by +0.05 ± 0.10°C per decade, but the global average surface temperature has increased significantly by +0.15 ± 0.05°C per decade. The difference in the warming rates is statistically significant. This difference occurs primarily over the tropical and subtropical regions.
- The lowest 8 km of the atmosphere and the surface are influenced differently by factors such as stratospheric ozone depletion, atmospheric aerosols, and the El Niño phenomenon.[7] Hence, it is physically plausible to expect that over a short time period (e.g., 20

FIGURE 9.1. Variations of the earth's surface temperature over the last 140 years and the last millennium. (a) The earth's surface temperature is shown year by year (red bars) and approximately decade by decade (black line, a filtered annual curve suppressing fluctuations below near decadal time-scales). There are uncertainties in the annual data (thin black whisker bars represent the 95% confidence range) due to data gaps, random instrumental errors and uncertainties, and uncertainties in bias corrections in the ocean surface temperature data and also in adjustments for urbanization over the land. Over both the last 140 years and 100 years, the best estimate is that the global average surface temperature has increased by 0.6 ± 0.2°C. (b) Additionally, the year-by-year and 50-year average variations of the average surface temperature of the northern hemisphere for the past 1,000 years have been reconstructed from proxy data calibrated against thermometer data (see the list of the main proxy data in the diagram). The 95% confidence range in the annual data is represented by the grey region. These uncertainties increase in more distant times and are always much larger than in the instrumental record due to the use of relatively sparse proxy data. Nevertheless, the rate and duration of warming of the twentieth century has been much greater than in any of the previous nine centuries. Similarly, it is likely that the 1990s have been the warmest decade and 1998 the warmest year of the millennium.
Source: Intergovernmental Panel on Climate Change, 2001. *Climate Change 2001: Working Group I: The Scientific Basis*, Cambridge University Press, Cambridge, United Kingdom and New York, NY.

years) there may be differences in temperature trends. In addition, spatial sampling techniques can also explain some of the differences in trends, but these differences are not fully resolved.

Snow cover and ice extent have decreased:

- Satellite data show that there are very likely to have been decreases of about 10% in the extent of snow cover since the late 1960s, and ground-based observations show that there is very likely to have been a reduction of about two weeks in the annual duration of lake and river ice cover in the mid- and high latitudes of the Northern Hemisphere, over the twentieth century.
- There has been a widespread retreat of mountain glaciers in non-polar regions during the twentieth century.
- Northern Hemisphere spring and summer sea-ice extent has decreased by about 10–15% since the 1950s. It is likely that there has been about a 40% decline in Arctic sea-ice thickness during late summer to early autumn in recent decades and a considerably slower decline in winter sea-ice thickness.

Global average sea level has risen and ocean heat content has increased:

- Tide gauge data show that global average sea level rose between 0.1 and 0.2 meters during the twentieth century.
- Global ocean heat content has increased since the late 1950s, the period for which adequate observations of subsurface ocean temperatures have been available.

The total emissions of greenhouse gases in the United States in 1997 grew by 1.4% from the previous year. Overall, U.S. emissions are now about 10% higher than they were in 1990. The expansion in 1997 is a return to earlier trends after the unusual growth in 1996 emissions (up by a revised 2.8% from the 1995 level), which was caused primarily by severe weather in 1996 (see Table 9.1). Since 1990, U.S. emissions have increased at a compounded annual rate of about 1.3%, slightly faster than population growth (1.1%) but more slowly than increases in energy consumption (1.7%), electricity consumption (2.0%), or gross domestic product (2.3%). Carbon dioxide produced by burning fossil fuels accounts for the lion's share of greenhouse gas emissions in the United States (see Figure 9.2).

Table 9.1 shows trends in emissions of the principal greenhouse gases, measured in million metric tons of gas. Every gas has a unique impact on the greenhouse effect. For example, Table 9.2 shows the weighted value, the global warming potential (GWP), which is a measure of "radiative forcing" for some important gases. This concept, developed by IPCC, allows for com-

TABLE 9.1
Estimated U.S. emissions of greenhouse gases by gas, 1990–1997 (million metric tons of gas; see Appendix 10).

Gas	*1990*	*1991*	*1992*	*1993*	*1994*	*1995*	*1996*	*P1997*
Carbon Dioxide	4,971.7	4,916.3	4,988.8	5,109.8	5,183.9	5,236.4	5,422.3	5,503.0
Methane	30.2	30.4	30.4	29.7	29.9	30.0	29.1	29.1
Nitrous Oxide Halocarbons and Other Gases	1.0	1.0	1.0	1.0	1.1	1.0	1.0	1.0
CFC-11, CFC-12, CFC-113	0.2	0.2	0.1	0.1	0.1	0.1	0.1	*
HCFC-22	0.1	0.1	0.1	0.1	0.1	0.1	0.1	0.1
HFCs, PFCs, and SF_6	*	*	*	*	*	*	*	*
Methyl Chloroform	0.2	0.2	0.1	0.1	0.1	*	*	*
Carbon Monoxide	87.4	89.2	86.2	86.3	90.3	81.3	80.4	NA
Nitrogen Oxides	21.6	21.5	21.9	22.2	22.5	21.7	21.3	NA
Nonmethane VOCs	18.9	19.1	18.7	18.9	19.5	18.6	17.2	NA

*Less than 50,000 metric tons of gas.
P = preliminary data.
NA = not available.
Sources: U.S. Department of Energy, 1997. *Emissions of Greenhouse Gases in the United States 1996*, DOE/EIA-0573(96), Washington, D.C. U.S. Environmental Protection Agency, 1998. *Inventory of U.S. Greenhouse Gas Emissions and Sinks, 1990–1996*, Review Draft, Washington, D.C. U.S. Environmental Protection Agency, 1997. Office of Air Quality Planning and Standards, *National Air Pollutant Emission Trends, 1900–1996*, EPA-454-R-97-011, Research Triangle Park, NC.

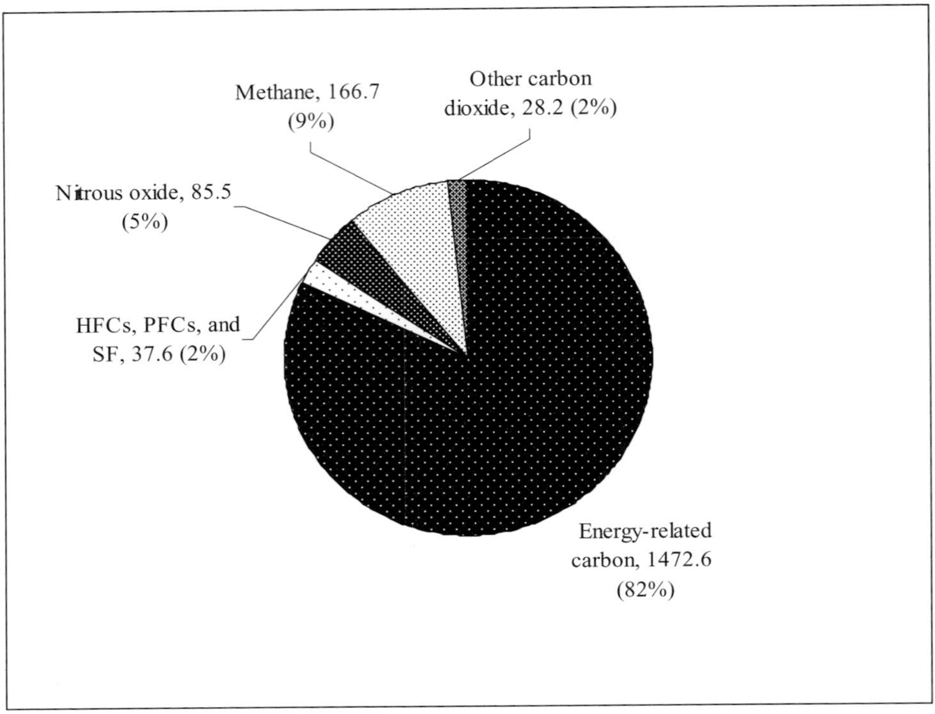

FIGURE 9.2. U.S. greenhouse gas emissions, 1997, in metric tons carbon equivalent. *Source*: U.S. Department of Energy, 1998. Energy Information Administration.

TABLE 9.2
U.S. emissions of greenhouse gases, based on global warming potential, 1990–1997 (million metric tons of carbon equivalent; see Appendix 10).

Gas	*1990*	*1991*	*1992*	*1993*	*1994*	*1995*	*1996*	*1997*^{P}
Carbon	1,356	1,341	1,361	1,394	1,414	1,428	1,479	1,501
Methane	173	174	174	170	171	172	167	167
Nitrous Oxide	82	83	85	86	91	88	86	85
HFCs PFCs and SF_6	22	22	23	23	26	31	35	38
Total	1,633	1,620	1,643	1,673	1,702	1,719	1,767	1,791

P = preliminary data.
Source: Revised from data from U.S. Department of Energy, 1997. *Emissions of Greenhouse Gases in the United States 1996*, DOE/EIA-0573(96), Washington, D.C.

parisons of the impacts of different greenhouse gases on global warming, with the effect of carbon dioxide being equal to 1.[8] The GWPs for other greenhouse gases are considerably higher.

Over 80% of U.S. greenhouse gas emissions are caused by the combustion of fossil fuels, especially coal, petroleum, and natural gas. Consequently, U.S. emissions trends are largely caused by trends in energy consumption. In recent years, national energy consumption, like emissions, has grown relatively slowly, with year-to-year fluctuations caused (in declining order of importance) by weather-related phenomena, business cycle fluctuations, and developments in domestic and international energy markets.

Other U.S. emissions include carbon dioxide from noncombustion sources (2% of total U.S. greenhouse gas emissions), methane (9%), nitrous oxide (5%), and other gases (2%). Methane and nitrous oxide emissions are caused by the biological decomposition of various waste streams, fugitive emissions from chemical processes, fossil fuel production and combustion, and many smaller sources. The other gases include hydrofluorocarbons (HFCs), used primarily as refrigerants; perfluorocarbons (PFCs), released as fugitive emissions from aluminum smelting and also used in semiconductor manufacture; and sulfur hexafluoride, used as an insulator in utility-scale electrical equipment.

The Kyoto Protocol, drafted in December 1997, raised the public profile of climate change issues in the United States in general, and of emissions estimates in particular. Emissions inventories are the yardstick by which the success or failure in complying with the Kyoto Protocol would be measured.

Carbon Dioxide

Carbon dioxide (CO_2) is the most prevalent greenhouse gas being emitted in the United States, accounting for 84% of its greenhouse gas emissions. Most CO_2 emissions originate from fossil fuel combustion and are influenced by the interaction of three factors:

- Consumption of energy-using services, such as transportation, heating and cooling, and industrial manufacturing
- Energy intensity of energy-using services; that is, the amount of energy used for each type of service
- Carbon intensity of the energy sources; that is, the amount of carbon released per unit of energy used to provide the services, usually in the form of electricity

Emissions per dollar of GDP and emissions per capita are crude measures of the carbon intensity of the use of energy services. United States emissions per capita, which declined in the early 1980s, have risen in the

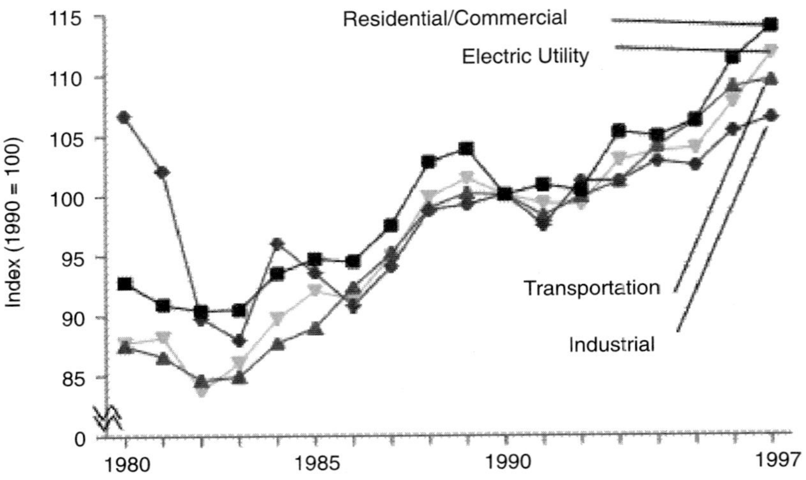

FIGURE 9.3. Carbon dioxide emissions in the United States by economic sector. *Source*: Department of Energy, 1998. Environmental Information Agency.

1990s, although at a relatively low rate. Emissions per dollar of GDP have declined almost every year. Conversely, some of the indicators of carbon intensity have begun to increase, especially emissions per kilowatt-hour of electric power generation. During the early 1990s, several unrelated factors combined to lower the carbon intensity of power generation, including the expansion of natural-gas-fired generation caused by relatively low natural gas prices and better nuclear power plant operating rates. Over the past two years, however, the trends for some of those factors have reversed. Several nuclear power plants have been shut down since 1995, and nuclear generation declined by about 7% between 1996 and 1997; natural gas prices have risen, with the result that utilities have turned increasingly to existing coal plants for power generation.

The trends in carbon dioxide emissions by energy consumption sector are shown in Figure 9.3. Emissions from the industrial sector dropped substantially in the early 1980s, corresponding to energy prices that induced industry to adopt energy-efficient technologies. Emissions from other sectors also dropped slightly in the early 1980s. In the late 1980s, however, emissions rose consistently as energy prices dropped dramatically and the economy grew. In 1990, somewhat higher energy prices induced an economic slowdown that was felt most strongly in 1991, with the result that emissions fell. Since 1991, emissions have grown consistently in all sectors, with the largest increases in the transportation and electric power sectors. Emissions in the industrial sector have grown relatively slowly, even during a vigorous economic expansion, due to energy efficiency improvements and low growth in energy-intensive industries.

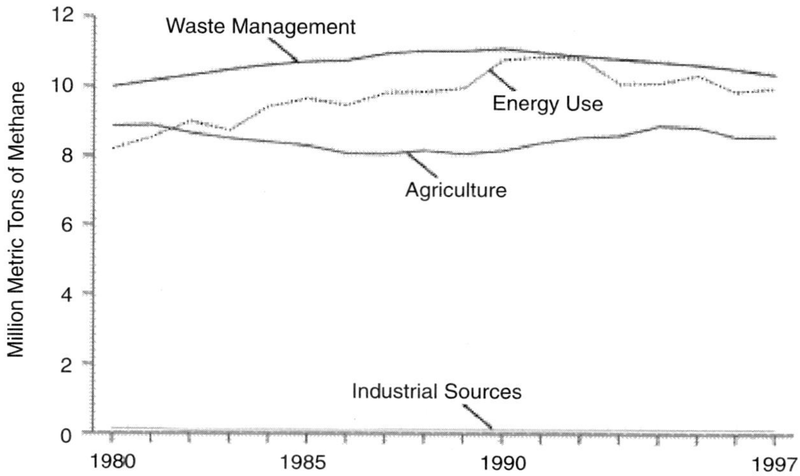

FIGURE 9.4. Methane emissions in the United States by type of source.
Source: Department of Energy, 1998. Environmental Information Agency.

Methane

Methane (CH_4) accounts for about 9% of U.S. GWP-weighted greenhouse gas emissions. Emissions of CH_4 appear to have remained roughly constant through the 1990s, or perhaps to have declined slightly. Methane emissions estimates are more uncertain than those for carbon dioxide, however, and correspondingly less confidence can be placed in the apparent trends. Methane emissions come from three categories of sources, each accounting for approximately one-third of U.S. methane emissions, or about 3% of the nation's total greenhouse gas emissions. The largest of the three sources is the anaerobic decomposition of municipal solid waste in landfills (see Figure 9.4). Emissions from this source are declining (although very slowly) as a consequence of a reduction in the volume of waste landfilled and a gradual increase in the volumes of landfill gas captured for energy or flared. Methane is also a byproduct of fossil energy production and transport when it leaks from natural gas production and distribution systems and when mine gas is released during coal production. Farm animal management also contributes, as a result of anaerobic decomposition of animal waste.

Nitrous Oxide

Nitrous oxide (N_2O) makes up approximately 5% of U.S. GWP-weighted greenhouse gas emissions. Emissions estimates for N_2O are more uncertain than those for either carbon dioxide or methane. Estimated nitrous oxide emissions have been roughly constant in the 1990s, without an obvious trend. The revised estimates of nitrous oxide emissions include one large

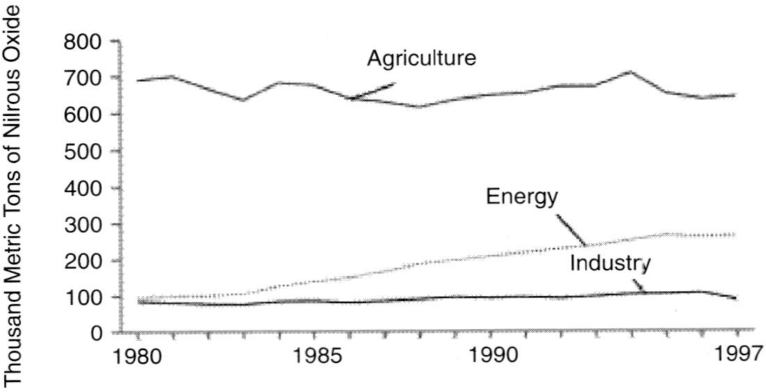

FIGURE 9.5. Nitrous oxide emissions in the United States by type of source. *Source*: Department of Energy, 1998. Environmental Information Agency.

class of sources and two small classes (see Figure 9.5). Agriculture is the principal source, dominated by emissions from nitrogen fertilization of agricultural soils. Secondary N_2O emissions from nitrogen in agricultural runoff into streams and rivers have been incorporated. Motor vehicles equipped with catalytic converters also emit significant amounts of N_2O.[9] Chemical processes, fuel combustion, and wastewater treatment plants are comparably small emitters of N_2O.

Halocarbons and Other Gases

The Kyoto Protocol specifies that emissions of several classes of engineered gases be limited: hydrofluorocarbons (HFCs), perfluorocarbons (PFCs), and sulfur hexafluoride (SF_6). Emissions of these three classes of gases account for about 2% of U.S. GWP-weighted emissions. There are several other categories of chemicals that also qualify as greenhouse gases but are excluded from the Framework Convention on Climate Change and the Kyoto Protocol because they are already controlled under the Montreal Protocol on Ozone-Depleting Substances. They include chlorofluorocarbons (CFCs), hydrochlorofluorocarbons (HCFCs), and several solvents. Emissions of the gases included in the Kyoto Protocol have increased rapidly in the 1990s, but emissions of all of them are very small (at most a few thousand metric tons). On the other hand, many of the gases have atmospheric lifetimes measured in the hundreds or thousands of years, and consequently they are potent greenhouse gases with global warming potentials hundreds or thousands of times higher than that of carbon dioxide per unit of molecular weight.

Land Use and Forestry

Forest lands in the United States are net sinks of carbon dioxide from the atmosphere. According to U.S. Forest Service, U.S. forest land stores about

200 million metric tons of carbon, equivalent to almost 15% of U.S. carbon dioxide emissions. Extensive deforestation of the United States occurred in the late nineteenth and early twentieth centuries. Since then, millions of hectares of formerly cultivated land been abandoned and returned to forest. The regrowth of forests is sequestering (i.e., storing in its tissues) carbon on a large scale. The sequestration is diminishing, however, because the rate at which forests absorb carbon slows as the trees mature. The extent to which carbon sequestration should be included in emissions inventories generally, and the extent to which sequestration would "count" under the Kyoto Protocol, are still being determined. The Kyoto Protocol specifically limits countable effects for countries like the United States to anthropogenic afforestation, deforestation, and reforestation that has occurred since 1990, and only if it is measurable and verifiable. Each clause would probably limit the applicability of carbon sequestered as a result of land use changes and forestry.

Tree planting, reforestation, and protection of tropical rainforests are examples of "win-win" opportunities. Not only do they sequester CO_2, but they are "oxygen factories." Plus, they give numerous other benefits, not the least of which is their socio-psychological values. People of all cultures are connected to their flora.

Threats to the Stratospheric Ozone Layer

The incidence in skin cancer, especially the most virulent form, melanoma, increased throughout much of the twentieth century. Some of this can be attributed to the population migrations toward the tropics with the increases associated with greater exposures to ultraviolet radiation from the sun. By extension, observations of the thinning of the ozone layer in the stratosphere have been coincidental with the increase in skin cancer cases.

A simple representation of how ozone (O_3) is formed and destroyed in the stratosphere is indicated in two reactions.

Ozone Formation:

$$O_2 \xrightarrow[<200\text{nm}]{\text{Sunlight}} 2O \tag{9.1}$$

$$O + O_2 \rightarrow O_3 \tag{9.2}$$

Ozone Destruction:

$$O_3 \xrightarrow[200\text{–}300\text{nm}]{\text{Sunlight}} O_2 + O \tag{9.3}$$

$$O + O_3 \rightarrow 2O_2 \tag{9.4}$$

Note that this is a huge oversimplification of the numerous photochemical steps involved in either formation or destruction of O_3, but it demonstrates the overall result of molecular oxygen (O_2) absorbing a photon at a wavelength shorter than 200 nanometers. This absorbed energy splits the molecule into two O atoms. In turn, one of these O atoms can react

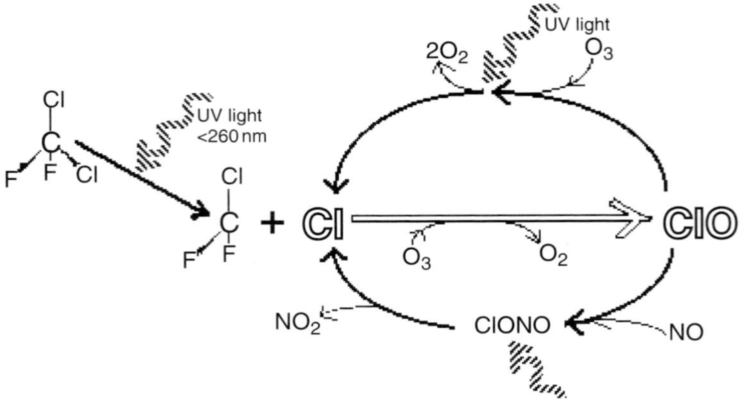

FIGURE 9.6. Halogen-catalyzed ozone depletion process.
Source: National Aeronautics and Space Administration, Advanced Supercomputing Division.

with another O_2, forming a new ozone molecule. Up to 98% of the sun's high-energy ultraviolet light (UV-B and UV-C) is absorbed by the destruction and formation of atmospheric ozone. The global exchange between ozone and oxygen is about 300 million tons per day.

The key is that both reactions are occurring in the stratosphere simultaneously. In fact, it is the energy expended in these two net reactions that uses up much of the incoming UV radiation. The problem is that in recent decades, the destruction has greatly outpaced the formation of ozone. A principal reason for the imbalance can be attributed to the presence of halogenated compounds that find their way to the upper levels of the troposphere, ultimately reaching the stratosphere. The destruction of ozone is sped up (catalyzed) by halogen atoms, notably chlorine (Cl) and fluorine (F). The process is depicted in Figure 9.6. One Cl molecule can degrade over 100,000 molecules of ozone before it is removed from the stratosphere or is transformed to a nonreactive species such as chlorine nitrate ($ClONO_2$). Even these nonreactive species, known as reservoirs, can release another active Cl when excited by sunlight, however.

Human activities contribute as much as 85% of the chlorine found in the atmosphere. That takes some doing, considering the oceans are loaded with chlorine in the form of salts, which aerosolize into the atmosphere. However, chlorine is also a common constituent of many modern products, such as plastics and solvents.

Coral Reef Destruction[10]

Two-thirds of the earth's coral reefs are dying. It is estimated that 10% of the earth's coral reefs have already been degraded beyond

FIGURE 9.7. Coral reef biodiversity.
Source: Ocean World Web site: http://oceanworld.tamu.edu/; accessed April 21, 2005. *Photo credit*: California Academy of Sciences.

recovery. A much larger percentage is now threatened. Human activities are among the major cause of reef decline.
—United States National Oceanic and Atmospheric Administration (NOAA).[11]

Besides air, planetary effects must also consider the other major environmental fluid, water. In oceans, for example, coral reefs (see Figure 9.7) are among the most diverse ecosystems on the planet and home to nearly a million different types of marine animals. Reefs serve to protect coastal communities from potentially destructive wave energy and beach erosion. Like so many ecosystems, the scientific research community is only beginning to identify the bounty of this species richness; for example, coral reefs have been the focus of promising biomedical studies. Ironically, as the appreciation of their value has become better understood, these delicate ecosystems are highly threatened.

Ten percent of the world's reefs have been completely destroyed, and some reefs are almost completely gone. For example, the Philippines reef system has lost more than 70% of its original reef, with but 5% in good conditions. Reefs are destroyed both indirectly and directly. When the general environmental conditions change, this has an indirect effect on reefs, which have a tight set of conditions to survive. For example, coral reefs can live only within a certain temperature and salinity range. So, even a very small mean water temperature elevation, for example 1°C, will not allow the coral organism to reproduce. Thus, the potential effects from global climate change could increase the rate and extent of reef loss worldwide.

The common symptom of reef condition is the amount of coral bleaching; the pigmented algae that live as a vital component of the reef ecosystem die, and when they do, the pigmentation is also diminished, leaving only the limestone shell visible through the transparent coral bodies. Coral bleaching has increased at an increasing rate since the 1980s.

Algal diversity is also affected by increased water temperatures, so that invasive, harmful algae grow rapidly on the upper layers of the coral, diminishing the incoming solar radiation and lowering the rate of photosynthesis, thereby killing the coral. Coral are also threatened by overfishing and ocean pollution.

A direct and intentional method of coral reef destruction is to destroy the reefs physically. This is particularly problematic in the Philippines, where divers fish the reefs for food and aquarium fish using highly destruction methods like explosives (see Figure 9.8) that stun the fish and make

FIGURE 9.8. Explosives used to stun fish that are also physical stressors to reef ecosystems.
Source: Ocean World Web site: http://oceanworld.tamu.edu/; accessed April 21, 2005. *Photo credit*: Thomas Heeger: Philippines.

collection much easier than with conventional fishing practices. In the process, the coral polyps that serve as the reef infrastructure are destroyed. Poisons, especially those containing cyanide (CN^-), are also used to stun the fish when applied onto the reef. These are dramatic and instantaneous illustrations of habitat loss; that is, a physical stressor and a chemical stressor, respectively.

The reefs may also fall victim to spreading diseases or natural predators, such as the Crown of Thorns starfish, which can completely dismantle a reef in little more than two weeks.[12] However, the partial destruction and death of the reefs that occurs naturally is small relative to the human-caused destruction.

"As the 'rain forests of the sea,' coral reefs provide services estimated to be worth as much as $375 billion annually, a staggering figure for an ecosystem covering less than one percent of the earth's surface," says the United States Coral Reef Task Force.[13] In fact, reef tourism is paradoxical. Because it brings in such substantial annual capital, tourism has the potential to save the dying reefs through awareness programs, creation of parks, and regeneration efforts. On the other hand, however, it is partly responsible for the reef destruction. Divers can kill an entire coral colony by touching a single polyp. Uneducated tourists can trample miles of coral in a week and their boat anchors may disturb the reef beyond recovery when they hit "bottom."[14]

When sewage or soil enters the water, the ocean's top layer becomes cloudy and blocks the coral's available sunlight. With the reduced light, photosynthesis cannot occur and the coral communities degenerate. Another consideration is the eutrophication (see Chapter 4) that occurs as a result of dumping. When waste materials are dumped into the ocean, kelps, seaweed, and other algal plants grow in the newly nutrient-rich environment.[15] The algal species then over take the coral, killing the polyps one by one.

A final consideration is over-harvesting. Because the reefs are so rich in fish diversity, they are prime catch sites. The problem arises when the delicate balance is disturbed by over-fishing larger animals, such as grouper or flounder. Without these members of the food web around, some smaller fish increase rapidly in population, then die due to lack of food, and other species die off almost immediately. Furthermore, explosives, poisons, and other questionable fishing tactics are often used in and around the reef; these throw the ecosystem out of sync, as well.

Thus, decisions currently made to protect or ignore the current coral reef problem can be displayed in a flow chart (see Figure 9.9).

Syllogisms for Coral Reef Destruction

A means of analyzing the ethical aspects of coral reef destruction is to make moral arguments using syllogisms. Here are a few:

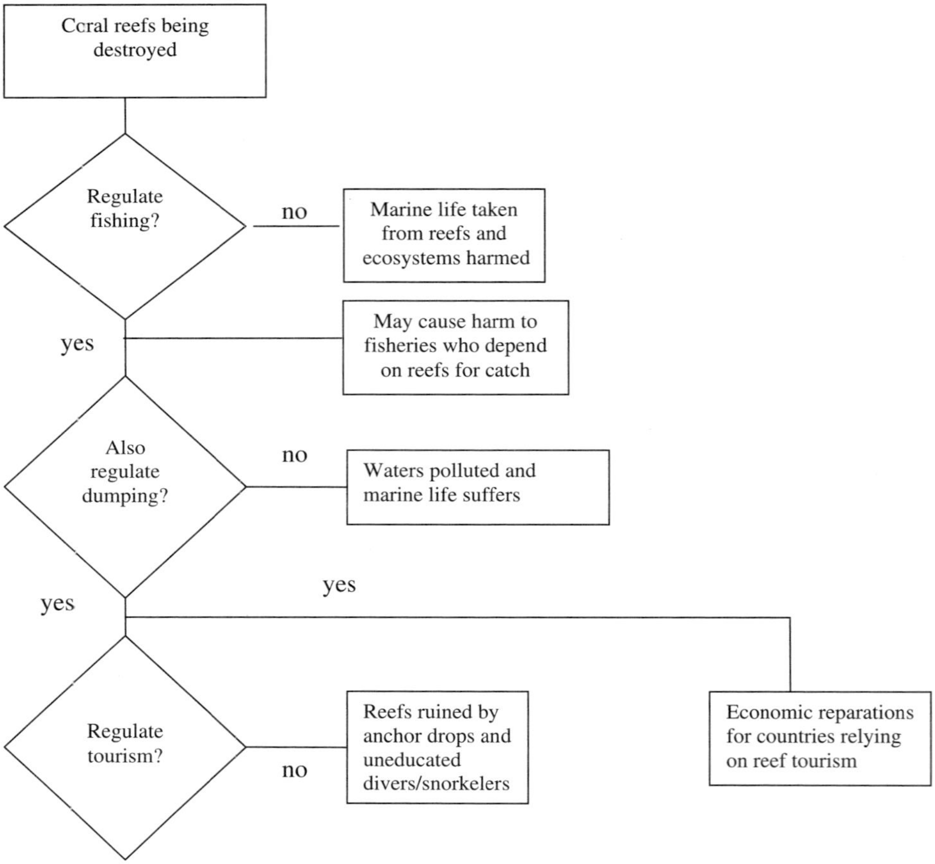

FIGURE 9.9. Coral reef decision chart.

Syllogism 1

Factual Premise: Corporations often get rid of byproducts in the most economical manner possible.

Connecting Premise: Corporations dump waste near coral reefs and subject the reefs to toxic chemicals that poison fish and eat away at the polyp structures.

Evaluative Premise: Poisoning the coral reefs is wrong.

Conclusion: Therefore, the corporations who are dumping near the coral reefs and destroying them are doing the wrong thing.

Syllogism 2

Factual Premise: Much of the revenue of tropical, coastal nations depends upon tourism.

Connecting Premise: Tourists love to walk on and touch the colorful coral structure, but are in fact damaging polyp structures and disturbing the delicate marine ecosystem.
Evaluative Premise: Allowing for excessive tourism in order to generate a profit, at the expense of the coral reefs, is irresponsible.
Conclusion: Therefore, the agencies involved with excessive tourism leading to the demise of the coral reefs are acting in an irresponsible manner.

Syllogism 3
Factual Premise: Anglers must fish to make a living.
Connecting Premise: Anglers love to fish in the diversely inhabited coral reef areas, but are in fact damaging polyp structures by dropping anchor and disturbing the delicate marine ecosystem by over-fishing.
Evaluative Premise: Over-fishing and unnecessarily dropping anchor is reckless.
Conclusion: Therefore, the anglers causing coral reef destruction are acting in a reckless manner.

The key lesson that needs to be heeded is that to address coral reef destruction the first step is to heighten public awareness of the critical situation. In addition, there is a strong need to impose legal dumping restrictions. Other possible mitigating measures include greater tourism regulations, fishing caps, and fewer anchor drops, so the coral reef communities can rebuild themselves and maintain their delicate balance for future generations.

The lesson from the 1960s was the need to take actions at home while considering the big picture. And the atmosphere and ocean are very big systems, indeed. Our planetary challenge has been succinctly boiled down to a very catchy and relevant motto:

Think globally and act locally!

Notes and Commentary

1. The source of this discussion is Goddard Institute of Space Studies, 2005. Research, New York, NY; http://www.giss.nasa.gov/research/chemistry/; accessed April 18, 2005.
2. For example, the phenomenon of global warming, although not embraced by all scientists, is generally accepted, in concept, by much of the atmospheric science community. However, I recall in the early 1970s a very intelligent and knowledgeable professor, Dr. Loran Marlow in the Department of Earth Sciences and Geography at Southern Illinois University, lecturing on the distinct likelihood of "global cooling." His reasoning and that of other respected

researchers at the time, was that the buildup of aerosols would block incoming solar radiation (insolation) and decrease the atmosphere's total energy budget. That is, there would be less short-wave and visible light entering the atmosphere, so there would be less conversion of these wavelengths to the infrared range (i.e., less heat formation) that are radiated into the atmosphere.

3. The principal source of this discussion is U.S. Department of Energy, 2004. Energy Information Agency, Emissions of Greenhouse Gases in the United States 2003, Report #: DOE/EIA-0573(2003), Washington, D.C. This is also the source of the estimates for global fluxes of the gases discussed in this section.
4. Intergovernmental Panel on Climate Change, 2001. *Climate Change 2001: Working Group I: The Scientific Basis*, Cambridge University Press, Cambridge, United Kingdom, and New York, NY.
5. Some years back, I received a telephone call from an angry citizen who rejected the concept of the greenhouse effect itself. I thought he was disagreeing with the methods of data collection or the interpretation of the data, so I began to give him an abbreviated version of a lecture I have given at Duke and other universities on the thermodynamic fundamentals of the greenhouse effect, including using a greenhouse and automobile on a cold, sunny day to make my point. He stopped me abruptly, telling me he thought that the whole concept was fallacious. I was left speechless for some minutes. For my lecture to be compelling, I need a modicum of acceptance of thermodynamics. I must add that this person is intelligent and knowledgeable. He considers himself to be a scientist and has written papers on the myth of the greenhouse effect. I realized then that we had two very different paradigms. The factual premises that I needed for my valid argument (see Chapter 1) were missing or simply rejected. I was unable to convince the caller of the existence of the greenhouse effect so, obviously, I was also unable to bring the conversation to the point of discussing which, if any, variables would be rate limiting on future climate changes. The episode was another reminder that assuming agreement on any issue, no matter how fundamental, at the outset of a public meeting can be perilous!
6. Intergovernmental Panel on Climate Change, 2001. *Climate Change 2001: Working Group I: The Scientific Basis*, Cambridge University Press, Cambridge, United Kingdom, and New York, NY.
7. El Niño is a description of large temperature fluctuations in the tropical regions of the Pacific Ocean. In 2002, the National Oceanic and Atmospheric Administration defined an El Niño event to be equal to or greater than 0.5°C averaged over three months.
8. Intergovernmental Panel on Climate Change, 1996. *Climate Change 1995: The Science of Climate Change*, Cambridge University Press, Cambridge, United Kingdom, and New York, NY.
9. This is another example of attempting to solve one problem (i.e., hydrocarbon emissions) and exacerbating another (i.e., greenhouse gas emissions).
10. The principal source for this presentation is a research report by Kristen Boswell, an undergraduate student in Duke University's Pratt School of Engi-

neering. The report from which this discussion was drawn was conducted as a requirement for the course EGR 108S, Ethics in Professions.
11. Available online at www.noaa.gov; accessed February, 2005.
12. Coral Reef Destruction: http://www.uwsp.edu/geo/courses/geog100/CoralReef.htm; accessed February, 2005.
13. T. Garrison, 2004. *Essentials of Oceanography.* Thomson Learning, Inc., Pacific Grove, CA.
14. It is not really "bottom", in fact it is the top of the reef that they are crushing.
15. This is the same process of eutrophications as that discussed in Chapter 4 for freshwaters.

CHAPTER 10

Myths and Ideology: Perception versus Reality

By far, most of the cases we have considered thus far are generally well documented insofar as the scientific community reaching a consensus that they have caused harm. In most instances, these cases are regarded as milestones, albeit often negative, in society's journeys toward environmental consciousness. Such milestones have served as drivers for environmental policy. Many perceived problems, however, do not enjoy such a consensus. The hypotheses linking anthropogenic activities to changes in global climate and the threats posed by long-term storage of radioactive wastes (and the whole issue of using fission to produce electricity, for that matter) are examples of major disagreements between policy makers, journalists, and lay people, as well as within the scientific research community.

In this chapter, we analyze cases where either scientific consensus was not reached or in which, in retrospect, we find that the requisite scientific underpinning either is questionable or does not, in fact, exist.

Solid Waste: Is It Taking over the Planet?

Municipal solid waste (MSW),[1] the trash or garbage collected by towns, cities, and counties, is made up of commonly used and disposed of items like lawn waste and grass clippings, boxes, plastics and other packaging, furniture, clothing, bottles, food scraps, newspapers, appliances, paint, and batteries. In 2001, U.S. residents, businesses, and institutions produced more than 229 million tons of MSW, which is approximately 4.4 pounds of waste per person per day, up from 2.7 pounds per person per day in 1960 (see Figure 10.1).

The amount of solid waste generated in the United States has grown steadily (see Figure 10.2), and almost every local authority since the 1990s has implemented management practices to stem the burgeoning amounts of solid waste being generated and needing disposal. These measures have included source reduction, recycling and composting, and prevention or

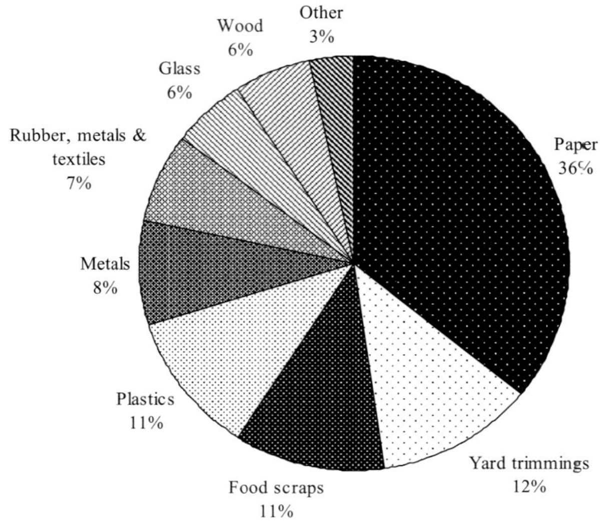

FIGURE 10.1. Composition of waste generated in the United States in 2001. *Source*: U.S. Environmental Protection Agency, 2005. Municipal Solid Waste, http://www.epa.gov/epaoswer/non-hw/muncpl/facts.htm; accessed April 5, 2005.

diversion of materials from the waste stream. Source reduction involves altering the design, manufacture, or use of products and materials to reduce the amount and toxicity of what gets thrown away. Recycling averts items from reaching the landfill or incinerator. Such items include paper, glass, plastic, and metals. These materials are sorted, collected, and processed and then manufactured, sold, and bought as new products. Composting is the microbial decomposition of the organic fraction of wastes, such as food and yard trimmings. The microbes, mainly bacteria and fungi, produce a substance that is valuable as a soil conditioner and fertilizer, which is sold or given away by the local authorities, often at the landfill site itself.

Engineers and environmental professionals have also improved waste handling once the items find their way to the landfill. For example, landfills must be engineered in a manner and in areas where waste is placed into the land (see Table 10.1). Landfills usually have liner systems and other safeguards to prevent groundwater contamination. Combusting solid waste is another practice that has helped reduce the amount of landfill space needed.

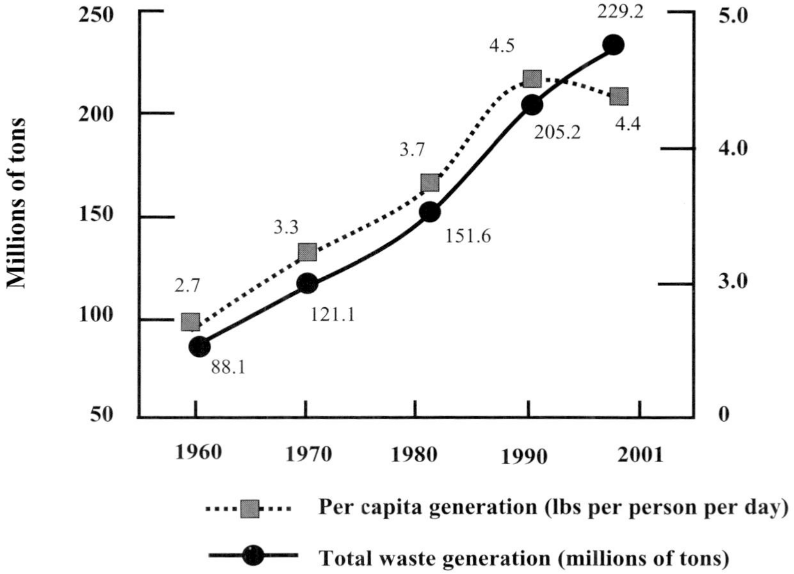

FIGURE 10.2. Trends in municipal solid waste generation. Up to the late 1980s, the rate of waste generated per capita was proportional to the total waste generated. After that time, in part due to waste reduction and recycling programs, the waste generated per capita dropped, but the total waste generated continued the upward pace.

TABLE 10.1
Summary of federal landfill standards as prescribed by the U.S. Environmental Protection Agency, 2005. http://www.epa.gov/epaoswer/non-hw/muncpl/landfill; accessed April 22, 2005.

Location restrictions ensure that landfills are built in suitable geological areas away from faults, wetlands, flood plains, or other restricted areas.
Liners are geomembrane or plastic sheets reinforced with two feet of clay on the bottom and sides of landfills.
Operating practices such as compacting and covering waste frequently with several inches of soil help reduce odor; control litter, insects, and rodents; and protect public health.
Groundwater monitoring requires testing groundwater wells to determine whether waste materials have escaped from the landfill.
Closure and postclosure care include covering landfills and providing long-term care of closed landfills.
Corrective action controls and cleans up landfill releases and achieves groundwater protection standards.
Financial assurance provides funding for environmental protection during and after landfill closure (i.e., closure and postclosure care).

Combustion facilities burn solid wastes at high temperatures, reducing waste volume and generating electricity.

This is all well and good, and demonstrates great progress in how we think about wastes. However, there is an argument that the original premise on which this progress is based is, in fact, flawed. Are we really facing a solid waste crisis? Talk show hosts and a recent Home Box Office show hosted by the comedy team, Penn and Teller, consider the solid waste problem to be a convenient myth. One of their postulations is that the issue is another way that the government interferes with privacy and freedoms. In fact, one of Penn and Teller's conclusions is that the recycling is okay, but the ends should not justify the means. They argue that it is unethical to control people's lives based on a flawed premise.

The controversy cuts both ways. Others believe that the progress being made is overstated and that measures to reduce waste, for example, looking at only total volume reduced, are inadequate because pockets of intractable problems exist. Take "disposable" diapers, for instance. A cloth diaper service, with an obvious vested interest, argues:

> An entire generation is growing up believing that the term "disposable diaper" is redundant: There's only one thing you put on babies' bottoms. They're plastic, you get them in huge bags and boxes at the grocery store or the convenience store, and you fold them up, and toss them in the trash when they're dirty. The product name itself is a misnomer, testament to the power of Madison Avenue and to our own Freudian neuroses surrounding our bodies and our wastes. For Huggies and Pampers and Luvs are not "disposable" at all. We throw about 18 billion of them away each year into trash cans and bags, believing they've gone to some magic place where they will safely disappear. The truth is, most of the plastic-lined "disposables" end up in landfills. There they sit, tightly wrapped bundles of urine and feces that partially and slowly decompose only over many decades. What started out as a marketer's dream of drier, happier, more comfortable babies has become a solid-waste nightmare of squandered material resources, skyrocketing economics, and a growing health hazard, set against the backdrop of dwindling landfill capacity in a country driven by consumption.[2]

So, even one of the most successful programs in terms of changing the attitude about waste is subject to controversy.

Alar and Apples

From the mid-1960s through the 1980s, most apples grown in U.S. orchards were sprayed with the compound butanedioic acid mono(2,2-dimethylhydrazine), also known as daminozide ($C_6H_{12}N_2O_3$), and best known by its tradename Alar. The compound is an amino acid derivative growth retardant that was formulated by the Uniroyal Chemical Company. The chemical was a valuable asset to orchard operations. When sprayed on apples, the growth process can be controlled to allow for nearly simultaneous ripening, allowing orchards to be harvested all at once. Thus, labor, machinery, and other expenses could be minimized. Laboratory testing conducted from 1973 through 1977 began to suggest a linkage between Alar and its degradation product unsymmetrical dimethyl hydrazine (UDMH) to cancer in rodents, especially mice and hamsters. This weight of evidence led to the 1984 designation by the National Toxicology Program that UDMH be considered a "probable human carcinogen." It is not unusual for a degradation product to be more toxic than the parent compound (see the discussion box, "Parent versus Progeny"). By extension, due to the widespread exposure to apples and apple products, some preliminary risk assessments predicted as many as 30 million additional cancer deaths from ingesting Alar in products like applesauce. Although this risk was subsequently lowered to 20 million, it was considered an extremely important public health issue, especially in light of the fact that one of the largest demographic groups ingesting apple products is young children. Such information is important if true.

Parent versus Progeny

Chemical kinetics is the description of the rate of a chemical reaction.[3] This is the rate at which the reactants are transformed into products. In environmental situations, this is the process by which chemicals are synthesized, such as in microbial processes, and degraded. Degradation is both good and bad. It is beneficial in situations where we are able to manufacture substances in a way that they are "biodegradable," meaning that they are relatively easily broken down to simpler compounds, such as by sunlight (photodegradation), by ubiquitous chemicals like carbonic acid (abiotic chemical degradation), or by organisms (biodegradation). It can be very harmful when the substances are changed to become more persistent, more bioaccumulating, or more toxic. It is not uncommon, for example, for a compound to become more toxic when it is metabolized, a process known as *bioactivation*.

The reaction rate is a function of the concentrations of the reacting substances (see Chapter 2). The mathematical expression of this function is known as the rate law.

The concentrations in the rate law are the concentrations of reacting chemical species at any specific point in time during the reaction. The rate is the velocity of the reaction at that time.

Vinclozolin Degradation

An example of a toxic substance degrading to a more toxic compound is that of the fungicide vinclozolin (see Figure 10.3). The dichlorobenzene in the structure engenders persistence to the molecule, and the carbamate ring provides reactivity. Vinclozolin usually breaks down to form principal degradation products when the carbamate ring opens: a butenoic acid (2-[(3,5-dichlorophenyl)-carbamoyl]oxy-2-methyl-3-butenoic acid), referred to as M1; and an enanilide (3′,5′-dichloro-2-hydroxy-2-methylbut-3-enanilide), M2. The M1 reaction is reversible; that is, the carbamate ring closes and returns to the vinclozolin structure, depending upon environmental conditions; whereas M2 is a nonreversible degradation product and does not return to the parent form. All three of the compounds are hormonally active and disrupt endocrine systems in mammals. However, M1 is even more disruptive than the parent compound.

Environmental conditions can profoundly affect the rate of degradation and the types of compounds formed, that is, the chemical speciation (see Figure 10.3). For example, pH of the soil, sediment, and water column is a principal factor in vinclozolin degradation rates.[4] The degradation is quite rapid at higher pH and much slower at low pH. At 35°C, the half-life of vinclozolin at pH 8.3 is less than one hour, and at pH 4.5 it is 530 hours. This difference can be explained in part by vinclozolin's increased resistance to hydrolysis at lower pH. The pH also determines the principal degradation pathway that vinclozolin will take, with higher pH values yielding greater amounts of M1 and lower pH values yielding more M2. A third degradation product, 3,5-dichloroaniline, has been detected after considerable time (672.3 h at pH 6.5, 1537 h at pH 5.5, and 505.8 h at pH 4.5). This points to important considerations for estimating the fate of vinclozolin. Not only does increased soil and solution acidity increase vinclozolin's persistence, but acidity also influences the degradation pathways and the appearance of secondary degradation products.

The type of solvent also influences degradation and bioavailability. For example, vinclozolin in water differs from vinclozolin in acetone. The acetone-vinclozolin solution is more persistent than the water suspension. This may result from the dichorophenyl group's influence on the lipophilicity of vinclozolin. The persistence observed in the field has been lower than in the laboratory studies, possibly the result of increased photodegradation and greater moisture gradients. Available moisture in the soil column would also affect the rate of degradation because it provides hydrogen ions and may render compounds more polar over time. So, many factors can influence the breakdown of compounds in the environment.

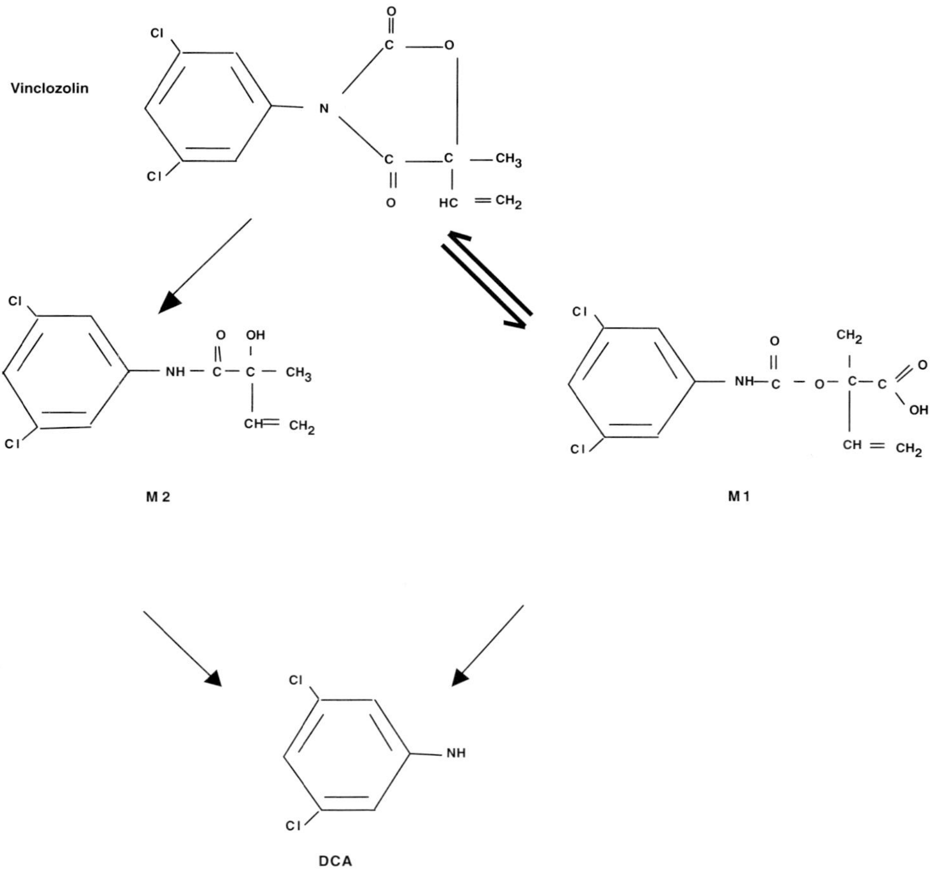

FIGURE 10.3. Structural formulae and degradation pathways of hazardous chemicals and its principal degradation products.
Source: S.Y. Szeto, N.E. Burlinson, J.E. Rahe, J.E., and P.C. Oloffs, 1989. "Kinetics of hydrolysis of the dicarboximide fungicide vinclozolin," *Journal of Agricultural and Food Chemistry*, 37: 523–529.

Timing was critical in the Alar case. The National Toxicology Program was established in 1978 by the U.S. Health, Education, and Welfare department (today known as the Department of Health and Human Services), right about the time that the Alar data were being released and evaluated. The new program was created because the government was being criticized for a lack of coordination of its myriad toxicology testing programs. A central program was seen as a way to strengthen the science base in toxicology, to develop and to validate improved testing methods. The program was also designed to provide more reliable information about potentially toxic chemicals to health, regulatory, and research agencies, scientific and medical communities, and the general public.

The 1970s saw an increasing amount of concern by scientists, politicians, journalists, and activists about the human health effects of chemical agents in our environment. A number of human diseases were thought to be directly or indirectly related to chemical exposures; so many argued that decreasing or eliminating human exposures to those chemicals would help prevent some human disease and disability. This sounds almost trite by contemporary standards, where chemical exposure and risks are logical, but such linkages were nascent and risk assessment was still in its formative stages.

Although many considered the whole problem of toxic substances to be much ado about nothing, others saw a cancer link to virtually any chemical other than water. The latter was brought home to me in 1977 in a discussion with a Regional Administrator of the U.S. EPA. She was trying to convince people in a meeting that "everything doesn't cause cancer." She was particularly irritated by some of the local cartoons that showed a rat consuming massive amounts of artificial sweeteners with bylines to the effect that any substance consumed to excess would lead to cancer. Her major point was that if a substance is not carcinogenic, it will not lead to cancer, no matter the dose. Again, through the prism of risk assessment, such a statement seems unnecessary, but in retrospect it was not only needed, but in many venues it was seen as pro-chemical or even anti-environmental! In fact, this EPA official was one of the most environmental of any I have met, but her objectivity as a scientist would not allow her to ignore the dose-response relationships. Vestiges of these concerns are still with us, as we try to have a common understanding of terms like organic, additives, natural, and holistic. Few chemicals, even those where data seem to support their safety, are considered acceptable for certain segments of society.

So, Alar became at the focal point for this controversy. Its structure was perceived to be menacing by many (see Figure 10.4). In February 1989, *60 Minutes* ran a story about a Natural Resources Defense Council (NRDC) report on Alar as a human carcinogen—one that posed particular risks for children. Public protest forced apple growers to stop using it and Uniroyal to pull it off the market. Although Uniroyal voluntarily canceled its use on fruits and vegetables, the chemical is still used for ornamental and bedding plants.

On the other hand, it appears that Alar's degradation product can be linked to cancer (see Table 10.2).

So then, what are the real effects? The problem with answering this question includes both the adjective and the noun; what is "real" and what is the "effect" in question? In some cases, for example, the answer is partially answered by animal studies; for example, mammals may experience skin or eye irritation. Also, Alar seems to be very rapidly distributed after

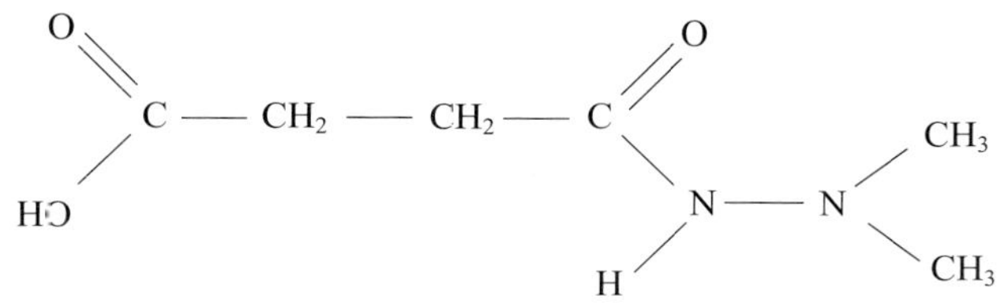

FIGURE 10.4. Alar structure.

TABLE 10.2
Results of cancer studies of Alar and its degradation product.

Animal	*Substance*	*Amount*	*Time*	*Result*
Rats	daminozide	5, 25, 250, or 500 mg/kg/day	2 years	No increase in tumor formation
Mice	daminozide	15, 150, 300, or 500 mg/kg/day	2 years	No increase in tumor formation
Rats	UDMH in water	0, 1, 50, or 100 ppm	2 years	Significant, but slight, dose-related increase in liver tumors in females
Rats	UDMH in water	100 ppm	2 years	Bile duct hyperplasia and inflammation of the liver in males
Rats	UDMH in water	50 and 100 ppm	2 years	Bile duct hyperplasia and inflammation of the liver in females
Mice	UDMH in water	0, 1, 5, or 10 ppm (males) 0, 1, 5, or 20 ppm (females)	2 years	Females exhibited decreased survival at the highest dose tested; also a significant increase in the incidence of lung tumors

mammalian exposure; for example, 96 hours after swine were exposed to a single oral dose ($5\,mg\,kg^{-1}$) daminozide, it was detected in all body tissues at concentrations as high as 73 ppb. The highest levels were found in the liver and kidney. Urinalysis showed that about 84% of the dose was eliminated in the urine and that 1% of the dose was metabolized to UDMH. The majority of daminozide residues ingested by animals is rapidly excreted in the urine and feces.

Science was never left alone to do its objective and careful research, however. The Natural Resources Defense Council (NRDC), a nonprofit environmental group, allegedly encouraged CBS's *60 Minutes* into running a story on the dangers of Alar. The broadcast was based largely on the NRDC report "Intolerable Risk: Pesticides in Our Children's Food," which identified 66 potentially carcinogenic pesticides in foods that a child might eat.

The NRDC's public relations firm, Fenton Communications, then convinced other major news organizations to feature the story. Actress Meryl Streep testified before Congress and on TV talk shows about the dangers of Alar.[5] The public panicked: school systems removed apples from their cafeterias, and supermarkets took them off their shelves. The scare cost apple growers over $100 million. The American Council on Science and Health, another advocacy group, paid newscaster Walter Cronkite $25,000 to narrate a TV documentary on the Alar scare entitled *Big Fears, Little Risks*.

In a previous two-year period, stocks rose an average 14% for companies negatively profiled on *60 Minutes*. Market insiders, aware of the upcoming story, exchanged a large number of shares in Uniroyal, which dropped in value precipitously after the story.

Some of the decisions that were made or should have been made by company managers and government officials in determining the safety of Alar, as well as actions that needed to be taken, are shown in the flowchart in Figure 10.5.

Another important lesson learned from the Alar episode is that of risk perception, especially as it pertains to children. Children are particularly sensitive and susceptible to many environmental pollutants. They are growing, so tissue development is highly prolific. Plus, society has stressed (as it certainly should!) special levels of protection for infants and children. For example, regulations under the Food Quality Protection Act mandate special treatment of children, evidenced by the so-called 10 X Rule, which recommends that, after all other considerations, the exposure calculated for children include 10 times more protection (thus, the exposure is multiplied by 10 when calculating risks) when children are exposed to toxic substances. The Federal Food Quality Protection Act[6] (FQPA) requires that risk assessments related to children include a safety factor regarding the potential for prenatal and postnatal effects. Several concepts of risk are introduced in Chapters 2 and 5, including the observation that environmental risk is a function of hazard and exposure. The 10 X Rule is an effort to ameliorate possible risks to children by addressing the potential exposures, even when the hazard is not completely understood. Frequently, the prenatal and postnatal toxicities are included when trying to establish some level of safety or precaution.

Recall that the concept of risk is expressed as the likelihood (statistical probability) that harm will occur when a receptor (e.g., human or a part

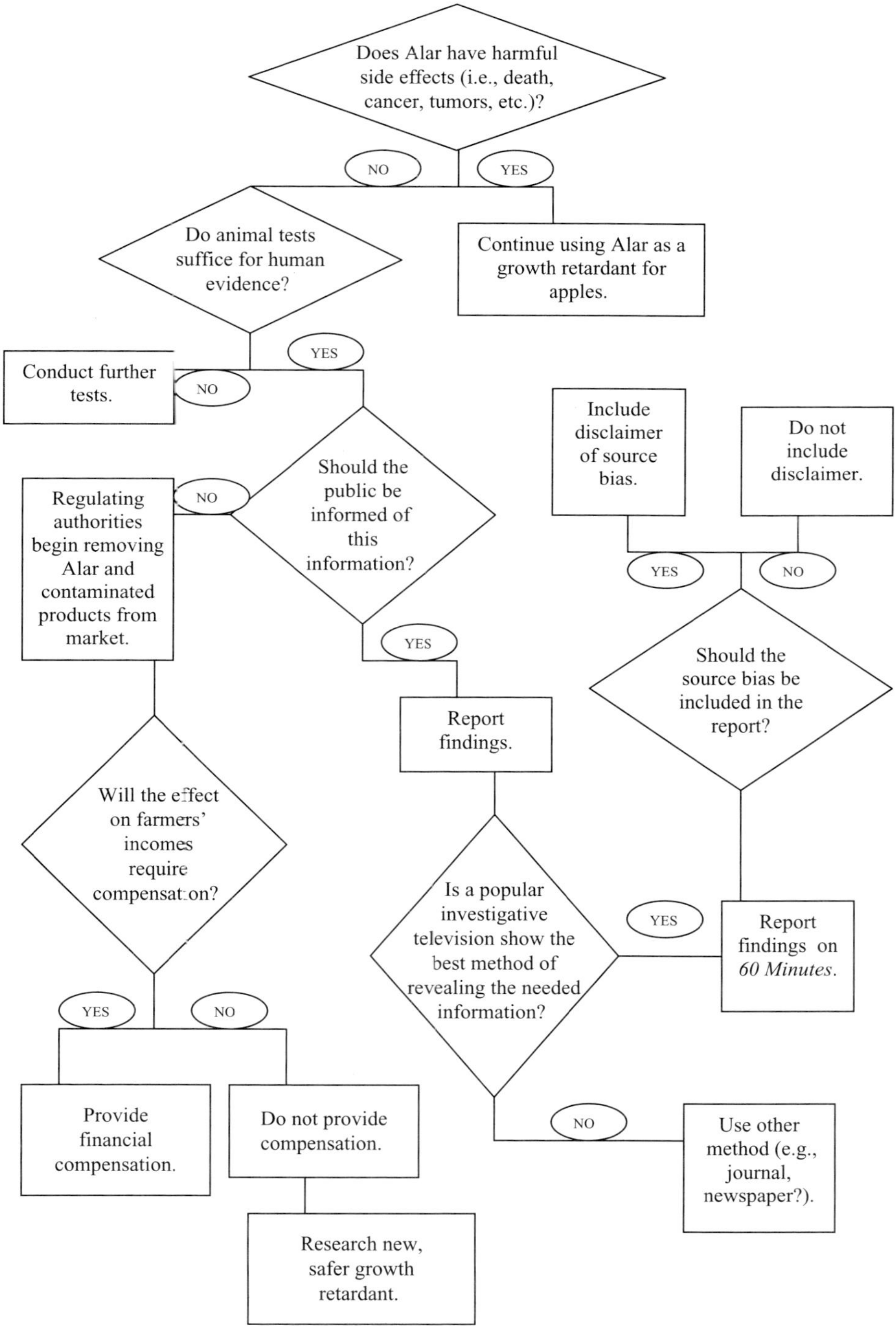

FIGURE 10.5. Flowchart of decisions available regarding the safety of Alar.

of an ecosystem) is exposed to that hazard. So, an example of a toxic hazard is a carcinogen (a cancer-causing chemical), and an example of a toxic risk is the likelihood that a certain population will have an incidence of a particular type of cancer after being exposed to that carcinogen (e.g., the population risk that one person out of a million will develop lung cancer when exposed to a certain dose of a chemical carcinogen for a certain period of time).

Dose is the amount, often mass, of a contaminant administered to an organism (so-called applied dose), the amount of the contaminant that enters the organism (internal dose), the amount of the contaminant that is absorbed by an organism over a certain time interval (absorbed dose), or the amount of the contaminant or its metabolites that reach a particular target organ (biologically effective dose or bio-effective dose), such as the amount of a neurotoxin (a chemical that harms the nervous system) that reaches the nerve or other nervous system cells. Theoretically, the higher the concentration of a hazardous substance that an organism contacts, the greater the expected adverse outcome. The classic demonstration of this gradient is the so-called dose-response curve (see Figure 5.11). If the amount of the substance is increased, a greater incidence of the adverse outcome would be expected.

Another aspect of the dose-response curve is that with increasing potency, the range of response decreases. A small increase in dose leads to a relatively large increase in biological response. As shown in Figure 10.6, a severe response represented by a steep curve will be manifested in greater rates of disease or mortality over a smaller range of dose. For example, an acutely toxic contaminant's dose that kills 50% of test animals (i.e., the LD_{50}) is closer to the dose that kills only 5% (LD_5) and the dose that kills 95% (LD_{95}) of the animals since the range between the LD_5 and LD_{95} has shrunk. The dose difference of a less acutely toxic contaminant will cover a broader range, with the differences between the LD_{50} and LD_5 and LD_{95} being more extended than that of the more acutely toxic substance.

There are a number of uncertainties associated with the data used to create a dose-response curve. Dose-response relationship based upon comparative biology from animal studies are usually high-dose, short duration (at least compared to a human lifetime) studies. From these animal data, models are constructed and applied to estimate the dose-response that may be expected in humans. Thus, the curve may be separated into two regions (see Figure 10.7). When environmental exposures do not fall within the range of observation, extrapolations must be made to establish a dose relationship. Generally, extrapolations are made from high to low doses, from animal to human responses, and from one route of exposure to another. The first step in establishing a dose-response is to assess the data from empirical observations. To complete the dose-response curve, extrapolations are made either by modeling or by employing a default procedure based upon information about the chemical and biochemical characteristics.

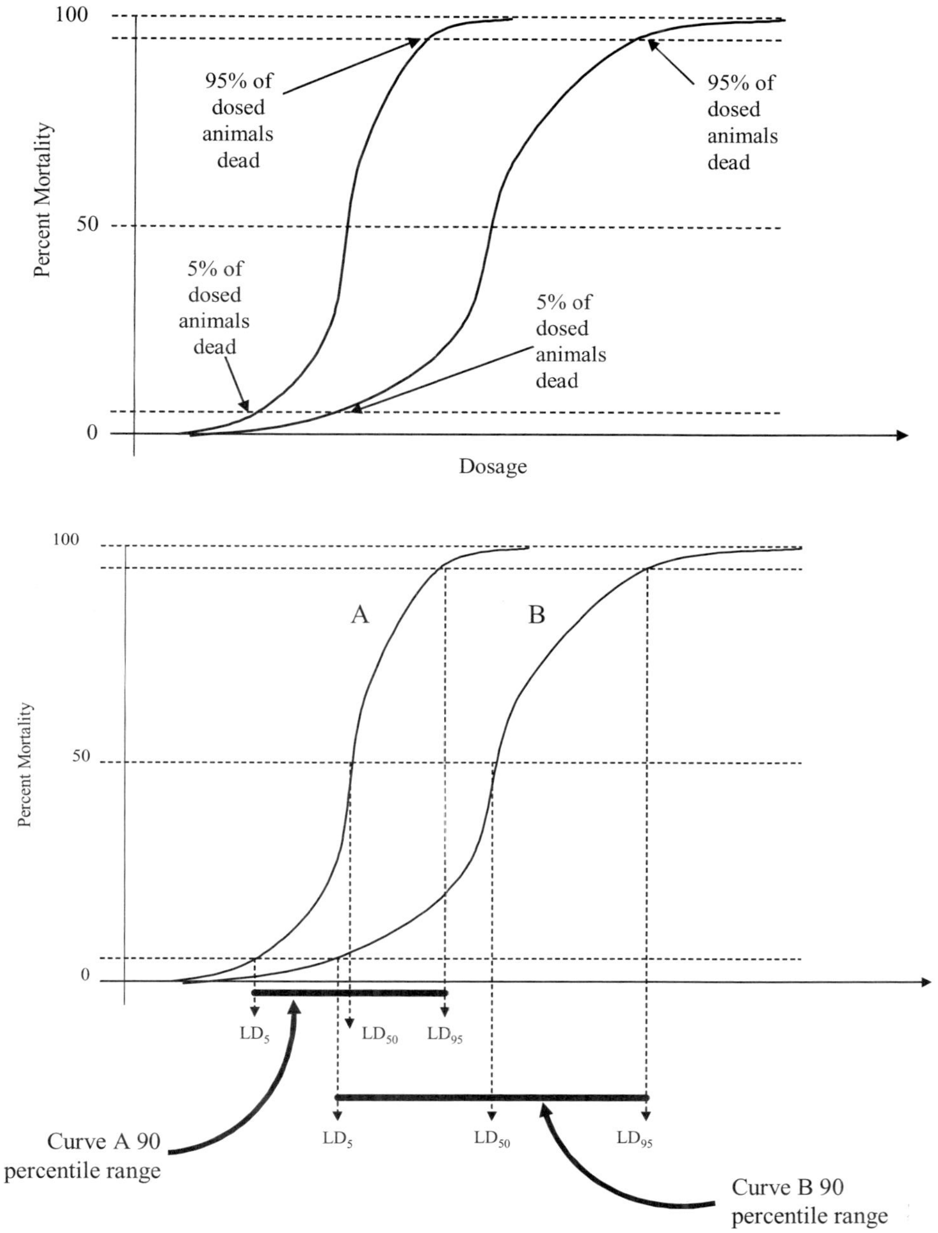

FIGURE 10.6. The greater the potency or severity of response (i.e., steepness of the slope) of dose-response curve, the smaller the range of toxic response (90 percentile range shown in bottom graph). Also, note that both curves have thresholds and that Curve B is less acutely toxic based upon all three reported lethal doses (LD_5, LD_{50}, and LD_{95}). In fact, the LD_5 for Curve A is nearly the same as the LD_{50} for Curve B, meaning that at about the same dose, contaminant A kills nearly half the test animals, but contaminant B has killed only 5%. Thus, contaminant A is much more acutely toxic.
Source: D.A. Vallero, 2004. *Environmental Contaminants: Assessment and Control*, Elsevier Academic Press, Burlington, MA.

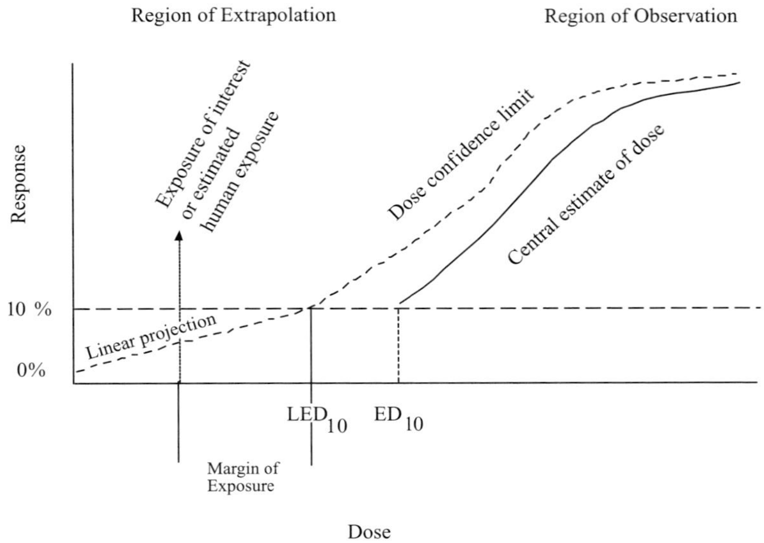

FIGURE 10.7. Dose-response curves showing the two major regions of data availability. Based upon discussions with the U.S. Environmental Protection Agency.

Based on the scientific weight-of-evidence available for the hazardous chemical, the U.S. EPA classifies its cancer-causing potential. Carcinogens fall into the following classifications (in descending order of strength of weight-of-evidence):

- A Carcinogen—The chemical is a human carcinogen.
- B Carcinogen—The chemical is a probable human carcinogen, with two subclasses:
  B1—Chemicals that have limited human data from epidemiological studies supporting their carcinogenicity
  B2—Chemicals for which there is sufficient evidence from animal studies but inadequate or no evidence from human epidemiological studies.
- C Carcinogen—The chemical is a possible human carcinogen.
- D Chemical—The chemical is not classifiable as to human carcinogenicity.
- E Chemical—There is evidence that the chemical does not induce cancer in humans.

As a precautionary principle, it generally is assumed in cancer risks assessments that there is no safe level of exposure; there is no threshold below which an exposure is acceptable. Such thresholds include the NOAEL, as shown in Figure 5.11, or the "lowest observed adverse effect level" (LOAEL), which is the area of the dose-response curve where studies have actually linked a dose to an effect. Thus, the precautionary principle renders the NOAEL and LOAEL irrelevant to cancer risk. Instead, cancer slope factors are used to calculate the estimated probability of increased cancer incidence over a person's lifetime (the so-called excess lifetime cancer risk, or ELCR). Like the reference doses, slope factors follow exposure pathways (see Appendix 3); for example, when calculating a Reference Dose (RfD) or Margin of Exposure (MOE) from the prenatal or postnatal adverse effects in the offspring and traditional uncertainty factors.

However, uncertainties or a elevated concern for children is not always sufficiently addressed using uncertainty factors in the RfD and MOE. Thus, the FQPA requires an additional evaluation of the weight of all relevant evidence. This involves examining the level of concern for how children are particularly sensitive and susceptible to the effects of a chemical, and determining whether traditional uncertainty factors already incorporated into the risk assessment adequately protect infants and children. As mentioned, this is accomplished mathematically in the exposure assessment. The U.S. EPA has prepared guidance on how data deficiency uncertainty factors should be used to address the FQPA childrens' safety factor. The final decision to retain the default 10X FQPA safety factor or to assign a different FQPA safety factor is made during the characterization of risk and not determined as part of the RfD process. The weight-of-the-evidence approach, therefore, includes both hazard and exposure considered together for the chemical being evaluated. The FQPA safety factor for a particular chemical must have the level of confidence in the hazard and exposure assessments and an explicit judgment of the possibility of other residual uncertainties in characterizing the risk to children.

By extension, other sensitive strata of the population also need protection beyond those of the general population.[7] The elderly and asthmatic members of society are more sensitive to airborne particles. Pregnant women are at greater risk from exposure to hormonally active agents, such as phthalates and a number of pesticides. Pubescent females undergo dramatic changes in their endocrine systems and, consequently, are sensitive to exposures during this time.

The Alar case illustrates that science is a necessary but wholly insufficient basis for decision making. People wanted to be sure that their children were not being exposed to a toxicant, particularly when the substrate in question was the apple and the exposure was ingestion of apple juice. The public reaction and any sensational media attention should not have come as a big surprise.

Agent Orange: Important If True

Agent Orange was a defoliant and weed-killing chemical used by the U.S. military in the Vietnam War. It was sprayed to remove leaves from trees behind which the enemy troops would hide. Agent Orange was applied by airplanes, helicopters, trucks, and backpack sprayers. In the 1970s, years after the tours of duty in Vietnam, some veterans became concerned that exposure to Agent Orange might cause delayed health effects. One of the ingredients in Agent Orange contained small amounts of the highly toxic compound tetrachlorodibenzo-*para*-dioxin (TCDD).

The U.S. Department of Veteran Affairs (VA) has listed a number of diseases that could have resulted from exposure to herbicides like Agent Orange. To receive consideration the law requires that some of these diseases be at least 10% disabling under VA's rating regulations within a deadline that began to run the day a veteran left Vietnam. If there is a deadline, it is listed in parentheses after the name of the disease:

- Chloracne or other acneform disease consistent with chloracne (must occur within one year of exposure to Agent Orange)
- Chronic lymphocytic leukemia
- Diabetes Mellitus, Type II
- Hodgkin's disease
- Multiple myeloma
- Non-Hodgkin's lymphoma
- Acute and subacute peripheral neuropathy (for purposes of this section, the term acute and subacute peripheral neuropathy means temporary peripheral neuropathy that appears within weeks or months of exposure to an herbicide agent and resolves within two years of the date of onset)
- Porphyria cutanea tarda (must occur within one year of exposure to Agent Orange)
- Prostate cancer
- Respiratory cancers (cancer of the lung, bronchus, larynx, or trachea)
- Soft-tissue sarcoma (other than osteosarcoma, chondrosarcoma, Kaposi's sarcoma, or mesothelioma)

The issue is international. After all, if it is true that U.S. soldiers exposed to Agent Orange show symptoms often associated with dioxin exposure, there is likely to be residual dioxin contamination in the treated areas of Vietnam. Dioxin is highly persistent, as we learned in Times Beach and other hazardous waste sites (see Chapter 5). On March 3–6, 2002, the Vietnamese and U.S. governments sponsored a scientific conference in Hanoi, Vietnam, regarding the health and environmental effects of Agent Orange and dioxin. Scientists from Vietnam, the United States, and 11 other countries discussed the state-of-the-science of research into the health effects of dioxin. The day after the conference ended, March 7, 2002, a select panel of international scientists identified data gaps in our understanding of the health and environmental effects of dioxin and recommended general areas of research in Vietnam that would help to fill these data gaps.

The next day, senior scientists from the Vietnamese Ministry of Science, Technology, and the Environment, the Vietnamese Ministry of Health, the U.S. National Institute of Environmental Health Sciences, the U.S. Environmental Protection Agency, and the U.S. Centers for Disease Control and Prevention met in Hanoi to establish an agreement for future research activities using the findings from the three-day conference and one-day workshop as a guide. The Vietnamese and U.S. government agencies agreed to a joint research plan, addressing the need for direct research on human health outcomes from exposure to dioxin and research on the environmental and ecological effects of dioxin and Agent Orange. The participants identified the highest priority areas.

The primary concerns in Vietnam from prolonged exposure to dioxin are for reproductive and developmental disorders that may be occurring in the general population. The key areas for research in Vietnam include miscarriage, premature birth, congenital malformations, endocrine disorders, neurological disorders, immunodeficiency, cancer, genetic damage, and diabetes mellitus. United States and Vietnamese scientists agreed to review the available literature and set priorities for areas where to determine that the literature is insufficient and to determine the presence or absence of a hazard and to identify where more research is needed.

Preliminary discussions have suggested two areas of research that should be further developed; research on existing populations with high exposures to dioxin relative to populations with low exposures (for example, people living near hotspots) and research on therapies to reduce dioxin body burdens in humans (such as some herbal therapies being proposed in Vietnam).

Dioxin contaminants of Agent Orange have persisted in the environment in Vietnam for over 30 years. In addition to a better understanding of outcomes of exposure, an improved understanding of residue levels and rates of migration of dioxin and other chemicals in the environment is needed. Hot spots containing high levels of dioxin in soil have been identified and others are presumed to exist but have yet to be located.

Dioxin has migrated through soil and has been transported through natural processes such as wind-blown dust and erosion into the aquatic environment. Contamination of soil and sediments provides a reservoir source of dioxin for direct and indirect exposure pathways for humans and wildlife. Movement of dioxin through the food web results in bioconcentration and biomagnification with potential ecological impacts and continuing human exposure. Research is needed to develop approaches for more rapid and less expensive screening of dioxin residue levels in soil, sediment, and biological samples that can be applied in Vietnam.

Improvements in this first step of analysis should be complemented by efforts to upgrade capabilities of laboratory facilities and equipment to international standards required by the research needs. This is no small task. Even in the United States, few labs have the capability to measure dioxins adequately. One of the most difficult aspects of measurement is the ability to extract the dioxin and similar compounds from the media where it is found. Tissue extraction is particularly difficult. Improved analytical capabilities can then be used to help to determine locations of highly contaminated areas, monitor remediation, and understand migration of dioxin in the natural environment. Monitoring efforts need to be linked to modeling efforts to understand fate and transport of dioxin in the environment. Innovative and cost-effective approaches to environmental remediation for application in Vietnam need to be developed, tested, and applied.

Preliminary discussions have suggested two areas of research that should be further developed: ecological and restoration research on a degraded upland forest (such as the Ma Da forest) and research on the identification, characterization, and remediation of hot spots (such as Da Nang Airport).

Actually, a number of defoliating agents were used in Vietnam, including those listed in Table 10.3. Most of the formulations included the two herbicides 2,4-D and 2,4,5-T. The combined product was mixed with kerosene or diesel fuel and dispersed by aircraft, vehicle, and hand spraying. An estimated 80 million liters of the formulation was applied in South Vietnam during the war.[8] The next step, linking environmental measurements of these contaminants to health and ecological harm, will be quite challenging.

TABLE 10.3
Formulations of defoliants used in the Vietnam War.

Agent	*Formulation*
PURPLE	2,4,-D and 2,4,5,-T used between 1962 and 1964
GREEN	Contained 2,4,5-T and was used 1962–1964
PINK	Contained 2,4,5-T and was used 1962–1964
ORANGE	A formulation of 2,4,-D and 2,4,5-T used between 1965 and 1970
WHITE	A formulation of Picloram and 2,4,-D
BLUE	Contained cacodylic acid
ORANGE II	2,4,-D and 2,4,5-T used in 1968 and 1969 (sometimes referred to as Super Orange)
DINOXOL	2,4,-D and 2,4,,5-T, small quantities were tested in Vietnam between 1962 and 1964
TRINOXOL	2,4,5-T; small quantities tested in Vietnam 1962–1964

Source: Agent Orange Web site: http://www.lewispublishing.com/orange.htm; accessed April 22, 2005.

The Snail Darter: A Threat to the Endangered Species Act?

One of the most divisive environmental issues occurred in the 1970s, when the snail darter (*Percina tanasi*), a small fish living in the Little Tennessee River, became the epicenter of a debate pitting the protection of threatened and endangered species against the development of water resources. More importantly, the tiny fish was perceived by many to represent a harbinger of ominous, ecological damage unless the government stepped in, and an equally passionate opposing group perceived the darter to lack much ecological value. This opposition saw the darter merely as a tactical means to achieve an already established end, to stop large water projects at any costs. The lines were drawn, with science straddling the two sides.

Preliminary scientific investigations found that *Percina* was threatened with extinction by the building of a dam; this led to an amendment allowing petitions for exemption from the requirements of the Endangered Species Act of 1973 (ESA).[9] More recently, critics have questioned the science behind ESA enforcement, arguing that healthy species are placed on the protected list. Finally, the judicial costs are enormous; lawsuits from both pro-environmental and pro-growth factions add greatly to the expense of enforcing the ESA.[10]

The 1973 ESA was passed to provide a comprehensive program to protect species at risk of extinction, and especially to preserve the habitat of such organisms. Protected species include plants and vertebrate and invertebrate animals that may be listed as either endangered or threatened based upon scientific assessments of the risk of their extinction. Distinct population segments of vertebrate species may also be listed as threatened or endangered. For example, certain fish populations of chinook, coho, chum, and sockeye salmon in the Pacific Northwest and California are presently protected by the ESA, but other populations of these same species in Alaska are not listed and can be commercially harvested, because the Alaskan populations are healthy and do not meet the baseline ESA requirements for special protection. Less stringent protection is available for plant species under the ESA.

Once a species is listed, the Fish and Wildlife Service is granted potent legal tools, including penalties, to aid the recovery of the species and the protection of its habitat. The law also facilitates citizens' ability to sue in court to do the same.[11] By the year 2002, a total of 1,072 species of animals and 748 species of plants had been listed as either endangered or threatened. The majority of these species (517 species of animals and 745 species of plants) occur in the United States and its territories and the remainder only in other countries. Of the U.S. species, 1,000 are covered in recovery plans.[12]

Protection and recovery of listed species can be controversial. Sometimes the threat indicates a larger habitat and ecosystem problem, such as dwindling essential physical, chemical, or biological resources or changes such as the construction of barriers like roads that impede species survival. When species are threatened, we must know whether they are the proverbial "canaries in the coal mine."[13]

The ESA has been a highly contentious environmental statute, stemming from its strict substantive provisions affecting the use of both federal and nonfederal lands. Once a species is listed, powerful legal tools are available to aid the recovery of the species and the protection of its habitat. The ESA is administered by the U.S. Fish and Wildlife Service (FWS) for terrestrial and freshwater species and some marine mammals, and by the National Marine Fisheries Service (NMFS, now NOAA Fisheries) for marine and anadromous species. The U.S. Geological Survey's Biological Resources Division conducts research on species for which the FWS has management authority.

Back to the snail darter: It can grow to as large as 89 mm (3.5 inches), spawning in the Tennessee River from mid-winter through mid-spring, in the shallower shoal areas over large, smooth gravel.[14] Water temperature during this period ranges from 5° to 16°C. Hatching takes place in about 18 to 20 days after spawning, with the larvae then drifting with the current to nursery areas farther downstream. After a nursery period of five to seven months, the juvenile darters begin to migrate back to the upstream spawning areas, where they spend the remainder of their lives. About one-fourth of the darters reach sexual maturity in their first year, and the remainder during the second year. The maximum lifespan appears to be four years. Food of larger snail darters includes aquatic insects and snails, but snails form the bulk of the diet.

The snail darter was discovered in August 1973 in the lower Little Tennessee River, Loudon County, Tennessee, by David A. Etnier. After further collections and study, Etnier published his findings in January 1976, indicating the snail darter to be a new species of percid fish. Before the construction of various impoundments, this fish probably was abundant in the main channel of the Tennessee River and possibly ranged from the Holston, French Broad, Lower Clinch, and Hiwassee Rivers, and downstream in the Tennessee drainage to northern Alabama.

The critical habitat for the snail darter in the Little Tennessee River was completely eliminated in 1979 by the closure of Tellico Dam. There is some evidence, however, that immediately downstream in the Tennessee River (headwater of Watts Bar Reservoir) a viable population still remains in the five- to 10-mile stretch of riverine habitat below Fort Loudon Dam. Another population, quite likely of natural origin, was discovered by Etnier in November 1980, in South Chickamauga Creek between creek mile 5.6 in Tennessee (Hamilton County) and creek mile 19.3 in Georgia (Catoosa County). Subsequent 1981 and 1982 surveys in the Tennessee River drainage have revealed snail darters in Sewee Creek (Meigs County), and a few darters have also been taken from the Tennessee River mainstream just below Chickamauga and Nickajack Dams, the Sequatchie River (Tennessee), and Paint Rock River (Alabama). The remaining distribution has resulted from transplants. Since 1975, snail darters have been transplanted in various Tennessee Rivers: Hiwassee (Bradley and Polk Counties), Nolichucky (Cocke/Greene Counties), Holston (Knox County), and Elk (Giles County). The Nolichucky transplant work was discontinued early, and there has been no definite evidence of a surviving population. In 1988, snail darters were found in the French Broad River upstream from its confluence with the Holston River. The population's status is unknown, but occurences probably stem from the Holston River transplants. The population in the Little Tennessee River was variously estimated at 5,000 to 20,000 prior to the onset of detrimental impacts from the construction of Tellico Dam.

Conflict was inevitable since the snail darter lived in the part of the Tennessee Valley Authority's nearly completed Tellico dam:

> Under section 7 of the Endangered Species Act, environmentalists filed a suit to prevent completion of the $116 million dam. After extensive legal battles, on June 15, 1978, the U.S. Supreme Court ruled in favor of the fish.[15]

In response, the United States Congress decided to revise the ESA to make it more realistic and flexible, especially to accommodate additional economic interests. Representative John Duncan of Tennessee argued that 3,000 Tellico workers were unemployed because of the Supreme Court decision. The way Congress decided to address this was to establish a seven-member committee of high-ranking federal officials who would be given the authority to exempt important projects from certain provisions of the Endangered Species Act. The so-called God Committee was quick to decide on the snail darter controversy, which entailed saving the darter and in the process, stopped the Tellico Dan project on January 23, 1979. Six months later, Rep. Duncan struck back with a short piece of legislation, a rider, and with 42 seconds of debate and a voice vote, that exempted the Tellico project from the requirements of the Endangered Species Act. The bill first failed when it came to the U.S. Senate, but the sponsor, Senator Howard Baker of Tennessee, forced a voice vote September 11, 1979, which led a four-vote victory. A presidential decision in favor of the minnow was the environmentalists' last hope, but political pressures prevented President Jimmy Carter's veto.

The scientific inertia continued into the 1980s, however, and the Little Tennessee River was the snail darter's only known spawning habitat when the species was listed as endangered. Although no populations now exist in the Little Tennessee River, the proposed and subsequent construction of Tellico Dam sparked reintroduction efforts and population surveys. New populations were either discovered or started in the main stem of the

Tennessee River and in six of its tributaries. Etnier found some darters living in the South Chickamauga Creek in November 1980, and subsequently discovered a single fish in the lower Sequatchie River, causing the Tennessee Valley Authority and the U.S. Fish and Wildlife Service to start studies in 1981 to characterize the species' habitat range. These investigations identified a population of snail darters in Sewee Creek, along with some darters in the Sequatchie and Paint Rock Rivers. As a consequence of these and other discoveries, members of the Snail Darter Recovery Team met with Fish and Wildlife Service biologists to recommend changes. As a result the Service decided:

1. To downlist the species from endangered to threatened
2. To keep the species on the federal list
3. To retain requirements for a federal permit to collect snail darters if downlisting to threatened occurs

A number of the recovery team members voted not to delist the species because the viability of its populations is still unknown. A substantial reestablishment effort was put into effect to move the threatened fish population to amenable habitats. The transplanting is still ongoing. In fact, some scientists believe it is but a matter of time before *Percina tanasi* is removed entirely from any ESA list.

The case is yet another lesson that neither politics nor science always has complete primacy in environmental decision making. Environmental professionals must be prepared to answer questions like "Is such a small fish worth all this effort, time, and expense?"

Hint #1: Sometimes it is the very small things that are among the most precious. Just ask researchers delving into nanotechnology!
Hint #2: Ask a coal miner if they would prefer a 100-kg methane-tolerant canary!

Seveso Plant Disaster

This case could appear in Chapter 3 as an example of a toxic cloud. It might also be included in Chapter 5 as a harbinger of the dioxin scares that paved the way for Superfund and hazardous waste legislation throughout the world. It could be another sword of Damocles in Chapter 6, emblematic of potential exposures to toxic substances to those millions of people living near industrial facilities. But it appears here because the disaster represents so much more than the event itself in terms of why it occurred, what were the real exposures, and consequently how risks from such incidents can be better managed in the future.

On July 10, 1976, an explosion at a 2,4,5-trichlorophenol (2,4,5-T) reactor at a manufacturing plant near Seveso, Italy, resulted in the highest concentrations of the most toxic chlorinated dioxin (TCDD) levels known in human residential populations[16] (see Chapter 5 for an explanation of dioxins and TCDD). Up to 30 kg of TCDD were deposited over the surrounding area of approximately 18 km^2. That is kg, as in kilograms! We usually are worried about dioxin, and especially TCDD, if a gram or less is released in a year. This release of 30,000 grams occurred instantaneously!

The plant was operated by Industrie Chemiche Meda Societa, Anonima (ICMSA), an Italian subsidiary owned by the Swiss company, Givaudan, which was owned by another Swiss company, Hoffman-LaRoche.[17] The release resulted from a ruptured disc on a batch plant. The 2.4,5-T was formulated from 1,2,4,5-tetrachlorobenzene and caustic soda (NaOH) in the presence of ethylene glycol. Normally, only a few molecules of dioxins form from several trillion molecules of the batch, but in the events leading up to the explosion, the reactor temperatures rose beyond the tolerances, leading to a "runaway reaction," bursting the disc. A large volume of hydrogen gas (H_2) was formed, which propelled the six tons of fluid in the reactor into the Seveso region. Actually, the discharge could have been worse, since upon hearing the noise being made at the beginning of the release, a plant foreman opened a valve to release cooling waters onto the heating coils of the reactor. This action likely reduced the propulsion.

The explosion is, in part, an example of doing the wrong thing for the right reason. The Italian legislature had passed a law requiring the plant to shut down on weekends, whether or not a batch was complete. On the weekend of the explosion, the plant was shut down after the formulation reaction, but before the complete removal of the ethylene glycol by distillation. This was the first time that the plant was shut down at this stage. Based upon chemistry alone, the operator had no reason to fear anything would go awry. The mixture at the time of shutdown was 158°C, but the theoretical temperature at which an exothermic reaction would be expected to occur was thought to be 230°C. Subsequent studies have found that exothermic reactions of these reagents can begin at 180°C, but these are very slow processes when below 230°C. The temperature in fact rose, mainly due to a temperature gradient from the liquid to the steam phases. The reactor wall in contact with the liquid was much cooler than the wall in contact with the steam, with heat moving from the upper wall to the surface of the liquid. The stirrer was switched off so that the top few centimeters of liquid rose in temperature to about 190°C, beginning the slow exothermic reaction. In seven hours the runaway reaction commenced. The reactions may also have been catalyzed by chemicals in residue that had caked on the upper wall that, with the temperature increase, was released into the liquid in the reactor. Thus, the runaway reaction and explosion could have been prevented if the plant had not had to be prematurely shut down and if the plant operators had been better trained to consider contingencies beyond the textbook conditions.

A continuous, slight wind dispersed the contents over the region that included 11 towns and villages. The precipitate looked like snow. No official emergency action took place the day of the explosion. People with chemical burns checked themselves into local hospitals. The mayor of Seveso was only told of the incident the next day.

The response to the disaster and pending dioxin exposure was based on the perceived potential risks involved. The contaminated area was divided into three zones based on the concentration of TCDD in the soil. The 211 families in Zone A, the most heavily contaminated area, were evacuated within 20 days of the explosion and measures were taken to minimize exposure to residents in nearby zones. In a preliminary study, the U.S. Center for Disease Control and Prevention (CDC) tested blood serum samples from five residents of Zone A who suffered from the dioxin-related skin disease known as chloracne, as well as samples from four from Zone A without chloracne, and three from outside the contaminated area. All samples had been collected and stored shortly after the accident. In Zone A, serum TCDD levels ranged from 1,772 to 10,439 parts per trillion (ppt) for persons without chloracne and from 828 to 56,000 ppt for persons with chloracne. The TCDD concentrations are the highest ever reported in humans. Interestingly, TCDD was detected in only one sample from the unexposed group, but it was comparably high, 137 ppt. The elevated TCDD level was thought to be due to misclassification or sample contamination. The CDC presently is evaluating several hundred historical blood samples taken from Seveso residents for TCDD. There are plans to determine half-life estimates and to evaluate serum TCDD levels for participants in the Seveso cancer registry.

As in the case of the snail darter controversy, scientific research has continued even after the sensational aspects of the disaster have waned. In fact, a recent study by Warner, et al.[18] found a statistically significant, dose-response-increased risk for breast cancer incidence with individual serum TCDD level among women in the Seveso population; that is, a dose-response relationship of a two-fold increase in the hazard rate associated with a 10-fold increase in serum TCDD. This result is an early warning because the Seveso cohort is relatively young, with a mean age at interview of less than 41 years. These findings are consistent with a 20-year follow-up study that showed that even though no increased incidence of breast cancer had been observed in the Seveso population 10 and 15 years after the incident, after 20 years, breast cancer mortality emerged among women who resided in heavily contaminated areas, and who were younger than 55 years at death [relative risk (RR) = 1.2 with a 95% confidence interval (CI) = 0.6–2.2], but not in those who were older.

The findings are not significant statistically, but they are scary. For example, it is very difficult to characterize a population for even a year, let alone 20 years, because people move in and out of the area. Epidemiological studies are often limited by the size of the study group in comparison to the whole population of those exposed to a contaminant. In the Seveso case, the TCDD exposure estimates have been based on zone of residence, so the study is deficient in individual-level exposure data. There is also variability within the exposure area; for example, recent analyses of individual blood serum TCDD measurements for 601 Seveso women suggest a wide range of individual TCDD exposure within zones.[19]

Some argue that the public's response to the real risks from Seveso were overblown, considering no one died directly from the explosion and the immediate aftermath, although hundreds were burned by the NaOH exposure and hundreds others did develop skin irritations. This may be true, but so much is unknown, especially when the chemical of concern is TCDD, which has been established to cause cancer, neurological disorder, endocrine dysfunction and other chronic diseases. The medical follow-up studies may indicate that the public response was plausible, given the latent connections to chronic diseases like breast cancer.

The toxicity, especially the very steep cancer slope for TCDD, is met with some skepticism. Exposures at Seveso were high compared to general environmental exposures. However, some high-profile cases, especially the recent suspected TCDD poisoning of the Ukrainian President Viktor Yushchenko in 2004 and the 1998 discovery of the highest blood levels of TCDD in two Austrians, have added to the controversy. In fact, both of these cases were precipitated by the adverse effects, especially chloracne. Usually, dioxins, PCBs, and other persistent, bioaccumulating toxicants are studied based on exposure, followed by investigations to see if any effects are present. This has been the case for Agent Orange, Seveso, Times Beach, and most other contamination instances. It is unusual to see such severe effects, but with the very high doses (see Figure 10.8) reported in the Yushchenko and Austrian cases, it would have come as a greater surprise not to have seen some adverse effect.

Poverty and Pollution

There is little debate on whether economics and environmental quality are related. Most agree that they are. The disagreement comes in two forms. The most common is the degree, extent, and type of relationships between economic development and environmental condition. Often, the perception is that greater development threatens environmental quality; more cars, more consumerism, more waste, and more releases of toxicants. In his 2003 book, *The Real Environmental Crisis* (University of California Press, Berkeley, CA), J.M. Hollander takes an opposite view:

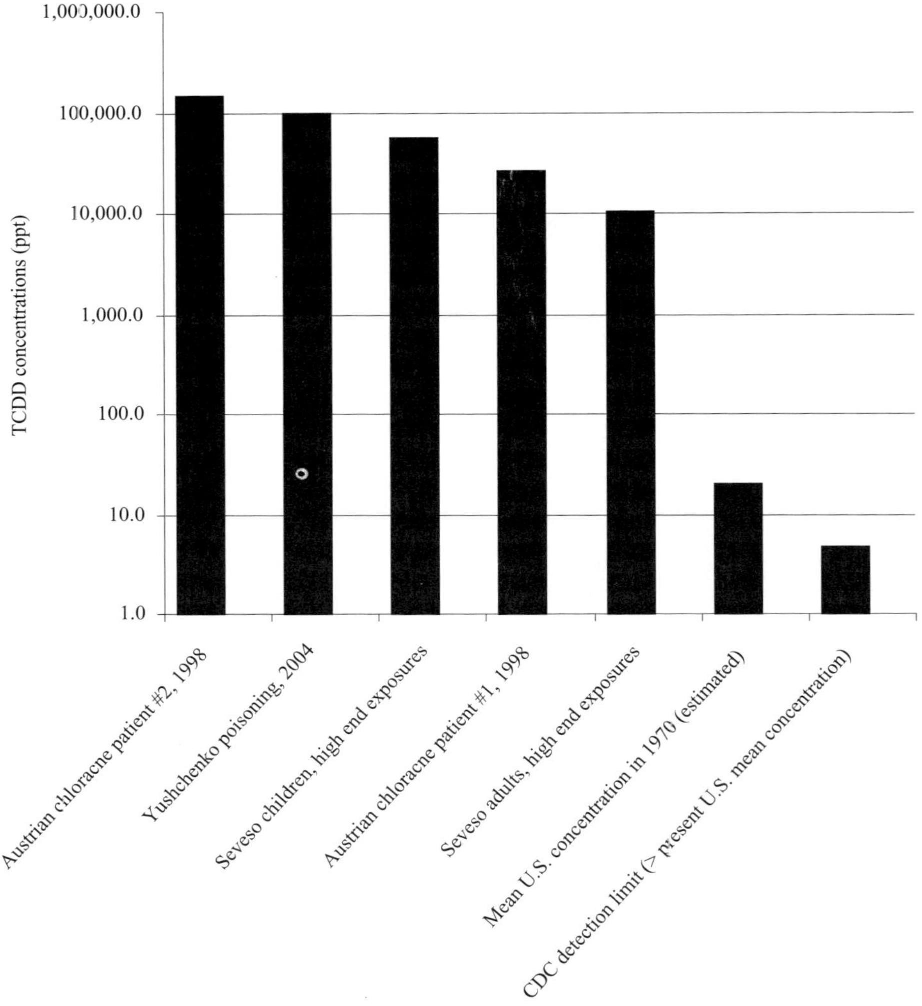

FIGURE 10.8. Tetrachlorodibenzo-*para*-dioxin blood serum concentrations (log scale) from various exposure scenarios.
Sources: Center for Disease Control and Prevention; and Dioxin facts.org: http://www.dioxinfacts.org/dioxin_health/dioxin_tissues/yushchenko.html; accessed April 24, 2005.

People living in poverty perceive the environment very differently from the affluent. To the world's poor—several billion people—the principal environmental problems are local, not global. They are not the stuff of media headlines or complicated scientific theories. They are mundane, pervasive and painfully obvious:

- Hunger—chronic undernourishment of a billion children and adults caused not only by scarcity of food resources but by poverty, war, and government tyranny and incompetence.
- Contaminated Water Supplies—a major cause of chronic disease and mortality in the third world.
- Diseases—rampant in the poorest countries. Most could be readily eradicated by modern medicine, while others, including the AIDS epidemic in Africa, could be mitigated by effective public health programs and drug treatments available to the affluent.
- Scarcity—insufficient local supplies of fuel, wood, and other resources, owing not to intrinsic scarcity but to generations of overexploitation and underreplenishment as part of the constant struggle for survival.
- Lack of Education and Social Inequality, Especially of Women—lack of education associated with high birthrates and increasing the difficulty for families to escape from the dungeons of poverty.[20]

Hollander argues that environmental quality and a healthy economy are highly compatible. His evidence is that most Western countries that developed economically coincidentally improved in most environmental aspects. A dramatic example is that of the island of Hispaniola. On the Dominican Republic side, there is much lush vegetation. But, on the Haiti side, the land is denuded. The Dominican Republic has embraced a more open, capitalistic marketplace, and Haiti has suffered the ravages of totalitarian regimes.

Is this anecdotal or can we expect that reducing poverty by enhancing economic development will always be associated with a cleaner environment? Perhaps the answer is not dichotomous. Perhaps, like so many issues in engineering and the environment, the answer is that there is a somewhat fluid range in which to optimize both conditions, economic well-being and environmental protection. Operating outside of this range—going too far toward commercialism on one end and stifling freedom on the other—may pit one against the other.

The cases in this chapter leave us with an uneasy feeling. But, they provide us with examples of the complexity of environmental problems. Perhaps, the predominant lesson is that complex environmental problems really are not solved, only managed.

Notes and Commentary

1. The source of background information in this section is U.S. Environmental Protection Agency, 2005. Municipal Solid Waste, http://www.epa.gov/epaoswer/non-hw/muncpl/facts.htm; accessed April 5, 2005.

2. Dy-Dee Diaper Service, 2005. http://www.dy-dee.com/; accessed April 22, 2005.
3. Although kinetics in the physical sense arguably can be shown to share many common attributes with kinetics in the chemical sense, for the purposes of this discussion, it is probably best to treat them as two separate entities. Physical kinetics, as discussed in Chapter 2, is concerned with the dynamics of material bodies and the energy in a body owing to its motions. Chemical kinetics address rates of chemical reactions. The former is more concerned with mechanical dynamics, the latter with thermodynamics.
4. S.Y. Szeto, N.E. Burlinson, J.E. Rahe, and P.C. Oloffs, 1989. "Persistence of the Fungicide Vinclozolin on Pea Leaves under Laboratory Conditions," *Journal of Agricultural Food Chemistry*, 37, 529–534.
5. I always find it interesting that reputable scientists can make credible arguments about an issue that falls within their areas of expertise, but those are largely ignored by the public. However, when a popular entertainer makes a statement, the public (at least a large part of it) considers this to be reliable. Go figure!
6. The FQPA was enacted on August 3, 1996, to amend the Federal Insecticide, Fungicide, and Rodenticide Act (FIFRA) and the Federal Food, Drug, and Cosmetics Act (FFDCA). Especially important to risk assessment, the FQPA established a health-based standard to provide a reasonable certainty of no harm for pesticide residues in foods. This new provision was enacted to assure protection from unacceptable pesticide exposure and to strengthen the health protection measures for infants and children from pesticide risks.
7. A very interesting development over the past decade has been the increasing awareness that health research has often ignored a number of polymorphs or subpopulations, such as women and children, and is plagued by the so-called "healthy worker" effect. Much occupational epidemiology has been based upon a tightly defined population of relatively young and healthy adult white males who had already been screened and selected by management and economic systems in place during the twentieth century. Moreover, health studies have tended to be biased toward adult white males even when the contaminant or disease of concern was distributed unevenly throughout the general U.S. population. For example, much of the cardiac and cancer risk factors for women and children have been extrapolated from studies of adult, white males. Pharmaceutical efficacy studies had also been targeted more frequently toward adult males. This has been changing recently, but the residual uncertainties are still problematic.
8. Agent Orange Web site. http://www.lewispublishing.com/orange.htm; accessed April 22, 2005.
9. 16 *U.S.C.* 1531–1543; P.L. 93–205, as amended.
10. The principal source for background information regarding the Endangered Species Act is Congressional Research Service, 2003. E.H. Buck, M.L. Corn, and P. Baldwin, Report RL31654, *Endangered Species: Difficult Choices*, Government Printing Office, Washington, D.C.

11. Congressional Research Service, 1993. R. Meltz, Report RL31654, *The Endangered Species Act: A Primer*, Government Printing Office, Washington, D.C.
12. For protection and recovery descriptions, see the U.S. Fish and Wildlife Service (FWS) at http://endangered.fws.gov/ and the National Marine Fisheries Service (NMFS, which recently changed its name to NOAA Fisheries) at http://www.nmfs.noaa.gov/endangered.htm.
13. The analogy of a canary in the coal mine represents an extrapolation of a real practice used by underground coal miners. In addition to the constant fear of cave-ins and explosions from coal dust, miners were threatened with the release of the highly toxic mine gas. The gas is actually methane, CH_4, which in the enclosed mine shaft conditions can combust when its concentrations reach 5% of the air or, if coal dust is also present, only 2%, so a single spark could ignite the shaft. The gas is also toxic when inhaled at a sufficiently high concentration, as is another gas that sometimes forms in mines, carbon monoxide (CO). A toxic response is a function of concentration, time, and body mass. Thus the smaller bird would exhibit symptoms, in this case, asphyxiation and death, hopefully before the larger miners did. The other advantage of the canary is that it is a loud and persistent singer. When low doses of CH_4 or CO entered the areas where miners were working, the miners would notice the bird would stop singing. And, before its collapse, the bird would begin to sway on its perch. This is a classic example of dose-response, that is, more intense and severe symptoms with increasing dose. The canary is an excellent example of a low-tech solution.

    The analogy for ecosystems is that losing a sensitive species indicates that with time larger problems loom. The time to change what we are doing is before the big problems begin. In other words, pay attention to the canary and get out of the coal mine now!
14. The sources of background information regarding the snail darter are D.A. Etnier, 1976. "A New *Percid* Fish From The Little Tennessee River, Tennessee," *Proceedings of the Biological Society*, 88:469–488. G.D. Hickman, and R.B. Fitz, 1978. A Report on the Ecology and Conservation of the Snail Darter (*Percina tanasi* Etnier) 1975–1977, Tennessee Valley Authority, Norris, Tennessee. W.C. Starnes, 1977. "The Ecology and Life History of the Endangered Snail Darter, *Percina* (*Imostoma*) *tanasi* Etnier)," Ph.D. dissertation, University of Tennessee, Knoxville, TN.
15. R.Nash, 1989. *The Rights of Nature: A History of Environmental Ethics*, The University of Wisconsin Press, Madions, WI, p. 125.
16. P. Mocarelli and F. Pocchiari, 1988. Preliminary report: 2,3,7,8-tetrachlorodibenzo-*p*-dioxin exposure to humans—Seveso, Italy. *Morbidity and Mortality Weekly Report*, 37:733–736. A. Di Domenico, V. Silano, G. Viviano, and G. Zappni, 1980. Accidental release of 2,3,7,8-tetrachlorodibenzo-*p*-dioxin (TCDD) at Seveso, Italy: V. Environmental persistence of TCDD in soil, *Ecotoxicology and Environmental Safety*, 4:339–345.

West Virginian coal miner holding a canary, *circa* 1900.
Photo credit: West Virginia Office of Miners' Health, Safety and Training.

17. The principal source for the engineering aspects of the industrial accident is T. Kletz, 2001. *Learning from Accidents*, 3e, Gulf Professional Publishing, Oxford, UK.
18. M. Warner, B. Eskenazi, P. Mocarelli, P.M. Gerthoux, S. Samuels, L. Needham, D. Patterson, and P. Brambilla, 2002. "Serum Dioxin Concentrations and Breast Cancer Risk in the Seveso Women's Health Study," *Environmental Health Perspectives*, 110 (7): 625–628.
19. *Ibid.*
20. J.M. Hollander, 2003. *The Real Environmental Crisis*, University of California Press, Berkeley, CA.

CHAPTER 11

Just Environmental Decisions, Please

No state shall . . . deny to any person . . . equal protection of the laws.
The Constitution of the United States of America

. . . with liberty and justice for all.
The Pledge of Allegiance

The fourteenth amendment to the U.S. Constitution states:

> All persons born or naturalized in the United States, and subject to the jurisdiction thereof, are citizens of the United States and of the state wherein they reside. No state shall make or enforce any law which shall abridge the privileges or immunities of citizens of the United States; nor shall any state deprive any person of life, liberty, or property, without due process of law; nor deny to any person within its jurisdiction the equal protection of the laws.

The equal protection clause and the conclusion of the *Pledge of Allegiance* is a fitting place to start a discussion of environmental justice. The *Pledge* underlines the principle of fairness articulated throughout the Constitution of the United States. In a way, every case in this book represents a form of injustice. People should not be exposed to harmful pollutants and those who are exposed are victims of wrongdoing, whether intentional or not. Destruction of the earth's ecological resources is actually denying future generations the benefits that previous people have enjoyed.

> *Man's capacity for justice makes democracy possible, but man's inclination to injustice makes democracy necessary.*
> Reinhold Neibuhr[1]

The reality articulated by Neibuhr indicates the complexities of the human condition in providing a public benefit, such as an environment that supports public health and ecosystems. The Declaration of Independence's unalienable rights of life, liberty, and the pursuit of happiness depend upon a livable environment. The government plays a central role of overcoming the forces that will militate against equity in environmental protection. Democracy and freedom are at the core of achieving fairness and Americans rightfully take great pride in these foundations of our Republic. The framers of our constitution wanted to make sure that life, liberty, and the pursuit of happiness were available to all; first with the protection of property rights and later, with the Bill of Rights, by granting human and civil rights to all the people.

The 1960s saw several social upheavals, not the least of which were those advocating for environmental protection and civil rights. Ironically, given the commonalities of the two movements, the environmental movement has been slow to recognize fully its role in justice. The movement has incrementally grown (see Table 11.1). This recognition is evidenced by the evolution of names given to the denial of justice in environmental decisions and projects. At first, groups expressed concerns about *environmental racism*, followed by the recognition for the need for greater *environmental equity*. These transitional designations reflect more than simple changes in jargon. When attention began to be paid to the particular incidents of racism, the focus was logically placed on eradicating the menace at hand—blatant acts of willful racism. This was a necessary, but not completely sufficient, component in addressing the environmental problems of minority communities and economically disadvantaged neighborhoods, so the concept of equity was employed more assertively. Equity implies the need not only to eliminate the overt problems associated with racism, but to initiate positive change to achieve more evenly distributed environmental protection. What has become evident only in the past few decades is that without a clean environment, life is threatened by toxic substances, liberty is threatened by the loss of resources, and happiness is less likely in an unhealthful and unappealing place to live.

The only way to protect public health and the environment is to ensure that *all* persons are adequately protected. In the words of Reverend Martin Luther King, "Injustice anywhere is a threat to justice everywhere."[2] By extension, if any group is disparately exposed to an unhealthy environment, then the whole nation is subjected to inequity and injustice. Put in a more positive way, we can work to provide a safe and livable environment by including everyone, leaving no one behind. This mandate has a name. It is called *environmental justice,* and equal protection can be extended intellectually (if not legally) to matters of public health and environmental quality.

Environmental Justice

The term *environmental justice* is presently the term of choice when addressing fairness in environmental protection. The term usually is

TABLE 11.1
Milestones in the environmental justice movement.

1971	Council on Environmental Quality annual report notes inequities in distribution of environmental hazards.
1982	500 citizens are arrested for demonstrating in opposition to the siting of a PCB disposal landfill in the predominantly black and poor Warren County, North Carolina.
1983	The General Accounting Office investigates the relationship between race and the siting of four commercial hazardous waste landfills in the Southeast. At three of the four landfills, African Americans made up the majority of the population living nearby. At least 26% of the population in all four communities was below the poverty level.
1987	United Church of Christ's Toxic Wastes and Race in the United States is released and concludes that race, not income, was the factor more strongly correlated to residence near a hazardous waste site. It found that the proportion of minorities in communities with a hazardous waste facility is nearly double that of communities without one. Where two or more such facilities are located, the proportion of minorities is more than triple.
1990	Michigan Conference on Race and the Incidence of Environmental Hazards is held, bringing together academics, activists, and policymakers around the issue of environmental justice.
1992	The Environmental Protection Agency's (EPA's) Environmental Equity Working Group, established in 1990, releases Environmental Equity: Reducing Risk for All Communities, which concludes that racial minorities and low-income people are disproportionately exposed to lead, selected air pollutants, hazardous waste facilities, contaminated fish, and agricultural pesticides in the workplace.
	EPA establishes the Office of Environmental Equity (renamed the Office of Environmental Justice in 1994).
1994	President Clinton issues Executive Order 12898, Federal Actions to Address Environmental Justice in Minority Populations and Low-Income Populations, requiring federal agencies to develop a comprehensive strategy for making environmental justice part of their daily operations.
	The Interagency Working Group on Environmental Justice is established, chaired by EPA Administrator Carol M. Browner and comprised of the heads of 11 agencies and several White House offices.
	Update of United Church of Christ report finds that minority populations in 1993 are more likely to live in ZIP codes where hazardous waste facilities are located than they were in 1980; race/ethnicity remains a stronger indicator of proximity to a facility than income.

Source: Council on Environmental Quality. *Environmental Quality 1994–1995*, Washington, D.C.

applied to social issues, especially as they relate to neighborhoods and communities. For example, the so-called environmental justice (EJ) communities possess two basic characteristics:

1. They have experienced historical (usually multigenerational) exposures to disproportionately high doses of potentially harmful substances (the *environmental* part). Exposure is preferred to *risk*, since risk is a function of the hazard and the exposure to that hazard. Even a substance with a very high toxicity (one type of hazard) that is confined to a laboratory of a manufacturing operation may not pose much of a risk due to the potentially low levels of exposure (see Chapter 2).
2. They have certain, specified socioeconomic and demographic characteristics. EJ communities must have a majority representation of low socioeconomic status (SES) and racial, ethnic, and historically disadvantaged people (the *justice* part).

The determination of disproportionate impacts (i.e., diseases and other health endpoints) is fundamental to environmental justice. Epidemiologists look at clusters and other indications of elevated exposures and effects in populations. Certain cancers, neurological and other chronic diseases have been found to be significantly higher in minority communities and in socioeconomically depressed areas. Acute diseases, as indicated by hospital admissions, may also be higher in certain segments of society, such as pesticide poisoning in migrant workers. In addition, each person responds to an environmental insult uniquely, and that person is affected differently at various life stages. For example, young children are at higher risk to neurotoxins. However, subpopulations also can respond differently than the whole population, meaning that genetic differences seem to affect people's susceptibility to contaminant exposure. Scientists are very interested in genetic variation, so that genomic techniques[3] (e.g., identifying certain polymorphisms) are a growing area of inquiry.[4]

In a sense, historical characteristics constitute the environmental aspects of EJ communities, and socioeconomic characteristics entail the justice considerations. The two sets of criteria are mutually inclusive, so for a community to be defined as an EJ community, both of these sets of criteria must be present.

A recent report by the Institute of Medicine[5] found that numerous EJ communities experience a "certain type of double jeopardy." The communities must endure elevated levels of exposure to contaminants, while being ill equipped to deal with these exposures because so little is known about the exposure scenarios in EJ communities. The latter problem is exacerbated by the disenfranchisement from the political process that is endemic to EJ community members. The report also found large variability among communities as to the type and amount of exposure to toxic substances.

This pointed to the need for improved approaches for characterizing human exposures to toxicants in EJ communities.

Old Paradigm:
  Technical professionals must participate only in science-based decisions.
Paradigm Shift:
  Managing the environmental risks of complex situations requires that multiple perspectives be considered, but always underpinned by sound science.

At first blush, this appears to be inconsistent with the warnings throughout this book regarding advocacy. The real problem with advocacy science arises when sound science is replaced by "junk science." At the other end of the spectrum, however, is "culturally ignorant science." Throughout most of the environmental movement, scientists have had to incorporate sound science into decisions. Few, if any, decisions can be made exclusively from physical scientific principles, no matter how much we engineers wish it to be. To visualize this complexity, let us consider a force field analogous to decision making, where a magnet is placed in each sector at a point equidistant from the center of the diagram (see Figure 11.1). If we assume that the decision being considered will be pulled in the direction of

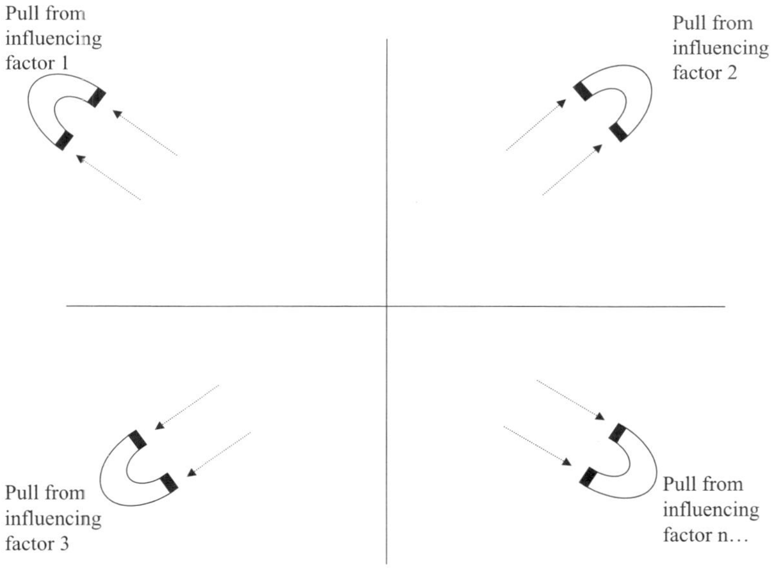

FIGURE 11.1. Decision force field.

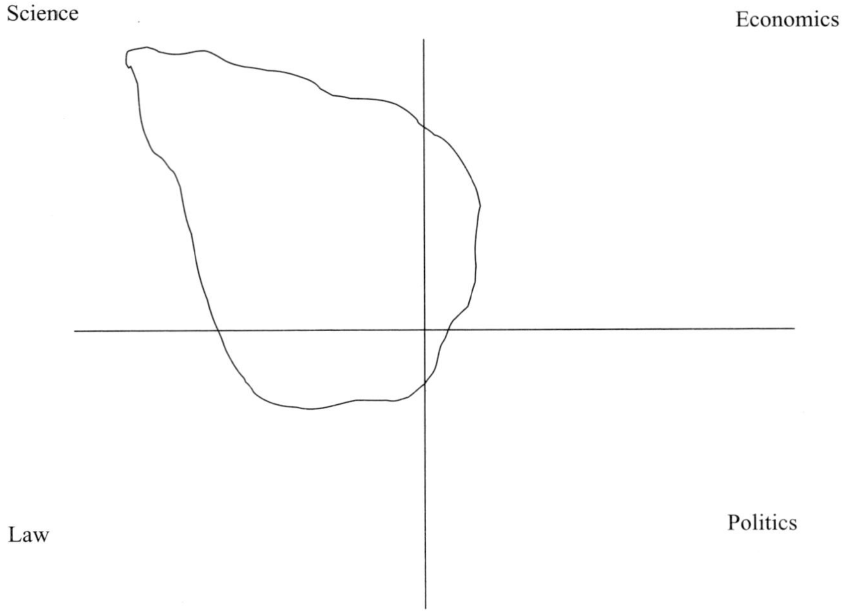

FIGURE 11.2. Decision force field for a scientifically based decsison.

the strongest magnetic force, the stronger the magnet the more the decision that actually will be made will be pulled in that direction.

A decision that is almost entirely influenced by strong science will appear something like the force field in Figure 11.2. However, if other factors exist, such as a plant closing or the possibility of new jobs being attracted at the expense of environmental, geological, or other scientific influences, the decision would migrate toward these stronger influences, as shown in Figure 11.3.

Environmental decision making is actually an exercise in environmental ethics. Engineers by their very nature are called to be risk managers. Risk management is the process of selecting ways to respond to risk. The risk management process is informed by the quantitative results of the risk assessment process. Health and environmental risk characterizations are based upon the toxicology, exposure, and effects results derived from the methods and approaches discussed in Chapters 1 and 2. These characterizations are essential, but wholly insufficient to manage risks. In addition to quantitative risk assessment findings, managing risks must consider social, economic, legal, and political realities (see Figure 11.4), which can include some quantitative information, but are often semiquantitative and even subjective. Engineers are usually more comfortable dealing with quantitative approaches, but it is imperative that we compel ourselves to con-

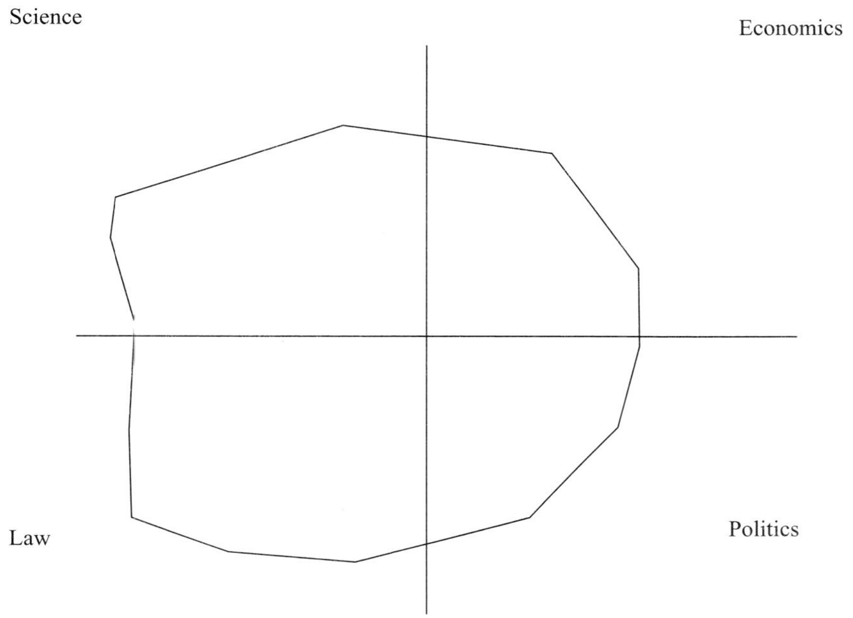

FIGURE 11.3. Decision force field for complex decision with multiple influences.

sider these other aspects of every engineering response to risks. The combination of quantitative, semiquantitative, and qualitative information is particularly crucial for managing risks in environmental justice situations, given the complex socioeconomic factors that can determine an engineering project's success or failure.

How Can Engineers Best Manage Risks in a Changing Environment?

When we ask how engineers can best manage risks, we are really asking engineers to do what they do best: anticipating, preventing, and solving problems. Engineering interventions to prevent and solve environmental problems and to address public health challenges have been approached in various ways.

The engineering community has come to recognize that our span of control has changed. For example, when considering the consequences of building collapse on environmental quality prior to 9/11, we almost exclusively considered it to be under the aegis of programs designed to protect people from contact with materials during and following building demolition. For example, Section 112 of the Clean Air Act envisioned the need to

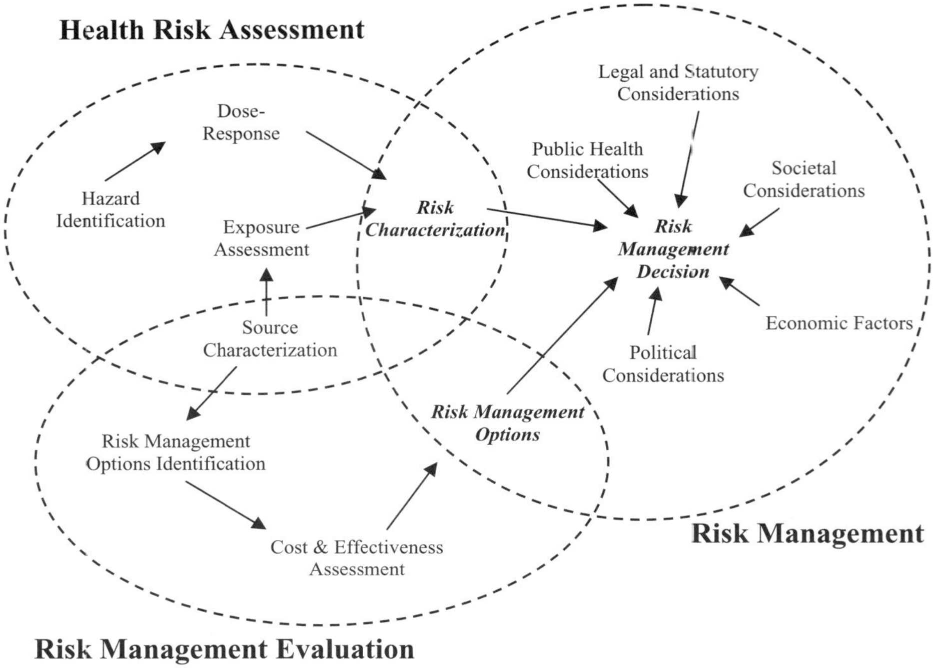

FIGURE 11.4. Risk assessment and risk management processes.
Source: Adapted from the National Research Council, 1983. *Risk Assessment in the Federal Government: Managing the Process*; National Academy Press, Washington, D.C. National Research Council, 1994. *Science and Judgment in Risk Assessment*, National Academy Press, Washington, D.C.

consider fugitive dust emissions during construction and deconstruction of structures as a mechanism of exposing people to toxic substances, especially asbestos and heavy metals.[6] States are often tasked with oversight of demolition projects. For example, in Wisconsin, disturbance of asbestos is regulated in part by Chapter NR 447 of the Wisconsin Administrative Code. The code states that before initiating any demolition or renovation project, the owner-operator of a structure must have the structure inspected for the presence of asbestos by an asbestos inspector licensed by the state's Department of Health and Family Services (DHFS). The inspection must identify the types of asbestos present in the structure, because different forms of asbestos have different toxicities and vary in their likelihood to be transported. As a result, some categories of asbestos may be allowed to remain on-site, but others must be removed before the demolition or renovation begins.[7] Other states have similar programs.

In its normal use, deconstruction describes a well-established process; that is, dismantling or removing materials from structures prior to, or instead of, demolition. When heavy equipment demolishes the structure, the building materials are aggregated into a mixture, so deconstruction is a way to separate out hazardous materials or to improve the likelihood that valuable materials will be recycled.[8] Recently, however, deconstruction has been used to mean intentional demolition, at least where potential exposures to asbestos are possible. Standard engineering practice and risk management have been reconsidered in light of terrorism. Following the terrorist attacks on the World Trade Center in New York and the Pentagon in Washington, new guidance has been formulated to address the increased likelihood that people can be exposed to these pollutants in scenarios other than the highly regulated and engineering controlled conditions of a planned demolition.

Old Paradigm:
Little can be known until after a substance is released into the environment.
Paradigm Shift:
New analytical and computational tools are available to model risks before products are released.

In addition to environmental science and engineering, other disciplines are key players in risk management. The chemical engineer must be cognizant of the risks associated with the synthesis of certain chemicals in reactors, and even the use of those chemicals following synthesis. Researchers are developing new tools, for example, computational toxicology and quantitative structure activity relationships (QSARs), to model the possible behavior of contaminants even before the molecules are synthesized.[9]

Optimization in Environmental Risk Management

By three methods we may learn wisdom: First, by reflection, which is noblest; Second, by imitation, which is easiest; and third by experience, which is the bitterest.

Confucius (circa 551–479 B.C.)

Engineers spend most of the preparation for the profession learning the scientific principles and applying them to problems; what Confucius might have called "reflection." Next, we observe and apply lessons from our mentors. And we hope not to experience the bitterness of direct failure by adopting practices that have worked for others in the past.

Environmental and health risk management takes the engineer into uncomfortable venues. The frustration for engineers lies in the fact that there is seldom a simple answer to the questions "How healthy is healthy enough?" and "How protected is protected enough?" These questions are actually probing into what is an acceptable risk. Managing risks consists of balancing among alternatives. Usually, no single solution to an environmental problem is available. Whether a risk is acceptable is determined by a process of making decisions and implementing actions that flow from these decisions to reduce the adverse outcomes or at least to lower the chance that negative consequences will occur.[10]

Risk managers can expect that whatever risk remains after their project is implemented, those potentially affected will not necessarily be satisfied with that risk. It is difficult to think of any situation where anyone would prefer a project with more risk than one with less, all other things being equal. It has been said that "acceptable risk is the risk associated with the best of the available alternatives, not with the best of the alternatives which we would hope to have available."[11]

Since risk involves chance, risk calculations are inherently constrained by three conditions:

1. The actual values of all important variables cannot be known completely and thus cannot be projected into the future with complete certainty.
2. The physical science of the processes leading to the risk can never be fully understood, so the physical, chemical, and biological algorithms written into predictive models will propagate errors in the model.
3. Risk prediction using models depend on probabilistic and highly complex processes that make it infeasible to predict many outcomes.[12]

The "go or no go" decision for most engineering designs or projects is based upon some sort of risk-reward paradigm, and should be a balance between benefits and costs.[13] This creates the need to have costs and risks significantly outweighed by some societal good. The adverb "significantly" reflects two problems: the uncertainty resulting from the three constraints described earlier and the margin between good *versus* bad. Significance is the province of statistics; it tells us just how certain we are that the relationship between variables cannot be attributed to chance. But when comparing benefits to costs, we are not all that sure that any value we calculate is accurate. For example, a benefit/cost ratio of 1.3 with confidence levels that give a range between 1.1 and 1.5 is very different from a benefit/cost ratio of 1.3 with a confidence range between 0.9 and 1.7. The former does not include any values below 1, but the latter does (i.e., 0.9). This value means that with all the uncertainties, our calculation shows that the project could be unacceptable (i.e., more costs than benefits). This situation is compounded by the second problem of not knowing the proper margin of safety. That is, we do not know the overall factor of safety to ensure that the decision is prudent. Even a benefit/cost ratio that appears to be mathematically high, that is, well above 1, may not provide an ample margin of safety given the risks involved.

The likelihood of unacceptable consequences can result from exposure processes, from effects processes, or from both processes acting together. So, four possible permutations can exist:

1. Probabilistic exposure with a subsequent probabilistic effect.
2. Deterministic exposure with a subsequent probabilistic effect.
3. Probabilistic exposure with a subsequent deterministic effect.
4. Deterministic exposure with a subsequent deterministic effect.[14]

A risk outcome is deterministic if the output is uniquely determined by the input. A risk outcome is probabilistic if they are generated by a statistical method, for example, randomly. Thus, the accuracy of a deterministic model depends on choosing the correct conditions, those that will actually exist during a project's life and correctly applying the principles of physics, chemistry, and biology. The accuracy of the probabilistic model depends on choosing the right statistical tools and correctly characterizing the outcomes in terms of how closely the subpopulation being studied (e.g., a community or an ecosystem) resembles those of the population (e.g., do they have the same factors or will there be sufficient confounders to make any statistical inference incorrect?). A way of looking at the difference is that deterministic conditions depend on how well we understand the science underpinning the system, whereas probabilistic conditions depend on how well we understand the chance of various outcomes (see Table 11.2).

Actually, the deterministic exposure/deterministic effect scenario is not really a risk scenario because there is no chance involved. It would be like saying that releasing a 50-kg steel anvil from 1 meter above the earth's surface runs the risk of falling toward the ground! The risk comes into play only when we must determine external consequences of the anvil falling. For example, if an anvil is suspended at the height of one meter by steel wire and used by workers to perform some task (i.e., a deterministic exposure) there is some probability that it may fall (e.g., studies have shown that the wires fail to hold one in 10,000 events; i.e., a failure probability of 0.0001), so this would be an example of a deterministic exposure followed by a probabilistic effect (wire failure), permutation number 2.

Estimating risk using a deterministic approach requires the application of various scenarios, for example, a very likely scenario, an average scenario, or a worst-case scenario. Very likely scenarios are valuable in some situations when the outcome is not life threatening or not one of severe effects, like cancer. For example, retailers may not care so much about the worst case (a customer is not likely under most scenarios to buy their product) in designing a store, but wants to avoid the risk of losing a likely customer (e.g., most potential Armani suit customers are not likely to want to hear loud heavy metal music when they walk in the store; but conversely loud heavy metal music may be a selling feature for most customers looking to buy tube amplifiers at a guitar store). The debate in the public health

TABLE 11.2
Exposure and effect process risk management.

	Probabilistic Exposure	*Deterministic Exposure*
Probabilistic Effect	Contracting the West Nile Virus. Although many people are bitten by mosquitoes, most mosquitoes do not carry the West Nile virus. There is a probability that a person will be bitten and another, much lower probability that the bite will transmit the virus. A third probability of this bitten group may be rather high that once bitten by the West Nile virus-bearing mosquito that the bite will lead to the actual disease. Another conditional probability exists that a person will die from the disease. So, death from a mosquito bite (probabilistic exposure) leads to a very unlikely death (probabilistic effect).	Occupational exposure to asbestos. Exposure to asbestos in vermiculite workers is deterministic because the worker chooses to work at a plant that processes asbestos-containing substances. This is not the same as the person choosing to be exposed to asbestos, only that the exposure results from an identifiable activity. The potential health effects from the exposures are probabilistic, ranging from no effect to death from lung cancer and mesothelioma. These probabilistic effects increase with increased exposures which can be characterized (e.g., number of years in certain jobs, availability of protective equipment, and amount of friable asbestos fibers in the air).
Deterministic Effect	Death from methylisocyanate exposure. Exposure to a toxic cloud of high concentrations of the gas methylisocyanate (MIC) is a probabilistic exposure, which is very low for most people. But, for people in the highest MIC concentration plume, such as those in the Bhopal, India, tragedy, death was 100% certain. Lower doses led to other effects, some acute (e.g., blindness) and others chronic (e.g., debilitation that led to death after months or years).	Generating carbon dioxide from combusting methane. The laws of thermodynamics dictate that a decision to oxidize methane, e.g., escaping from a landfill where anaerobic digestion is taking place, will lead to the production of carbon dioxide and water (i.e., the final products of complete combustion). Therefore, the engineer should never be surprised when a deterministic exposure (heat source, methane, and oxygen) lead to a deterministic effect (carbon

TABLE 11.2
Continued

	Probabilistic Exposure	*Deterministic Exposure*
	The chronic deaths may well be characterized probabilistically, but the immediate poisonings were deterministic (i.e., they were completely predictable based on the physics, chemistry, and biology of MIC).	dioxide release to the atmosphere). In other words, the production of carbon dioxide is 100% predictable from the conditions. The debate on what happens after the carbon dioxide is released (e.g., global warming) is the province of probabilistic and deterministic models of these effects.

arena is often between a mean exposure and a worst-case exposure (i.e., maximally exposed and highly sensitive individuals). The latter is more protective, but almost always more expensive and difficult to attain. For example, lowering the emissions of particulate matter (PM) from a power plant stack to protect the mean population of the state from the effects of PM exposures is much easier to achieve than lowering the PM emissions to protect an asthmatic, elderly person living just outside of the power plant property line (see Figure 11.5). Although the most protective standards are best, the feasibility of achieving them can be a challenge (see Figure 11.6).

Actual or realistic values are input into the deterministic model. For example, to estimate the risk of tank explosion from rail cars moving through and parked in a community, the number of cars, the flammability and vapor pressure of contents, the ambient temperature, the vulnerability of the tank materials to rupture, and the likelihood of derailment would be assigned numerical values from which the risk of release or explosion is calculated. A probabilistic approach would require the identification of the initiating events and the plant operational states to be considered; analysis of the adverse outcome using statistical analysis tools, including event trees; application of fault trees for the systems analyzed using the event trees (i.e., reliability analyses; see Chapters 1 and 2), collection of probabilistic data (e.g., probabilities of failure and the frequencies of initiating events), and the interpretation of results.

Human beings engage in risk management decisions every day. They must decide throughout whether the risk from particular behaviors is acceptable or whether the potential benefits of a behavior do not sufficiently outweigh the hazards associated with that behavior. In engineering terms, they are optimizing their behaviors based upon a complex set of variables that lead to numerous possible outcomes. A person wakes up and must

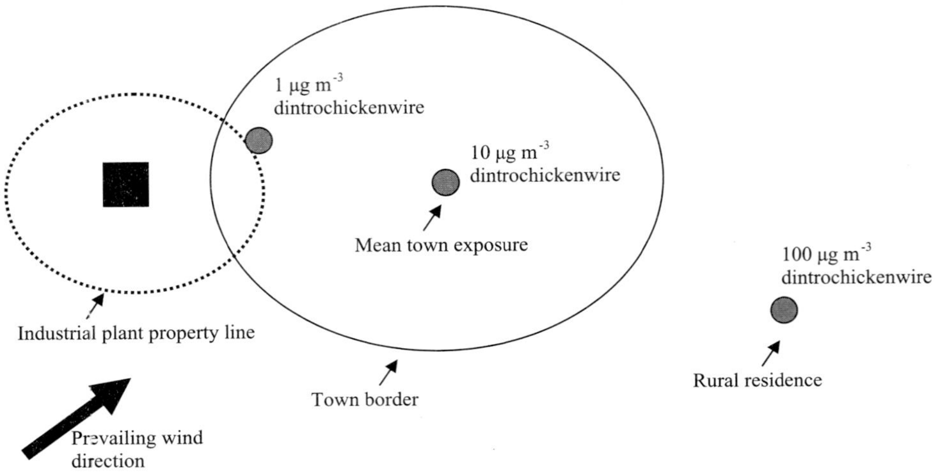

FIGURE 11.5. Difference in control strategies based on concentration of the allowable emissions of an air pollutant to reduce risks to maximally exposed versus mean population and a very low exposure scenario. Fictitious data for fictitious toxic substance, "dintrochickenwire." The concentration would be even lower if the risks are based on a highly sensitive subpopulation (e.g., elderly, infants, or asthmatics), depending upon the effects elicited by the emitted pollutant. For example, if dintrochickenwire is expected to cause cardiopulmonary effects in babies, an additional factor of safety may push the risk-based controls downward by a factor of 10 to achieve $0.1\,\mu g\,m^{-3}$ to protect the maximally exposed, sensitive population.

decide whether to drink coffee that contains the alkaloid caffeine. The benefits include the morning jump-start, but the potential hazards include induced cardiovascular changes in the short term, and possible longer-term hazards from chronic caffeine intake. The decision is also optimized according to other factors, such as sensory input (e.g., a spouse waking earlier and starting the coffee could be a strong positive determinate pushing the decision toward yes), habit (more likely to drink a cup if it is part of the morning routine, less likely if it is not), and external influences (e.g., seeing or hearing a commercial suggesting how nice a cup of coffee would be, or conversely reading an article in the morning paper suggesting a coffee-related health risk). This decision includes a no-action alternative, along with a number of other available actions. One may choose not to drink coffee or tea. Other examples may include a number of actions, with concomitant risk.

The no-action alternative is not always innocuous. For example, if a person knows that exercise is beneficial but does not act upon this knowledge, the potential for adverse cardiovascular problems is increased. If a person does not ingest an optimal amount of vitamins and minerals, disease

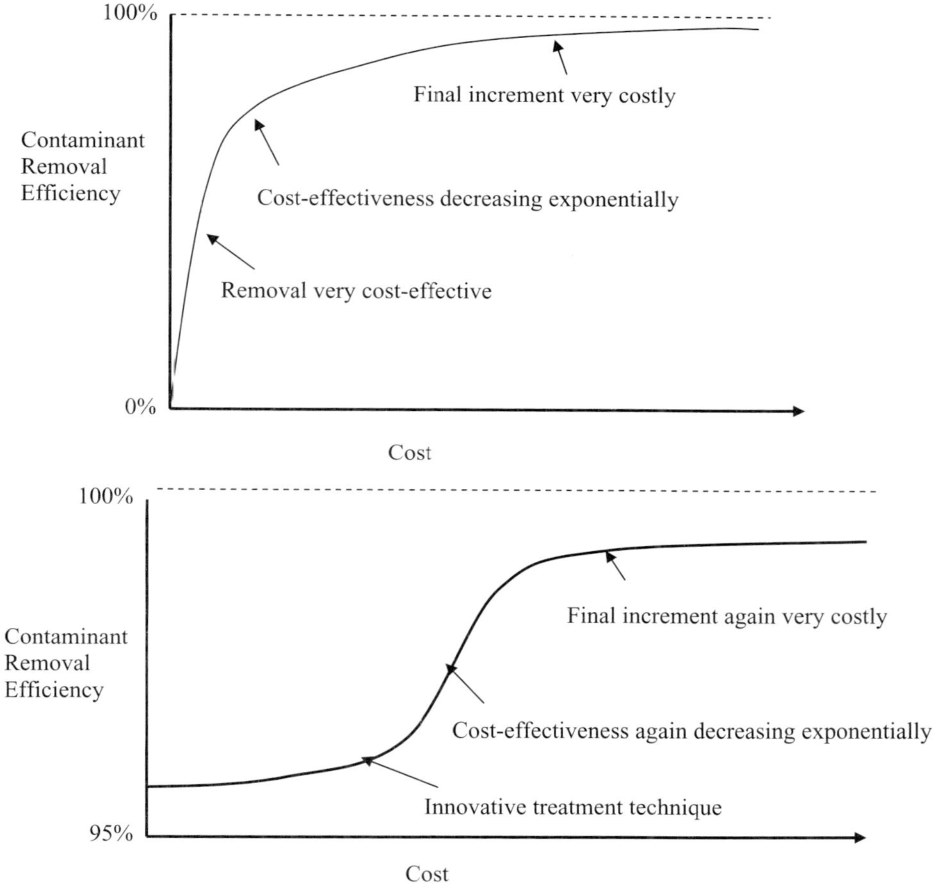

FIGURE 11.6. Prototypical contaminant removal cost-effectiveness curve. In the top diagram, during the first phase, a relatively large amount of the contaminant is removed at comparatively low costs. As the concentration in the environmental media decreases, the removal costs increase substantially. At an inflexion point, the costs begin to increase exponentially for each unit of contaminant removed, until the curve nearly reaches a steady state where the increment needed to reach complete removal is very costly. The top curve does not recognize innovations that, when implemented, as shown in the bottom diagram, can make a new curve that will again allow for a steep removal of the contaminant until its cost-effectiveness decreases. This concept is known to economists as the law of diminishing returns. *Source*: D.A. Vallero, 2004. *Environmental Contaminants: Assessment and Control,* Elsevier Academic Press, Burlington, MA.

resistance may be jeopardized. If a person always stays home to avoid the crowds, no social interaction is possible and the psyche suffers. The management decision in this case may be that the person decided that human-to-human contact, correctly, is a means of transmitting pathogens. But, implementing that decision carries with it another hazard, social isolation. Likewise, the engineer must take an action only if it provides the optimal solution to the environmental problem, while avoiding unwarranted financial costs without causing unnecessary disruption to normal activities, and in a manner that is socially acceptable to the community.

In addition, the engineer must weigh and balance any responsibility to represent the client with environmental due diligence. This diligence must be applied to designs and plans for manufacturing processes that limit, reduce, or prevent pollution, to ways to reduce risks in operating systems, to the assessment of sites and systems for possible human exposures to hazardous and toxic substances, and to the evaluation of design systems to reduce or eliminate these exposures. Ultimately, the engineer participates in means to remedy the problem, to ameliorate health, environmental, and welfare damages. The remedy process varies according to the particular environmental compartment of concern (e.g., water, air, or soil), the characteristics of contaminant of concern (e.g., toxicity, persistence in the environment, and likelihood to accumulate in living tissues), and the specific legislation and regulations covering the project. However, it generally follows a sequence of preliminary studies, screening of possible remedies, selecting the optimal remedy from the reasonable options, and implementing the selected remedy (see Figure 11.4). The evaluation and selection of the best alternative is the stuff of risk management.

Given this logical and seemingly familiar role of the engineer, why then do disasters and injustices occur on our watch? What factors cause the engineer to improperly optimize for the best outcome? In part, failures in risk decision making and management are ethical in nature. Sometimes organizational problems and demands put engineers in situations where the best and most moral decision must be made against the mission as perceived by management. Working within an organization has a way of inculcating the corporate culture into professionals. The process is incremental and can desensitize employees to acts and policies that an outsider would readily see to be wrong. Much like the proverbial frog placed in water that gradually increases to the boiling point, an engineer can work in gradual isolation, specialization, and compartmentalization that ultimately changes to immoral or improper behavior, such as ignoring key warning signs that a decision to locate the facilitate will have an unfair and disparate impact on certain neighborhoods, that health and safety are being compromised, and that political influence or the bottom line of profitability is disproportionately weighted in an engineer's recommendation.[15]

Another reason that optimization is difficult is that an engineer must deal with factors and information that may have not been adequately addressed during formal engineering training or even during career development. Although environmental and public health decisions must always give sufficient attention to the toxicity and exposure calculations, these quantitative results are tempered with feasibility considerations. Thus, engineers' strengths lie to the far left and right of Figure 5.1, but the middle steps—feasibility and selecting the best alternative—require information that is not "well behaved" in the minds of many engineers. For example, in 1980 the U.S. Congress passed the Comprehensive Environmental Response, Compensation, and Liability Act (CERCLA), commonly referred to as the Superfund.[16] The Superfund law authorizes the federal government to respond directly to releases or to the threat of releases of hazardous substances and enables the U.S. Environmental Protection Agency (EPA) to take legal action to force parties responsible for causing the contamination to clean up those sites or reimburse the Superfund for the costs of cleanup. If the responsible parties for site contamination cannot be found or are unwilling or unable to clean up a site, the EPA can use funds from the Superfund to clean up a site.

Engineering risk management is a type of *teleology*. The term is derived from the Greek, *telos*, meaning "far." So, what engineers do is based upon something "out there" that we would like to achieve—a desired outcome—or something we would like to avoid—an adverse outcome. In particular, engineering is utilitarian; that is, we are called upon "to produce the most good for the most people."[17] However, engineers are duty-bound to our codes of ethics, design criteria, regulations, and standards of practice. In this way, engineering is a type of *deontology* (Greek, *deon* for "duty"). Engineers must, to the best of their abilities, consider all possible design outcomes, planned or otherwise. But this is not easy because the most appropriate benchmarks for success in engineering projects are moving targets. There is no consensus about success at the outset, so that even if the design is implemented to meet all specifications, the project could be deemed a failure. For example, environmental decisions regarding the level of cleanup at a hazardous waste site can be based on the target health risk following cleanup. This remediation target has been based on what was called the residential standard. That is, immediately following the passage of key hazardous waste laws, regulators held the general view that a polluted site needed to be cleaned up to the point that no undue risk remained and the site could be returned to productive use, as if it had never been polluted. This closely follows the steps of hazard identification, dose-response relationships, exposure analysis, and effects assessment to characterize risks, as described in Chapters 1 and 2.[18]

The Superfund law requires that before the actual cleanup of a hazardous waste is conducted that a feasibility study be completed. These studies must address nine criteria that will determine the best approach for addressing the contamination, as well as the ultimate level of cleanup:

1. Overall protection of human health and environment
2. Compliance with applicable or relevant and appropriate requirements
3. Long-term effectiveness and permanence
4. Reduction of toxicity, mobility, or volume through treatment
5. Short-term effectiveness
6. Ease of implementation
7. Cost
8. State acceptance
9. Community acceptance

The first and fourth criteria are clearly the product of a sound, quantitative risk assessment. The other criteria must also include semiqualitative and qualitative information. So, although scientists and engineers devote much attention to quantitative values for toxicity and exposure, the actual level of protection to be achieved must factor in these other social and economic aspects. Frequently, for example, a site cleanup may start (and finish) as a removal action, where the contaminated material is simply dug up and taken away to be disposed of and treated off-site. The National Oil and Hazardous Substances Pollution Contingency Plan (NCP)[19] mandates that a removal action is needed when there is:

- Actual or potential exposure of humans, animals, or the food chain
- The presence of contained hazardous substances that pose a threat of release
- The threat of migration of the hazardous substances
- The threat of fire or explosion
- The availability of an appropriate federal or state response

The major purpose of a removal action is to eliminate exposures, or at least keep them at a minimum, so a removal action commonly is taken to address an imminent danger to human health or the environment, and usually takes a short period of time. Removals are classified as emergencies, time critical, or nontime critical. Emergency removals are undertaken when the danger is so large that little time is available to undertake a planning process. Time critical removals have less than six months before site activities must be initiated. Nontime-critical removals call for at least six months but not more than 12 months of planning before the removal action is undertaken. Removal actions normally occur as part of the initial response to a badly contaminated site that will later be the subject of a more formal and extensive remedial action.

Removal actions can be fairly broad and include actions needed to monitor, assess, and evaluate the release or threat of release of hazardous substances; the disposal of removed material; or measures to prevent, minimize, or mitigate damage to the public health or welfare of the United

States or to the environment, which may otherwise result from a release or threat of release. Removal also includes sundry actions like security fencing or other measures to limit access, provision of alternative water supplies, and temporary evacuation and housing of people with a potential of being exposed.

Thus, the desired level of protection has to be considered within the whole framework of engineering judgment and public good. For this reason, immediate actions like hazardous waste removals often begin with a higher target cleanup level; the target risk may be one in 10,000 (one additional cancer attributed to this level of pollution in a population of 10,000 exposed individuals) instead of one in a million. The removal action may be followed by a more stringent and more permanent cleanup (i.e., the remediation phase) or may be done in conjunction with remediation, but these decisions again are part of the risk management process.

The lack of hard and fast levels of detection extends to other environmental management decisions. For example, the Clean Air Act Amendments of 1990 (CAAA) require an "ample margin of safety" beyond maximum available control technologies (MACT).[20] In other words, the risks to public health that exist after a facility puts in place state-of-the-science technologies still may not adequately protect people from cancer and other adverse effects. This risk, known as residual risk, must be addressed in ways beyond the installation of control devices. The actual requirements differ by pollutant and by the type of emission (i.e., source category). The CAAA identified 188 hazardous air pollutants (known as air toxics) whose sources need to be addressed. In fact, 174 source categories have been identified in the regulations (40 CFR Part 63). The MACT standards applied to any source that emitted 10 tons per year of any single hazardous air pollutant, or 25 tons per year total hazardous air pollutants.

So, the CAAA provided a two-step process in protecting public health against hazardous air pollutants. First, technology-based standards required sources to meet specific emission limits based on emissions levels already being achieved by many similar sources in the country. Second, the EPA applies a risk-based approach to assess how well the technology-based standards have done in lowering risks. If the EPA finds that there are any significant remaining or residual health or environmental risks, the agency would write and implement additional standards. To provide the ample margin of safety to protect public health and prevent adverse environmental effects, the CAAA requires a human health risk and adverse environmental effects-based needs test. This residual risk standard setting considers the need for additional national standards on stationary emission sources following regulation to protect public health and the environment. However, when the EPA and states consider this safety margin, they must also weigh economics, occupational, safety, energy use, and other relevant considerations in any risk reduction decision. Thus, there is no single target for any source, but a range of what is considered to be acceptable risk (see

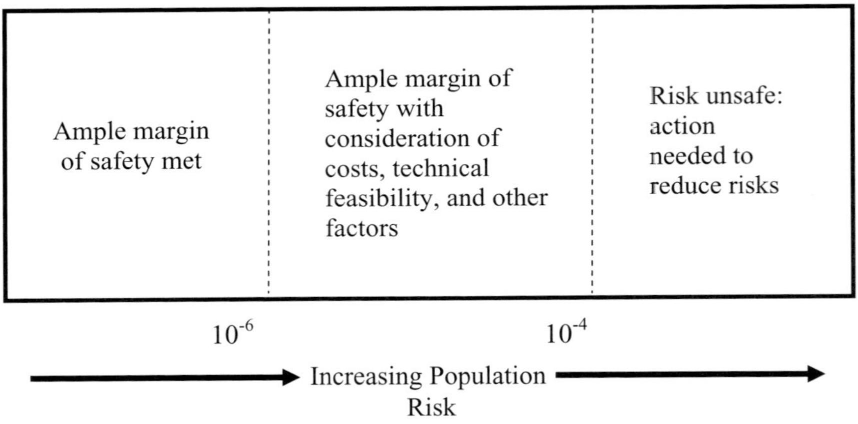

FIGURE 11.7. Ample margin of safety based on airborne contaminant's cancer risk.

Figure 11.7). The range between acceptable and unacceptable risk varies, but no residual cancer risk above one in 10,000 is acceptable; and risks below one additional cancer per million people in a population are generally considered to have met the safety requirements.

From this discussion, it may be easy to see how actual or perceived injustices can occur. People may wonder why their neighborhood cleanup level is an order of magnitude worse than their neighbors'. This may be justifiable, for example if the cleanup is not feasible given available technology due to a particular soil type, or the contaminant of concern in one neighborhood may be much more recalcitrant to treatment than a different contaminant in another neighborhood. Again, the key to dealing with such perceptions is in good risk communications. However, the engineer should be wary of differences and needs to be certain that the disparities in cleanup are truly justified. If the other neighborhood cleanup levels are stricter because the people have more savvy or are better politically connected than a neighborhood that may not be as sophisticated or is less of a "squeaky wheel," this might be an instance of injustice.

Although risk management is an example of optimization, it is not as straightforward for environmental justice as some of the other engineering applications, such as simplex or multiplex optimization routines. These optimization models often apply algorithms to arrive at a net benefit/cost ratio, with the selected option being the one with the largest value, or greatest quantity of benefits compared to costs. If this sounds like a utilitarian approach, it is. There are numerous challenges when using such models in environmental decisions. Steven Kelman of Harvard University was one of the first to articulate the weaknesses and dangers of taking a purely utilitarian approach in managing environmental, safety, and health risks.[21]

Kelman asserts that in such risk management decisions, a larger benefit/cost ratio does not always point to the correct decision. He also opposes the use of dollars—monetization of nonmarketed benefits or costs—to place a value on environmental resources, health, and quality of life. He uses the logical technique of *reductio ad absurdum* (from Greek, *ἥεις το αδυνατον απαγωγη*, "reduction to the impossible") where an assumption is made for the sake of argument and a result found, but it is so absurd that the original assumption must have been wrong.[22] For example, the consequences of an act, whether positive or negative, can extend far beyond the act itself. Kelman gives the example of telling a lie. Using the pure benefit/cost ratio, if the person telling the lie has much greater satisfaction (however that is quantified) than the dissatisfaction of the lie's victim, the benefits would outweigh the cost and the decision would be morally acceptable. At a minimum, the effect of the lie on future lie-telling would have to be factored into the ratio, as would other cultural norms. Another of Kelman's examples of flaws of utilitarianism is the story of two friends on an Arctic expedition, wherein one becomes fatally ill. Before dying, he asks that the friend return to that very spot in the Arctic ice in 10 years to light a candle in remembrance. The friend promises to do so. If no one else knows of the promise and the trip would be a great inconvenience, the benefit/cost approach instructs him not to go (i.e., the costs of inconvenience outweigh the benefit of the promise because no one else knows of the promise). These examples point to the fact that benefit/cost information is valuable, but care must be taken in choosing the factors that go into the ratio, properly weighing subjective and nonquantifiable data, ensuring that the views of those affected by the decision are properly considered, and being mindful of possible conflicts of interest and undue influence of special interests.

One possible means of enhancing the risk management process while maintaining the rigorous risk assessment process is to approach each project from a site-wide perspective that combines health and ecological risks with land use considerations. In other words, determining the residual risk that should be allowed to remain is based on both traditional risk outcomes (disease, endangered species) *and* future land uses (see Figure 11.8). The attractive aspect of adding the future land use perspective is that it requires that a project be at least somewhat sustainable. Even a very attractive near-term project may not be so good when viewed from a longer term perspective. Conversely, a project with seemingly large initial costs may in the long run be the best approach. Even projects with larger risks upfront may be the best of available alternatives. This has been the case for some asbestos and lead remedies, where the workers involved in removing the contaminants are subjected to the threat of elevated concentrations of toxicants, but the overall benefits of the action were deemed necessary to protect children. Plus, in a well-planned engineering project, the risk is transformed from a risk that is widely distributed in space and time (i.e., numerous buildings with the looming threat to children's health for decades to come) to a

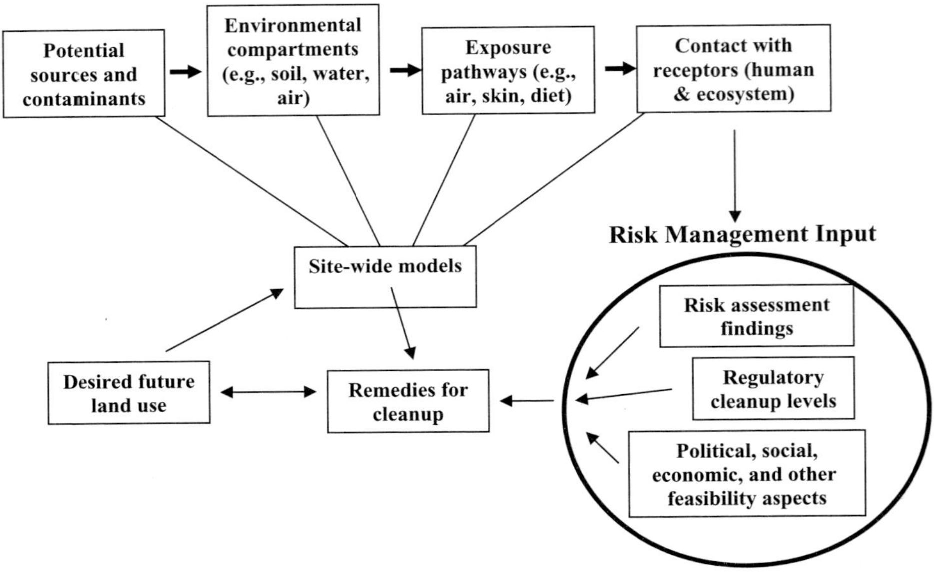

FIGURE 11.8. Site-wide cleanup model based upon targeted risk and future land use. *Source*: Adapted from J. Burger, C. Powers, M. Greenberg, and M. Gochfeld, 2004. "The Role of Risk and Future Land Use in Cleanup Decisions at the Department of Energy," *Risk Analysis*, 24 (6), 1539–49.

concentrated risk that can be controlled (e.g., safety protocols, skilled workers, protective equipment, removal and remediation procedures, manifests and controls for contaminated materials, and ongoing monitoring of fugitive toxicant releases).

This combined risk and land use approach also helps to moderate the challenge of "one size fits all" in environmental cleanup. That is, limited resources may be devoted to other community objectives if the site does not have to be cleaned to the level prescribed by a residential standard. This does not mean that the site can be left to be "hazardous," only that the cleanup level can be based on a land use other than residential, where people are to be protected in their daily lives. For example, if the target land use is similar to the sanitary landfill common to most communities in the United States, the protection of the general public is achieved through measures beyond concentrations of a contaminant. These measures include allowing only authorized and adequately protected personnel in the landfill area, barriers and leachate collection systems to ensure that contamination is confined within certain areas within the landfill, and security devices and protocols (fences, guards, and sentry systems) to limit the opportunities for exposures and risks by keeping people away from more hazardous areas.

This can also be accomplished in the private sector. For example, turnkey arrangements can be made so that after the cleanup (private or governmental) meets the risk/land use targets, a company can use the remediated site for commercial or industrial uses. Again, the agreement must include provisions to ensure that the company has adequate measures in place to keep risks to workers and others below prescribed targets, including periodic inspections, permitting, and other types of oversights by governmental entities to ensure compliance with agreements to keep the site clean (i.e., so-called closure and post-closure agreements).

Even within the risk assessment component (left-hand side of Figure 11.4), balancing is needed. The balance between human health and ecological protection is one such situation. For example, an endangered species or an on-site worker can be used as the "risk driver." This is an example of a human *versus* ecological risk balance.[23] Even if "human" is chosen as the risk driver, the decision to protect the general public is very different (and more difficult in most cases) than protecting workers. The decision on risk drivers involves trade-offs. For example, the very decision to clean up a site may impose risks to a group of people who would have not had any risks related to the site, the cleanup crew. Similarly, trade-offs are involved if ecological risk drivers are used, such as between the risks of one species to another. A sensitive species, such as wildflowers, may be more greatly harmed by removal due to soil disturbances than if the soil had not been disturbed. Also, scientists have found that if left to their own devices, many soil microbes adapt and break down even highly toxic substances, a process known as *natural attenuation* (see Figure 11.9). In fact, the microbes may not do as well if disturbed, plus other possible risks can be avoided that would result from soil removal, tranport, and *ex situ* treatment.

Risk trade-offs can also involve the difference between one community and another. This is a common consideration in environmental justice decision making. For example, is the area to be cleaned up at the expense of another community, for example a neighborhood near an incinerator where contaminated soil will be stored and treated? Historically, communities of low socioeconomic status have been more likely to be near such treatment facilities.

Another consideration is the temporality of the risk abatement. For example, are remedies being considered a benefit to the present population, but will introduce added risk to future human populations or ecosystems? For instance, storage of hazardous or radioactive wastes in facilities designed for 100 years may sound okay today, but what happens when their useful life is ended? Timing is also important in terms of sequencing. In some cases, the best approach is to take deliberate and thoughtful analysis before removing material, but in others it may be feasible and the best approach to begin remediation (e.g., incineration of the top 15 cm of soil for the whole site) at once. Other aspects of removal and remediation may take place "out of sequence" but with a better overall improvement.

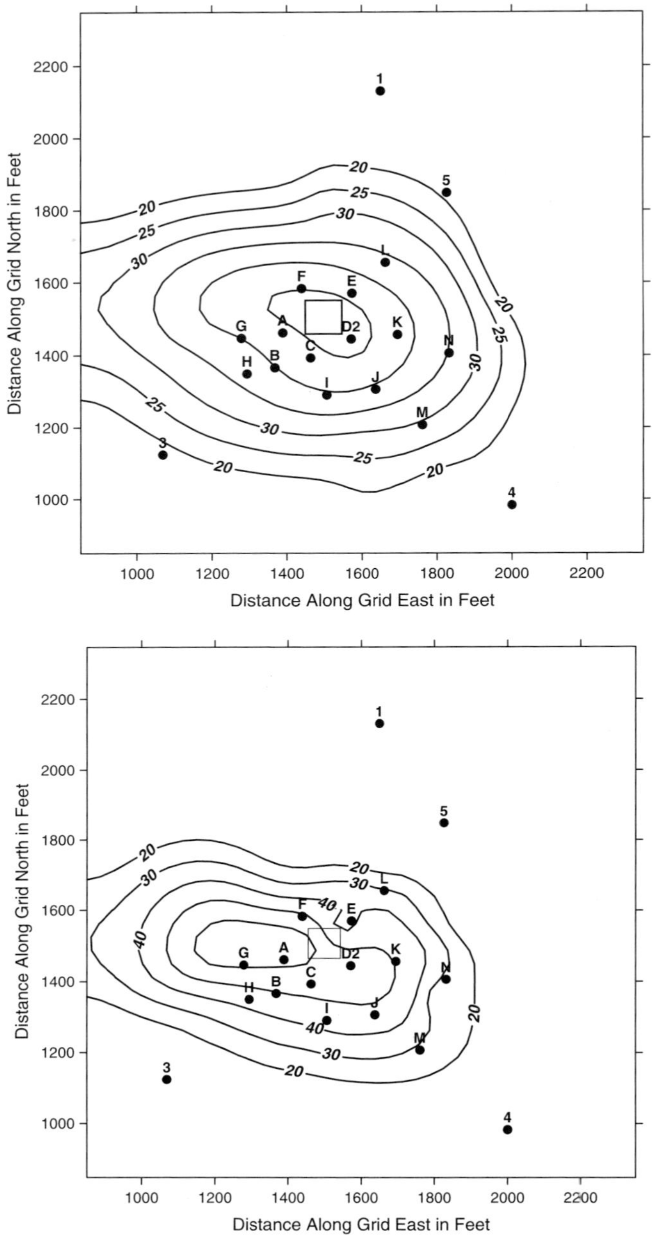

FIGURE 11.9. Duke Forest Gate 11 Waste Site in North Carolina. Top map: Modeled paradioxane plume after 50 years of natural attenuation. Bottom map: Paradioxane plume modeled after 10 years of pump and recharge remediation. Numbered points are monitoring wells. The difference in plume size from intervention versus natural attenuation is an example of the complexity of risk management decisions; i.e., does the smaller predicted plume justify added costs, possible risk tradeoffs from pumping (e.g., air pollution), and disturbances to soil and vegetation?
Source: M.A. Medina, Jr., W. Thomann, J.P. Holland, and Y-C. Lin, 2001. "Integrating Parameter Estimation, Optimization and Subsurface Solute Transport," *Hydrological Science and Technology* 17, 259–282. Used with permission from first author.

Unlike the risk assessment component, the feasibility and social aspects of risk management are often ill-defined and have no consensus even among professionals within the same discipline on how they should be approached. Recently, educators have recognized the difficulty of introducing students to complex and highly emotional risk management decisions. For example, Debra Satz, director of Stanford University's Ethics in Society Program has characterized the mismatch of real ethical dilemmas and abstract moral principles:

> . . . I have been struck by how ill equipped much moral and political philosophy is to deal with the "limits of the possible." By the limits of the possible, I have in mind the non-ideal aspects of our world: that people don't always do the right thing, that there can be very high costs to doing the right thing when others do not, that information is imperfect, resources are limited, interests are powerful, the best options may not be politically or materially feasible, and that collective action problems are everywhere. We need a moral and political philosophy that integrates theoretical reflection on values with practical knowledge about how the world sets limits on what we can do and what we can hope for.[24]

Satz's commentary is particularly germane to engineers, who must always work within the "limits of the possible," as we translate theory from the laboratory to the field.

The complexity of balancing all these factors to manage risks is also tied inextricably to social justice. A concise yet eloquent description of the balance needed to achieve justice has been put forth by two professors of religion, James B. Martin-Schramm and Robert L. Stivers:

> For the Greeks justice meant "treating equals equally and unequals unequally." This simple statement of the norm of justice hides the complexities of determining exactly who is equal and who is not and the grounds for justifying inequality. It leads in modern interpretations of justice, however, to freedom and equality as measures of justice. It also leads to the concept of equity, which is justice in actual situations where a degree of departure from freedom and equality is permitted in the name of achieving other social goods. So, for example, most societies give mentally and physically impaired individuals extra resources and justify it in the name of greater fairness.[25]

Socially responsible risk management requires a complicated balance of many factors, with a strong consideration of how the actions stemming from the decisions will affect people now and in the future, especially those who may not be heard or who may not completely grasp the risks. The proper modicum of precaution and safety is not formulaic, but depends on the unique synthesis of factors in every project.

Precautionary Principle and Factors of Safety in Risk Management

Like nearly everything else in risk management, the concepts of due diligence, precaution, and factors of safety are not static, but depend on a dynamic relationship of internal factors that vary from project to project. Also, there have been enough disasters and near catastrophes that prudence is required. Such calamities do not usually occur in a nice neat linear projection, but may be the consequence of an iteration of unlikely occurrences (see the discussion box, "The Butterfly Effect" in Chapter 7). But, on the other hand, engineers cannot simply add up all the factors of safety to arrive at an overall factor of safety for their project. This over-conservatism is not only expensive in terms of time and resources; it may well not be protective. In fact, this approach could lead to unjust, unsafe, and unacceptable outcomes.

The risk assessment process provides information that has "built-in" factors of safety, such as the reference dose (RfD), which includes uncertainty factors and modifying factors in determining the "safe" dose of toxicants.

Likewise, other data being used in risk management decisions include their own factors of safety, such as those of designs of treatment structures (landfills, treatment and storage facilities, and lagoons); transportation of contaminated materials; propagation of error in air, soil, and water models (e.g., atmospheric dispersion model assumptions about ranges and variances in the vertical and horizontal axes, as shown in Figure 11.10); activity patterns of the population; and other socioeconomic models and data. Complicating this further is that the types of data from these other sources are highly variable. Some are quantitative and even of a ratio or interval type, such as average body weight, miles driven, and median residential lot size. Others are semiquantitative or qualitative, such as ordinal data about people's preferences, such as those regarding aesthetics or willingness to pay (e.g., data derived from a survey using a Likert scale from 0 to 4, i.e., 0 = unwilling to pay *versus* 4 = willing to pay a very large amount). And, data can also be nonquantitative, but useful, such as nominal data (e.g., identification numbers, street names, and ethnic origin). The way these data can be handled is dependent on the type. For example, statistical measures of central tendencies (e.g. means, medians, and modes) are only properly applied to ratio or interval data. Data aggregation, grouping, and reduction techniques must be matched to data type. Even within quantitative data sets, the manner in which the data are combined with other data sets is not always straightforward. For example, if the source of an asbestos-laden rock stratum is characterized with a high degree of certainty (i.e., type of

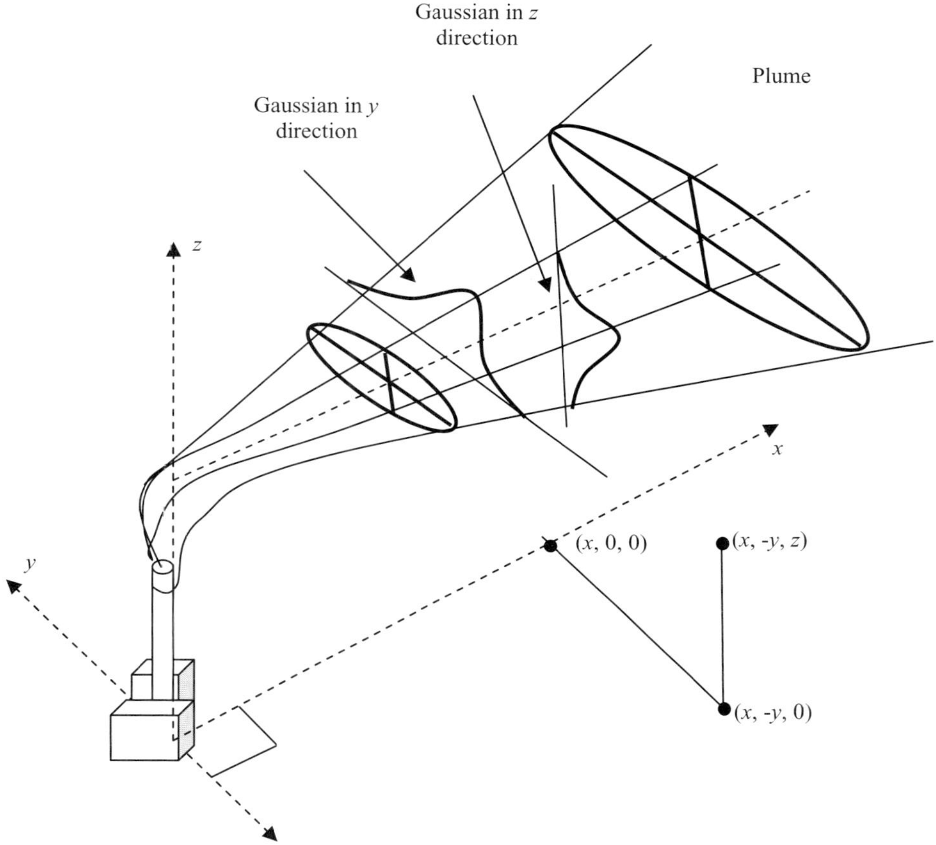

FIGURE 11.10. Atmospheric Gaussian plume model based upon random distributions in the horizontal and vertical directions. Uncertainties regarding the width and height of the plume would be depicted by error lines in the *x*-axis and *z*-axis, respectively.
Source: Adapted from D. Turner, 1970. Workbook of Atmospheric Dispersion Estimates. Office of Air Programs Publication No. AP-26 (NTIS PB 191 482). U.S. Environmental Protection Agency, Research Triangle Park, NC.

asbestos, concentration of asbestos per volume of rock, friability of asbestos, and variability of asbestos concentrations within the stratum), but the manner in which the asbestos is transported and disposed after being extracted is unknown, this creates a challenge for risk manager. For example, assumptions about stochastic distributions from the source would be incorrect; that is, the amount of asbestos cannot be predicted by distance from the source, so even a location very near to the source may have low or no asbestos, whereas a source at some distance, even thousands of kilometers, may have very high concentrations of the asbestos extracted from the site (e.g., shipped by railcar, disposed by a manufacturer, or used by a consumer).

Addressing the myriad individual, factor-specific uncertainties and factors of safety is, again, a matter of optimization. A recent example of the complexity of optimizing risk targets and processes is that of paper recycling. Quantitative measures of success are available and used frequently, such as the utilization rates for wastepaper. These metrics, however, are insufficient to determine whether wastepaper recycling on the whole is a success or failure in a community, in a nation, or worldwide. In fact, in many communities recycling has been emblematic of the commitment to environmental protection. It is not unusual in even the most profit-oriented offices to hear someone exclaim that they did not make copies of a given document in the interest of "saving a tree" (granted, this may have been a convenient excuse for not wanting to spend time at the copy machine, but it is indicative of the remarkable change in environmental awareness in the corporate culture).

Nations report recovery and utilization rates as indicators of environmental success. What complicates such indicators is that numerous socioeconomic factors are behind these results, not just environmental considerations. For example, costs and economic decisions may either be important or relatively ignored factors. The problem with ignoring these factors is that the programs may be less sustainable than the glowing report of utilization rates may indicate. If costs are always "in the red," the programs may be more vulnerable to market forces and may eventually fail. Other socioeconomic factors, like community pride and the desire to "think globally but act locally" can buoy local environmental programs, even if they are not cost-effective (see the discussion box, "Market versus Nonmarket Valuation: Uncle Joe the Junk Man").

Market versus Nonmarket Valuation: Uncle Joe the Junk Man

Many factors drive decisions about how to manage environmental risks. Environmental protection is now a commonly held value in most contemporary cultures, but the manner of assigning value to environmental resources varies considerably. This is challenging since selecting decisions is based in part on the extent to which benefits outweigh costs. If costs are greater than benefits from a given engineering project, the project is usually not acceptable. The problem is that some benefits do not lend themselves well to economical analyses, for example, value of life, aesthetics, and transgenerational effects (i.e., costs and problems well into the future).

Recycling has become a generally accepted practice in many modern cultures, including those in the most economically developed countries. The problem is that from a purely market-based standpoint, many recycling programs do not pay for themselves. The good news, however, is that costs are not the only socioeconomic factors. Let me give a personal example of both market-based and nonmarket valuation approaches.

When I was a teenager, I worked for my uncle in a salvage business. Uncle Joe had a keen business acumen and knew how to turn a profit. He proudly referred to himself as a "junk man" and in fact often used the gerund "junking" to describe his chosen calling. I happened to be one of his favorite nephews (I heard from my Dad that it was because Uncle Joe observed that "Danny wasn't afraid to get dirty"—which I wore as a badge of honor, I must say!), so he hired me to work with him in his salvage business on weekends and during the summer. My uncle was an honorable man, but I would not characterize him as "environmentally sensitive." Neither had I yet to gain much of an environmental *pathos*, except from what I learned from camp outings and later from a high school earth sciences course. My uncle's concern for what was to become known as recycling was almost purely profit-oriented. He developed a knack for finding value in what others discarded. He was particularly fond of metals. Copper was his favorite. In fact, he could tell before breaking open various devices, such as armatures, what type of copper he would find. The highest grade was known as Number 1 copper, which if memory serves was worth about 60 cents per pound. Number 2 copper could fetch considerably less depending on its purity. The same ranking system was used for some other valuable metals, such as aluminum, lead, and zinc. These metals, however, were worth considerably less than copper. Every once in a while, someone might slip a load of armatures by my uncle, which by their outward appearance included copper wiring, but only after dissembling them did we find, to Uncle Joe's dismay, that they were spun with aluminum wires. At that time, aluminum was worth only about a dime per pound. Other metals, especially ferrous alloys, cast iron, steel, and brass were worth considerably less. In fact, they sold by the ton rather than the pound (a few dollars per ton).

The profit-driven recycling was actually not environmentally friendly in the least. We commonly were confronted with wires that were insulated with polyvinyl chloride (PVC) and other plastic coatings. The PVC had to be removed before the copper or aluminum could be sold to salvage companies. I believe that these companies would have burned off the coatings, but local or state regulations probably prohibited this activity, so the salvage companies would sell them to private entrepreneurs like my uncle, knowing that they would take the load of wires to remote locations and burn off the coatings. I vividly recall the odors of the dense black plumes that rose from these fires. I am certain that the particulate matter (PM), vinyl chloride (VC), and polycyclic aromatic hydrocarbons (PAHs) in the plumes were at levels far above anything seen in industry today. The closest would be a building fire. The freed chlorine atoms from the VC are highly toxic, and VC and a number of the PAHs are known carcinogens. This open-burning practice has since been banned nationally, for good reason. I also remember breaking apart fluorescent light ballasts and other electrical equipment to retrieve the wires. In the process, I recall oils being released. I subsequently learned that polychlorinated biphenyls (PCBs) were used in this equipment as a dielectric fluid. Finally, in search of lead and zinc, we would break apart automotive radiators and batteries. The fluids from these devices were also very toxic. The batteries were filled with sulfuric acid (H_2SO_4) that contained extremely high concentrations of lead and zinc. The radiators contained the toxic antifreeze ingredient, ethylene glycol, as well as water into which lead, cadmium, nickel, or other toxic heavy metals had leached. I recall that the sections of the salvage yards where we retrieved and dismantled these devices were ankle deep in muck that must have contained heavy metals, ethylene glycol, asbestos (from brake systems), and other toxic substances in astonishingly high concentrations. My exposures to toxic substances, by today's standards, would have been almost off the charts.

I am not describing this type of recycling as an indictment of the salvage business in the 1960s. Hindsight is always better than what we see at the time. I point this out to show that recycling had taken place long before we had a symbol for it and before it was the topic of public service announcements. Also, junkyards and salvage businesses generally are found nearest lower socioeconomic neighborhoods. To sell his recycled materials, my uncle and I would travel to East St. Louis, Illinois (incidentally my birthplace), where junkyards were ubiquitously adjacent to the rail yards next to the Mississippi River. We also would go to salvage businesses in St. Louis, Missouri, if the price were right. In addition, we would occasionally visit a private metal refinery in the middle of a residential neighborhood! I recall one that was constructed of refractory materials shaped like a wigwam, in one of my uncle's "client's" backyard. I was not yet familiar with the concept of zoning and land use regulations, but it is hard to imagine how his neighbors allowed smelting of metals to occur within a few feet of where they lived. It suffices to say that the tolerance of pollution was much higher back then.

Interestingly, junkyards adjacent to or even in lower socioeconomic communities are still a problem. And, many of the toxic substances mentioned here are still likely to be found in the soil and groundwater around these sites. I am aware of recent public meetings where the presence of junkyards was listed as one of the biggest environmental concerns in selected neighborhoods. The challenge for local decision makers is whether the aesthetic displeasure or the nuisance of these sites is really what the community is concerned about. The only sure way of determining whether the junkyards are actual sources of contamination is by conducting sound studies that provide reliable information about environmental quality.

So, if we use only a measure of success as the amount of metals and other materials recycled, my recycling efforts of the 1960s would have to be considered highly successful. But, if other factors like environmental quality, land use, and health are thrown into the benefit/cost assessment, the overall utility of my recycling experience is unacceptable. The challenge is how to account for nonmarket costs and benefits, which are often difficult to quantify, let alone to convert to monetary values. A number of intrepid economists have attempted to help environmental decision makers by developing new "nonmonetized" approaches to valuation. Let us wish them success!

No environmental risk management decision is made on the basis of a single factor, no matter how important that factor is. And the manner of arriving at the decision is highly variable and seldom involves a linear solution. Even what seems to be a universally accepted practice such as waste minimization is not the best approach in every situation.

The public policy and process for citizen participation in the decision can be simple in theory but complex in reality. The U.S. Environmental Protection Agency (EPA) is charged with implementing Executive Order 12898, calling for equal consideration of all health and environmental effects in federal actions. The agency established an official Office of Environmental Justice and issued a formal definition of environmental justice:

> The fair treatment and meaningful involvement of all people regardless of race, color, national origin, or income with respect to the development, implementation, and enforcement of environmental laws, regulations, and policies. Fair treatment means that no group of people, including racial, ethnic, or socioeconomic group, should bear a disproportionate share of negative environmental consequences resulting from industrial, municipal, and commercial operations or the execution of federal, state, local, and tribal programs and policies.[26]

Environmental injustice does not necessarily require nefarious intent. It may or may not result from willful decisions in the "back room" or "boardroom" to shift environmental burdens to those with little power to resist. In fact, the very identification of environmental justice cases has come about under a veil of skepticism, criticism, and even denial that injustices exist. The unwillingness to accept many cases as "environmental injustices" has taken at least three different forms; denial that there is such discrimination, the argument that discrimination is beneficial, and that whatever does occur, it is not racially motivated.[27]

The first argument questions whether injustice is actually occurring.[28] It may be argued that the appearance of disparate treatment is merely anecdotal, and that rigorous studies have not been done. This is a typical criticism of scientific problems. Another line of reasoning is that regulatory agencies have not won many cases for complainants, that a large number of complaints have been dismissed, and that the decisions of the Supreme Court and various Appeals Courts suggest that there are no legal grounds for such a case. The problem here is that there is no correlation between numbers of winning cases and the existence of environmental injustice. It may simply mean that old paradigms are still in place or that significant legal findings do not yet constitute precedence. Applying such logic to other social movements, for example suffrage or racial equality, would lead to the invalid conclusion that women were not denied access to the vote and that there was no racial discrimination until the courts said there was!

A second denial argument centers on demographics and statistics. First, since it is nearly impossible to determine if there is such a thing as race, it ought to be impossible then to argue for discrimination. If we are able to somehow define race, then (the argument goes) we have to show that statistically people of identifiable racial characteristics are being discriminated against. In order to illustrate these arguments, let's define a town, with say, four neighborhoods, with racial characteristics as shown:[29]

Neighborhood	*Percent Minority*
A	10
B	5
C	35
D	95

Which of these neighborhoods is considered a "minority" neighborhood? That is, where is the line drawn? Simply because there is no clear threshold constituting a minority neighborhood does not obviate the fact that Neighborhood D is a minority neighborhood.

A related problem with defining a minority neighborhood is in choosing the size of the land area. The seminal report by the United Church of Christ Commission for Racial Justice, for example, used zip codes to identify locations. This is a blunt tool at best, since so many small communities have only one zip code. The community discussed earlier, for example, might have a single ZIP code, and thus the uneven distribution of minorities would never be evident. A wastewater treatment plant located in neighborhood D would not even be on the environmental justice radar screen. And if this town had, say, 45% minority population, it would not show up as a minority community, and any suggestion of environmental discrimination would disappear.

Another argument against environmental justice is that it is not the percent of people in a neighborhood that matters, but rather the number of people affected. For example, suppose neighborhood D in the earlier community is sparsely populated, and even though it is 95% minority, there are only a few people in that neighborhood.

The point of these arguments is to deny problems with disparate distribution of pollution because we cannot conduct well-documented studies. This is the classic "head in the sand" approach.

A third denial argument is that the undesirable land use facilities often are not constructed in lower socioeconomic neighborhoods, but rather the neighborhoods grow up around such facilities since the land there is affordable. Experience has shown that this point has some validity. The location of airports often has caused major shifts in population because of the noise from the airplanes. People with fewer resources are able to afford homes in areas vacated by the more affluent, even though they are aware of the high noise levels.

The need of some citizens to seek more economical housing is not, of course, an excuse for exposing them to higher levels of environmental contaminants. People do not move to less expensive neighborhoods to be nearer to contamination and unhealthy conditions. They move there because this is all they can afford.

Finally, some reports claim that all socioeconomic groups resist the siting of undesirable facilities and land use in their neighborhoods, and the final siting of these facilities in lower socioeconomic class neighborhoods is as a result of the inability or unsophistication in being able to fight off such decisions.[30] The argument goes that sites are initially equally and equitably distributed, and the lower socioeconomic neighborhoods are not very good at protecting their communities.

Although this may be true, it does not provide justification for disparate treatment. The inability to summon the sophisticated resistance to unfair siting practices demonstrates a type of second-order discrimination that results from first-order practices that have led to lower education levels and histories of discrimination in these neighborhoods. In the interest of diversity, let us consider a range of environmental justice cases, beginning with one of the most famous.

The Warren County, North Carolina, PCB Landfill

In the early 1980s, a predominantly African American community in Warren County held a large demonstration to oppose the siting of a landfill designed to store toxic polychlorinated biphenyls (PCBs). Following the protest, additional studies and public attention raised concerns about the fairness and equal protection afforded under environmental regulations and policies. This protest has been recognized as one of the key galvanizing events of the environmental justice movement.

The Warren County PCB Landfill was constructed in 1982 to contain soil that was contaminated by the illegal spraying of oil containing PCBs from over 340 km of highway shoulders. Over 100,000 liters of contaminated oil were sprayed illegally along roadsides in 14 North Carolina counties.

Like all environmental problems, or any problem for that matter, the first step is to understand the scientific facts of the case. To begin, what are PCBs and why do they elicit such concern? PCBs all have the structure of $C_{12}H_{(10-n)}Cl_n$, where n is within the range of 1 to 10:

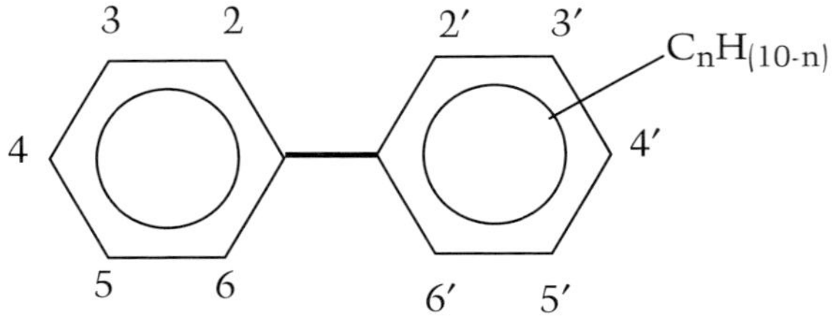

Polychlorinated Biphenyl Structure

All PCBs have this arrangement, but differ from each other by the number and location of chlorine atoms at each of the numbered positions. These different arrangements are known as *congeners*, a single, unique, well-defined chemical compound in the PCB category. The name of a congener specifies the total number of chlorine substituents and the position of each chlorine. For example, 4,4′-dichlorobiphenyl is a congener comprising the biphenyl structure with two chlorine substituents, one on each of the two carbons at the "4" (also known as "para") positions of the two rings. PCBs can exist as 209 possible chlorinated biphenyl congeners, although only about 130 of these generally are found commercially. PCBs were first marketed in 1929 and were manufactured in various countries and with different trade names (e.g., Aroclor, Clophen, and Phenoclor). See Table 11.3 for a listing of the different trade names used for PCBs.

Most PCBs were manufactured by combining a biphenyl compound with ferrous or ferric iron (Fe^{+2} or Fe^{+3}, respectively), yielding a complex mixture of different congeners of PCBs. The name used by the manufacturer, Monsanto, for these mixtures was Arochlor. The three most common PCB mixtures were Aroclor 1242, Arochlor 1254, and Arochlor 1260. The "12" indicates that it contains 12 carbons (i.e., a biphenyl arrangement). The next two numbers indicate the percentage of chlorine represented in the molecular mass, that is, 42%, 54%, and 60%, respectively. Thus, a congener can, to some extent, be identified by its molecular weight, which obviously differs because chlorine (Cl) atoms are substituting for hydrogen (H) atoms on the rings, a difference in atomic weight of about 34.5 for each substitution. That is, each Cl weighs about 35.5 and each hydrogen weighs 1, so every substitution results in a net increase of 34.5 in molecular weight. However, compounds with the same formula can also have different arrangements, or isomers. A homolog is a subcategory of PCB congeners, all with the same number of chlorine substituents. For example, the hexachlorobiphenyls, or hexa-PCBs, hexa-CBs, or simply HxCBs, are all PCB congeners with exactly six chlorine substituents that may be in any arrangement. See Table 11.4 for a complete list of PCB homologs.

In this case, for example, a hexachlorobiphenyl (HxCB) will have six chlorines, but they can be at different locations on the molecule. This is important in that isomeric differences can lead to differences in persistence, bioaccumulation, and toxicity. Thus, the best way to describe an individual congener, as for other halogenated aromatic compounds, is to use the numbering system in the molecule shown earlier. For example, consider one arrangement of a six-chlorine PCB. 2,2′,4,4′,5,5′-hexachlorobiphenyl has the arrangement of:

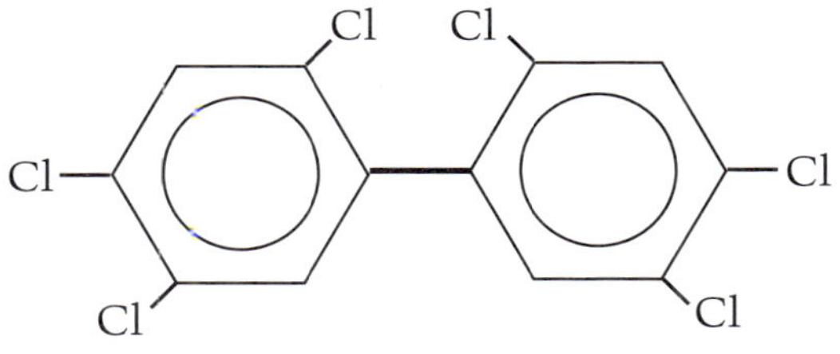

This compound is also known as the 153rd PCB congener, or CB-53.

As shown in Table 11.5, PCBs are very persistent; they resist breakdown in the environment.

Because of their chemical stability and heat resistance, they were used worldwide as dielectric fluids in electrical equipment, especially transformers and capacitors, as well as in hydraulic and heat exchange fluids and

TABLE 11.3
PCB trade names and synonyms.

Aceclor	Diaclor	Orophene
Adkarel	Dicolor	PCB
ALC	Diconal	PCB's
Apirolio	Diphenyl, chlorinated	PCBs
Apirorlio	DK	Pheaoclor
Arochlor	Duconal	Phenochlor
Arochlors	Dykanol	Phenoclor
Aroclor	Educarel	Plastivar
Aroclors	EEC-18	Polychlorinated biphenyl
Arubren	Elaol	Polychlorinated biphenyls
Asbestol	Electrophenyl	Polychlorinated diphenyl
ASK	Elemex	Polychlorinated diphenyls
Askael	Elinol	Polychlorobiphenyl
Askarel	Eucarel	Polychlorodiphenyl
Auxol	Fenchlor	Prodelec
Bakola	Fenclor	Pydraul
Biphenyl, chlorinated	Fenocloro	Pyraclor
Chlophen	Gilotherm	Pyralene
Chloretol	Hydol	Pyranol
Chlorextol	Hyrol	Pyroclor
Chlorinated biphenyl	Hyvol	Pyronol
Chlorinated diphenyl	Inclor	Saf-T-Kuhl
Chlorinol	Inerteen	Saf-T-Kohl
Chlorobiphenyl	Inertenn	Santosol
Chlorodiphenyl	Kanechlor	Santotherm
Chlorphen	Kaneclor	Santothern
Chorextol	Kennechlor	Santovac
Chorinol	Kenneclor	Solvol
Chorinol	Leromoll	Sorol
Clophen	Magvar	Soval
Clophenharz	MCS 1489	Sovol
Cloresil	Montar	Sovtol
Clorinal	Nepolin	Terphenychlore
Clorphen	No-Flamol	Therminal
Decachlorodiphenyl	NoFlamol	Therminol
Delor	Non-Flamol	Turbinol
Delorene	Olex-sf-d	

Source: U.S. Environmental Protection Agency, 2005. PCB ID—Definitions: http://www.epa.gov/toxteam/pcbid/defs.htm; accessed April 4, 2005.

TABLE 11.4
PCB homologs.

PCB Homolog	*Cl Substituents*	*PCB Congeners*
Monochlorobiphenyl	1	3
Dichlorobiphenyl	2	12
Trichlorobiphenyl	3	24
Tetrachlorobiphenyl	4	42
Pentachlorobiphenyl	5	46
Hexachlorobiphenyl	6	42
Heptachlorobiphenyl	7	24
Octachlorobiphenyl	8	12
Nonachlorobiphenyl	9	3
Decachlorobiphenyl	10	1

TABLE 11.5
Physical, chemical, and toxicological properties of polychlorinated biphenyls (PCBs).

Properties	*Persistence/Fate*	*Toxicity**
Water solubility decreases with increasing chlorination: 0.01 to 0.0001 $\mu g L^{-1}$ at 25°C; vapor pressure: 1.6–0.003 × 10-6 mm Hg at 20°C; log K_{ow}: 4.3–8.26.	Most PCB congeners, particularly those lacking adjacent unsubstituted positions on the biphenyl rings (e.g., 2,4,5-, 2,3,5- or 2,3,6-substituted on both rings) are extremely persistent in the environment. They are estimated to have $T_{1/2}$ ranging from three weeks to two years in air and, with the exception of mono- and dichlorodiphenyl, more than six years in aerobic soils and sediments. PCBs also have extremely long $T_{1/2}$ in adult fish; for example, an eight-year study of eels found that the $T_{1/2}$ of chlorobiphenyl 153 was more than 10 years.	Acute toxicity (50% lethal concentration = LC_{50}) for the larval stages of rainbow trout is 0.32 $\mu g L^{-1}$ with a threshold of no observed adverse effect level (NOAEL) of 0.01 $\mu g L^{-1}$. The acute toxicity of PCB in mammals is generally low and LD_{50} values in rats of 1 $g kg^{-1}$ body weight. The International Agency for Research on Cancer (IARC) has concluded that PCBs are carcinogenic to laboratory animals and probably also for humans. They have also been classified as substances for which there is evidence of endocrine disruption in an intact organism.

*Chemical half-life = $T_{1/2}$
Lethal dose to 50% of tested organism = LD_{50}
Lethal concentration to 50% of tested organism = LC_{50}
Bioconcentration factor = BCF
No observed adverse effect level = NOAEL
No observed adverse effect concentration = NOAEC
Source: United Nations Environmental Programme, 2002. Chemicals: North American Regional Report, Regionally Based Assessment of Persistent Toxic Substances, Global Environment Facility.

lubricating and cutting oils. Most PCB congeners, especially the ones that do not have adjacent unsubstituted positions on the benzene rings (e.g., 2,4,5-, 2,3,5- or 2,3,6-substituted on both rings), have estimated half-lives ($T_{1/2}$) ranging from three weeks to two years in air and, with the exception of those with only one or two chlorines (i.e., mono- and di-chlorobiphenyls), more than six years in aerobic soils and sediments. PCBs also have large $T_{1/2}$ values in fish and other aquatic fauna.[31] All PCBs also readily accumulate in tissue, including mammalian and human tissue, and are associated with a wide range of health effects, including cancer, endocrine disruption, immunity disorders, reproductive and developmental effects, and nervous system damage. There is convincing evidence that PCBs cause cancer in animals. In addition, a number of epidemiological studies of workers exposed to PCBs have been performed. Results of human studies raise concerns for the potential carcinogenicity of PCBs. Studies of PCB workers found increases in rare liver cancers and malignant melanoma. The presence of cancer in the same target organ (liver) following exposures to PCBs both in animals and in humans and the finding of liver cancers and malignant melanomas across multiple human studies adds weight to the conclusion that PCBs are probable human carcinogens. There is also a great deal of concern about health effects on yet-to-be-born, newborn, and infant children because PCBs can effectively migrate through the placental barrier, which can lead to myriad problems, including low birth weight[32] and delayed central nervous system function following *in utero* PCB exposure.[33]

Concern over the toxicity and persistence in the environment of PCBs caused the U.S. Congress in 1976 to enact a specific section, 6(e), of the Toxic Substances Control Act (TSCA) to address PCB contamination. This included prohibitions on the manufacture, processing, and distribution in commerce of PCBs. This is the so-called "cradle to grave" (i.e., from manufacture to disposal) management of PCBs in the United States. Similar prohibitions and management measures were adopted worldwide.

Back to Warren County:

The landfill was located on a 142-acre tract about three miles south of the town of Warrenton, and held about 60,000 tons of contaminated soil collected solely from the contaminated roadsides. The U.S. EPA permitted the landfill under the Toxic Substances Control Act, which is the controlling federal regulation for PCBs. The state owns approximately 19 acres of the tract and Warren County owns the remaining acreage surrounding the state's property. The containment area of the landfill cell occupied approximately 3.8 acres, enclosed by a fence. The landfill surface dimension was approximately 100 m × 100 m with a depth of approximately 8 m of contaminated soil at the center. The landfill was equipped with both polyvinyl chloride and clay caps and liners, with a dual leachate collection system. The landfill was never operated as a commercial facility.

The site is located in the Shocco Township of the county, which has a population of approximately 1,300. Sixty-nine percent of the township res-

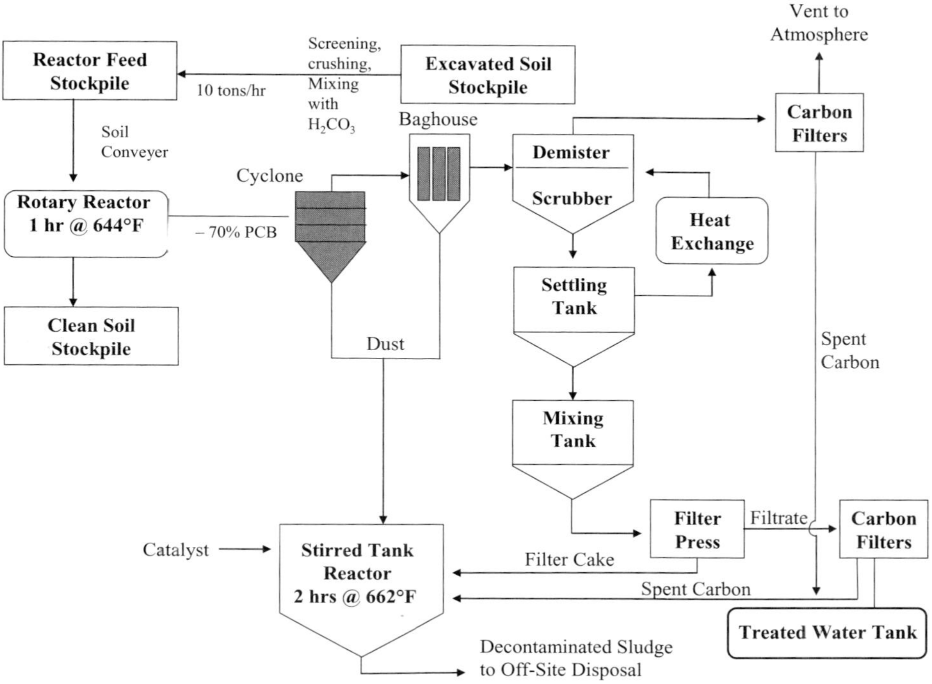

FIGURE 11.11. Base Catalyzed Decomposition (BCD). This is the process recommended to treat PCB-contaminated soil stored in Warren County, North Carolina. *Source*: Federal Remediation Technologies Roundtable, 2002. *Screening Matrix and Reference Guide*, 4e, Washington, D.C.

idents are nonwhite and 20% of the residents have incomes below the federal poverty level.

Residents of Warren County and civil rights leaders passionately protested the location of the landfill in Warren County. These protests are considered the watershed event that brought environmental justice to the national level. In 1982, during the construction of the landfill, then-Governor Jim Hunt made a commitment to the people of Warren County. He stated that if appropriate and feasible technology became available, the state would explore detoxification of the landfill.

In 1994, a Working Group, consisting of members of the community and representatives from the state, began an in-depth assessment of the landfill and a study of the feasibility of detoxification. Tests using landfill soil and several treatment technologies were conducted. In 1998, the working group selected base catalyzed decomposition (BCD) as the most appropriate technology (see Figure 11.11). Approximately $1.6 million in state funds had been spent by this time. In 1999, the Working Group fulfilled its mission and was reformed into a community advisory board. In the BCD process, PCBs are separated from the soil using thermal desorption. Once separated, the PCBs are collected as a liquid for treatment by the BCD process. BCD is a nonincineration, chemical dechlorination process that transforms PCBs, dioxins, and furans into nontoxic compounds. In the process, chlorine atoms are chemically removed from the PCB and dioxin/furan molecules and replaced with hydrogen atoms. This converts the compounds to biphenyls, which are much less hazardous. Treated soil is returned to the landfill, and the organics from the BCD process are recycled as a fuel or disposed off-site as nonhazardous waste.

A cleanup target of 200 parts per billion (ppb) was established by the working group for the landfill site and was made a statutory requirement by the N.C. General Assembly. The EPA cleanup levels for high occupancy usage is 1 part per million (ppm). EPA's examples of high-occupancy areas include residences, schools, and day-care centers; thus the target is five times lower than the EPA requirement. The removal of PCBs from the soil will eliminate further regulation of the site and permit unrestricted future use.

A public bid opening was held on December 22, 2000, for the site detoxification contract. Site preparation work was completed in December 2001. Work included the construction of concrete pads and a steel shelter for the processing area, the extension of county water, an upgrade of electrical utilities, and the establishment of sediment and erosion control measures.

The treatment equipment was delivered in May 2002. An open house was held on-site the next month so community members could view the site and equipment before start-up. Initial tests with contaminated soil started at the end of August 2002. The EPA demonstration test was performed in January of 2003. An interim operations permit was granted in March based on the demonstration test results. Soil treatment was completed in October of 2003. A total of 81,600 tons of material was treated from the landfill site. The treated materials included the original contaminated roadside soil and soil adjacent to the roadside material in the landfill that had been cross-contaminated. The original plan specified using the BCD process to destroy the PCBs after thermal desorption separated them from the soil. With only limited data available to estimate the quantity of liquid PCBs that would be collected, conservative estimates were used to design the BCD reactor. In practice, the quantity of PCBs recovered as liquid was much less than anticipated. The BCD reactor tanks were too large to be used for the three-run demonstration test required under TSCA to approve the BCD process. As an alternative, one tankload of liquid containing PCBs was shipped to an EPA permitted facility for destruction by incineration. Most of the equipment was decontaminated and demobilized from the site by the end of 2003. Site restoration will be complete in the spring . . . once vegetation has become established. The total cost of the project was \$17.1 million.

The Orange County, North Carolina, Landfill[34]

Sometimes the best cases to illustrate environmental justice are close to home. For example, every community has a landfill, wastewater treatment plant, and other environmental facilities. And unlike some of the more prominent environmental justice cases, such as the PCB landfill in Warren County, NC, many injustices take place in small neighborhoods or even around a public facility that is rather isolated. So, it is important to analyze a case that represents this broader, more latent problem that is less likely to garner national and international headlines. One such case took place in Chapel Hill, North Carolina. Chapel Hill has become a burgeoning community, but was once a charming village hosting the University of North Carolina, the flagship university in the North Carolina higher education system. The town has a rich history, especially as the home to the first state institution to open its doors to students and as a survivor of the invasion of the Union troops during the Civil War.

Chapel Hill remained a village until the 1960s when the expansion of the university caused a surge in population and pressure for new developments. During that time Chapel Hill also was becoming a mecca for retired people, with its mild climate, great golf courses, beautiful gardens, and of course the advantages of a first-rate university drawing people from the Northeast. The village was becoming a city of over 56,000 people.

During the 1960s progressive era, Chapel Hill organized the first truly integrated school system in North Carolina, carving out the central section of town in a way that essentially integrated all schools. This forward-looking liberal attitude carried through in the election of municipal officers, and it was no wonder that Chapel Hill was the first town in North Carolina to elect an African American as mayor.

Howard Lee was a talented and hard-working mayor who went on to become a state senator. During his tenure as mayor, he had to grapple with intense development pressures that necessitated the organization of many municipal services, including the creation of a bus service.

At that time the town was using a small landfill owned by the university for the disposal of its solid waste, but this landfill was rapidly running out of space and the university wanted to close it, so in 1972 a search commenced for a new landfill site. Searches then were not nearly as intense as they are today, and the entire process was quite informal. The town council decided that it wanted to buy a piece of land to the north of the town and construct the new landfill there. This land seemed like a good choice since it was between Chapel Hill and Hillsborough, the county seat of Orange County. It was also a convenient location for Carrboro, a small community next to Chapel Hill. There were no new housing developments near the proposed landfill site and it was off a paved road, Eubanks Road, and this would facilitate the transport of refuse to the landfill.

There was, however, a vibrant African American community, the Rogers Road neighborhood, that abutted the intended landfill area, and these people expressed their dissatisfaction with the choice of a landfill site and went to Mayor Lee for help. The mayor talked them into accepting the decision and allegedly promised them that this would be the one and only landfill that would be located near their neighborhood, and if they could endure this affront for ten years, the finished landfill would be made into a neighborhood park. Most importantly, they were told that the next landfill for Chapel Hill would be somewhere else and that their area would not become a permanent dumping site. The citizens of the Rogers Road neighborhood grudgingly accepted this deal and promise and then watched as the Orange County Regional Landfill was built near their community.

The site for the landfill was 202 acres, cut into two sections by Eubanks Road, and abutting Duke Forest, a research and recreational facility owned by Duke University. On one side of the site was the Rogers Road neighborhood.

The landfill, which had no liner or any other pollution control measures, was opened in 1972. The three communities contributing to the landfill, Chapel Hill, Carrboro, and Hillsborough, along with Orange County, formed a quasi-governmental body called the Landfill Owners Group (LOG) to operate the landfill. The LOG was composed of elected officials from the four governmental bodies. One of the early actions by this group was to establish a sinking fund that would eventually pay for the expansion of this landfill or a new site when this became necessary.

As the population of Orange County exploded in the 1970s, it became quite clear that this landfill would not last very long and that a new landfill would be needed fairly soon. LOG, using money from tipping fees, purchased a 168-acre tract of land next to the existing landfill, called the Green Tract, with the apparent intent of using it when the original landfill became full, but without actually publicly declaring that this was the intended use for this land.

In the early 1980s, it became quite apparent that a new landfill would be necessary, but by that time the Green Tract was considered to be too small. This would not be a long-term solution, and a need was apparent for a larger site that would accommodate the needs on a long term. The four governmental agencies asked LOG to initiate proceedings to develop a new landfill, which could be opened in the mid-1990s.

The LOG set up a landfill selection committee (LSC) to oversee the selection of the new landfill and asked Eddie Mann, a local respected banker and civic-minded citizen, to chair the LSC. The LOG directed the LSC to seek technical help with the selection process, and as a result, Joyce Engineering, a Virginia firm that had assisted other communities in the selection of landfills, was hired to conduct the search.

After a study of Orange County, Joyce Engineering selected 16 locations as potential landfill sites, using criteria established by the LSC such as proximity to cities, airports, and environmentally sensitive areas. One of the 16 sites chosen by Joyce was the Green Tract, which became known as OC-3.

The next step was to hold public hearings and then to cull the list of 16 down to a smaller list for final discussion. As the 16 sites were being considered, each was placed in one of three categories: 1) to be considered further, 2) to be placed in reserve for possible consideration later, or 3) not to be considered further.

The public hearings were classic "Not in My Back Yard" (so called, "NIMBY") exercises. Neighbors who lived around their sites hired lawyers and environmental scientists or were fortunate enough to have lawyers, physicians, and engineers as neighbors to persuade the LSC that their site simply was inappropriate. In other cases the members of the LSC themselves had a reason to eliminate a specific site from consideration. Often the classification of a site into the third (not to be considered further) category was on what appeared to be flimsy evidence. In one case, a member of the LSC who happened to live near a site said that this was nice farmland and sheep would graze on the hillside. This apparently was a sufficient reason for eliminating this site from further consideration. There was no overt collusion or visible trading of votes, but it became quite clear to observers that many decisions had already been made far in advance of the public hearing.

The Rogers Road neighborhood (and the Green Tract, which was one of the possible sites being considered) was represented on the LSC by a graduate student who did not live in the neighborhood and who appeared to have little interest in the outcome.

Following these hearings, the LSC pared down the original 16 sites to five, one of which still was the Green Tract. The argument that the former mayor of Chapel Hill had promised the residents in that neighborhood that future landfills would be elsewhere was not considered persuasive by the LSC. Since Howard Lee, the former mayor of Chapel Hill, did not represent Carrboro, Hillsborough, or Orange County, the well-intentioned promise was not considered binding by the other governmental entities. In addition, although Lee acknowledged making this promise, this was never found on any written document.

One of the problems with the Green Tract was that it was too small to afford a long-term solution, a source of encouragement to the Rogers Road neighborhood. But this was all changed when, late in the process and well after the public hearings, Eddie Mann introduced a new site, named OC-17. This site abutted the existing landfill and the Rogers Road neighborhood, and included a large tract of land in Duke Forest, a section called the Blackwood Mountain region.

The introduction of this site and its acceptance by the LSC as a finalist was a case of local politics at their worst. The people who were least able to resist the backdoor expansion of the existing landfill, the Rogers

Road neighborhood, were told that the promises made by elected officials were null and void because the new politicians could not be held to these promises. The effect of this argument was to suggest that any promise made by one administration does not need to be kept by another. This is analogous to buying savings bonds from the federal government with no guarantee that it will be redeemed in 10 years since a new administration will be in Washington. Or, in environmental parlance, all political decisions are unsustainable.

The opponents of these two tracts, OC-3 (the original Green Tract) and OC-17 (the new Blackwood Mountain area) began to fight the selection process, aided by many Chapel Hillians who saw the inequity in this process. The resisters packed the LSC committee meetings, printed T-shirts ("WE HAVE DONE OUR SHARE"), wrote letters to the newspaper, and did everything they could to keep the inevitable from happening.

In 1995 the LSC approved the selection of OC-3 and OC-17 as the new landfill, but suggested that some form of compensation be made to the citizens in the Rogers Road neighborhood. The decision next went to the LOG for their consideration.

The vote in the LOG was 6–3 in favor of the selected site. Two of the negative votes were by the representatives from Carrboro. The town of Carrboro would not be directly affected by the location of the landfill in the Eubanks Road area, and thus Carrboro ought to have had a clear selfish motive for choosing this site. But the two Carrboro representatives on LOG, Mayor Mike Nelson and Alderwoman Jacquelyn Gist, based their negative vote on the promise made by Howard Lee to the Rogers Road neighborhood, and announced that they would fight the selection of this site.

Nevertheless, having been approved by the LOG, the decision next went to the four governmental bodies for approval. Chapel Hill, Hillsborough, and Orange County approved the site with little debate. In the meeting of the Chapel Hill Town Council the previous promise by Mayor Lee was not even brought up. But Mayor Nelson and Alderwoman Gist convinced the Carrboro council to delay the approval until compensation could be worked out *in advance* of the decision, citing the previous broken promises as loss of trust in politicians.

This delay by Carrboro allowed Duke University to marshal its forces and to hire appropriate lawyers and scientists to come to the defense of Duke Forest. The university trustees voted unanimously to fight the siting, and the president of Duke, Nan Keohane, wrote a strong letter to the LOG and the four governmental bodies threatening legal action if the land in Duke Forest was to be taken. Using his knowledge of the area, Jud Edeburn, the manager of the Duke Forest, quickly located areas with endangered species and several wetland locations, thus reducing the available acreage for the landfill. A historic African American cemetery was discovered in the forest and placed on the protected National Registry, further reducing the availability of land. But Joyce Engineering found ways to redesign the landfill so as to accommodate these restrictions and to still use the major part of the tract for burial of solid waste. Demands for public hearings and more tests did not change the decision, and a year after the vote, OC-17 remained the first choice of the LOG and the three governments. The government of Carrboro was under increasing pressure to cave in.

Then, in 1997, Duke University announced that it had deeded the Blackwood Mountain section of Duke Forest to NASA for conducting experiments. The federal government now controlled this land and the fight was over. It took clever legal work, the effective battle fought by the citizens of the Rogers Road neighborhood, and the courage of Carrboro's Mayor Nelson and Alderwoman Gist to stop the landfill from being sited at a location where the people had already done their share.

If It Does Occur, It Is Not Bad[35]

The second line of argument advanced against the environmental justice movement is to first admit that there might be environmental discrimination, but that this discrimination is not harmful. Suppose the price of land in our imaginary community is as shown:

Neighborhood	*Percent Minority*	*Land Cost, \$/Acre*
A	10	20,000
B	5	15,000
C	35	10,000
D	95	3,000

Now suppose a municipal facility like a solid waste incinerator would need 100 acres. The cost to the municipality would be \$2,000,000 if the facility is built in neighborhood A, and \$300,000 if it were built in neighborhood D. This is a large savings that would be passed on to the taxpayers in the community. Hence (the argument goes) it is more advantageous to build the plant in neighborhood D, because all members of the community, including the residents of neighborhood D, would share in the cost savings by having lower taxes. In addition, the lower property values in neighborhood D caused by the undesirable facility will also represent a savings to the property owners since the property taxes will be lower.

Thus, the greatest savings in taxes would be for those who pay the most, so the argument that everyone shares in the economics of using lower-priced land is not persuasive. In addition, lower taxes should not be an excuse for unfair and unequal treatment of all citizens.

Another utilitarian (greatest good) argument for using the lower socioeconomic neighborhood is that people who live there would most likely have high unemployment, and a facility located there would provide jobs. These critics point to the influx of less-advantaged people around such urban facilities as incinerators and suggest that the facilities were actually beneficial to the neighborhood.

But this is not the point. By disparate distribution of environmental contamination we are still unfairly treating some less advantaged part of our population.

Finally, another argument advanced against environmental justice is to say that whatever the situation might be, the cure is far worse than the disease. For example, the elimination of pollution would be extremely costly, and in the long run, impossible. We could not as a nation set a goal of zero pollution and still hope to pay for other social needs.

This is true, but not germane. The issue is the fair distribution of environmental costs and benefits.

If It Does Occur and It Is Bad, It Is Not Racially Motivated

The third argument against environmental injustice accepts that there certainly seems to be disparity in the siting of pollution-producing facilities, and acknowledges that often the less-advantaged residents do not have much choice in the placement of such facilities—in other words, discrimination does occur. But, the argument goes, this unfairness is not racially motivated.

Proponents of environmental justice argue, as does Kristen Shrader-Frechette, that there seems unlikely to be any other reason for such injustice.

> If the area closest to a noxious facility tends to have a population of nonwhites rather then whites, then regardless of what ZIP codes (or any other system of aggregation) reflect, there is likely to be environmental racism.[36]

Racism is a difficult and often contentious issue. It is like any other –ism; it is a belief in something that is taken on faith and cannot be proven. Even further, sometimes rational proof does not change the thinking of a true believer. Racism is the belief that some racially identifiable group is inferior to some other group, or does not deserve equal moral or legal protection. This belief then leads to racial discrimination, or the actions that are manifested based on this belief.

In the case of environmental justice, racial discrimination can be defined as:

> . . . those institutional rules, regulations, and policies of government or corporate decisions that deliberately target certain communities

> for least desirable land uses, resulting in the disproportionate exposure of toxic and hazardous waste on communities based upon prescribed biological characteristics. Environmental racism is the unequal protection against toxic and hazardous waste exposure and the systematic exclusion of (disadvantaged groups) from decisions affecting their communities.[37]

But saying that this occurs and proving it are two different issues. There seems to be no doubt as to the fact that racially identifiable socioeconomic groups have been unfairly treated in terms of environmental contamination. But is this due to racial discrimination? If such discrimination can be shown to be racially motivated, then the Civil Rights Act, Title VI §601, which makes racial discrimination illegal, can be used to correct the injustice. Proving racial discrimination is, however, quite difficult. What has to happen in order to prove that racial discrimination has occurred is for the person responsible to admit that he or she intentionally discriminated on the basis of race. This is an admission of guilt and thus unlikely to occur.

Also, much of the alleged discrimination is corporate; that is, there is not a single person or small group of persons engaging in the acts of discrimination. It is more a manifestation of company or other corporate policies and actions. Finding the guilty parties can be a tortuous process.

The next section in the Civil Rights Act, Title VI §602, makes it easier to prove racial discrimination by stating that if there is unfair treatment of an identifiable minority, then there is *de facto* proof that racial discrimination has occurred. But this is the section of the act that was basically nullified by a 5–4 Supreme Court decision.

Environmental injustice may seem intractable, but progress is being made. The facts are that environmental inequality exists, and that often it is the minority populations in our country who bear the brunt of the pollution.

Is Environmentalism a Middle-Class Value?

An environmental response is often precipitated first by a complaint. But to complain, one must have a "voice." If a certain group of people has had little or no voice in the past, they are likely to feel and be disenfranchised. Although there have been recent examples to the contrary, African American communities have had little success in voicing concerns about environmentally unacceptable conditions in their neighborhoods. Hispanic Americans may have even less voice in environmental matters since their perception of government, the final arbiter in many environmental disagreements, is one of skepticism and outright fear of reprisal in the form of being deported or being profiled. Many of the most adversely affected communities are not likely to complain.

Land use is always a part of an environmental assessment. However, justice issues are not necessarily part of these assessments. Most environmental impact assessment handbooks prior to the late 1990s contained little information and guidelines related to fairness issues in terms of housing and development. They were usually concerned about open space, wetland and farmland preservation, housing density, ratios of single- *versus* multiple-family residences, owner occupied housing *versus* rental housing, building height, signage and other restrictions, designated land for public facilities like landfills and treatment works, and institutional land uses for religious, health care, police, and fire protection.

When land uses change (usually to become more urbanized), the environmental impacts may be direct or indirect. Examples of direct land use effects include eminent domain, which allows land to be taken with just compensation for the public good. Easements are another direct form of land use impacts, such as a 100-m right-of-way for a highway project that converts any existing land use (e.g., farming, housing, or commercial enterprises) to a transportation use. Land use change may also come about indirectly, such as so-called secondary effects of a project that extend, in time and space, the influence of a project. For example, a wastewater treatment plant and its connected sewer lines will create accessibility, which spawns suburban growth.[38] People living in very expensive homes may not even realize that their building lots were once farmland or open space and, had it not been for some expenditure of public funds and the use of public powers like eminent domain, there would be no subdivision.

Environmentalists are generally concerned about increased population densities, but housing advocates may be concerned that once the land use has been changed, environmental and zoning regulations may work against affordable housing. Even worse, environmental protection can be used as an excuse for some elitist and exclusionary decisions. In the name of environmental protection, certain classes of people are economically restricted from living in certain areas. This problem first appeared in the United States in the 1960s and 1970s in a search for ways to preserve open spaces and green areas. One measure was the minimum lot size. The idea was that rather than having the public sector securing land through easements or outright purchases (i.e., fee simple) to preserve open spaces, developers could either set aside open areas or require large lots in order to have their subdivisions approved. Thus, green areas would exist without the requisite costs and O&M entailed by public parks and recreational areas. Such areas have numerous environmental benefits, such as wetland protection, flood management, and aesthetic appeal. However, minimum lot size translates into higher costs for residences. The local rules for large lots that result in less affordable housing is called *exclusionary zoning*. One value (open space and green areas) is pitted against another (affordable housing). In some cases, it could be argued that preserving open spaces is simply a tool for excluding people of lesser means or even people of minority races.[39]

Habitat for Humanity

A recent case reflects the common problem of competing values in land use. The housing advocacy group, Habitat for Humanity, proposed a development of affordable houses in a college town in the Southeast. The cost of housing in this town is well above the state average, so a number of advocates supported the Habitat model, where potential homeowners invest in their own homes through "sweat equity" and receive voluntary support. But some groups formed in opposition to the plan. In an early meeting, one neighbor stated a desire that the homes be like a nearby high-cost subdivision (houses costing much more than even the already expensive town average). She recommended that they be "single family homes with a nice, friendly, college town look and feel."[†] Another later said that "From day one, we have said that that parcel is not suited for a high-density project." That may be the case, but the result of such thinking is, in the end, exclusionary. People are very passionate and protective about their neighborhoods, well beyond concern about property values. This is a form of "NIMBY," so common to the environmental engineer who must balance science and social needs to site unpopular facilities like landfills and treatment plants.

[†]The quotes have been changed to protect anonymity, but the meanings are maintained.

Engineers, in order to best serve their clients, must be sensitive to the fact that most of us want to protect the quality of our neighborhoods but, at the same time, engineers and land use planners must take great care that their ends (environmental protection) are not used as a rationale for unjust means (unfair development practices). Environmental laws and policies, like zoning ordinances and subdivision regulations, should not be used as a means to keep lower socioeconomic groups out of privileged neighborhoods.

Exclusionary zoning touches on the larger issue of justice. It has everything to do with fairness and how certain groups may bear a disproportionate burden. Historically, certain groups have borne disproportionate exposure to pollution and such exposures have led to disproportionate adverse health effects, not to mention their having to live in an unpleasant environment. Some of this occurred unintentionally, as a result of migrations near pollution sources. Some highly polluted industrial sites near U.S. cities were home to working class, predominantly white workers and their families for several decades in the first half of the twentieth century. In fact, in the United States, even well-paid plant managers lived near their factories. Much of this was dictated by available transportation. Plant owners often lived well away from the plant, since they had the means and the flexibility to visit as needed. Before air and water pollution controls became prevalent, people living near steel mills, coke ovens, chemical processing facilities, and power plants were exposed to visible plumes of pollutants. And surface waters commonly showed visible signs of pollution. As the result of personal automobile ownership and increased wages, workers were able to move to nicer neighborhoods. They were replaced by lower income people even though many of these lower income families were not employed at the polluting plant. Over time, many of these undesirable neighborhoods became populated largely by African Americans. In some cases, the neighborhoods continued to change from predominantly African American populations to recently arrived immigrants.

This incidental injustice is associated with higher than average pollutant exposures. Intentional injustice, however, is even worse. That is, some industries have been accused of intentionally locating polluting facilities in areas inhabited by lower socioeconomic groups. The default in environmental matters is that complaints start the process of assessment, and, ultimately, action. This often works best in higher income neighborhoods. In that sense, this complaint system is really a middle- or even upper-class process.

How can the engineering profession help people who have historically had no voice or who have not been taught how to make it heard? For environmental fairness, everyone potentially being affected needs a voice and a place at the table from the earliest planning stages of a project. The default seems to be changing, but some argue that environmental quality still is being used, knowingly or innocently, to work against fairness. And people who are likely to be exposed to the hazards brought about by land use decisions need to be aware of options well before decisions are made. This should be applied not only to land development and civic decisions, but to everything engineers do that may have an impact on the health, safety, and welfare of the public.

Sometimes we need to remind ourselves of just how far we have come in battling pollution in the past 50 years. We also need to realize that the general perception of environmental quality and the public's expectations have grown substantially in a relatively short time. The revolution in thinking and the public acceptance of strong measures to regulate the actions of private industry has been phenomenal. So, what may have previously been considered to be simply the "cost of doing business" (brown haze, smelly urban areas, obviously polluted water, and disposal of pollutants in pits, ponds, lagoons, and by land burial) are now considered to be inappropriate and even immoral activities. However, it is not automatic that all private and public entities have gotten that message (for example, see the case study, "Carver Terrace"). Engineers can help continue to raise their client's appreciation of fairness and justice, as well as the improvement to the bottom line that can result from strong environmental programs.

Carver Terrace, Texas[40]

In the 1970s the citizens of Carver Terrace, in Texacana, Texas, a predominantly African American community, began to see dark, vile-smelling gunk oozing out of their lawns. They could not interest the local authorities in the problem, even when they started to believe that their community was experiencing a higher than usual number of medical problems.

In 1978 the problems at Love Canal emerged and hazardous waste became a national issue and problem. A year later, after Congress ordered large chemical companies to identify what chemicals they had disposed of and at what locations, the discovery was made that Koppers Company of Pittsburgh had operated a creosote plant in the area now known as Carver Terrace, and when Koppers closed the plant they bulldozed everything into the ground, including the vats and pond holding creosote, a known human carcinogen.

Actually, creosote is a complex mixture of several highly toxic compounds (see Table 11.6) that can elicit numerous health effects including burns, skin irritation, convulsions, and kidney and liver problems. Because the land was inexpensive, poorer families eagerly bought lots and built homes in what became known as Carver Terrace. About 25,000 people, 85% a racial minority, lived within four miles of the former creosote plant.

When it became known that the ground contained large quantities of creosote, the U.S. EPA sent a team of hazardous waste experts in moon suits who concluded that there was no problem. The citizens knew better and soon found out that the EPA had done other studies and reports that clearly showed this area to be a candidate for the Superfund cleanup, but that these were not made available to the citizens of Carver Terrace.

In retrospect, the extent of pollution was large; 170 million liters of shallow groundwater were contaminated, along with 1,650 cubic meters of soil, contaminated to a depth of one foot. The situation was so bad that the citizens' group urged the government to "buy out" the residents of Carver Terrace just as they were currently buying out the

TABLE 11.6
Major components of wood creosote.

phenol
methylhydroxycyclopentenone
ortho-cresol
dimethylhydroxycyclopentenone
para-cresol
guaiacol
2,6-xylenol
3,4-xylenol
6-methylguaiacol
3,5-xylenol
2,4-xylenol
2,5-xylenol
2,3-xylenol
3-methylguaiacol
5-methylguaiacol
4-methylguaiacol
2,4,6-trimethylphenol
2,3,6-trimethylphenol
4-ethylguaiacol
4-ethyl-5-methylguaiacol
4-propylguaiacol
Other compounds (unknown peaks in chromatogram)

Source: Agency for Toxic Substances and Disease Registry, 2002, Toxicological Profile for Creosote, Washington, D.C.

people who lived around Love Canal. They could not see why they were being treated differently from the people around Love Canal. The only obvious difference, at least to the residents, was that Love Canal was predominantly European American and Carver Terrace was mostly African American. Eventually, through energetic and determined activism, the residents of Carver Terrace were also bought out.

Public decisions have brought lower socioeconomic communities into environmental harms way. Although public agencies, such as housing authorities and public works administrations, do not have a profit motive, *per se*, they do need to address budgetary and policy considerations. If open space is cheaper and certain neighborhoods are less likely to complain (or by extension, vote against elected officials), the "default" for unpopular facilities, such as landfills and hazardous waste sites, may be to locate them in lower income neighborhoods. Also, elected and appointed officials and bureaucrats may be more likely to site other types of unpopular projects, such as public housing projects, in areas where complaints are less likely to be put forth or where land is cheaper (see the case study, "West Dallas Lead Smelter").

West Dallas Lead Smelter[41]

In 1954 the Dallas, Texas, Housing Authority built a large public housing project on land immediately adjacent to a lead smelter. The project had 3,500 living units and became a predominantly African American community. During the 1960s, the lead smelter stacks emitted over 200 tons of lead annually into the air. Recycling companies had owned and operated the smelter to recover lead from as many as 10,000 car batteries per day. The lead emissions were associated with blood lead levels in the housing project's children and these were 35% higher than in children from comparable areas. Lead is a particularly insidious pollutant because it can result in developmental damage. Study after study showed that the children at this project were in danger of higher lead levels, but nothing was done for over 20 years. Finally, in the early 1980s, the city brought suit against the lead smelter, and the smelter immediately initiated control measures that reduced its emissions to allowable standards. The smelter also agreed to clean up the contaminated soil around the smelter and to pay compensation to people who had been harmed.

This case illustrates two issues of environmental racism and injustice. First, the housing units should never have been built next to a lead smelter. The reason for locating the units there was justified on the basis of economics. The land was inexpensive and thus this saved the government money. The second issue was the foot dragging by the city in insisting that the smelter clean up the emissions. Once the case had been made, within two years the plant was in compliance. As of 2003, blood lead levels in West Dallas are now below the national average. Why did it take 20 years for the city to do the right thing?

Lessons Applied: The Environmental Justice Movement

In spite of the general advances in environmental protection in the United States, the achievements have not been evenly disseminated throughout our history. Environmental science and engineering, like much of the rest of our culture for the past three centuries, has not been completely just and fair.

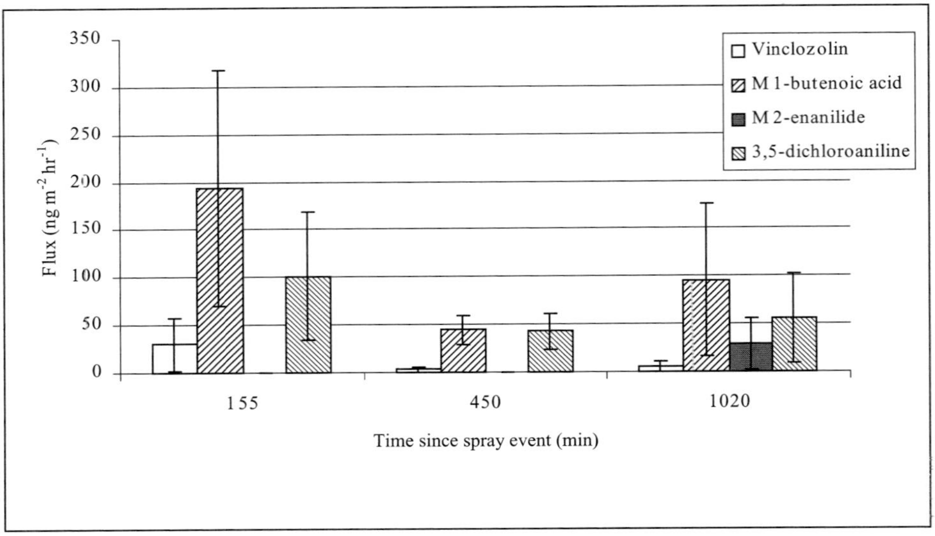

FIGURE 11.12. The results of a laboratory chamber study where an agricultural fungicide, vinclozolin (5 mL of 2,000 $mg L^{-1}$ suspended in water), is sprayed onto soil. The bars show the time-integrated atmospheric flux of organic compounds from nonsterile North Carolina Piedmont soil (aquic hapludult) with pore water pH = 7.5, following a 2.8-mm rain event and soil incorporation. Error bars indicate 95% confidence intervals. Vinclozolin (3-(3,5-dichlorophenyl)-5-methyl-5-vinyl-oxzolidine-2,4-dione), M1 (2-[3,5-dichlorophenyl)-carbamoyl]oxy-2-methyl-3-butenoic acid), and M2 (3′,5′-dichloro-2-hydroxy-2-methylbut-3-enanilide) are all suspected endocrine disrupting compounds; that is, they have been shown to affect hormone systems in mammals. This indicates that workers are not only potentially exposed to the parent compound, the pesticide that is actually applied, but to degradation products as the product is broken down in the soil.
Source: D.A. Vallero and J.J. Peirce, 2002. Transport and Transformation of Vinclozolin from Soil to Air, *Journal of Environmental Engineering*, 128 (3), 261–268.

The history of environmental contamination has numerous examples where certain segments of society are exposed inordinately to chemical hazards. This has been particularly problematic for communities of low socioeconomic status. For example, the landmark study by the Commission for Racial Justice of United Church of Christ[42] found that rate of landfill siting and the presence of hazardous sites in a community were disproportionately higher in African American communities than is non-minority community. Occupational exposures may also be disproportionately skewed in minority populations. For example, Hispanic and other minority workers can be exposed to higher concentrations of chemicals where they live and work, in large part due to the nature of their work (e.g., agricultural chemical exposures can be very high when and shortly after fields are sprayed, as shown in Figure 11.12. In fact, the metabolites or degradation products may be more toxic than the applied pesticide, e.g. M1-butenoic acid is more hormonely active than the parent, vinclozolin.). All too often these exposures can affect whole families of workers, who may join them in the fields, or who are exposed when the workers return to their homes.

The concept of environmental justice (EJ) began in the early 1980s and the movement has spread rapidly since. Advocacy groups in EJ communities throughout the United States have formed in response to environmental problems, particularly those that may be associated with human exposures and risks. Such groups often originated in neighborhoods in response to a specific event that had occurred or was being proposed, such as the siting of an incinerator. In such cases, the EJ aspects were discovered after the fact following studies that indicated statistical relationships between racial, ethnic, and socioeconomic status and environmental and public health risks. As a result, in 1992, the U.S. Environmental Protection Agency (EPA) created the Office of Environmental Justice to coordinate the agency's EJ efforts, and in 1994, President Clinton signed Executive Order 12898, "Federal Actions to Address Environmental Justice in Minority and Low-Income Populations." This order directs that federal agencies attend to the environment and human health conditions of minority and low-income communities, and requires that the agencies incorporate EJ into their missions. In particular, EJ principles must be part of the federal agency's day-to-day operation by identifying and addressing "disproportionately high and adverse human health and environmental effects of programs, policies, and activities on minority populations and low-income populations."[43]

Environmental Justice and the Catalytic Converter

Executive Order 12898 calls upon the agencies of the federal government to determine up front whether their projects and decisions will burden any particular groups unfairly.

Can we liken this to the debate that took place in the 1970s about how best to remove pollutants found in car exhaust that were contributing to smog? Decision makers in the government and in automobile companies were assessing the best means of reducing hydrocarbons that were being released from mobile sources (i.e., cars, trucks, buses, and trains). The options boiled down to whether to redesign and retool the internal combustion engines to improve efficiency and consequently reduce emissions, or whether to find a way to retrofit or "add-on" a product without making major changes to the engines.

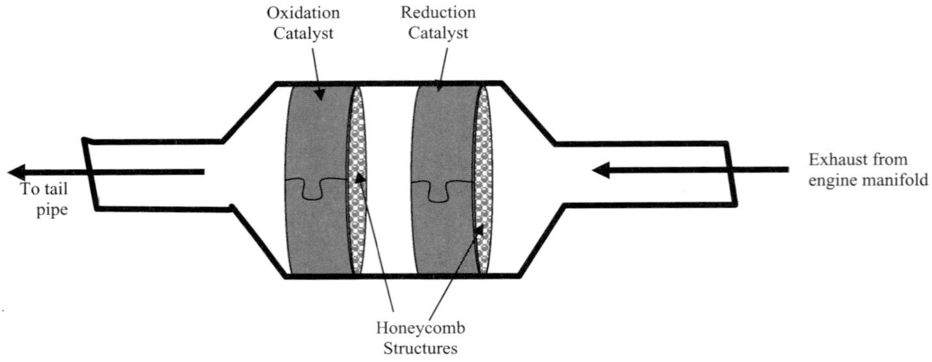

FIGURE 11.13. Catalytic converter.

The domestic automobile companies decided to retain the basic engines and to add a catalytic converter to the exhaust system. A catalytic converter uses metal catalyst pellets (e.g., platinum) to oxidize hydrocarbons to water and CO_2 and convert carbon monoxide, CO, to CO_2. (see Figure 11.13). Some perceived these as "band-aids" on inefficient engines but the automotive engineers insisted that this would be the most cost-effective way to reduce emissions. These same engineers were shocked when the Japanese automobile manufacturers, most notably Honda, chose the second alternative, to produce highly efficient engines that did not require catalytic converters. They decided that the reduction of emissions was an essential and integral part of engine design. American engineers, on the other hand, decided to use a bolt-on device to solve the problem without attacking the fundamental source of the emissions.

Is this comparable to what agencies are now doing about environmental justice? Each federal agency has a mission. Agencies are evaluated by the American people and their elective representatives on how well the missions are accomplished. That is one of the purposes of Congressional oversight, for example. So, when a new initiative comes along, agencies are more likely to see it as an ancillary objective, or worse, as an obstacle in the way of the "real" business of achieving its mission. We saw this in the early days of the National Environmental Policy Act (NEPA), which called upon federal agencies to rethink their missions regarding the environment. Like the EJ executive order, NEPA defined a new ethic in how the federal government does business, making environmental policy a leading priority. Specifically, NEPA created the environmental impact statement (EIS) that

income citizens in their regions. The decision also established remedies to meet this objective, especially by giving three judges responsibility to rule on exclusionary zoning cases.

40. M. Ritzdorf, 1997. "Locked Out of Paradise: Contemporary Exclusionary Zoning, the Supreme Court, and African Americans, 1970 to the Present," in J.M. Thomas and M. Ritzdorf, eds., 1997. *Urban Planning and the African American Community: In the Shadows*, SAGE Publications, Thousand Oaks, CA.
41. K. Shrader-Frechette, 2002. *Environmental Justice: Creating Equality, Reclaiming Democracy*, Oxford University Press, New York, NY.
42. Commission for Racial Justice, United Church of Christ, 1987. *Toxic Wastes and Race in the United States.*
43. Presidential Executive Order 12898, 1994. "Federal Actions to Address Environmental Justice in Minority Populations and Low-Income Populations," February 11, 1994.
44. Section 102 of NEPA, 42 U.S.C. § 4321 *et seq.*, Public Law No. 91–190, 83 Stat. 852.
45. Although I have been unable to locate the citation, this is very close to a mid-1970s quote shared by Timothy Kubiak, who at the time was an EIS reviewer for the U.S. EPA and who had studied NEPA under Lynton K. Caldwell, a strong advocate for a national environmental policy, at the University of Indiana.

Part IV

What Is Next?

As mentioned in the beginning of this book, analyzing cases not only tells us what went wrong, but what can be done about it. The case analyses give the opportunity to construct scenarios for the future. The stockbrokers' advice that "past performance does not necessarily predict future results" holds equally true for environmental predictions. However, the information from the analyses gives us an opportunity to assess possible outcomes given certain assumptions.

CHAPTER 12

Bottom Lines and Top of the Head Guesses

There are many methods for predicting the future. For example, you can read horoscopes, tea leaves, tarot cards, or crystal balls. Collectively, these methods are known as "nutty methods." Or you can put well-researched facts into sophisticated computer models, more commonly referred to as "a complete waste of time."

Scott Adams, U.S. cartoonist[1]

Any book with case studies calls for an assessment of the lessons learned, suggestions for improving things, and predictions of what will happen under various present and future scenarios. These considerations may be either explicit or implicit, but they are needed all the same. I am a bit more sanguine than Adams about the success of predictions and models. I have indeed seen some fairly good predictions in the last three decades of what would happen if things did not change or if certain measures were taken. The problem from an objective and scientific perspective is that we never know completely whether the measures indeed were the cause of improvements, or whether some third or n^{th} variable was the major contributor.

The best we can usually do is provide some estimation, based upon weight-of-evidence, that a set of actions seems to have led to an outcome. For example, there has been some recent data that seem to show that the ozone hole in the stratosphere above the North Pole is shrinking. Stratospheric ozone depletion has been widely reported in the scientific and lay literature and is of great concern as an indication of increased ultraviolet light exposure to human populations and wildlife. So, if the "hole" is indeed getting smaller, that is wonderful news. But, can we attribute this to measures brought about by international treaties and legislative bans on chlorofluorocarbons (CFCs) and other ozone depleting substances? Or, would something similar have happened anyway? Although the former is always a possibility, it would be difficult to dismiss the latter. Actually, this brings to mind another quote by Adams:

TABLE 12.1
Percentage decrease in ambient concentrations of national ambient air quality standard pollutants from 1985 through 1994.

Pollutant	*Decrease in Concentration*
CO	28%
Lead	86%
NO_2	9%
Ozone	12%
PM_{10}	20%
SO_2	25%

Source: U.S. Environmental Protection Agency.

The creator of the universe works in mysterious ways. But he uses a base ten counting system and likes round numbers.

Although I cannot be certain, I believe the quote is more about Adam's perception of humans, especially scientists and engineers, than about God. We must always "round off" and make assumptions. This makes the math easier, but makes the assessments and predictions more uncertain. Chaos theory has taught us that ill-posed problems, like predicting environmental consequences, seldom follows a linear path. Indeed, the predictions are usually not amenable to singular solutions, although it is human nature to try to fit them to such conveniences.

In spite of the uncertainties, it is possible to extract some important lessons and even to predict outcomes. Overall, there have been remarkable environmental progress and significant lessons learned in a relatively short amount of time. Much of this progress has been in the marked reductions in conventional pollutants. For example, the ambient atmospheric concentrations of six major air pollutants measured at over 4000 monitoring sites across the United States have continuously decreased, as shown in Table 12.1

The significant drop is directly related to tougher emission standards. The decrease in lead can be attributed directly to the phase-out of tetraethyl lead and other lead gasoline additives that began in 1973. Likewise, conventional surface water quality in developed nations has also enjoyed significant improvement. For example, in the early 1970s, 36% of the stream miles in the United States were safe for fishing and swimming, but 60% have now been designated as fishable and swimmable. This is a relatively rapid occurrence since the first permit limiting the discharge of pollutants to surface waters was issued in 1973. There has also been significant progress in addressing toxic pollutants. The bald eagle was taken off the endangered species list in 1996, which many saw as a direct result of the DDT ban since 1972.

Unfortunately, there remain significant environmental challenges and the global view is certainly less hopeful than that of the West.

According to the United Nations Environmental Programme (UNEP),[2] poverty in much of the global population continues to lead to increased disparities both within and among nations. Rapid globalization—particularly through developments in information technology, transport, and trade regimes—is observed. In many countries, there are trends toward decentralization of environmental responsibilities from national to subnational authorities, an increasing role for the transnational corporations in environmental stewardship and policy development, and a move toward integrated environmental policies and management practices. Increased willingness by governments to cooperate on a global basis is witnessed by the multitude of world summits in the last decade. The question arises, however, as to how this willingness is translated into concrete and effective actions. There are greater recognition and popular insistence that the wealth of nations and the well-being of individuals lie not just in economic capital, but also in social and natural resources.

Fundamental global environmental trends are emerging from the diverse regional accounts of priority environmental concerns—global and regional, current and future—recently summarized by the UNEP:[3]

- The use of renewable resources—land, forest, fresh water, coastal areas, fisheries, and urban air—is beyond their natural regeneration capacity and therefore is unsustainable.
- Greenhouse gases are still being emitted at levels higher than the stabilization target internationally agreed upon under the United Nations Framework Convention on Climate Change.
- Natural areas and the biodiversity they contain are diminishing due to the expansion of agricultural land and human settlements.
- The increasing, pervasive use and spread of chemicals to fuel economic development is linked to major health risks, environmental contamination, and disposal problems.
- Global developments in the energy sector are unsustainable.
- Rapid, unplanned urbanization, particularly in coastal areas, is putting major stress on adjacent ecosystems.
- The complex and often little understood interactions among global biogeochemical cycles are leading to widespread surface water acidification, climate variability, changes in the hydrological cycles, and the loss of biodiversity, biomass, and bioproductivity.

There are also widespread social trends, intrinsically linked to the environment, that also harm environmental trends, notably:

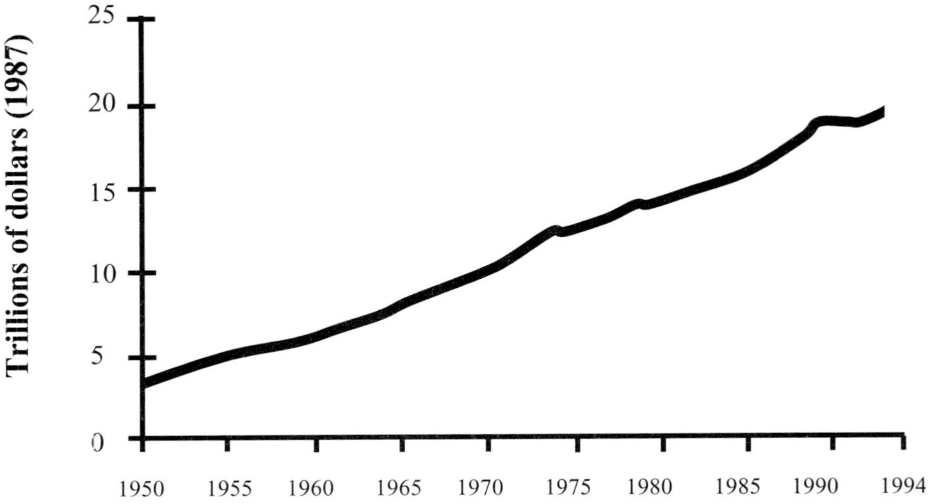

FIGURE 12.1. Gross world product.
Source: UNEP.

- An increase in inequality, both among and within nations, in a world that is generally healthier and wealthier (see Figure 12.1)
- A continuation, at least in the near future, of hunger and poverty despite the fact that globally enough food is available
- Greater human health risks resulting from continued resource degradation and chemical pollution

Current patterns of energy use need to be addressed, otherwise the loss and damage to land and natural resources, climate, air quality, rural and urban settlements, and human health and well-being will continue in many regions of the world. A promising area is to develop alternative energy sources, a step that has begun but needs to be accelerated and enhanced. Energy efficiency needs to be improved in industrial, domestic, or agricultural sectors, along with reduced emissions of pollutants, including greenhouse gases. Pollution is always an indication of inefficiency.

Technology can and will improve, giving rise to better uses of natural resources, less waste, and fewer ancillary pollutants. This will occur in all sectors, especially in industry, agriculture, transportation, and infrastructure development. Sharing of technologies among nations will help, such as coal cleaning and pollution control approaches that have been tested in developed countries. Low-tech solutions need to be developed and shared with developing countries, where possible, since these are more likely to be successful, even when resources are scarce and education levels limited.

Water quality and quantity will be problematic in many parts of the world and will limit economic development and environmental protection efforts. Combined with topsoil loss and diminishing arable lands, water will be a limiting factor for food self-sufficiency in several regions, forcing a dependence on food trade.

The environmental science community will need to invest in improved ways to share data and information on environmental quality, including consolidation and harmonization of national datasets, and in the acquisition of global datasets, where appropriate and where it does not threaten national security. Scientists will have to communicate in ways that can be understood by non-scientists who must make well-informed environmental decisions.

Again, scientifically sound advice needs to accompany our encouragement that people think globally and act locally.

The Future of Environmental Science and Engineering

Environmental engineers and scientists are often defined by what they do. They are called upon to handle and treat chemical, biological, and thermal waste, to find ways to clean up water, air, and land that have been polluted. They are also heavily involved in techniques to prevent pollution in the first place. However, this is not a complete picture, because the environment is an aggregate of all conditions, not simply the physical conditions that comprise the surroundings of an organism or a community of organisms. So, when we are talking about the human environment, the environmental engineer must apply not only the physical and natural sciences, but also the social and economic sciences. In fact, the environmental engineer is not just an applier of sciences, but also of the humanities, politics, religion, and any other human endeavor. Thus, environmental engineers must apply physical, chemical, and biological principles to prevent and to solve problems. They must also be highly sensitive to how these principles are acceptable and feasible to the human populations in which the principles are being applied. Even the most efficient and scientifically rigorous engineering project will fail unless it fully accounts for these human variables. We can design and build a most efficient treatment facility or landfill, but if it does not account for the societal needs, it will be ineffective. Since society is comprised of an amalgam of systems, and chemical engineering is all about systems, we can look to the chemical engineers to help to provide insights.

The Systematic Approach

One of the key lessons learned from the past three decades is the need for a comprehensive, systematic approach to environmental problems. The chemical engineering discipline's systematic perspective is instructive to

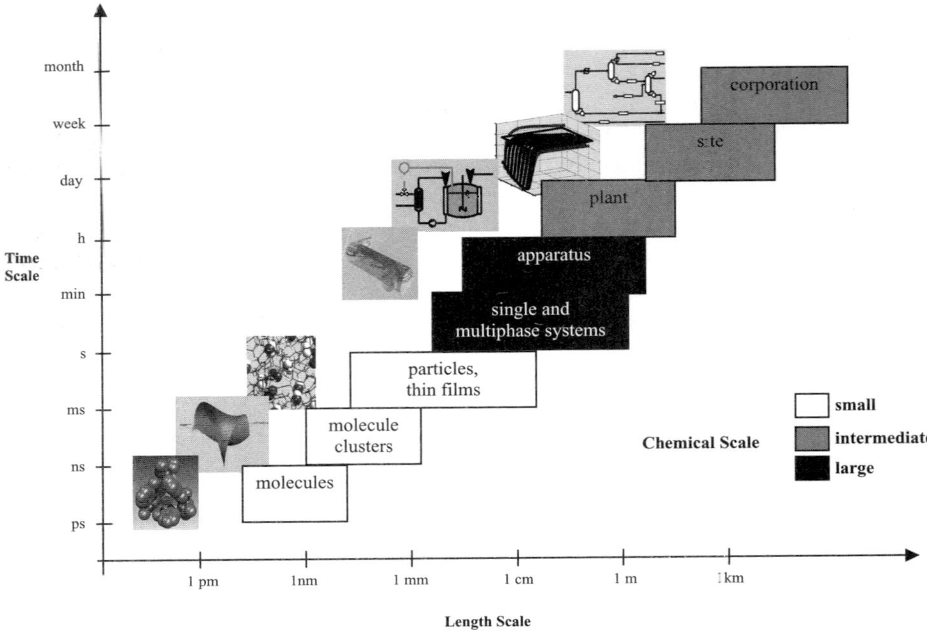

FIGURE 12.2. Scales and complexities of reactors.
Source: W. Marquardt, L. von Wedel, B. Bayer, "Perspectives on Lifecycle Process Modeling," 2000, in M.F. Malone, J.A. Trainham, B. Carnahan, eds., *Foundations of Computer-Aided Process Design*, AIChE Symp. Ser. 323, Vol. 96, 192–214. Note: ms = millisecond; ns = nanosecond; ps = picosecond.

environmental problem solving. Arguably, chemical engineering is, or should be, incorporated into any environmental engineering endeavor. After all, chemical engineering is "a broad discipline dealing with processes (industrial and natural) involving the transformation (chemical, biological, or physical) of matter or energy into forms useful for mankind, economically and without compromising environment, safety, or finite resources."[4] In fact, one of the first chemical engineering concepts to come to mind is the reactor. Next are mass and energy balances. Thus, it is impossible to address environmental problems without making use of chemical engineering concepts. The most common reactors are those that operate at an industrial scale. They include tanks and vats that have certain things go in and certain, but different, things come out. In environmental engineering, we do the same thing but cover a vast scale from subcellular to global (see Figure 12.2). For example, the processes that lead to a contaminant moving and changing in a bacterium may be very different from those processes at the lake or river scale, which in turn are different from those processes that cause the contaminant's fate as it crosses the ocean. This is simply a

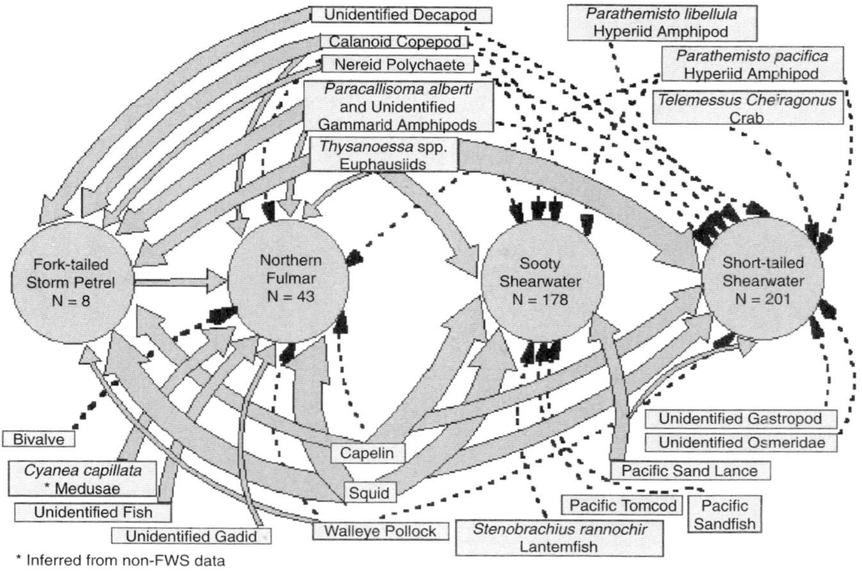

FIGURE 12.3. Flow of energy and mass among invertebrates, fish, and seabirds (*Procellariform*) in the Gulf of Alaska. The larger the width of the arrow, the greater the relative flow. Note how some species prefer crustaceans (e.g., copepods and euphausiids), but other species consume larger forage species like squid.
Source: G.A. Sanger, 1983. Diets and food web relationships of seabirds in the Gulf of Alaska and adjacent marine areas. U.S. Department of Commerce, National Oceanic and Atmospheric Administration, OCSEAP Final Report # 45, 631–771.

manifestation of the First Law of Thermodynamics—energy (or mass) is neither created or destroyed, only altered in form. This means that energy and mass within a system must be in balance, what comes in must equal what goes out. Environmental scientists and engineers must measure and account for these energy balances within a region in space through which a fluid travels. Scale and complexity can vary by orders of magnitude. So the bottom line is that environmental science and engineering must make use of the tools that chemical engineers provide, especially the thermodynamics of mass and energy balances.

Understanding the interrelationships among the abiotic (nonliving) and biotic (living) environments is paramount to solving environmental problems. Without scientifically characterizing it as such, human societies have been taking advantage of the concept of "trophic state" for much of our history. Organisms, including humans, live within an interconnected network or web of life (see Figure 12.3). In a way this is not any different from the energy and mass budgets of the chemical reactors familiar to chemical engineers. Of course, living things are more complex and complicated, but that is something to which any successful environmental professional will have to adapt. For example, the ecologist may be perfectly happy to understand the complex interrelationships shown in Figure 12.3, but in the event of designing an off-shore oil rig or following an oil spill, the environmental engineer must append this web to another one that includes humans as consumers. Also, the rig or the spill may change the abundance and richness of species, so the whole web is changed. These feedbacks are the stuff of environmental engineering. The engineer will be called upon to optimize the constructed project and to preserve (limit the effects of) the energy and mass balances. Sometimes, the environmental engineer must decide that there is no way to optimize both. At times, the ethical engineer must recommend the no-build option. Usually, though, the engineer must help the client navigate through numerous permutations and optimize solutions for more than two variables (e.g., species diversity, productivity and sustainability, costs and feasibility, and oil extraction efficiencies).

Since environmental engineering takes the systematic and ecological view, and because this view is scale and complexity dependent, there is a growing focus on human health and an evolving role of environmental engineers in assessing and managing risks. This is particularly important to the risk assessment process, especially exposure assessment at the microenvironmental scale. There is no clear consensus on many risk assessment terms, and exposure assessment is one of those where researchers, engineers, and the general public use varying definitions. The World Health Organization (WHO) defines exposure assessment as:

> The process of estimating or measuring the magnitude, frequency, and duration of exposure to an agent, along with the number and characteristics of the population exposed. Ideally, it describes the sources, pathways, routes, and the uncertainties in the assessment.[5]

Unfortunately, the definition twice embeds the word exposure, the very word we hope to define. We can further define exposure as:

> The contact between an agent and a target. Such contact occurs at an exposure surface over an exposure period.[6]

This definition uses engineering vernacular in expressing exposure as a function of surface (implying initial and boundary conditions) and time. So, we are ready to consider the concept of microenvironment (μE). At least three groups of engineers, upon seeing this term, will have unique perspectives. Those who work in emerging fields of nano-technologies and air pollution are thinking about length, and consider anything with the micro prefix to mean, in their world, pretty big stuff (e.g., particles with diameters >100 nanometers (nm) and <10 μm). The same is generally true for another group, the biomedical and microbial engineers, who consider micro to mean organisms about the size of bacteria and other single-celled creatures, but probably larger than viruses. The third group is the so-called exposure scientists, who have actually coined this term to deal with the difference in scale between a person (i.e., a single human being, known as "personal exposure") and the ambient environment (i.e., outside of that person's home). The concept of μE is to address pollution "where we live." So, examples of μE include one's home or a room or garage in that home, a bar, a church, inside of a car, even a cubicle and lab bench.

So, applying mass and energy balances, for example, to a home (see Figure 5.14) or even the globe (see Figure 7.2), helps to determine the amount of contaminant that finds its way to where it can do damage. Engineers have been increasingly involved and, I expect, will be called upon even more to address contamination at the μE scale. Recent examples, including the radon (Rn) permeation into homes where the public concern crested in the 1980s, asbestos and lead (Pb) exposures in home renovations and demolitions, and the current national concern over toxic molds (engineers refer to them as examples of bioaerosols), call for proper design and operations of residential, occupational, and business microenvironments.

New Thinking

Ross McKinney, an engineer who has advanced the state of biological treatment knowledge, is fond of saying that the solutions to many environmental problems are often "under our feet." His statement can be taken both literally and figuratively. Literally, many of the microbes that we use to treat pollutants are found in the soil, under our feet. Environmental engineers have been quite clever in finding ways to enrich these microbes' environments and to acclimate them to want to use our wastes as food sources. McKinney has found the soil bacteria and fungi to be particularly adaptable to highly concentrated waste environments, such as those in wastewater treatment tanks and hazardous waste reactors. In fact, it is a joy to observe McKinney's awe of the efficiency and elegance of the microbe and its capacity to be a miniscule chemical factory. McKinney has played a large part in changing the view of what environmental engineering is all about, having been trained in microbiology at MIT. However, he was more interested in how these "bugs" work rather than trying to find and name a new microbe (or more correctly, new to us, since they knew they were around the whole time!). Now, most university environmental engineering programs have a cadre of experts in microbiology and biochemistry. Even those in the more physical realms of environmental engineering, like system design, ultraviolet and ozonization disinfection controls, and exposure assessment, have a working knowledge of microbiology. This trend should only increase in the decades ahead.

McKinney is being a bit figurative in his advice to look under our feet for answers to the most perplexing environmental problems. Engineers are highly creative people. Aarne Vesilind, another environmental engineering pioneer, often reminds us that engineers "do things."[7] Both McKinney and Vesilind are telling us that in the process of our doing things we should be observant to new ways of doing those things. The answer can be right under our feet.

This book highlights a number of important changes in ethos that have taken place in recent decades. Unfortunately, there is a common human failing that we tend to keep doing things the way we have always done things. Maybe this is not so much a failing as an adaptive skill of the survival of the species. If we suffered no ill effects after eating that berry, we will eat more of that particular species. But even a similar looking species may be toxic, so we are careful to eat only the species that did not kill us. So, we have a paradox as environmental professionals. We do not want to expose our clients or ourselves to unreasonable risks, but we must, to some extent, push the envelope to find better ways of protecting the environment. So, we tend to suppress new ways of looking at problems. However, facts and theories may be so overwhelmingly convincing that we must change our world view. Thomas S. Kuhn refers to this as a "paradigm shift."[8] Scientists are often very reluctant to accept these new ways of thinking (Kuhn said such resistance can be "violent"). In fact, even if we do accept them they are often not dramatic reversals (revolutions), but modifications of existing designs (evolutions). Some say that the bicycle was merely a mechanical rerendering of the horse (e.g., the saddle seat, the linear and bilateral symmetrical structure, and the handle bars); as was the automobile. Often, as the great philosophers tell us, "we might look but we do not see!"

Combining the advice of Kuhn, McKinney, and Vesilind as it pertains to the lessons learned from the cases in this book gives us something on the order of:

> Yes, go with what works, but be aware of even the most subtle changes in what you are doing today *versus* what you did successfully yesterday. And, do not disregard the importance of common sense and rationality in environmental engineering.

The answers are often readily available, cheap, and feasible, but it takes some practice and a willingness to admit that there *is* a better way to do it. McKinney's advice that we look under our feet also tells us that natural systems are our allies. This observation, which may be intuitively obvious to this generation of environmental professionals, was not fully accepted in the 1950s and 1960s. In fact, there was a growing preference toward abiotic, chemical solutions as opposed to the biological approaches. The Second World War precipitated a petrochemical revolution. Modern society at that time placed a premium on synthetic, plastic solutions. Toward the end of the decade of the 1960s, the concept of using *passé* techniques like acclimated bacteria to treat wastes increasingly was seen as old fashioned. We searched for a miracle chemical to do this in a shorter time and in a more efficient manner.[9] The contemporary environmental professional has yet another challenge—sorting through what is hype and what is truly a technological breakthrough. I liken this to the physician who is daily inundated with literature from the pharmaceutical industry on all the new drugs that will allow her to be a more effective doctor, the seemingly endless series of visits from pharmaceutical reps, and the sharing of success stories with colleagues in person or virtually on Web sites. How do we separate the wheat from the chaff? The environmental engineer will be confronted with similar ill-posed problems. What is the best software for hazardous waste design? How different, really, is the new genetically altered species from those grown from native soils? What is the value-added of an early warning system for the drinking water plant? What are the added risks of intervention *versus* letting nature take its course (i.e., natural attenuation)?

Another emerging challenge has to do with the changing working environment itself. Many environmental professionals have been employed by firms, agencies, and institutions. The engineering codes of ethics recognize this by helping to remind the engineer that he or she serves numerous interests (i.e., the so-called "conflict of interest" clauses, as well as the "faithful agent" provisions). The new conflicts of commitment and interests will become increasingly more complicated in the coming decades. The workforce has undergone significant change since the 1980s, with greater numbers of contractors and fewer actual employees in many organizations. This means that the new engineer will need to be abundantly more self-sufficient than even a decade ago. The Future Engineer (FE) and Professional Engineer (PE) certification processes will become even more important. The professional will need a whole host of mentors beyond the PE. The interdisciplinary nature of environmental engineering requires mentoring in each of the disciplines and perspectives. The actual amount of tutelage will vary considerably. If an engineer seeks to design and oversee wetland restoration projects, hands-on experience with wetland ecologists is vital. If the engineer is more concerned about hazardous waste remediation, some time in the laboratory of an environmental analytical chemist would be worthwhile. And, in both cases, after the initial experience, a career-long relationship with these mentors should be maintained. The mix of inputs from trusted mentors could make for a solution very different from one where only handbooks are consulted. This mentor-learner model also helps to ensure that the knowledge and wisdom of this generation are passed on to the next, a means of providing a way to preserve corporate memory in the ever-changing field of environmental engineering.

The future environmental professional will have to produce products that are more transparent and coherent than those of the generation before him.[10] Products will be transparent because clients and the affected public can watch them as they are being developed because of the openness of the Internet, the increased technical savvy of larger numbers of nonengineers, and the growing comfort with what had heretofore been the sole domain of the scientists, engineers, and technicians. Products will be coherent in that the public will see that they have been articulated and structured by rational thinking, and will need to result in concrete products (plans and programs are merely means to the end), like the restored ecosystems (e.g., constructed wetlands), the structures (e.g., innovative drinking water systems), and improved quality of life (e.g., reduced risks from abandoned waste sites). Coherence, then, calls for engineers to design monitoring and ongoing assessment into any project; that is, how well is it doing over the life (and even after the life) of the project.

The sheer amount and complexity of data and information is enormous at present and will continue to grow. Environmental professionals must be comfortable with ambiguity, since every environmental decision is made under uncertainty, often a great deal of it. A lot of what scientists and engineers do does not always seem logical. And explaining the meaning of data can be very challenging. That is, in part, due to the incompleteness of our understanding of the methods used to gather data. Even well established techniques like chromatography have built-in uncertainties. I liken these scientific and methodological uncertainties to Plato's Allegory of the Cave. Recall that Plato argued that when we are ignorant of how things work (actually Plato was referring to people untutored in the Theory of Forms, but let us not worry about that), we are like prisoners chained inside of a cave, who are not allowed to turn our heads. So, what we see, exclusively, is the wall of the cave. Behind us burns a fire, and between the fire and us is a parapet, on which puppeteers are walking. The puppeteers behind us periodically hold up puppets that cast shadows on the wall of the cave. We cannot see these puppets (i.e., reality). What we do see and hear are the shadows cast and echoes from the objects (i.e., measurements).[11] Recall from statistics that the definition of accuracy is how close we are to the true value, or reality. Thus our instruments and other methods are to some extent what we see on the cave wall and not the puppets themselves. In chromatography, for example, we are fairly certain that the peaks we are seeing represent the molecule in question, but actually depending on the detector, all we are seeing is the number of carbon atoms (e.g., flame ionization detection) or the mass to charge ratios of molecular fragments (e.g., mass spectrometry), but not the molecule itself. Add to this instrument and operator uncertainties, and we can see that even the more accepted scientific approaches are biased and inaccurate, let alone an approach like mathematical modeling, where assumptions about initial and boundary conditions, values given to parameters, and the propagation of error render our results even more uncertain.

The first ethical principle of the engineering profession is to hold paramount the health, safety, and welfare of the general public. However, the public itself is changing, as exemplified by two seemingly contradictory trends. The first is that there seems to be a great divergence between the technologically literate and those not conversant in technical matters. Those who are trained in the technical fields appear to be becoming more highly specialized and steeped in techno-jargon, leaving the majority of people dependent upon whatever the technologists say. Simultaneously, there seems to be a trend of greater fluency of what were formerly specialized, engineering terms in the larger public arena.

The first trend, the technological literacy gap, as it applies to environmental engineering is not the same as the so-called digital divide; that is, the difference in access and use of information technology by groups of different socioeconomic status, race, and gender. The literacy gap is more fundamental than any single issue. Some fear that the future citizenry is ill-prepared and undereducated to participate fully in an increasingly complex, technology-rich future. In the United States, this concern in part is manifested by lack of preparation in math and science of students, and their entry into the present and future workforce. This is the mirror image of the problem of engineers' inadequate training in and appreciation for the humanities and social sciences. Engineers will definitely have to enhance their reach to include a greater number of perspectives in their projects and, simultaneously, we need to help the general public increase its appreciation for things technological. This confluence will not be easy. For example, the National Center for Education Statistics[12] reports that in 1999, the United States lagged behind much of the developed world, and even a number of developing nations, in its middle school and high school students' achievement in mathematics and science (see Table 12.2). Interestingly, the tests used in these comparisons place heavy emphasis on environmental sciences and the math and science underpinning environmental engineering. The Trends in International Mathematics and Science Study (TIMSS) measures aptitude in fractions and number sense, algebra, geometry, data representation, analysis, and probability, measurement, earth sciences, life sciences, physics, chemistry, environmental science, scientific inquiry, and the nature of science.

The other trend is that previously arcane and highly technical concepts and jargon are becoming increasingly mainstreamed. So, although many students do not seem motivated to participate fully in the increasing technological demands of society, they somehow are gaining a large repertoire of scientific expertise in their everyday lives.[13]

Few things have changed environmental science and engineering more than the computer. In fact, engineering has traditionally been among the

TABLE 12.2
Mathematics and science achievement of eighth-graders in 1999 Trends in International Mathematics and Science Study (TIMSS), formerly known as the Third International Mathematics and Science Study. TIMSS was developed by the International Association for the Evaluation of Educational Achievement to measure trends in students' mathematics and science achievement. The regular four-year cycle of TIMSS allows for comparisons of students' progress in mathematics and science achievement.

Mathematics		*Science*	
Nation	*Average*	*Nation*	*Average*
Singapore	604	Chinese Taipei	569
Korea, Republic of	587	Singapore	568
Chinese Taipei	585	Hungary	552
Hong Kong SAR	582	Japan	550
Japan	579	Korea, Republic of	549
Belgium-Flemish	558	Netherlands	545
Netherlands	540	Australia	540
Slovak Republic	534	Czech Republic	539
Hungary	532	England	538
Canada	531	Finland	535
Slovenia	530	Slovak Republic	535
Russian Federation	526	Belgium-Flemish	535
Australia	525	Slovenia	533
Finland	520	Canada	533
Czech Republic	520	Hong Kong SAR	530
Malaysia	519	Russian Federation	529
Bulgaria	511	Bulgaria	518
Latvia-LSS	505	**United States**	**515**
United States	**502**	New Zealand	510
England	496	Latvia-LSS	503
New Zealand	491	Italy	493
Lithuania	482	Malaysia	492
Italy	479	Lithuania	488
Cyprus	476	Thailand	482
Romania	472	Romania	472
Moldova	469	Israel	468
Thailand	467	Cyprus	460
Israel	466	Moldova	459
Tunisia	448	Macedonia, Republic of	458
Macedonia, Republic of	447	Jordan	450
Turkey	429	Iran, Islamic Republic of	448
Jordan	428	Indonesia	435
Iran, Islamic Republic of	422	Turkey	433
Indonesia	403	Tunisia	430
Chile	392	Chile	420
Philippines	345	Philippines	345
Morocco	337	Morocco	323

first adopters of many new technologies, and this has certainly been the case for information technology (IT). But, like any tool, it can be either well used or misused. Two quotes seem to capture the range of thinking on whether the computer is a blessing or a curse:

> In a way not seen since Gutenberg's printing press that ended the Dark Ages and ignited the Renaissance, the microchip is an epochal technology with unimaginably far-reaching economic, social, and political consequence.
>
> —Michael Rothchild[14]

> While all this razzle-dazzle connects us electronically, it disconnects us from each other, having us "interfacing" more with computers and TV screens than looking in the face of our fellow human beings. Is this progress?
>
> —Jim Hightower[15]

At one extreme we have information technology being a panacea and at the other end the bane of modern civilization. Engineering is not immune to this debate. Arguably, most engineers and environmental technicians fall closer to Rothchild's perspective, as strong advocates for the application of IT in every aspect of environmental engineering. After all, engineers need to apply the sciences, so we must avail ourselves of the best tools and methods to accomplish this. And the adoption by society of these new technologies has been phenomenally rapid and the rate of adoption has increased with each new technology. For example, it took only 16 years for the personal computer to be adopted by 25% of U.S. households and merely seven years for them to accept Internet access (see Table 12.3).

TABLE 12.3
Years needed for 25% of U.S. households to adopt new technologies.

Technology	*Years to Reach 25% of U.S. Households*
Automobiles	55
Electricity	45
Telephone	35
Radio	27
Television	25
Personal computers	16
Cellular phones	13
Internet (World Wide Web)	7

Source: S. Baase, 2003. *A Gift of Fire: Social, Legal, and Ethical Issues for Computers and the Internet*, 2e, Prentice Hall, Upper Saddle River, NJ.

There are also downsides, such as the temptation for environmental engineering students to observe the world through their computer screens, rather than actually getting out into the environment that needs protecting! But there is no arguing that computer modeling has become an essential part of almost any environmental assessment, design, or cleanup.

Is this an oxymoronic situation? The future environmental professional will have to reconcile any technical deficiencies indicated by the math and science gaps with the creeping technological savvy of the general public. Perhaps the best strategy is to be ready to explain complicated engineering concepts in a straightforward manner, but at the same time, be prepared for a public that expects high-tech solutions to their problems.

The Morning Shows the Day

Major challenges are ahead, but there is much reason for optimism. Not the least of which is the strong preparation of the next generation of environmental scientists and engineers. They have the skills, knowledge, and perspective to address these complex problems. The environmental movement that began some decades ago preceded the rigorous science needed to address these problems, but science has been catching up. I believe most of my colleagues will agree that the present generation of scientists and engineers possesses the creativity and capabilities that will be needed. We see this in our classrooms, our laboratories, our agencies, and our corporations. They stand ready to take the profession to the next level. Or, as Milton might have put it, there is a new environmental day dawning:

> The childhood shows the man,
> As morning shows the day.
> *Paradise Regained. Book iv. Line 220.*

Notes and Commentary

1. S. Adams, 1998. *The Dilbert Future: Thriving on Business Stupidity in the 21st Century*, HarperBusiness, New York, NY.
2. United Nations Environmental Program, Division of Early Warning and Assessment, 1997 and 2002. Global Environmental Outlook, Nairobi, Kenya.
3. United Nations Environmental Program, 1997. Global Environmental Outlook—1, Global State of the Environment, 1997.
4. This definition comes from the Worcester Polytechnic Institute (http://www.wpi.edu/Academics/Depts/CHE/About/definition.html).
5. World Health Organization, 2002. Harmonization of Approaches to the Assessment of Risk from Exposure to Chemicals, Final Report of the International

Programme on Chemical Safety Harmonization Project, Exposure Assessment Planning Workgroup,Terminology Subcommittee.

6. *Ibid.*
7. I first heard Vesilind publicly share this profundity during an engineering conference. At first blush, the statement sounds like a truism or even silly unless you think about it. There are many, and I would add a growing number of, enterprises that do not *do* anything (or at least it is hard to tell what they do). They think about things, they come up with policies, they review and critique the work of others, but their value-added is not so "physical." I have to say that I envy my family members and friends in construction who, at the end of every working day, see a difference because of what they did that day. This can take many forms, such as a few more meters of roadway, a new roof, or an open lot where a condemned structure once stood. The great thing about environmental engineering is that we can do both. We can plan and do. I should say we *must both plan and do!* In the words of the woodworker, we must "measure twice and cut once." The good news (and the responsibility) is that the environmental engineer's job is not done when the blueprints are printed. The job is not even done when the project is built. The job continues for the useful life of the project. And, since most environmental projects have no defined end, but are continuously in operation, the engineers get to watch the outcomes indefinitely. Engineers must get out there and observe (and oversee) the fulfillment of their ideas and the implementation of their plans. That is why engineers are called to "do things."

   This reminds me of a conversation I had with a boilermaker, who happens to be my brother-in-law, that points out that engineers need to be aware of the knowledge and wisdom all around them (not just from texts, manuals, or even old professors). He has been installing, welding, and rigging huge boiler systems for power plants and refineries for decades and has a reputation among his fellow boilermakers as highly intelligent and skilled at his craft. He recently shared with me that he likes to work with young engineers, mainly because they listen. They are not concerned about hierarchies or chain of command so much as some of the more senior engineers or managers. Perhaps it is because they know so little about the inner workings of complex and large systems like those needed in coal-fired combustion. Plus, they seem to know how to have fun. He contrasts this with the engineer who shows up on the job and lets everyone know that he is the pro. My brother-in-law recounts one memorable occasion when one such arrogant professional chose not or did not know to ask the boilermakers about what happens when a multiton boiler tank is rigged. Had he asked, the boilermaker would have shared the extent of his knowledge about "stretching." In other words, the height of the superstructure had to be sufficiently taller than the boiler to account for the steel alloy elasticity due to the tremendous weight. As it was designed, the superstructure was too short, so the boiler stretched all the way to the ground surface and the whole thing had to be redesigned and retrofitted. Had the professional asked, he would have known early on to modify the design. My brother-in-law surmises that the young guys would have asked. My guess is that, out of respect, even if they hadn't asked, he would have warned them simply because they put him in a place where he could communicate with them. The moral of this story is that leadership often comes from places other than the top. Another moral is that as you mature, don't forget what made you successful in the first place.

8. T.S. Kuhn, 1962. *The Structure of Scientific Revolutions*, University of Chicago Press, Chicago, IL. Kuhn actually changed the meaning of the word "paradigm," which had been almost the exclusive province of grammar (a fable or parable). Kuhn extended the term to mean an accepted specific set of scientific practices. The scientific paradigm is made up of what is to be observed and analyzed, the questions that arise pertaining to this scientific subject matter, to whom such questions are to be asked, and how the results of the investigations into this subject matter will be interpreted. The paradigm can be harmful if it allows incorrect theories and information to be accepted by the scientific and engineering communities. Such erroneous adherences can result from "groupthink," a term coined by Irving Janis, a University of California, Berkeley psychologist. Groupthink is a collective set of systematic errors (biases) held by and perpetuated by a group. See I. Janis, 1982. *Groupthink: Psychological Studies of Policy Decisions and Fiascoes*, 2e, Houghton Mifflin Company, Boston, MA.
9. McKinney was my mentor at the University of Kansas. Interestingly, from our conversations about the then nascent area of genetic engineering, if memory serves, he shared the same skepticism that he held for abiotic chemistry as the new paradigm. In a sense, McKinney argued that engineers had been doing genetic engineering all along and that we should be wary of the sales pitches for new super genes. Again, it appears that he has been proven generally correct. I believe that he would be among the first to use an organism that would do a better job, no matter whether it was achieved through natural acclimation or through contemporary genetic engineering. Before spending precious funds on transgenic bacteria, first "look under your feet" for microbes that have the natural potential to do the same thing.
10. Sometimes the best resources come from the strangest places. The discussion on transparency and coherence must be credited to an article by columnist George Will that I read in the September 8 2004 edition of the Durham (NC) *Herald Sun* newspaper, page A-11. Will was particularly curious as to the success of the Entertainment and Sports Programming Network (ESPN). One reason, according to Will, is that unlike most endeavors in contemporary society, sports provide closure almost immediately. We know who won and lost. This is coherence. The other is that the spectator is included in the process. What is going on is completely transparent. There are 4 minutes and 15 seconds left in the game and the Chiefs are ahead by 11 points. The Cardinals are leading the Cubs 4 to 2, with two outs in the top of the ninth inning, with Albert Pujols on deck. Engineers will increasingly be working in such environments as a result of the Internet, sunshine laws, and, I believe, an

increased technical literacy of greater numbers of people. Not everyone wants to know the score of the Cardinals game, but they can find it immediately if they wish. They can even find the box scores of every game played this year, or the statistics of every minor league player on the Cardinals' triple-A farm club. Likewise, not everyone wants to know the details of the pump station you may be designing in the southwestern part of your city, but they can find them almost immediately if they so choose. So, unlike only a few years ago, it behooves the engineer to expect at least one person to have as much or more information than you do about the proposed project in the neighborhood, and the well-advised engineer should be ready for this *before* the public hearing, zoning appeal, or the environmental assessment meeting.

Unfortunately, the technical literacy is not evenly distributed throughout the population. For example, the so-called "digital divide" does exist, but it is rapidly changing. For example, according to Sara Base (2003, *A Gift of Fire: Social, Legal, and Ethical Issues for Computers and the Internet*, 2e, Prentice Hall, Upper Saddle River, NJ), at the beginning of the 1990s, about 10% of Internet users were women, but by the end of the decade about half the users were women. However, in 1999, lower socioeconomic status (SES) groups were half as likely as the general population to have Internet access in their residences. And there is also a racial divide, as African American and Hispanic households were half as likely as the general population to have a personal computer. Sometimes, even against such odds, people will have access. For instance, Alaska has the highest percentage of homes with Internet access of all the states in the United States. In this case, with its greater isolation and large expanses, the need for remote access overrides even some very large cultural and SES barriers.

11. B. Jowett (translator), *The Republic by Plato*, Oxford University Press, Oxford U.K. Another way of looking at the incompleteness of our understanding of how things work is St. Paul's likening of the way that we see the world to our ". . . seeing through a glass, darkly. . . ." (1 Corinthians 13:12). As humans, we will always be limited in our understanding of even the most fundamental aspects of the tools we use.
12. U.S. Department of Education, National Center for Education Statistics, 2003. J.D. Sherman, S.D. Honegger, and J.L. McGivern, Comparative Indicators of Education in the United States and Other G8 Countries: 2002. Report No. NCES 2003–026, Washington, D.C.
13. In a discussion, Aarne Vesilind argued that there is a problem in terminology, especially in what is labeled "technology." For example, just because you play a video game or watch television does not make you technologically advanced. On the other hand, if you write codes for video games or understand the circuitry of the TV, this is an indication of technological literacy.
14. M. Rothchild, 1995. "Beyond Repair: The Politics of the Machine Age Are Hopelessly Obsolete," *The New Democrat*, July/August, 8–11.
15. J. Hightower, 1995. Quoted in R. Fox, "Newstrack," *Communications of the ACM*, 38(8), 8–11.

APPENDIX 1

Equilibrium

Chemical reactions depend on colligative (collective) relationships between reactants and products. Colligative properties are expressions of the number of solute particles available for a chemical reaction. So, in a liquid solvent like water, the number of solute particles determines the property of the solution. This means that the concentration of solute determines the colligative properties of a chemical solution. These solute particle concentrations for pollutants are expressed as either mass-per-mass (e.g., $mg\,kg^{-1}$) or, most commonly, as mass-per-volume (e.g., $mg\,L^{-1}$) concentrations. In gas solutions, the concentrations are expressed as mass-per-volume ($mg\,m^{-3}$). Colligative properties may also be expressed as mole fractions, where the sum of all mole fractions in any solution equals 1.

Equilibrium Example

Consider the equilibrium involved in dissolving sucrose in water. The given sugar solution contains 240 g sucrose per 1,000 g water. What are the mole fractions of the solute and the solvent?

Solution

Sucrose is our *solute.* Water is our *solvent.* The gram molecular weight of sucrose ($C_6H_{12}O_6$) is 180 g, so we would have 240/180 = 1.3 moles sucrose 1000 g water.

Since the molecular weight of H_2O is 18, the mole fraction of our sugar solvent equals

$$\frac{moles(solute)}{moles(solute)+moles(solvent)}=\frac{240/180}{240/180+1000/18}=1.3/56.9=0.02.$$

and the mole fraction of water is $\frac{1000/18}{1000/18+240/180}=0.98.$

Thus, the mole fraction (expressed as mole-percent) of our solute is approximately 2% and the mole-percent of our solvent is about 98%. The sum of all mole-percentages is 100% because the sum of all mole fractions is 1.

Colligative properties depend directly upon concentration. One important property is vapor pressure, which is decreased with increased temperature. This is why water will require higher temperatures to boil when a solvent is present. For example, pure water will boil at 100°C because of one atmosphere (760 mm Hg) of pressure, and escape as water vapor. In this case, all of the molecules are water (100% mole fraction). By adding solvent to the pure water, we change the mole fraction. For example, if we heat the example solution (2% sucrose), our vapor pressure is lowered by 2%, so that rather than 760 mm Hg, our vapor pressure = (0.98 water mole fraction) (760 mm Hg) = 745 mm Hg. Thus, the vapor pressure of the solvent (P) in any solution is found by:

$$P = X_A P^0 \tag{A.1.1}$$

where

X_A = Mole fraction of solvent
P^0 = Vapor pressure of 100% solvent

Solution Equilibria

A body is considered to be in thermal equilibrium if there is no heat exchange within the body and between that body and its environment. Analogously, a system is said to be in chemical equilibrium when the forward and reverse reactions proceed at equal rates. Again, since we are looking at finite space and time, such as a spill or an emission, or movement through the environment, reactions within that time and space may be either nonequilibrium ($xA + yB \rightarrow zC + wD$) or equilibrium ($xA + yB \Leftrightarrow zC + wD$) chemical reactions. The x, y, z, and w terms are the *stoichiometric* coefficients, which represent the relative number of molecules of each reactant (A and B) and each product (C and D) involved in the reaction. So, to have chemical equilibrium, the reaction must be reversible, so that the concentrations of the reactants and the concentrations of the products are constant with time.

The Law of Concentration Effects states that the concentration of each reactant in a chemical reaction dictates the rate of the reaction. So, using our equilibrium reaction ($xA + yB \Leftrightarrow zC + wD$), we see that the rate of the forward reaction, that is, the rate that the reaction moves to the right, is most often dictated by the concentrations of A and B. So, we can express the forward reaction as:

$$r_1 = k_1[A]^x[B]^y \tag{A.1.2}$$

The brackets indicate molar concentrations of each chemical species (i.e., all products and reactants). Further, the rate of the reverse reaction can be expressed as:

$$r_2 = k_2[C]^z[D]^w \tag{A.1.3}$$

Since at equilibrium, $r_1 = r_2$ and $k_1[A]^x[B]^y = k_2[C]^z[D]^w$, we can rearrange the terms to find the equilibrium constant K_{eq} for the reversible reaction:

$$\frac{k_1}{k_2} = \frac{[C]^z[D]^w}{[A]^x[B]^y} = K_{eq} \tag{A.1.4}$$

The equilibrium constant for a chemical reaction depends upon the environmental conditions, especially temperature and ionic strength of the solution. An example of a thermodynamic equilibrium reaction is chemical precipitation water treatment process.[1] This is a heterogeneous reaction in that it involves more than one physical state. For an equilibrium reaction to occur between solid and liquid phases the solution must be saturated and undissolved solids must be present. So, at a high hydroxyl ion concentration (pH = 10), the solid phase calcium carbonate ($CaCO_3$) in the water reaches equilibrium with divalent calcium (Ca^{2+}) cations and divalent carbonate (CO_3^{2-}) anions in solution. So, when a saturated solution of $CaCO_3$ contacts solid $CaCO_3$, the equilibrium is:

$$CaCO_3(s) \Leftrightarrow Ca^{2+}\,(aq) + CO_3^{2-}(aq) \tag{A.1.5}$$

The (s) and (aq) designate that chemical species are in solid and aqueous phases, respectively.

Thus, applying the equilibrium constant relationship in Equation A.1.4, the dissolution (precipitation) of calcium carbonate is:

$$K_{eq} = \frac{[Ca^{2+}]+[CO_3^{2-}]}{[CaCO_3]} \tag{A.1.6}$$

The solid phase concentration is considered to be a constant K_s. In this instance, the solid $CaCO_3$ is represented by K_s, so:

$$K_{eq}K_s = [Ca^{2+}] + [CO_3^{2-}] = K_{sp} \tag{A.1.7}$$

K_{sp} is known as the solubility product constant. These K_{sp} constants for inorganic compounds are published in engineering handbooks (e.g., in

Part 1, Appendix C of the *Handbook of Environmental Engineering Calculations*). Other equilibrium constants, such as the Freundlich Constant (K_d) discussed in Chapter 3, are also published for organic compounds (e.g., in Part 1, Appendix D of the *Handbook of Environmental Engineering Calculations*).

Gas Equilibria

For gases, the thermodynamic "equation of state" expresses the relationships of pressure (p), volume (V), and thermodynamic temperature (T) in a defined quantity (n) of a substance. For gases, this relationship is defined most simply in the ideal gas law:

$$pV = nRT \tag{A.1.8}$$

where R = the universal gas constant or molar gas constant = $8.31434\,\text{J}\,\text{mol}^{-1}\,{}^{\circ}\text{K}^{-1}$.

It should be noted that the ideal gas law applies only to ideal gases, those that are made up of molecules taking up negligible space, with negligible spaces between the gas molecules. So, for real gases, the equilibrium relationship is:

$$(p + k)(V - nb) = nRT \tag{A.1.9}$$

where

K = factor for the decreased pressure on the walls of the container due to gas particle attractions
nb = volume occupied by gas particles at infinitely high pressure

Further, the van der Waals equation of state is:

$$k = \frac{n^2 a}{V^2} \tag{A.1.10}$$

where a is a constant.

The van der Waals equation generally reflects the equilibria of real gases. It was developed in the early twentieth century and has been updated, but these newer equations can be quite complicated.

Gas reactions, therefore, depend upon partial pressures. The gas equilibrium K_p is the quotient of the partial pressures of the products and reactants, expressed as:

$$K_p = \frac{p_C^z p_D^w}{p_A^x p_B^y} \tag{A.1.11}$$

and from Equations 4-1, 5, 6, and 7, K_p can also be expressed as:

$$K_p = K_{eq}(RT)^{\Delta v} \qquad (A.1.12)$$

where Δv is defined as the difference in stoichiometric coefficients.

Free Energy

Equilibrium constants can be ascertained thermodynamically by employing the Gibbs free energy (G) change for the complete reaction. Free energy is the measure of a system's ability to do work, in this case to drive the chemical reactions. This is expressed as:

$$G = H - TS \qquad (A.1.13)$$

where G is the energy liberated or absorbed in the equilibrium by the reaction at constant T. H is the system's enthalpy and S is its entropy.

Enthalpy is the thermodynamic property expressed as:

$$H = U + pV \qquad (A.1.14)$$

where U is the system's internal energy.

Entropy is a measure of a system's energy that is unavailable to do work. Numerous handbooks[2] explain the relationship between Gibbs free energy and chemical equilibria. However, the relationship between a change in free energy and equilibria can be expressed by:

$$\Delta G^{\star} = \Delta G_f^{\star 0} + RT \ln K_{eq} \qquad (A.1.15)$$

where

$\Delta G_f^{\star 0}$ = Free energy of formation at steady state (kJ gmol^{-1})

Importance of Free Energy in Microbial Metabolism

Metabolism is the cellular process that derives energy from a cell's surroundings. Energy to do chemical work is exemplified by cellular processes. Microbes, like bacteria and fungi, are essentially tiny chemical, efficient factories that mediate reactions at various rates (kinetics) until they reach equilibrium. These simple organisms (and complex organisms alike) need to transfer energy from one site to another to power their machinery needed to stay alive and reproduce. Microbes play a large role in degrading pollutants, whether in natural attenuation, where the available microbial populations adapt to the hazardous wastes as an energy source, or in engineered

TABLE A.1.1
Genera of microbes able to degrade a persistent organic contaminant; that is, crude oil.

Bacteria	*Fungi*
Achromobacter	*Allescheria*
Acinetobacter	*Aspergillus*
Actinomyces	*Aureobasidium*
Aeromonas	*Botrytis*
Alcaligenes	*Candida*
Arthrobacter	*Cephaiosporium*
Bacillus	*Cladosporium*
Beneckea	*Cunninghamella*
Brevebacterium	*Debaromyces*
Coryneforms	*Fusarium*
Erwinia	*Gonytrichum*
Flavobacterium	*Hansenula*
Klebsiella	*Helminthosporium*
Lactobacillus	*Mucor*
Leucothrix	*Oidiodendrum*
Moraxella	*Paecylomyces*
Nocardia	*Phialophora*
Peptococcus	*Penicillium*
Pseudomonas	*Rhodosporidium*
Sarcina	*Rhodotorula*
Spherotilus	*Saccharomyces*
Spirillum	*Saccharomycopisis*
Streptomyces	*Scopulariopsis*
Vibrio	*Sporobolomyces*
Xanthomyces	*Torulopsis*
	Trichoderma
	Trichosporon

Source: U.S. Congress, Office of Technology Assessment, 1991. Bioremediation for Marine Oil Spills, Background Paper, OTA-RP-O-70, U.S. Government Printing Office, Washington D.C.

systems that do the same in a more highly concentrated substrate (see Table A.1.1). Free energy is an important factor in microbial metabolism.

The reactant and product concentrations and pH of the substrate affect the observed ΔG^* values. If a reaction's ΔG^* is a negative value, the free energy is released and the reaction will occur spontaneously, and the reaction is *exergonic*. If a reaction's ΔG^* is positive, the reaction will not occur spontaneously. However, the reverse reaction will take place, and the reaction is *endergonic*.

Time and energy are limiting factors that determine whether a microbe can efficiently mediate a chemical reaction, so catalytic processes

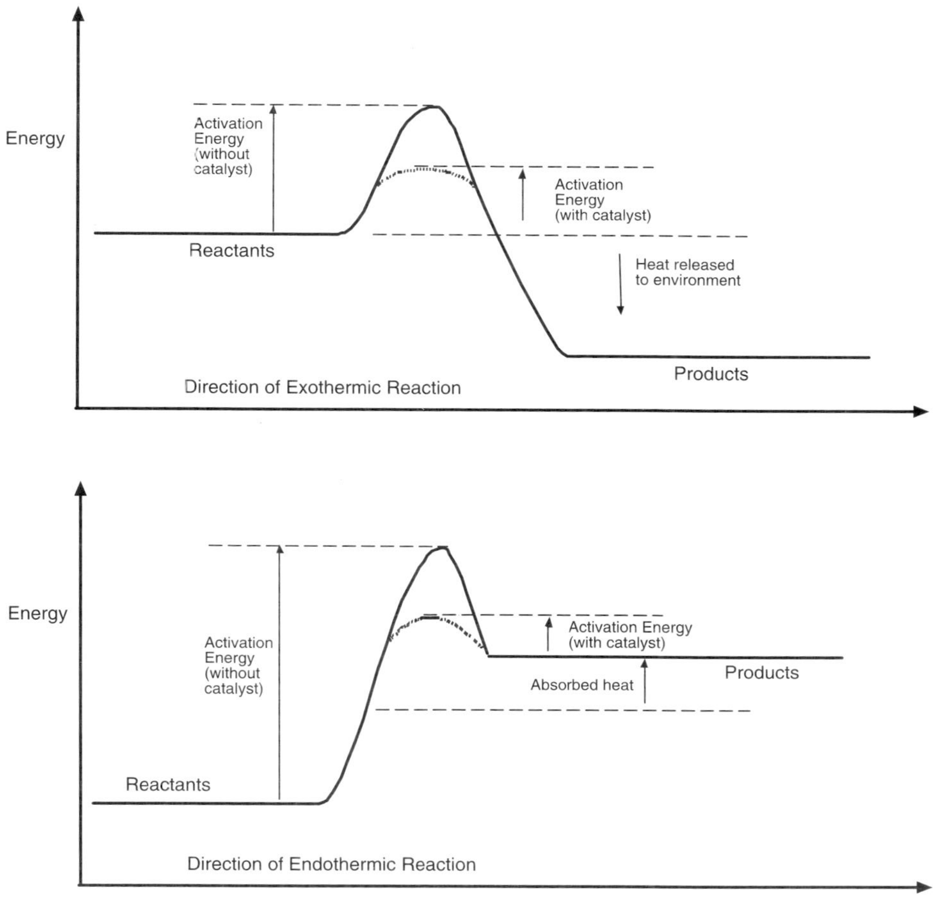

FIGURE A1.1. Effect of a catalyst on an exothermic reaction (top) and on an endothermic reaction (bottom).

are usually needed. Since an *enzyme* is a biological *catalyst*, these compounds (proteins) speed up the chemical reactions of degradation without themselves being used up. They do so by helping to break chemical bonds in the reactant molecules (see Figure A1.1). Enzymes play a very large part in microbial metabolism. They reduce the reaction's *activation energy*, which is the minimum free energy required for a molecule to undergo a specific reaction. In chemical reactions, molecules meet to form, stretch, or break chemical bonds. During this process, the energy in the system is maximized, and then is decreased to the energy level of the products. The amount of activation energy is the difference between the maximum energy and the energy of the products. This difference represents the energy barrier that must be overcome for a chemical reaction to take place. Catalysts (in this case, microbial enzymes) speed up and increase the likelihood of a reaction by reducing the amount of energy (i.e., the activation energy) needed for the reaction.

The most common microbial coupling of exergonic and endergonic reactions by means of high-energy molecules to yield a net negative free energy is that of the nucleotide adenosine triphosphate (ATP) with $\Delta G^{*} = -12$ to $-15\,\text{kcal mole}^{-1}$. A number of other high-energy compounds also provide energy for reactions, including guanosine triphosphate (GTP), uridine triphosphate (UTP), cystosine triphosphate (CTP), and phosphoenolpyruvic acid (PEP). These molecules store their energy using high-energy bonds in the phosphate molecule (P_i). An example of free energy in microbial degradation is the possible first step in acetate metabolism by bacteria:

$$\text{Acetate} + \text{ATP} \rightarrow \text{acetyl-coenzyme A} + \text{ADP} + P_i \qquad \text{(A.1.16)}$$

In this case, the P_i represents a release of energy available to the cell. Conversely, to add the phosphate to the two-P_i structure ADP to form the three-P_i ATP requires energy (i.e., it is an *endothermic process*). Thus, the microbe stores energy for later use when it adds the P_i to the ATP.

Notes and Commentary

1. For the calculations and discussions of solubility equilibrium, including this example, see C.C. Lee and S.D. Lin, eds., 2000. *Handbook of Environmental Engineering Calculations*, pp. 1.368–1.373.
2. See Michael LaGrega, Phillip Buckingham, and Jeffrey Evans, 2001. *Hazardous Waste Management*, 2e, McGraw-Hill, Boston, MA.

APPENDIX 2

Government Reorganizations Creating the U.S. Environmental Protection Agency and the National Oceanic and Atmospheric Administration[1]

Reorganization Plan No. 3 of 1970
July 9, 1970

Special Message from the President to the Congress about Reorganization Plans to Establish the Environmental Protection Agency and the National Oceanic and Atmospheric Administration.

To the Congress of the United States

As concern with the condition of our physical environment has intensified, it has become increasingly clear that we need to know more about the total environment—land, water, and air. It also has become increasingly clear that only by reorganizing our Federal efforts can we develop that knowledge, and effectively ensure the protection, development and enhancement of the total environment itself.

The Government's environmentally-related activities have grown up piecemeal over the years. The time has come to organize them rationally and systematically. As a major step in this direction, I am transmitting today two reorganization plans: one to establish an Environmental Protection Agency, and one to establish, with the Department of Commerce, a National Oceanic and Atmospheric Administration.

Environmental Protection Agency (EPA)

Our national government today is not structured to make a coordinated attack on the pollutants which debase the air we breathe, the water we drink, and the land that grows our food. Indeed, the present governmental structure for dealing with environmental pollution often defies effective and concerted action.

Despite its complexity, for pollution control purposes the environment must be perceived as a single, interrelated system. Present assignments of departmental responsibilities do not reflect this interrelatedness.

Many agency missions, for example, are designed primarily along media lines—air, water, and land. Yet the sources of air, water, and land pollution are interrelated and often interchangeable. A single source may pollute the air with smoke and chemicals, the land with solid wastes, and a river or lake with chemical and other wastes. Control of the air pollution may produce more solid wastes, which then pollute the land or water. Control of the water-polluting effluent may convert it into solid wastes, which must be disposed of on land.

Similarly, some pollutants—chemicals, radiation, pesticides—appear in all media. Successful control of them at present requires the coordinated efforts of a variety of separate agencies and departments. The results are not always successful.

A far more effective approach to pollution control would:

- Identify pollutants.
- Trace them through the entire ecological chain, observing and recording changes in form as they occur.
- Determine the total exposure of man and his environment.
- Examine interactions among forms of pollution.
- Identify where in the ecological chain interdiction would be most appropriate.

In organizational terms, this requires pulling together into one agency a variety of research, monitoring, standard-setting and enforcement activities now scattered through several departments and agencies. It also requires that the new agency include sufficient support elements—in research and in aids to State and local anti-pollution programs, for example—to give it the needed strength and potential for carrying out its mission. The new agency would also, of course, draw upon the results of research conducted by other agencies.

Components of the EPA

Under the terms of Reorganization Plan No. 3, the following would be moved to the new Environmental Protection Agency:

- The functions carried out by the Federal Water Quality Administration (from the Department of the Interior).
- Functions with respect to pesticides studies now vested in the Department of the Interior.
- The functions carried out by the National Air Pollution Control Administration (from the Department of Health, Education, and Welfare).
- The functions carried out by the Bureau of Solid Waste Management and the Bureau of Water Hygiene, and portions of the functions carried out by the Bureau of Radiological Health of the Environmental Control Administration (from the Department of Health, Education, and Welfare).
- Certain functions with respect to pesticides carried out by the Food and Drug Administration (from the Department of Health, Education, and Welfare).
- Authority to perform studies relating to ecological systems now vested in the Council on Environmental Quality.
- Certain functions respecting radiation criteria and standards now vested in the Atomic Energy Commission and the Federal Radiation Council.
- Functions respecting pesticides registration and related activities now carried out by the Agricultural Research Service (from the Department of Agriculture).

With its broad mandate, EPA would also develop competence in areas of environmental protection that have not previously been given enough attention, such, for example, as the problem of noise, and it would provide an organization to which new programs in these areas could be added.

In brief, these are the principal functions to be transferred:

[Extensive discussion of specific functions proposed to be transferred has been omitted.]

Advantages of Reorganization

This reorganization would permit response to environmental problems in a manner beyond the previous capability of our pollution control programs. The EPA would have the capacity to do research on important pollutants irrespective of the media in which they appear, and on the impact of these pollutants on the total environment. Both by itself and together with other agencies, the EPA would monitor the condition of the environment—biological as well as physical. With these data, the EPA would be able to establish quantitative "environmental baselines"—critical if we are to measure adequately the success or failure of our pollution abatement efforts.

As no disjointed array of separate programs can, the EPA would be able—in concert with the States—to set and enforce standards for air and water quality and for individual pollutants. This consolidation of pollution control authorities would help assure that we do not create new environmental problems in the process of controlling existing ones. Industries seeking to minimize the adverse impact of their activities on the environment would be assured of consistent standards covering the full range of their waste disposal problems. As the States develop and expand their own pollution control programs, they would be able to look to one agency to support their efforts with financial and technical assistance and training.

In proposing that the Environmental Protection Agency be set up as a separate new agency, I am making an exception to one of my own principles: that, as a matter of effective and orderly administration, additional new independent agencies normally should not be created. In this case, however, the arguments against placing environmental protection activities under the jurisdiction of one or another of the existing departments and agencies are compelling.

In the first place, almost every part of government is concerned with the environment in some way, and affects it in some way. Yet each department also has its own primary mission—such as resource development, transportation, health, defense, urban growth or agriculture—which necessarily affects its own view of environmental questions.

In the second place, if the critical standard-setting functions were centralized within any one existing department, it would require that department constantly to make decisions affecting other departments—in which, whether fairly or unfairly, its own objectivity as an impartial arbiter could be called into question.

Because environmental protection cuts across so many jurisdictions, and because arresting environmental deterioration is of great importance to the quality of life in our country and the world, I believe that in this case a strong, independent agency is needed. That agency would, of course, work closely with and draw upon the expertise and assistance of other agencies having experience in the environmental area.

Roles and Functions of the EPA

The principal roles and functions of the EPA would include:

- The establishment and enforcement of environmental protection standards consistent with national environmental goals.
- The conduct of research on the adverse effects of pollution and on methods and equipment for controlling it, the gathering of information on pollution, and the use of this information in strengthening environmental protection programs and recommending policy changes.
- Assisting others, through grants, technical assistance and other means in arresting pollution of the environment.

- Assisting the Council on Environmental Quality in developing and recommending to the President new policies for the protection of the environment.

One natural question concerns the relationship between the EPA and the Council on Environmental Quality, recently established by Act of Congress.

It is my intention and expectation that the two will work in close harmony, reinforcing each other's mission. Essentially, the Council is a top-level advisory group (which might be compared with the Council of Economic Advisers), while the EPA would be an operating, "line" organization. The Council will continue to be a part of the Executive Office of the President and will perform its overall coordinating and advisory roles with respect to all Federal programs related to environmental quality.

The Council, then, is concerned with all aspects of environmental quality—wildlife preservation, parklands, land use, and population growth, as well as pollution. The EPA would be charged with protecting the environment by abating pollution. In short, the Council focuses on what our broad policies in the environmental field should be; the EPA would focus on setting and enforcing pollution control standards. The two are not competing, but complementary—and taken together, they should give us, for the first time, the means to mount an effectively coordinated campaign against environmental degradation in all of its many forms.

National Oceanic and Atmospheric Administration

[Discussion of the establishment of NOAA has been omitted.]

An On-Going Process

The reorganization which I am here proposing affords both the Congress and the Executive Branch an opportunity to re-evaluate the adequacy of existing program authorities involved in these consolidations. As these two new organizations come into being, we may well find that supplementary legislation to perfect their authorities will be necessary. I look forward to working with the Congress in this task.

In formulating these reorganization plans, I have been greatly aided by the work of the President's Advisory Council on Executive Organization (the Ash Council), the Commission on Marine Science, Engineering and Resources (the Stratton Commission, appointed by President Johnson), my special task force on oceanography headed by Dr. James Wakelin, and by the information developed during both House and Senate hearings on proposed NOAA legislation.

Many of those who have advised me have proposed additional reorganizations, and it may well be that in the future I shall recommend further changes. For the present, however, I think the two reorganizations transmitted today represent a sound and significant beginning. I also think that in practical terms, in this sensitive and rapidly developing area, it is better to proceed a step at a time—and thus to be sure that we are not caught up in a form of organizational indigestion from trying to rearrange too much at once.

As we see how these changes work out, we will gain a better understanding of what further changes—in addition to these—might be desirable.

Ultimately, our objective should be to insure that the nation's environmental and resource protection activities are so organized as to maximize both the effective coordination of all and the effective functioning of each.

The Congress, the Administration and the public all share a profound commitment to the rescue of our natural environment, and the preservation of the Earth as a place both habitable by and hospitable to man. With its acceptance of the reorganization plans, the Congress will help us fulfill that commitment.

Richard Nixon
The White House
July 9, 1970

Notes and Commentary

1. This condensation of the President's Special Message to Congress transmitting Reorganization Plan No. 3 of 1970 was prepared by the U.S. Environmental Protection Agency in conjunction with the twentieth anniversary of the agency's creation. The omission references with regard to specific functions to be transferred to EPA and to the establishment of NOAA were added by EPA; the full text appears in *Public Papers of the Presidents: Richard Nixon, 1970,* on pages 578 through 586.

APPENDIX 3

Reliability in Environmental Decision Making

Reliability, like risk, is an expression of likelihood, but rather than conveying something bad, it tells us the probability of a good outcome. Reliability is the extent to which something can be trusted. A system, process, or item is reliable to the extent that it performs the designed function under the specified conditions during a certain time period. Thus, reliability means that something will not fail prematurely. Or, stated more positively, reliability is expressed mathematically as the probability of success. Thus reliability is the probability that something that is in operation at time *0* (t_0) will still be operating until the end of the designed life (time $t = (t_t)$). People in neighborhoods near the proposed location of a proposed facility want to know if it will work and will not fail. This is especially true for those facilities that may affect the environment, such as landfills and power plants. Likewise, when environmental cleanup is being proposed, people want to know how certain the environmental practitioners are that the cleanup will be successful.

The probability of a failure per unit time is the *hazard rate,* a term familiar to environmental risk assessment; many engineers may recognize it as a *failure density,* or $f(t)$. This is a function of the likelihood that an adverse outcome will occur, but note that it is not a function of the severity of the outcome. The $f(t)$ is not affected by whether the outcome is very severe (such as pancreatic cancer and loss of an entire species) or relatively benign (muscle soreness or minor leaf damage). The likelihood that something will fail at a given time interval can be found by integrating the hazard rate over a defined time interval:

$$P\{t_1 \leq T_f \leq t_2\} = \int_{t_1}^{t_2} f(t)dt \qquad \text{(A.3.1)}$$

where T_f = time of failure.

Thus, the reliability function $R(t)$, of a system at time t is the cumulative probability that the system has not failed in the time interval from t_0 to t_t:

$$R(t) = P\{T_f \geq t\} = 1 - \int_0^t f(x)dx \qquad \text{(A.3.2)}$$

One major point worth noting from the reliability equations is that everything we design will fail. Environmental practitioners can improve reliability by extending the time (increasing t_t); this is done by making the system more resistant to failure. For example, proper engineering design of a landfill barrier can decrease the flow of contaminated water between the contents of the landfill and the surrounding aquifer, for example, a velocity of a few microns per decade. However, the barrier does not completely eliminate failure (i.e., $R(t) = 0$); it simply protracts the time before the failure occurs (increases T_f).

Equation A.3.2 illustrates that if we have built-in vulnerabilities, such as unfair facility siting practices or the inclusion of inappropriate design criteria, like cultural bias, the time of failure is shortened. Like pollution, environmental injustice is a type of inefficiency. If we do not recognize these inefficiencies upfront, we will pay by premature failures (e.g., law suits, unhappy clients, and a public that has not been well served in terms of our holding paramount their health, safety, and welfare).

A discipline within engineering, reliability engineering, looks at the expected or actual reliability of a process, system, or piece of equipment to identify the actions needed to reduce failures, and once a failure occurs, how to manage the expected effects from that failure. Thus, reliability is the mirror image of failure. Since risk is really the probability of failure (i.e., the probability that our system, process, or equipment will fail), risk and reliability are two sides of the same coin. Recall from our discussion in Chapter 2 that the five types of failure may come in many forms and from many sources. Injustice is a social failure. A tank leaking chemicals into groundwater is an engineering failure, as is exposure to carcinogens in the air, water, and food. A system that protects one group of people at the expense of another is a type of failure. So, if we are to have reliable engineering we need to make sure that whatever we design, build, and operate is done so with fairness. Otherwise, these systems are, by definition, unreliable.

Note that in environmental engineering and other empirical sciences there is another connotation of reliability, which is an indication of quality, especially for data derived from measurements, including environmental and health data. In this use, reliability is defined as the degree to which measured results are dependable and consistent with respect to the study objectives (e.g., stream water quality). This specific connotation is sometimes called test reliability in that it indicates how consistent measured values are over time, how these values compare to other measured values, and how they differ when other tests are applied. Test reliability, like engineering reliability, is a matter of trust. As such, it is often paired with test validity; that is, just how near to the true value (as indicated by some type of known standard) the measured value is. The less reliable and valid the results, the less confidence scientists and engineers have in interpreting and using them. This is very important in engineering communications generally, and in risk communications specifically.

The environmental practitioner needs to know just how reliable and valid the data are. And the environmental practitioner must properly communicate this to clients and the public. This means, however discomfiting, that we must "come clean" about all uncertainties. Uncertainties are ubiquitous in risk assessment. The Chinese word for risk, *weij-ji*, is a combination of two characters, one representing danger and the other opportunity. *Weij-ji* indicates that risk is always an uncertain balance between benefit and cost; between gain and loss. The environmental practitioner should take care not to be overly optimistic, nor overly pessimistic about what is known and what needs to be done. Full disclosure is simply an honest rendering of what is known and what is lacking for those listening to make informed decisions. But, remember, a word or phrase can be taken many ways. Environmental practitioners should liken themselves to physicians writing prescriptions. Be completely clear, otherwise confusion may result and lead to unintended, negative consequences.

APPENDIX 4

Principles of Environmental Persistence

Persistent, bioaccumulating toxicants (PBTs) are a worldwide concern. Some of the most notoriously toxic chemicals are also very persistent. The concept of persistence elucidates the notion of tradeoffs that are frequently needed as part of many responses to environmental insults. It also underlines the importance that good science is necessary but never sufficient to provide an acceptable response to environmental justice issues. Let us consider the pesticide DDT (1,1,1-trichloro-2,2-bis-(4-chlorophenyl)-ethane ($C_{14}H_9Cl_5$)). DDT is relatively insoluble in water (1.2–5.5 mg L^{-1} at 25°C) and is not very volatile (vapor pressure: 0.02×10^{-5} mm Hg at 25°C).[1] Looking at the water solubility and vapor pressures alone may lead us to believe that people and wildlife are not likely to be exposed in the air or water. However, the compound is highly persistent in soils, with a $T_{1/2}$ of about 1.1 to 3.4 years, so it still may end up in drinking water in the form of suspended particles or in the air sorbed to fine particles. DDT also exhibits high bioconcentration factors (in the order of 50,000 for fish and 500,000 for bivalves), so once organisms are exposed, they tend to increase body burdens of DDT over their lifetimes. In the environment, the parent DDT is metabolized mainly to DDD and DDE.[2]

The physicochemical properties of a substance determine how readily it will move among the environmental compartments—to and from sediment, surface water, soil, groundwater, air, and in the food web, including humans. So, if a substance is likely to leave the water, it is not persistent in water. However, if the compound moves from the water to the sediment, where it persists for long periods of time, it must be considered environmentally persistent. This is an example of how terminology can differ between chemists and engineers. Chemists often define persistence as an intrinsic chemical property of a compound, whereas engineers see it as both intrinsic and extrinsic (i.e., a function of the media, energy and mass balances, and equilibria). So, engineers usually want to know not only about the molecular weight, functional groups, and ionic form of the compound, but also whether it is found in the air or water, and what is the condition of the media (e.g., pH, soil moisture, sorption potential, and microbial populations). The movement among phases and environmental compartments is known as partitioning. Many toxic compounds are semi-volatile (i.e., at 20°C and 101 kPa atmospheric pressure, vapor pressures = 10^{-5} to 10^{-2} kPa) under typical environmental conditions. The low vapor pressures and low aqueous solubilities mean they will have will low fugacities; that is, they lack a strong propensity to flee a compartment, for example, to move from the water to the air. The most common water-to-air fugacity measure is the *Henry's Law constant*. Henry's law states that the concentration of a dissolved gas is directly proportional to the partial pressure of that gas above the solution:

$$p_a = K_H[c] \tag{A.4.1}$$

where

K_H = Henry's Law constant
p_a = partial pressure of the gas
$[c]$ = molar concentration of the gas

or,

$$p_a = K_H \, C_W \tag{A.4.2}$$

where C_W is the concentration of gas in water.

Henry's law expresses the proportionality between the concentration of a dissolved contaminant and its partial pressure in the open atmosphere at equilibrium. That is, the Henry's Law constant is an example of an *equilibrium constant*, which is the ratio of concentrations when chemical equilibrium is reached in a reversible reaction, the time when the rate of the forward reaction is the same as the rate of the reverse reaction. Most of the time, when a partitioning coefficient is given, it is assumed to be an equilibrium constant. For environmental partitioning, the amount of chemical needed to reach equilibrium is usually very small, that is, very dilute solutions and other mixtures.

A direct partitioning between the air and water phases is the air-water partitioning coefficient (K_{AW}):

$$K_{AW} = \frac{C_A}{C_W} \tag{A.4.3}$$

where C_A is the concentration of a gas in the air.

The relationship between the air/water partition coefficient and Henry's Law constant for a substance is:

$$K_{AW} = \frac{K_H}{RT} \quad \text{(A.4.4)}$$

where R is the gas constant (8.21×10^{-2} L atm mol^{-1} K^{-1}) and T is the temperature (°K).

Under environmental conditions most toxic substances have very low K_H values since K_H is proportional to the concentration of a dissolved contaminant and its partial pressure in the atmosphere at equilibrium. There are many exceptions, however, such as the relatively water soluble compounds with high vapor pressures, like alcohols, benzene, toluene, and many organic solvents. Since Henry's law is a function of aqueous solubility and vapor pressure, estimating the tendency for a substance's release in vapor form, K_H is a good indicator of the fugacity from the water to the atmosphere.

Another common expression of partitioning is the octanol-water coefficient (K_{ow}). The K_{ow} value indicates a compound's likelihood to exist in the organic versus aqueous phase. A rule to keep in mind is that "like dissolves like." The configuration of a molecule determines whether it is polar or nonpolar. Polar compounds are electrically positive at one end and negative at the other. The water molecule, for example, is highly electronegative at the hydrogen atoms and electronegative at the oxygen atom. Other molecules, like fats, are not polar (i.e., do not have strong differences between the positive and negative ends). If a relatively nonpolar compound is dissolved in water and the water comes into contact with another substance, like octanol, the nonpolar compound will move from the water to the octanol. Its K_{ow} reflects just how much of the substance will move from the aqueous and organic solvents (phases) until it reaches equilibrium. For example, if at a given temperature and pressure a chemical is at equilibrium when its concentration in octanol is 100 mg L^{-1} and in water is 1000 mg L^{-1}, its K_{ow} is 100 divided by 1000, or 0.1. Since the range is so large among various environmental contaminants, it is common practice to express log K_{ow} values. So, for example, in a spill of equal amounts of the PCB, decachlorobiphenyl ($\log K_{ow} = 8.23$) and the pesticide chlordane ($\log K_{ow} = 2.78$), the PCB has much greater affinity for the organic phases than does the chlordane (more than five orders of magnitude). This does not mean that a greater amount of either of the compounds is likely to stay in the water column, since they are both hydrophobic, but it does mean that they will vary in the time and mass of each contaminant moving between phases. The time it takes to reach equilibrium (i.e., the kinetics) is different.

Even low K_H and K_{AW} compounds, however, can be transported long distances in the atmosphere when sorbed to particles. Fine particles can behave as colloids and stay suspended for extended periods of time, explaining in part why low K_H compounds can be found in the most remote locations relative to their sources, such as in the Arctic regions. This is important, for example, when explaining to indigenous populations why they may be exposed to contaminants that are not produced near them.

Sorption is the partitioning of a substance from the liquid to the solid phase, and is an important predictor of a chemical's persistence. If the substrate has sufficient sorption sites, such as many clays and organic matter, the substance may become tightly bound and persistent. The properties of the compound and those of the water, soil, and sediment determine the rate of sorption. The soil partition coefficient (K_d) is the experimentally derived ratio of a contaminant's concentration in the solid matrix to the contaminant concentration in the liquid phase at chemical equilibrium. Another frequently reported liquid to solid phase partitioning coefficient is the organic carbon partitioning coefficient (K_{oc}), which is the ratio of the contaminant concentration sorbed to organic matter in the matrix (soil or sediment) to the contaminant concentration in the aqueous phase. Thus, the K_{oc} is derived from the quotient of a contaminant's K_d and the fraction of organic matter (OM) in the matrix:

$$K_{oc} = \frac{K_d}{OM} \tag{A.4.5}$$

Many toxic substances are expected to be strongly sorbed, but K_{oc} varies from substrate to substrate.

It is important to keep in mind the difference between chemical persistence and environmental persistence. For example, we can look at Henry's law, solubility, vapor pressure, and sorption coefficients for a compound and determine that the compound is not persistent. However, in real life scenarios, this may not be the case. For example, there may be a repository of a source of a nonpersistent compound that leads to a continuous, persistent exposure of a neighborhood population (see the case study, "'Cancer Alley' and Vinyl Chloride" in Chapter 5). Or, a compound that is ordinarily not very persistent may become persistent under the right circumstances, for example, a reactive pesticide that is tracked into a home and becomes entrapped in carpet fibers. The lower rate of photolysis (degradation by light energy) indoors and the sorptive characteristics of the carpet twill can lead to dramatically increased environmental half-lives of certain substances.

Notes and Commentary

1. The source for the physicochemical properties of DDT and its metabolites is United Nations Environmental Programme, 2002. "Chemicals: North American Regional Report," Regionally Based Assessment of Persistent Toxic Substances, Global Environment Facility.

2. The two principal isomers of DDD are *p,p'*-2,2-bis(4-chlorophenyl)-1,1-dichloroethane; and *o,p'*-1-(2-chlorophenyl)-1-(4-chlorophenyl)-2,2-dichloroethane. The principal isomer of DDE is *p,p'*-1,1'-(2,2-dichloroethenylidene)-bis[4-chlorobenzene].

APPENDIX 5

Cancer Slope Factors

Slope factors are expressed in inverse exposure units since the slope of the dose-response curve is an indication of risk per exposure. Thus, the units are the inverse of mass per mass per time, usually $(mg\,kg^{-1}\,day^{-1})^{-1} = kg\,day^{-1}\,mg^{-1}$. This means that the product of the cancer slope factor and exposure, or risk, is unitless. The SF is the toxicity value used to calculate cancer risks. SF values are contaminant-specific and route-specific (e.g., via inhalation, through the skin, or via ingestion). Inhalation and oral cancer slope factors are shown in Table A.5.1. Note that the more potent the carcinogen, the larger the slope factor (i.e., the steeper the slope of the dose-response curve). For example, arsenic and benzo(a)pyrene are quite carcinogenic, with slope factors of 1.51 and 3.10, respectively. Their cancer potency is three orders of magnitude greater than aniline, bromoform, and chloromethane, for example. Also note that the SF is based on the linear portion of the curve.

The route of exposure can greatly influence the cancer slope. Note, for example, that the carcinogeniety of 1,2-dibromo-3-chloropropane is three orders of magnitude steeper via the oral route than from breathing vapors. Conversely, the cancer slope factor for chloroform is more than an order of magnitude greater from inhalation than from oral ingestion. Such information is important in deciding how to protect populations from exposure to contaminants. For example, if an industrial facility is releasing vinyl chloride, both inhalation and oral ingestion must be considered as possible routes of exposures for people living nearby. Both the inhalation and oral slope factors are high—3.00×10^{-1} and $1.90\,kg\,day^{-1}\,mg^{-1}$, respectively. In addition, if the vinyl chloride finds its way to the water supply, not only the amount in food and drinking water must be considered, but also indirect inhalation routes (e.g., showering), since vinyl chloride is volatile and can be released and inhaled. The physical and chemical characteristics (e.g., vapor pressure and Henry's Law constants) of vinyl chloride, coupled with its marked toxicity via multiple routes of exposure, make it a particularly onerous contaminant.

The table also indicates that the structure of a compound greatly affects its biological activity. For example, comparing halogen substitutions

TABLE A.5.1
Cancer slope factors for selected environmental contaminants.[1]

Contaminant	*Inhalation Slope Factor (kg day^{-1} mg^{-1})*	*Oral Slope Factor (kg day^{-1} mg^{-1})*
Acephate	1.74×10^{-2}	8.70×10^{-3}
Acrylamide	4.55	4.50
Acrylonitrile	2.38×10^{-1}	5.40×10^{-1}
Aldrin	1.71×10^{1}	1.70×10^{1}
Aniline	5.70×10^{-3}	5.70×10^{-3}
Arsenic	1.51×10^{1}	1.50
Atrazine	4.44×10^{-1}	2.22×10^{-1}
Azobenzene	1.09×10^{-1}	1.10×10^{-1}
Benzene	2.90×10^{-2}	2.90×10^{-2}
Benz(a)anthracene	3.10×10^{-1}	7.30×10^{-1}
Benzo(a)pyrene	3.10	7.30
Benzo(b)fluoranthene	3.10×10^{-1}	7.30×10^{-1}
Benzo(k)fluoranthene	3.10×10^{-2}	7.30×10^{-2}
Beryllium	8.40	Not given
Benzotrichloride	1.63×10^{1}	1.30×10^{1}
Benzyl chloride	2.13×10^{-1}	1.70×10^{-1}
Bis(2-chloroethyl)ether	1.16	1.16
Bis(2-chloroisopropyl)ether	3.50×10^{-2}	1.10×10^{-2}
Bis(2-ethyl-hexyl)phthalate	1.40×10^{-2}	7.00×10^{-2}
Bromodichloromethane	6.20×10^{-2}	6.20×10^{-2}
Bromoform	3.85×10^{-3}	7.90×10^{-3}
Cadmium	Not given	6.30
Captan	7.00×10^{-3}	3.50×10^{-3}
Chlordane	3.50×10^{-1}	3.50×10^{-1}
Chlorodibromomethane	8.40×10^{-2}	8.40×10^{-2}
Chloroethane (Ethylchloride)	2.90×10^{-3}	2.90×10^{-3}
Chloroform	8.05×10^{-2}	6.10×10^{-3}
Chloromethane	3.50×10^{-3}	1.30×10^{-2}
Chromium (VI)	3.50×10^{-3}	Not given
Chrysene	3.10×10^{-3}	7.30×10^{-3}
DDD	2.40×10^{-1}	2.40×10^{-1}
DDE	3.40×10^{-1}	3.40×10^{-1}
DDT	3.40×10^{-1}	3.40×10^{-1}
Dibenz(a,h)anthracene	3.10	7.30
Dibromo-3-chloropropane,1,2-	2.42×10^{-3}	1.40
Dichlorobenzene,1,4-	2.20×10^{-2}	2.40×10^{-2}
Dichlorobenzidine,3,3-	4.50×10^{-1}	4.50×10^{-1}
Dichloroethane,1,2-	9.10×10^{-2}	9.10×10^{-2}
Dichloroethene (mixture),1,1-	1.75×10^{-1}	6.00×10^{-1}
Dichloromethane	7.50×10^{-3}	1.64×10^{-3}
Dichloropropane,1,2-	6.80×10^{-2}	6.80×10^{-2}
Dichloropropene,1,3-	1.30×10^{-1}	1.75×10^{-1}
Dieldrin	1.61×10^{1}	1.61×10^{1}
Dinitrotoluene,2,4-	6.80×10^{-1}	6.80×10^{-1}

TABLE A.5.1
Continued

Contaminant	*Inhalation Slope Factor (kg day^{-1} mg^{-1})*	*Oral Slope Factor (kg day^{-1} mg^{-1})*
Dioxane,1,4-	2.20×10^{-2}	1.11×10^{-2}
Diphenylhydrazine,1,2-	7.70×10^{-1}	8.00×10^{-1}
Epichlorohydrin	4.20×10^{-3}	9.90×10^{-3}
Ethyl acrylate	6.00×10^{-2}	4.80×10^{-2}
Ethylene oxide	3.50×10^{-1}	1.02
Formaldehyde	4.55×10^{-2}	Not given
Heptachlor	4.55	4.50
Heptachlor epoxide	9.10	9.10
Hexachloro-1,3-butadiene	7.70×10^{-2}	7.80×10^{-2}
Hexachlorobenzene	1.61	1.60
Hexachlorocyclohexane, alpha	6.30	6.30
Hexachlorocyclohexane, beta	1.80	1.80
Hexachlorocyclohexane, gamma (lindane)	1.30	1.30
Hexachloroethane	1.40×10^{-2}	1.40×10^{-2}
Hexahydro-1,3,5-trinitro-1,3,5-traizine (RDX)	2.22×10^{-1}	1.11×10^{-1}
Indeno(1,2,3-cd)pyrene	3.10×10^{-1}	7.30×10^{-1}
Isophorone	9.50×10^{-4}	9.50×10^{-4}
Nitrosodi-*n*-propylamine, *n*-	7.00	7.00
Nitrosodiphenylamine, *n*-	4.90×10^{-3}	4.90×10^{-3}
Pentachloronitrobenzene	5.20×10^{-1}	2.60×10^{-1}
Pentachlorophenol	1.20×10^{-1}	1.20×10^{-1}
Phenylphenol,2-	3.88×10^{-3}	1.94×10^{-3}
Polychlorinated biphenyls (Arochlor mixture)	3.50×10^{-1}	2.00
Tetrachlorodibenzo-*p*-dioxin,2,3,7,8	1.16×10^{5}	1.50×10^{5}
Tetrachloroethane,1,1,1,2-	2.59×10^{-2}	2.60×10^{-2}
Tetrachloroethane,1,1,2,2-	2.03×10^{-1}	2.03×10^{-1}
Tetrachloroethene (PCE)	2.00×10^{-3}	
Tetrachloroethylene	2.03×10^{-3}	5.20×10^{-2}
Tetrachloromethane	5.25×10^{-2}	1.30×10^{-1}
Toxaphene	1.12	1.10
Trichloroethane,1,1,2-	5.60×10^{-2}	5.70×10^{-2}
Trichloroethene (TCE)	6.00×10^{-3}	1.10×10^{-2}
Trichlorophenol,2,4,6-	1.10×10^{-2}	1.10×10^{-2}
Trichloropropane,1,2,3-	8.75	7.00
Trifluralin	3.85×10^{-3}	7.70×10^{-3}
Trimethylphosphate	7.40×10^{-2}	3.70×10^{-2}
Trinitrotoluene,2,4,6- (TNT)	6.00×10^{-2}	3.00×10^{-2}
Vinyl chloride	3.00×10^{-1}	1.90

Sources: U.S. Environmental Protection Agency, 2002. Integrated Risk Information System. U.S. Environmental Protection Agency, 1994. Health Effects Summary Tables, 1994.

indicates that the greater the number of chlorine atoms on a molecule, the steeper the slope of the dose-response curve. Unsubstituted ethane is not carcinogenic (no slope factor). A single chlorine substitution in chloromethane renders the molecule carcinogenic, with a slope factor of 2.90×10^{-3}. Adding another chlorine atom to form 1,2-dichloroethane increases the slope to 9.10×10^{-2}. Completely halogenated ethane, hexachloroethane, has seen its cancer slope factor increase to 1.40×10^{-2}. Also, where the chlorine or bromine substitutions occur on the molecule will affect the cancer potential. For example, the isomers of tetrachloroethane have different slope factors; 1,1,1,2-tetrachloroethane's slope factor is 1.40×10^{-2}, but 1,1,2,2-tetrachloroethane's slope factor is 2.03×10^{-1}. This seemingly small difference in molecular structure leads to an order of magnitude greater cancer potency.

Dermal exposures are generally extrapolated from the other two major routes. For example, the dermal slope factor for Aroclor 1254, the polychlorinated biphenyl (PCB) mixture (21% $C_{12}H_6Cl_4$, 48% $C_{12}H_5Cl_5$, 23% $C_{12}H_4Cl_6$, and 6% $C_{12}H_3Cl_7$), the dermal exposure to soil or food is $2.22\,kg\,day^{-1}\,mg^{-1}$. Keep in mind that this is the dose-response slope associated with the handling or other skin contact with the contaminant, not the actual ingestion. The Aroclor 1254 dermal slope factor for exposure to water is $4.44\,kg\,day^{-1}\,mg^{-1}$. Both of these dermal slopes have been extrapolated from a gastrointestinal absorption factor of 0.9000.[2] The dermal slope factors shown in Table A.5.2 have been extrapolated from other routes. The GI tract absorption rate is also given, since these are often used to extrapolate slope factors for dermal and other routes of exposure. Note that the larger the GI absorption decimal, the more completely the contaminant is absorbed. For complete absorption, the value equals 1.

The absorption factor is not only important for extrapolating slope factors, but it is a variable in calculating certain exposures. As we shall see later, the air (both particle and gas) and water exposure equations include an absorption factor. The dermal exposure equation does not include an absorption factor, but since dermal cancer slope factors are extrapolated from the inhalation or ingestion slopes, by extension, the absorption factor is part of the dermal risk calculations. So, all other factors being equal, a contaminant with a larger absorption factor will have a larger risk. This is evident when considering the pathway taken by a chemical after it enters an organism. As shown in Figure A.5.1, the potential dose in a dermal exposure is what is available before coming into contact with the skin, but after this contact (i.e., the applied dose), it crosses the skin barrier and is absorbed. The absorption leads to the biological effectiveness of the contaminant when the chemical reaches the target organ, where it may elicit the effect (e.g., cancer). The absorption factor is the first determinant of the amount of the contaminant that reaches the target organ. For example, although the dermal slope factors for 1,4-dioxane and 1,4-dichlorobenzene are nearly the same (2.20×10^{-2} and 2.40×10^{-2}, respectively), all the

TABLE A.5.2
Gastrointestinal absorption rates and dermal cancer slope factors for selected environmental contaminants.[3]

Contaminant	*GI Absorption*	*Inhalation Slope Factor (kg day^{-1} mg^{-1})*
Acephate	0.5	1.74×10^{-2}
Acrylamide	0.5	9.00
Acrylonitrile	0.8	6.75×10^{-1}
Aldrin	1	1.72×10^{1}
Aniline	0.5	1.14×10^{-3}
Arsenic	0.95	1.58×10^{1}
Atrazine	0.5	4.44×10^{-1}
Azobenzene	0.5	2.20×10^{-1}
Benzene	0.9	3.22×10^{-2}
Benz(a)anthracene	0.5	1.46
Benzo(a)pyrene	0.5	1.46×10^{1}
Benzo(b)fluoranthene	0.5	1.46
Benzo(k)fluoranthene	0.5	1.46×10^{-1}
Beryllium	0.006	Not given
Benzotrichloride	0.8	1.63×10^{1}
Benzyl chloride	0.8	2.13×10^{-1}
Bis(2-chloroethyl)ether	0.98	1.13
Bis(2-chloroisopropyl)ether (DEHP)	0.8	8.75×10^{-2}
Bis(2-ethyl-hexyl)phthalate	0.5	2.80×10^{-2}
Bromodichloromethane	0.98	6.37×10^{-2}
Bromoform	0.75	1.05×10^{-2}
Cadmium	0.044	Not given
Captan	0.5	7.00×10^{-3}
Chlordane	0.8	4.38×10^{-1}
Chloroethane (Ethylchloride)	0.8	1.28
Chloroform	1	6.10×10^{-3}
Chloromethane	0.8	1.63×10^{-2}
Chromium(VI)	0.013	Not given
Chrysene	0.5	1.46×10^{-2}
DDD,4,4-	0.8	3.00×10^{-1}
DDE,4,4-	0.8	4.25×10^{-1}
DDT,4,4-	0.8	4.25×10^{-1}
Dibenz(a,h)anthracene	0.5	1.46×10^{1}
Dibromo-3-chloropropane,1,2-	0.5	1.12×10^{-1}
Dichlorobenzene,1,4-	1	2.40×10^{-2}
Dichlorobenzidine,3,3-	0.5	9.00×10^{-1}
Dichloroethane,1,2- (EDC)	1	9.10×10^{-2}
Dichloroethene,1,1-	1	6.00×10^{-1}
Dichloropropane,1,2-	1	6.80×10^{-2}
Dichloropropene,1,3-	0.98	1.84×10^{-1}
Dieldrin	1	1.60×10^{1}
Dinitrotoluene,2,4-	1	6.80×10^{-1}

TABLE A.5.2
Continued

Contaminant	*GI Absorption*	*Inhalation Slope Factor (kg day^{-1} mg^{-1})*
Dioxane,1,4-	0.5	2.20×10^{-2}
Diphenylhydrazine,1,2-	0.5	1.60
Epichlorohydrin	0.8	1.24×10^{-2}
Ethyl acrylate	0.8	6.00×10^{-2}
Ethylene oxide	0.8	1.28
Formaldehyde	0.5	Not given
Heptachlor	0.8	5.63
Heptachlor epoxide	0.4	2.28×10^{1}
Hexachloro-1,3-butadiene	1	7.80×10^{-2}
Hexachlorobenzene	0.8	2.00
Hexachlorocyclohexane, alpha	0.974	6.47
Hexachlorocyclohexane, beta	0.907	1.99
Hexachlorocyclohexane, gamma (lindane)	0.994	1.31
Hexachloroethane	0.8	1.75×10^{-2}
Hexahydro-1,3,5-trinitro-1,3,5-traizine (RDX)	0.5	2.22×10^{-1}
Indeno(1,2,3-cd)pyrene	0.5	1.46
Isophorone	0.5	1.90×10^{-3}
Nitrosodi-*n*-propylamine, *n*-	0.475	1.47×10^{1}
Nitrosodiphenylamine, *n*-	0.5	9.80×10^{-3}
Pentachloronitrobenzene	0.5	5.20×10^{-1}
Pentachlorophenol	0.5	2.40×10^{-1}
Phenylphenol,2-	0.5	3.88×10^{-3}
Polychlorinated biphenyls (Arochlor mixture)	0.85	2.35
Tetrachlorodibenzo-*p*-dioxin,2,3,7,8	0.9	1.68×10^{5}
Tetrachloroethane,1,1,1,2-	0.8	3.25×10^{-2}
Tetrachloroethane,1,1,2,2-	0.7	2.86×10^{-1}
Tetrachloroethene (PCE)	1	5.20×10^{-2}
Tetrachloromethane	0.85	1.53×10^{-1}
Toxaphene	0.63	1.75
Trichloroethane,1,1,2-	0.81	7.04×10^{-2}
Trichloroethene (TCE)	0.945	1.16×10^{-2}
Trichlorophenol,2,4,6-	0.8	2.20×10^{-2}
Trichloropropane,1,2,3-	0.8	8.75
Trifluralin	0.2	3.85×10^{-3}
Trimethylphosphate	0.5	7.40×10^{-2}
Trinitrotoluene,2,4.6- (TNT)	0.5	6.00×10^{-2}
Vinyl chloride	0.875	2.17

Sources: U.S. Environmental Protection Agency, 2002. Integrated Risk Information System. U.S. Environmental Protection Agency, 1994. Health Effects Summary Tables, 1994.

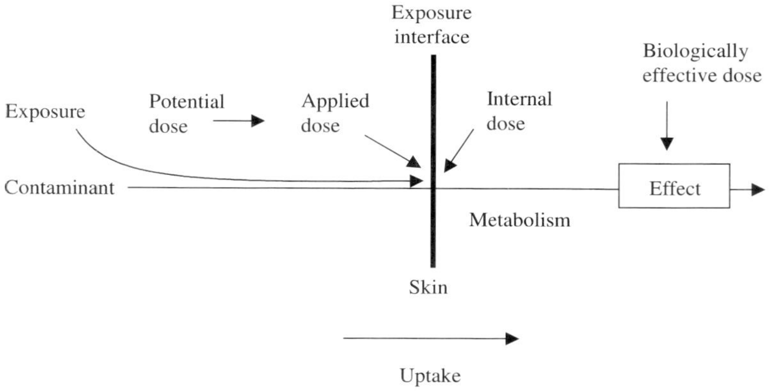

FIGURE A.5.1. Pathway of a contaminant from ambient exposure through health effect.
Source: U.S. Environmental Protection Agency, and D. Vallero, 2003. *Engineering the Risks of Hazardous Wastes*, Butterworth-Heinemann, Boston, MA.

dichlorobenzene is expected to be absorbed (i.e., absorption = 1), but only half the dioxane will be absorbed (i.e., absorption = 0.5). This means that if all other factors are equal, the risk from dichlorobenzene is twice that of dioxane.

Exposure route can influence the steepness of the slope. Note, for example, that the very lipophilic PCBs have a dermal slope that is an order of magnitude steeper than their inhalation slope. The absorption rate and, hence, the dermal slope is also affected by the contaminant's chemical structure. For example, trichloroethene (TCE) with its double bond between the carbon atoms has an absorption rate of 0.945, whereas 1,1,2-trichloroethane with only single bonds has a much lower absorption rate of 0.81, even though both have three chlorine substitutions.

Notes and Commentary

1. These values are updated. If a carcinogen is not listed in the table, visit http://risk.lsd.ornl.gov/tox/rap_toxp.shtml.
2. This information was obtained from the Risk Assessment Information System of the Oak Ridge National Laboratory, 2003.
3. These values are updated. If a carcinogen is not listed in the table, visit http://risk.lsd.ornl.gov/tox/rap_toxp.shtml.

APPENDIX 6

Equations for Calculating Lifetime Average Daily Dose (LADD) for Various Routes of Exposure[1]

Route of Exposure	*Equation* *LADD (in mg kg⁻¹ d⁻¹) =*	*Definitions*
Inhaling aerosols (particulate matter)	$\frac{(C)\cdot(PC)\cdot(IR)\cdot(RF)\cdot(EL)\cdot(AF)\cdot(ED)\cdot(10^{-6})}{(BW)\cdot(TL)}$	C = concentration of the contaminant on the aerosol/particle ($mg\,kg^{-1}$) PC = particle concentration in air ($mg\,m^{-3}$) IR = inhalation rate ($m^{-3}\,h^{-1}$) RF = respirable fraction of total particulates (dimensionless, usually determined by aerodynamic diameters, e.g., 2.5 μm) EL = exposure length ($h\,d^{-1}$) ED = duration of exposure (d) AF = absorption factor (dimensionless) BW = body weight (kg) TL = typical lifetime (d) 10^{-6} is a conversion factor (kg to mg)

Route of Exposure	*Equation LADD (in mg kg⁻¹ d⁻¹) =*	*Definitions*
Inhaling vapor phase contaminants	$\frac{(C)\cdot(IR)\cdot(EL)\cdot(AF)\cdot(ED)}{(BW)\cdot(TL)}$	C = concentration of the contaminant in the gas phase ($mg\,m^{-3}$) Other variables the same as above
Drinking water	$\frac{(C)\cdot(CR)\cdot(ED)\cdot(AF)}{(BW)\cdot(TL)}$	C = concentration of the contaminant in the drinking water ($mg\,L^{-1}$) CR = rate of water consumption ($L\,d^{-1}$) ED = duration of exposure (d) AF = portion (fraction) of the ingested contaminant that is physiologically absorbed (dimensionless) Other variables are the same as above
Contact with soil-borne contaminants	$\frac{(C)\cdot(SA)\cdot(BF)\cdot(FC)\cdot(SDF)\cdot(ED)\cdot(10^{-6})}{(BW)\cdot(TL)}$	C = concentration of the contaminant in the soil ($mg\,kg^{-1}$) SA = skin surface area exposed (cm^{-2}) BF = bioavailability (percent of contaminant absorbed per day) FC = fraction of total soil from contaminated source (dimensionless) SDF = soil deposition, the mass of soil deposited per unit area of skin surface ($mg\,cm^{-2}\,d^{-1}$) Other variables are the same as above

How to Apply LADD to Calculate Chronic Exposure

In the process of synthesizing pesticides over an 18-year period, a polymer manufacturer has contaminated the soil on its property with vinyl chloride. The plant closed two years ago but vinyl chloride vapors continue to reach the neighborhood surrounding the plant at an average concentration of 1 mg m^{-3}. Assume that people are breathing at a ventilation rate of 0.5 m^3 h^{-1} (about the average of adult males and females over 18 years of age[2]). The legal settlement allows neighboring residents to evacuate and sell their homes to the company. However, they may also stay. The neighbors have asked for advice on whether to stay or leave, since they have already been exposed for 20 years.

Solution and Discussion

Vinyl chloride is highly volatile, so its phase distribution will be mainly in the gas phase rather than the aerosol phase. Although some of the vinyl chloride may be sorbed to particles, we will use only a vapor phase LADD equation, since the particle phase is likely to be relatively small. Also, we will assume that outdoor concentrations are the exposure concentrations. This is unlikely, however, since people spend very little time outdoors compared to indoors, so this may provide an additional factor of safety. To determine how much vinyl chloride penetrates living quarters, indoor air studies would have to be conducted. For a scientist to compare exposures, indoor air measurements should be taken.

Find the appropriate equation and insert values for each variable. Absorption rates are published by the EPA and the Oak Ridge National Laboratory (http://risk.lsd.ornl.gov/cgi-bin/tox/TOX_select?select=nrad). Vinyl chloride is well absorbed, so to be conservative we can assume that $AF = 1$. We will also assume that the person stays in the neighborhood and is exposed at the average concentration 24 hours a day ($EL = 24$), and that a person lives the remainder of an entire typical lifetime exposed at the measured concentration. These assumptions all err on the side of safety (i.e., higher expected exposures).

Although the ambient concentrations of vinyl chloride may have been higher when the plant was operating, the only measurements we have are those taken recently. Thus, this is an area of uncertainty that must be discussed with the clients. The common default value for a lifetime is 70 years, so we can assume the longest exposure would be 70 years (25,550 days). Table 5.10 gives some of the commonly used default values in exposure assessments. If the person is now 20 years of age and has already been exposed for that time, and lives the remaining 50 years exposed at $1\,mg\,m^{-3}$:

$$\begin{aligned}\text{LADD} &= \frac{(C)\cdot(IR)\cdot(EL)\cdot(AF)\cdot(ED)}{(BW)\cdot(TL)}\\ &= \frac{(1)\cdot(0.5)\cdot(24)\cdot(1)\cdot(25{,}550)}{(70)\cdot(25{,}550)}\\ &= 0.2\,mg\,kg^{-1}\,day^{-1}\end{aligned}$$

If the 20-year-old leaves today, the exposure duration would be for the 20 years that the person lived in the neighborhood. Thus, only the *ED* term would change, from 25,550 days to 7300 days (i.e., 20 years).

Thus, the LADD falls to 2/7 of its value:

$$\text{LADD} = 0.05\,mg\,kg^{-1}\,day^{-1}$$

Notes and Commentary

1. M. Derelanko, 1999. "Risk Assessment," in *CRC Handbook of Toxicology*, M.J. Derelanko and M.A. Hollinger, eds., CRC Press, Boca Raton, FL.
2. U.S. Environmental Protection Agency, 1997. Exposure Factors Handbook, Report No. EPA/600/P-95/002Fa, Washington, D.C.

APPENDIX 7

Characterizing Environmental Risk

Risk can be quantitatively determined from hazard and exposure calculations. Two general risk characterization approaches are used in environmental problem solving: direct risk assessments and risk-based cleanup standards.

Direct Risk Calculations

In its simplest form, risk is the product of the hazard and exposure, but assumptions can greatly affect risk estimates. For example, cancer risk can be defined as the theoretical probability of contracting cancer when continually exposed for a lifetime (e.g., 70 years) to a given concentration of a substance (carcinogen). The probability usually is calculated as an upper confidence limit. The maximum estimated risk may be presented as the number of chances in a million of contracting cancer.

Two measures of risk are commonly reported. One is the individual risk; that is, the probability of a person developing an adverse effect (e.g., cancer) due to the exposure. This is often reported as a residual or increased probability above background. For example, if we want to characterize the contribution of all the power plants in the United States to increased cancer incidence, the risk above background would be reported. The second way that risk is reported is population risk; that is, the annual excess number of cancers in an exposed population. The maximum individual risk might be calculated from exposure estimates based upon a maximum exposed individual, or MEI. The hypothetical MEI lives an entire lifetime outdoors at the point where pollutant concentrations are highest. Assumptions about exposure will greatly affect the risk estimates. For example, the cancer risk from power plants in the United States has been estimated to be 100- to 1000-fold lower for an average exposed individual than that calculated for the MEI.

For cancer risk assessments, the hazard is generally assumed to be the slope factor and the long-term exposure is the lifetime average daily dose:

$$\text{Cancer risk} = \text{SF} \times \text{LADD} \qquad \text{(A.7.1)}$$

Cancer Risk Calculation Example

Using the lifetime average daily dose value from the vinyl chloride exposure calculation in the Appendix 6 example, estimate the direct risk to the people living near the abandoned polymer plant. What advice would you give the neighbors?

Solution and Discussion

Inserting the calculated LADD values from Appendix 6 and the vinyl chloride inhalation slope factor of 3.00×10^{-1} from Appendix 5 for the two LADD values under consideration, the cancer risk to the neighborhood exposed for an entire lifetime (exposure duration = 70 years) gives us $0.2\,\text{mg}\,\text{kg}^{-1}\,\text{day}^{-1} \times 0.3\,(\text{mg}\,\text{kg}^{-1}\,\text{day}^{-1})^{-1} = 0.06$. This is an incredibly high risk! The thresholds for concern are often 1 in 10,000 additional cancer risk as an initial exposure reduction number (e.g., emergency response cleanup) or 1 in a million (e.g., site remediation), whereas this is a probability of 6%.

Even at the shorter duration period (20 years of exposure instead of 70 years), the risk is calculated as $0.05 \times 0.3 = 0.017$ or nearly a 2% risk. The combination of a very steep slope factor and very high lifetime exposures leads to a very high risk. Vinyl chloride is a liver carcinogen, so unless corrective action significantly lowers the ambient concentrations of vinyl chloride the prudent course of action is that the neighbors accept the buyout and leave the area.

Incidentally, vinyl chloride has relatively high water solubility and can be absorbed to soil particles, so ingestion of drinking water (e.g., people on private wells drawing water from groundwater that has been contaminated) and dermal exposures (e.g., children playing in the soil) are also conceivable. The total risk from a single contaminant like vinyl chloride is equal to the sum of risks from all pathways (e.g., vinyl chloride in the air, water, and soil):

$$\text{Total Risk} = \Sigma \text{ risks from all exposure pathways} \qquad \text{(A.7.2)}$$

Requirements and measures of success are seldom if ever as straightforward as the vinyl chloride example. In fact, the engineer would be ethically remiss if the only advice given is to the local community, whether or not to accept the buyout. Of course, one of the canons is to be a "faithful agent" to the clientele. However, as we discussed in the previous chapter, the first engineering canon is to hold paramount the health and safety of the public. Thus, the engineer must balance any proprietary information that the client wants to be protected with the need to protect public health. In this case, the engineer must tell the client and prime contractors, for example, that the regulatory agencies need to know that even though the neighbors are moving, others, including future populations, are threatened. In other words, just because one's clients are taken out of harm's way does not obviate the need for remediation to reduce the vinyl chloride concentrations to acceptable levels.

The risk of adverse outcome other than cancer (so-called noncancer risk) is generally called the hazard quotient (HQ). It is calculated by dividing the maximum daily dose (MDD) by the acceptable daily intake (ADI):

$$\text{Noncancer risk} = \text{HQ} = \frac{\text{MDD}}{\text{ADI}} = \frac{\text{Exposure}}{\text{RfD}} \qquad \text{(A.7.3)}$$

Note that this is an index, not a probability, so it is really an indication of relative risk. If the noncancer risk is greater than 1, the potential risk may be significant, and if the noncancer risk is less than 1, the noncancer risk may be considered to be insignificant. As shown in Equation A.7.3, the reference dose, RfD, is one type of ADI.

Noncancer Risk Calculation Example

The chromic acid (Cr^{6+}) mists dermal chronic RfD of $6.00 \times 10^{-3}\,\text{mg}\,\text{kg}^{-1}\,\text{day}^{-1}$. If the actual dermal exposure of people living near a metal processing plant is calculated (e.g., by intake or LADD) to be $4.00 \times 10^{-3}\,\text{mg}\,\text{kg}^{-1}\,\text{day}^{-1}$, calculate the hazard quotient for the noncancer risk of the chromic acid mist to the neighborhood near the plant and interpret the meanings.

Solution and Discussion

From Equation A.7.3, $\frac{\text{Exposure}}{\text{RfD}} = \frac{4.00 \times 10^{-3}}{6.00 \times 10^{-3}} = 0.67$. Since this is less than 1, we would not expect people chronically exposed at this level to show adverse effects from skin contact. However, at this same chronic exposure, 4.00×10^{-3} mg kg^{-1} day^{-1}, to hexavalent chromic acid mists via oral route, the RfD is 3.00×10^{-3} mg kg^{-1} day^{-1}, meaning the HQ = 4/3 or 1.3. The value is greater than 1, so we cannot rule out adverse noncancer effects.

If a population is exposed to more than one contaminant, the hazard index (HI) can be used to express the level of cumulative noncancer risk from pollutants 1 through *n*:

$$HI = \sum_{1}^{n} HQ \qquad (A.7.4)$$

The HI is useful in comparing risks at various locations, for example, benzene risks in St. Louis, Cleveland, and Los Angeles. It can also give the cumulative (additive risk) in a single population exposed to more than one contaminant. For example, if the HQ for benzene is 0.2 (not significant), toluene is 0.5 (not significant), and tetrachloromethane is 0.4 (not significant), the cumulative risk of the three contaminants is 1.1 (potentially significant).

It is desirable to have realistic estimates of the hazard and the exposures in such calculations. However, precaution is the watchword for risk. Estimations of both hazard (toxicity) and exposure are often worst-case scenarios, because the risk calculations can have large uncertainties. Models usually assume effects to occur even at very low doses. Human data usually are gathered from epidemiological studies that, no matter how well they are designed, are fraught with error and variability (science must be balanced with the rights and respect of subjects, populations change, activities may be missed, and confounding variables are ever present). Uncertainties exist in every phase of risk assessment, from the quality of data, to limitations and assumptions in models, to natural variability in environments and populations.

APPENDIX 8

Risk-Based Contaminant Cleanup Example

Deciding on how "clean is clean" is one of the major controversies that have grown out of hazardous waste cleanup cases. Let us consider a fictitious example of how a risk-based cleanup can be applied to a contamination problem:

Problem

A well is the principal water supply for the town of Apple Chill. A study has found that the well contains 80 mg L^{-1} *tetrachloromethane (*CCl_4*) as a result of illegal dumping by Dump and Run Industries into a field above the aquifer from which the town's water is being drawn. Dump and Run was sued and required to ensure that the town's residents are not exposed to inordinate amounts of the contaminant. As a first step, the judge has ordered a review of the difference between current exposures and those needed to protect public health. Assuming that the average adult in the town drinks 2 L d*$^{-1}$ *of water from the well and lives in the town for an entire lifetime, what is the lifetime cancer risk to the population if no treatment is added? What concentration is needed to ensure that the population cancer risk is below* 10^{-6}*?*

Solution

The lifetime cancer risk added to Apple Chill's population can be estimated using the lifetime average daily dose (LADD) and slope factor for CCl_4. In addition to the assumptions given, we will use default values from the U.S. Environmental Protection Agency's Exposure Factors Handbook. We will also assume that people live in the town for their entire lifetimes, and that their exposure duration is equal to their typical lifetime.

Thus, ED and TL terms cancel, leaving the abbreviated (See Appendix 6 for definitions of terms):

$$\text{LADD} = \frac{(\text{C}) \cdot (\text{CR}) \cdot (\text{AF})}{(\text{BW})}$$

Since we have not specified male or female adults, we will use the average body weight, assuming that there are about the same number of males as females. We look up the absorption factor for CCl_4 and find that it is 0.85 (i.e., 85% of what is consumed stays in the body), so the adult lifetime exposure is:

$$\text{LADD} = \frac{(80)\cdot(2)\cdot(0.85)}{(65)} = 4.2\,\text{mg}\,\text{kg}^{-1}\,\text{day}^{-1}$$

Using the midpoint value between the default values $\left(\frac{15+40}{2} = 27.5\,\text{kg}\right)$ for body weight and default CR values ($1\,\text{L}\,\text{d}^{-1}$) the children lifetime exposure is:

$$\text{LADD} = \frac{(80)\cdot(1)\cdot(0.85)}{(27.5)} = 2.5\,\text{mg}\,\text{kg}^{-1}\,\text{day}^{-1}$$

for the first 13 years, and the adult exposure of $4.2\,\text{mg}\,\text{kg}^{-1}\,\text{day}^{-1}$ thereafter.

The oral cancer slope factor (SF) for CCl_4 is $1.30 \times 10^{-1}\,\text{kg}\,\text{day}^{-1}$, so the added adult lifetime cancer risk from drinking the water is the product of the SF and LADD:

$$4.2 \times (1.30 \times 10^{-1}) = 5.5 \times 10^{-1}$$

and the added risk to children is:

$$2.5 \times (1.30 \times 10^{-1}) = 3.3 \times 10^{-1}$$

However, for children, environmental and public health agencies recommend an additional factor of safety beyond what would be used to calculate risks for adults. This is known as the 10X rule; children need to be protected 10 times more than adults because they are more vulnerable, have longer life expectancies (so latency periods for cancer need to be accounted for), and their tissue is developing prolifically and changing. So, in this case, with the added risk, our reported risk would be 3.3. This is statistically impossible (i.e., we cannot have a probability greater than one because it would mean that the outcome is more than 100% likely, which of course is impossible!). However, what this tells us is that the combination of a very high slope of the dose response curve and a very high LADD leads to much needed protections, such as removal of either contaminant from the water or the provision of a new water supply. The city engineer or health department should mandate bottled water immediately.

The cleanup of the water supply to achieve risks below 1 in a million can also be calculated from the same information and reordering the risk equation to solve for C:

$$\text{Risk} = \text{LADD} \times \text{SF}$$

$$\text{Risk} = \frac{(\text{C}) \cdot (\text{CR}) \cdot (\text{AF}) \cdot (\text{SF})}{(\text{BW})}$$

$$\text{C} = \frac{(\text{BW}) \cdot \text{Risk}}{(\text{CR}) \cdot (\text{AF}) \cdot (\text{SF})}$$

Based on adult LADD, the well water must be treated so that the tetrachloromethane concentrations are below:

$$\text{C} = \frac{(65) \cdot 10^{-6}}{(2) \cdot (0.85) \cdot (0.13)} = 2.9 \times 10^{-4}\,\text{mg}\,\text{L}^{-1} = 290\,\text{ng}\,\text{L}^{-1}$$

Based on children's LADD and the additional 10X, the well water must be treated so that the tetrachloromethane concentrations are below:

$$\text{C} = \frac{(27.5) \cdot 10^{-7}}{(1) \cdot (0.85) \cdot (0.13)} = 2.5 \times 10^{-5}\,\text{mg}\,\text{L}^{-1} = 25\,\text{ng}\,\text{L}^{-1}$$

The town will have to remove the contaminant so that the concentration of CCl_4 in the finished water must be treated to a level six orders of magnitude less than the untreated well water; that is, lowered from $80\,\text{mg}\,\text{L}^{-1}$ to $25\,\text{ng}\,\text{L}^{-1}$.

Cleanup standards are part of the risk management available to decision makers and environmental professionals. However, other considerations need to be given to a contaminated site, such as how to monitor the progress in lowering levels and how to ensure that the community stays engaged and is participating in the cleanup actions, where appropriate. Even when the engineering solutions are working well, the engineer must allot sufficient time and effort to these other activities, otherwise skepticism and distrust can arise.

APPENDIX 9

Shannon-Weiner Index Example

Problem

In 1999, you conducted an environmental assessment of microbes in a small stream at your client's factory. You found seven species of these critters. Your actual number count of each microbial species in this stream community was 16, 49, 69, 124, 212, 344, and 660 mL^{-1}.

Find the diversity of this stream community using the Shannon-Weiner index:

$$D = -\sum_{i=1}^{m} P_i \log_2 P_i \tag{A.9.1}$$

or

$$D = -1.44\sum_{i=1}^{m}(n_i/N)\ln(n_i/N) \tag{A.9.2}$$

where

D = index of community diversity
P_i = n_i/N
n_i = number (i.e., density) of the ith genera or species
N = total number (i.e., density) of all organisms in the sample
i = 1, 2, . . ., m
m = number of genera or species

Solution

Construct a table to derive the values needed to find D, using Equation A.9.2:

$$D = -1.44\sum_{i=1}^{m}(n_i/N)\ln(n_i/N).$$

i	n_i	n_i/N	$-1.44 \ln(n_i/N)$	$-1.44(n_i/N) \ln(n_i/N)$
1	16	0.010855	6.513331	0.070701
2	49	0.033243	4.901637	0.162945
3	69	0.046811	4.408745	0.20638
4	124	0.084125	3.564653	0.299876
5	212	0.143826	2.792374	0.401617
6	344	0.233379	2.095335	0.489006
7	660	0.447761	1.157033	0.518075
Σ	1,474	1		2.148599

Thus, the diversity index is 2.1. The index is limited in its absolute meaning and is most useful when comparing different ecosystems. In the event that this system's Shannon-Weiner index is 2.1 and surrounding streams are all around 4, this system has a poorer biodiversity. Generally, D values range from about 1.5 to 4.5.

What would happen if all the numbers of species were ten times greater? Nothing, the index would still be 2.1. Thus total species abundance does not affect diversity. Thus, diversity is not affected, at least mathematically, but there is an overall increase in abundance, so long as the interspecies ratios stay the same.

You conducted a follow up study in 2004 and found that the density of these same species had changed to 2000, 25, 17, 18, 21, 40, and 11 microbes L^{-1}. How had the numbers and diversity changed in five years?

Solution and Discussion

Again, calculate D by constructing a table:

i	n_i	n_i/N	$-1.44 \ln(n_i/N)$	$-1.44(n_i/N) \ln(n_i/N)$
1	2,000	0.93809	0.09204	0.08634
2	25	0.01173	6.40215	0.07507
3	17	0.00797	6.95751	0.05548
4	18	0.00844	6.8752	0.05805
5	21	0.00985	6.65322	0.06553
6	40	0.01876	5.72535	0.10742
7	11	0.00516	7.58437	0.03913
Σ	2,132	1		0.48701

This shows that in five years, the actual number of microbes is increasing, but the diversity is far less ($D = 0.5$ versus 2.1 five years later). This may indicate that conditions favorable to one species and unfavorable to most of the others, for example, the presence of a toxic chemical, is detrimental to the other six species. Thus, the Shannon-Weiner index is a valuable tool for temporal comparisons of biodiversity within the same ecosystem.

However, a key question to ask is whether the two studies are truly comparable. For example, were the 1999 and the 2004 studies conducted in the same season (some microbes grow better in warmer conditions, whereas others may compete more effectively in cooler waters)? If the studies are comparable, this certainly is an indication that biodiversity is decreasing. In addition, since Shannon-Weiner values usually range from about 1.5 to 4.5, an index of 0.5 indicates a problem.

APPENDIX 10

Useful Conversions in Atmospheric Chemistry[1]

Permutations of SI Units

1 gC = 1 gram carbon (C)
1 GgC = gigagram carbon (C) = 1,000 metric tons carbon (C)
1 TgC = 1 teragram carbon (C) = 1 million metric tons carbon (C)
1 PgC = 1 petagram carbon (C) = 1 billion metric tons carbon (C)
1 ppmv = 1 part per million by volume in the atmosphere
1 ppbv = 1 part per billion by volume in the atmosphere
1 pptv = 1 part per trillion by volume in the atmosphere

Density

1 thousand cubic feet of methane = 42.28 pounds
1 thousand cubic feet carbon dioxide = 115.97 pounds
1 metric ton natural gas liquids = 11.6 barrels
1 metric ton unfinished oils = 7.46 barrels
1 metric ton alcohol = 7.94 barrels
1 metric ton liquefied petroleum gas = 11.6 barrels
1 metric ton aviation gasoline = 8.9 barrels
1 metric ton naphtha jet fuel = 8.27 barrels
1 metric ton kerosene jet fuel = 7.93 barrels
1 metric ton motor gasoline = 8.53 barrels
1 metric ton kerosene = 7.73 barrels
1 metric ton naphtha = 8.22 barrels
1 metric ton distillate = 7.46 barrels
1 metric ton residual oil = 6.66 barrels
1 metric ton lubricants = 7.06 barrels
1 metric ton bitumen = 6.06 barrels
1 metric ton waxes = 7.87 barrels
1 metric ton petroleum coke = 5.51 barrels

1 metric ton petrochemical feedstocks = 7.46 barrels
1 metric ton special naphtha = 8.53 barrels
1 metric ton miscellaneous products = 8.00 barrels

Alternative Measures of Greenhouse Gases

1 pound methane, measured in carbon units (CH_4-C) = 1.333 pounds methane, measured at full molecular weight (CH_4)
1 pound carbon dioxide, measured in carbon units (CO_2-C) = 3.6667 pounds carbon dioxide, measured at full molecular weight (CO_2)
1 pound carbon monoxide, measured in carbon units (CO-C) = 2.333 pounds carbon monoxide, measured at full molecular weight (CO)
1 pound nitrous oxide, measured in nitrogen units (N_2O-N) = 1.571 pounds nitrous oxide, measured at full molecular weight (N_2O)

Weight

1 kilogram = 2.205 pounds
1 short ton = 0.9072 metric tons
1 metric ton = 1.1023 short tons = 2,204.6 pounds
1 cubic meter = 35.3147 cubic feet
1 cubic centimeter = 3.531×10^{-5} cubic feet

Area

1 acre = 0.40468724 hectare (ha) = 4,047 m^2
1 hectare (ha) = 10,000 m^2 = 2.47 acres
1 kilometer = 0.6214 miles

Energy

1 joule = 947.9×10^{-21} quadrillion Btu
1 exajoule = 10^{18} joules = 0.9479 quadrillion Btu
1 quadrillion Btu = 1.0551 exajoule

Notes and Commentary

1. U.S. Department of Energy, 1997. *Emissions of Greenhouse Gases in the United States 1996*, DOE/EIA-0573(96), Washington, D.C.

Index

1,1,1-trichloro-2,2-bis-(p-chlorophenyl)ethane (DDT) 10, 55, 62–65, 216–217, 235, 302, 304, 305, 315, 316, 317, 319, 521, 524–525
10 X Policy 398
2,4,5-T 406
2,4-D, See dichlorophenoxyacetic acid
60 Minutes 396, 398

Aaron 163
absorption, mechanical 76
accuracy xvi
acid mine drainage 346–348
acid precipitation; acid rain 348–351
Adams, Scott 483–484
adenosine triphosphate 510
adsorption 76
advocacy science xx
AEC, See anion exchange capacity
affordable housing 466–468
Agent Orange 10, 404–406
air/water partition coefficient (K_{AW}) 86
Alar 392–403
Allegory of the Cave 494
Allied Chemical Company 235–237
aluminum 349
American Society of Civil Engineers (ASCE) 40, 330–332, 334
American Society of Mechanical Engineers (ASME) 328–330, 334
Amish 197
Amoco Cadiz 175
amphibole 138–140
ample margin of safety 437–438
anaerobe; anaerobic digestion 116, 233, 237, 305
anion exchange capacity (AEC) 76
antimony (Sb) 350
Appalachian Regional Development Act 169
aquaculture 288
Areas of Concern 170–174
areawide waste treatment management plan, See Section 208 plan
Aristotle 31, 32
arsenic xxvi, 116, 349–350, 356–360
asbestos xxii, xxviii, 138–140, 360–364, 426–427, 430, 439, 444–445, 448
ASCE, See American Society of Civil Engineers

Asian shore crab (*Hemigrapsus sanguineus*) 286–289
ASME, See American Society of Mechanical Engineers
asthma 10
Atlantic Empress 174
Atlantic menhaden 192
Aurelia spp. xxiii

Babylon 109
BACT, See best available control technology
Baez, Joan 339
Baker, Howard 409
Baltimore, Maryland 123
Bangladesh 356–360
base catalyzed decomposition 457–458
Base, Sara 501
Bay of Campeche, Mexico 174
BCF, See bioconcentration factor
Beggiatoa spp. 142
benchmark 10, 202, 265
Bennett, Edward H. 169
benzo(a)pyrene 302–303
Bernoulli, Jacob xxi, 69
Bernoulli's equation 106–107
best available control technology (BACT) 438
beta-naphthylamine (2-naphthylamine) 240, 265–266
Bhopal, India 15, 16, 34, 144–150, 430–431
bioaccumulation 89, 234–235
bioavailability 89
bioconcentration factor (BCF) 74, 89–91, 234
biodegradation 206, 233, 236
biodiversity 293, 327, 381, 547–549
biology, systems 3–4
biomarker 238–240
biosolids, See sludge
bitumen 109
black body radiation 300
Blackbird Mine, Idaho 346
bladder cancer 358
Bliss, Russell 213–214
Borland, Hal 163
Brittany, United Kingdom 174, 175
Brundtland Commission 41
bubble 2
bulk modulus 103
Bullitt County, Kentucky 216
Burnham, Daniel 169
Butterfly Effect 332–333
byssinosis 129

cadmium (Cd) xxi, xxvii, 116
Calculus, The 100
Cancer Alley 232–234
cancer classification system 402
cancer slope factor, See slope factor
cap and trade 2
Cape Mendocino xxiii
capillarity 103–105
carbon dioxide (CO_2) xxiii, 15, 16, 112, 209–210, 233, 243, 300–302, 311, 312, 368, 372–376, 377, 430–431, 474
carbon monoxide (CO) 16, 233, 243, 474
Carson, Rachel 14–15, 62
Carver Terrace, Texas 469–470
Caspian Sea 290
Castillo de Bellver 174
catalysis 5–6, 473–474, 508–509,
catalytic converter 473–474
catastrophes 16
categorical imperative xx
cation exchange capacity (CEC) 76
CEC, See cation exchange capacity
cesium (Cs) 321, 324, 325
CFCs, See chlorofluorocarbons
chemisorption 76
Chernobyl, Ukraine xxii, 15, 16, 321
Chesapeake Bay 192
Chester, Pennsylvania 204–213
Chicamacomico River 192
Chlorobiaceae spp. 142

chlorofluorocarbons (CFCs) 368, 373, 378, 483
Chromatiaceae spp. 142
chromic acid 541–542
chromium (Cr) 116, 244–246, 541–542
chrysotile 138–140
Circular A-95 170
Ciudad del Carmen, Mexico 174
civil rights 14
Civil Rights Act 465
Clark Fork River, Montana 346
Clean Air Act (and Amendments) 15, 16, 437–438
Clean Air Act of 1956 (United Kingdom) 111
coal mining 417
Coeur d'Alene Valley, Idaho 346
coherence 494, 500
Colburn, Theo 197
Collinsville, Illinois xxiii
combustion 209–211
command and control 1
Commoner, Barry 14
computational toxicology 3–8, 475–477
conditions, boundary 490
conditions, initial 490
conflict of interest 493
Confucius 429
congener 210, 214, 242
Construction Grants program 169
continuum fluid mechanics 98
Cool Hand Luke 21
copper (Cu) 116
Copper Basin, Tennessee 345
coral reefs 380–385
crayfish 291
credat emptor 33
creosote 469–470
Cronkite, Walter 398
Crown of Thorns starfish 381
crude oil 176
cryptosporidium 10
Cuyahoga River 163, 168, 171–174
cyanide 116, 209, 243
cyclone 131–134
cytochrome P450 239

Dallas, Texas 471
daminozide, See Alar
DDT, See 1,1,1-trichloro-2,2-bis-(p-chlorophenyl)ethane
decision force field 423–425
Denora, Pennsylvania 15, 139–140
density 100–102
deontology 435
deoxyribonucleic acid (DNA) 302, 324
depuration 90, 177–178
Design for Disassembly (DFD)
Design for Recycling (DFR) 2, 28
Design for the Environment (DFE) 2, 28
destruction removal efficiency (DRE) 136–137
DFD, See Design for Disassembly 2, 28
DFE, See Design for the Environment
DFR, See Design for Recycling
dibromochloropropane (DBCP) 241–242
dichlorophenoxyacetic acid (2,4-D) 302, 406
Dickens, Charles 110
diffusion 311
digital divide 495, 501
dinoflagellate 192–193
dioxane (*para*) 442
dioxin 15, 16, 88–89, 201, 207, 210–215, 227, 239, 249, 253–254, 270, 404–406, 411–413, 414
Dirty Dozen 91
disaster 226–231
discharge, stream 99
dissolution 80–85
distribution coefficient (K_D) 74, 78
District of Columbia (Washington, DC) 149

DNA, See deoxyribonucleic acid
Dominican Republic 415
Donora, Pennsylvania 138–140
dose-response 57, 248–255, 400–401
Dow Chemical Company 241
Drake Chemical Company (Lock Haven, Pennsylvania) 265–266, 272
DRE, See destruction removal efficiency
drum, freshwater (*Aplodinotus grunniens*) 291
Duke Forest, North Carolina 442, 462
Duke University 32
Duncan, John 409
Dylan, Bob 339

East St. Louis, Illinois 448
Eckenfelder, W. Wesley 327
ecological risk assessment 292–295
ecology 326
Edeburn, Jud 462
effluent 91
electronegativity 82–85
electrostatic precipitator 135
emission 91
Endangered Species Act 407–410
endergonic reaction 508–510
endocrine disruptor 316–319
energy balance 491
energy, kinetic 100
energy, potential 100
enthalpy 82–85
environmental justice 420–422, 471–475
enzyme 238, 508–509
epidemiology 197, 249–251, 265, 267–268
equilibrium 70, 503–510, 523
EROD, See ethoxyresorufin-*O*-deethylase
ethics 14, 38–42, 424–425, 495
ethike aretai 33
ethoxyresorufin-*O*-deethylase (EROD) 239
Etnier, David 408, 410
eutrophic lake 165–167
eutrophication 164–167
Eve of Destruction 339
Evelyn, John 110
Everglades 293
exergonic reaction 508–510
Exodus 163
exposure 255–263
Exposure Factors Handbook 61
Exxon Valdez xxii, 16, 174, 185–189

fabric filter 133–135
failure 21–28, 517–519
faithful agent 493
faithful agents xx
fate (pollutant) 96
Federal Food Quality Protection Act 398, 403, 416
federal reference method (FRM) 137
Federal Water Pollution Control Act 56
fiber 129–130, 138–139
FID, See flame ionization detection
filtration 135–136
fission, nuclear 320–321
flame ionization detection (FID) 208
fluids 96–107
fossil fuels 112
free energy, Gibbs 503–510
Freundlich Sorption Isotherm 79
FRM, See federal reference method
fugacity 75
Fuller, Buckminster 367
furan 210–213, 270
future engineer 493

gas chromatography (GC) 207
Gavia immer 90
GC, See gas chromatography
genetic engineering 320

genetically modified organism (GMOs) 16
Genoa, Italy 174
Gist, Jacquelyn 462, 463
global climate change 10, 299–302
Global Invasive Species Database 276–285
GMO, See genetically modified organism
Goiania, Brazil 247
gold rush 342
Goyer, Robert 358
Grace, W.R., Company xxviii
Graniteville, South Carolina 149
grass carp 278, 292
gray goo scenario 17
green chemistry 1–2
green engineering 2
greenhouse effect 300–301
greenhouse gases (in addition, see specific gases) 10, 485
groupthink 500
Guanabara Bay, Brazil 174

Habitat for Humanity 467
Haita 415
half-life ($T_{1/2}$) 87
halocarbons 373, 378
Hardin, Garrett 298–299
harm principle xx
hazard index 542
Hazard Ranking System (HRS) 200
hazard 57–61
hazardous waste 198, 199–202, 204, 212, 215–216, 219, 223–224, 245–246, 251, 252, 254, 263–264
Hazelwood, Joe 186
healthy worker syndrome 416
Henry's Law constant (K_H) 74, 85–87, 522–523
Henry's Law 85–86, 522
HEPA filter, See high efficiency particle air filter
Heritage Foundation 358
HI, See hazard index
high efficiency particle air (HEPA) filter 136
high performance liquid chromatography (HPLC) 207
Hightower, Jim 497
Hill, Bradford 35–36
Hill's Causal Criteria 35–36
Hispaniola 415
Hit 109
Hollander, Jack 413–415
Honda Motors, Inc. 474
Hooker Chemical Company 198
Hopewell, Virginia 235–237
hormonally active agent, See endocrine disruptor
Housing and Community Development Act 169
HPLC, See high performance liquid chromatography
HRS, See Hazard Ranking System
hydrochlorofluorocarbons 378
hydrogen sulfide (H_2S) 109, 116, 140–142
hydrolysis 310

ideal gas law 100
incidence 250–251
Indian River 192
Industrial Revolution 109
Industrie Chemiche Meda Societa, Anonima 411
information technology (IT) 496–498
intermedia transfer 326
Inuit 231–232, 302–305
Invasive Species Specialist Group 275, 285
invasive species, definition 275
inversion, thermal 119, 121, 139, 140, 143–144, 153–154, 158
ion exchange 76
ionization 117
Iron Gates Dam xxi, xxii–xxiii
Iron Mountain, California 340–346

Isles of Scilly, United Kingdom 175, 179
isomer; isomerization 211, 240, 311–312
itai itai disease 10
IXTOC I 174

Janus, Irving 500
Joyce Engineering Company 460–462
junk science xx
junkyards 446–448
justice, environmental 231

Kant, Immanuel xx
K_{AW}, See air/water partition coefficient
K_D, See distribution coefficient
Kepone (chlordecone) 235–237
K_H, See Henry's Law constant
kinetics 70–74, 507
King Dionysius 297
King Tikulti 109
King, Martin Luther 420
Knopfler, Mark 15
K_{oc}, See organic carbon-water coefficient
K_{ow}, See octanol-water coefficient
kudzu (*Pueraria* spp.) 283, 286, 292
Kuhn, Thomas xiv, 492, 500
Kuwait oil fires xxi
Kyoto Protocol 301

La Coruna, Spain 174, 190
LADD, See lifetime average daily dose
Lake Apopka, Florida 316–319
Lake Erie 163–168
Lake Michigan 287, 290
laminar flow 98
landfill siting 459–463
landfill xv
LCA, See life cycle analysis
LD, See lethal dose
lead (Pb) xxvi, 10, 91, 116, 351–353, 471, 484
Leadville, Colorado 345
Leaking Underground Storage Tank (LUST) program 92
Lee, Howard 462
Legionnaire's Disease 10
lethal dose (LD) 400–401
LEV, See low emission vehicle 56
Libby, Montana xxi, xxviii
life cycle analysis (LCA) 2, 28
Life Sciences Products 235–237
lifetime average daily dose (LADD) 129, 262, 264, 537–539, 543–545
liver cancer 345
LOAEL, See lowest observed adverse effect level
Lois, George 275
London, England xv, 15, 109–111, 142–144
Lorenz, Edward 332–333
Love Canal, New York xv, 15, 198–199, 223–224, 469–470
Love, William T. 198
low emission vehicle (LEV) 56
lowest observed adverse effect level (LOAEL) 249, 267–268, 403
lung cancer 358, 361
LUST, See Leaking Underground Storage Tank program

macro-ethics 332, 334
MACT, See maximum available control technology
Mangouras, Apostolos 190
Manokin River 192
margin of exposure (MOE) 403
Marine Pollution Control Unit 180
Martin-Schramm, James 443
mass balance 491
mass spectrometry (MS) 8, 208
maximum available control technology (MACT) 438
maximum contaminant level (MCL) 357

McDowell, B.D. 170
McGuire, Barry 339
McKinney, Ross xxviii–xxix, 327, 491–492, 500
MCL, See maximum contaminant level
McLuhan, Marshall 367
Mega Borg 175
mercury (Hg) xxvi–xxvii, 10, 91, 116, 306, 349, 351, 353–356
mesotrophic lake 165
methane 368, 373, 374, 375, 377, 430–431
methemoglobinemia 118
methyl tert-butyl ether (MTBE) 17–19, 65
methylisocynate (MIC) 34, 144–148, 430–431
Meuse Valley, Belgium 119–122
micro-ethics 332, 334
Milford Haven, Wales, United Kingdom 174
Mill, John Stuart xx, 298–299
Milton, John xiii, 498
Minamata Bay, Japan 10, 353–355, 364
Mnemiopsi spp. xxiii
MOE, See margin of exposure
Mogulof, Melvin 170
mole fraction 105–106
Moody Blues, The 339
MS, See mass spectrometry
MTBE, See methyl tert-butyl ether
Müller, Paul E. 62
Multiflora rose (*Rosa multiflora* (Thunb. ex Murr.)) xxiii–xxiv

nanotechnology 6, 16–17
National Contingency Plan (NCP) 180–182
National Environmental Policy Act 474–475
National Oceanic and Atmospheric Administration, creation of 95
National Oil and Hazardous Substances Pollution Contingency Plan 182
National Priority Listing (NPL) 200–201
National Research Council 358
natural attenuation 441–442
natural experiment 316, 493
Natural Resources Defense Council 398
NCP, See National Contingency Plan
Neibuhr, Reinhold 419
Nelson, Mike 462, 463
neo-Luddite 320
NEPA, See National Environmental Policy Act
net primary productivity 293
Neuse River 192
New York City, New York (Also see World Trade Center) 144
Newton, Isaac xxi, 100
Niagara Falls, New York 198–199
nickel (Ni) 116
NIMBY (not in my backyard) 461, 467
nitrogen compounds 112–115, 125
nitrogen cycle 117
nitrous oxide 368, 373, 374, 375, 377
no observed adverse effect level (NOAEL) 248–252, 267–268, 403
NOAEL, See no observed adverse effect level
nonpoint source 168
NPL, See National Priority Listing

O&M, See operation and maintenance
Occidental Chemical Company 198, 199
octanol-water coefficient (K_{ow}) 76, 81, 88, 90–91, 235, 523
OILPOL 182
oligotrophic lake 164–165

omics 4
one-hit model 251–252
operation and maintenance (O&M) 2, 28, 327
Orange County, North Carolina 459–463
organic acid 243
organic carbon-water coefficient (K_{oc}) 74
organochlorine pesticides 10
organophosphate pesticides 10
oxidation-reduction (redox) reactions 115–116, 310–311
ozone (O_3) 95, 114, 116, 118, 120–121, 123–124, 152–157, 158, 379–380

PA/SI, See preliminary assessment/site inspection
PAH, See polycyclic aromatic hydrocarbon
Palmerton, Pennsylvania 345
Paracelsus 57
particulate matter (PM) 117, 123–138, 431, 448
partition coefficient 69–70, 521–525
PBT, See persistent organic pollutant
PCBs, See polychlorinated biphenyls
pearls 291
Peirce, J. Jeffrey 36
Pender, Mike 339
perfluorocarbons 378
periodic table of elements 349–350
Persian Gulf 174
persistence 87, 90, 234–235, 314–316, 521–525
persistent organic pollutant (POP) 91, 302–306
persistent, bioaccumulating toxic (PBT) 91, 302, 305, 308, 309, 310, 521
Pfiesteria piscicida 191–194
photochemical oxidant smog 154–157
physicochemical properties 307–314
PIC, See product of incomplete combustion
Plato 494
PM, See particulate matter
point source 168
polarity 80
pollution, definition 55–56
polychlorinated biphenyls (PCBs) 10, 20–21, 107, 207, 208, 209, 211, 216, 227, 228, 231, 232, 239, 242, 302, 304, 305, 308, 313, 314, 317, 318, 452–458, 523
polycyclic aromatic hydrocarbon (PAH) 16, 208, 210, 239, 301–302, 314–318, 448
POP, See persistent, bioaccumulating toxic
Posner, Richard 16
potentially responsible party (PRP) 201, 218, 219
Poverty 413–415
Poza Rica, Mexico 140
precautionary principle 251, 268, 297, 299, 334, 444–446
precision xvi
preliminary assessment/site inspection (PA/SI) 200
pressure 99–100
Prestige 188–190
pretreatment 218
prevalence 250–251
Prieto, Robert 150
Prince William Sound, Alaska xxii, 16, 174, 188
product of incomplete combustion (PIC) 16
productivity (biological) 293, 327
professional engineer 493
PRP, See potentially responsible party
pyrite 342

Quayle, J. Danforth 56
Quicksilver Messenger Service 14

radioisotopes 246–247
rainforests xxiii
Rawls, John xx, 299
RD/RA, See remedial design/ remedial action
reaction rate 70–74
Record of Decision (ROD) 200, 201, 214, 353
reductio ad absurdum 439
Rees, Martin 16–17
reference concentration (RfC) 249, 267, 272
reference dose (RfD) 248–249, 267, 272, 403
Reilly, William xxii
reliability 38, 65–69, 223–224, 518–519
Remedial Action Plan 170, 172
remedial design/remedial action (RD/RA) 200, 201
remedial investigation/field study (RI/FS) 200, 201, 202
Reorganization Plan No. 3 of 1970 511–516
residential standard 435, 440
retention time (RT) 208
RfC, See reference concentration
RfD, See reference dose
RI/FS, See remedial investigation/ field study
Richter, Daniel 93
Right to Know 15
Rio de Janeiro, Brazil 297
risk analysis 223
risk assessment 223, 225, 230–231, 268–269
risk characterization 539–542
risk communication 225, 231
risk management 223, 425–444, 543–545
risk perception 225–226–231
risk tradeoff 61–65
risk 10, 37, 59–65, 398, 400
risk, cumulative 542
risk, definition 37
River Thames 110
Rocque, John 110
ROD, See Record of Decision
Rothchild, Michael 497
route of exposure 533
Royal Air Force 180
Royal Navy 179
RT, See retention time

safety (factor of safety) 224, 248–250, 403, 444–446
salt water intrusion 101–102
San Antonio, Texas 149
Sandman, Peter 229
sanitary engineering 328
Santa Barbara, California 182–185
Santayana, George 13–14
Santillian, Jesica 32–33
Satz, Debra 443
Schlosser, Paul xviii–xx
Section 208 plan 168–169
selenium (Se) 349
Seneca, the Younger 109
September 11, 2001 16, 144, 214
Seveso, Italy 15, 410–413, 414
SF, See slope factor
Shannon-Weiner Index 547–549
Shell Oil Company 241
Shetland Islands, United Kingdom 174
Sierra Club 358
Silent Spring 62
silver (Ag) 116
site-wide cleanup 439–440
slope factor (SF) 214, 232, 240, 242, 249, 252, 264, 265,527–533, 544–545
sludge 326–327
snail darter (*Percina tanasi*) 407–410
solid waste, municipal 389–392
solubility 87, 176, 302, 307–310, 312, 315

sorption 75–80, 307, 309, 310, 315–316
Spaceship Earth 14, 367
specific volume 101
specific weight 101
St. John's River 192
St. Paul (the Apostle) 501
St. Petersburg, Russia 188
Standard Fruit Company 241
standard, environmental (cleanup) 263–264, 543–545
standards 1, 438–441
Stivers, Robert 443
Stockholm Conference on the Human Environment 41
Stokes diameter 125
storytelling xv
stratospheric ozone 10, 379–380, 483
stretching 499–500
Stringfellow, California acid pits 216–219
sulfur compounds 111–112
sulfur dioxide (SO_2) 109, 112, 116, 139, 125
sulfur hexafluoride 378
Superfund 198, 200–201, 209, 214–215, 224, 265, 271, 435, 469
surface tension 103–104
sustainability 1, 40–41, 293
Sword of Damocles 297
syllogism xvi, xx, 383–385
system 326, 487–491

$T_{1/2}$, See half-life
Tar-Pamlico River 192
TCDD, See tetrachlorodibenzo-*para*-dioxin
technology 486, 490, 493, 495, 497, 501
teleology 435
Tellico Dam 407–409
Tennessee River 407–409
Tennessee Valley Authority 407–410
terrorism 325
tetrachlorodibenzo-*para*-dioxin (TCDD) 88–89, 210, 211, 214, 253–254, 404, 411–413, 414
tetrachloromethane (carbon tetrachloride) 543–545
thermodynamics, first law of 489
Thioploca spp. 142
Thiotrix spp. 142
Three Mile Island, Pennsylvania xxi, 15, 16
threshold 224, 248–249, 251–252, 264, 267–268
thyroid cancer 324–325
Times Beach, Missouri 10, 15, 198, 201, 213–215, 223
TIMSS, See Trends in International Mathematics and Science Study
Tobago 174
Torch Lake, Michigan 345
Torrey Canyon 175, 178–182
toxic cloud 144–152
Toxic Release Inventory (TRI)
toxicology 30
Tragedy of the Commons 298–299
transformation (pollutant) 96
transparency 494, 500
transport (pollutant) 96
Trends in International Mathematics and Science Study (TIMSS) 495–496
TRI, See Toxic Release Inventory
turbulent flow 98

U.S. Environmental Protection Agency, creation of 95
UDMH, See unsymmetrical dimethyl hydrazine
ULEV, See ultra-low emission vehicle 56
ultra-low emission vehicle (ULEV) 56
ultraviolet light 379–380

uncertainty xv–xvii
Union Oil Company 183
Uniroyal Chemical Company 393
United Church of Christ Commission for Racial Justice 451, 472
United Nations Educational, Scientific, and Cultural Organization 182
unsymmetrical dimethyl hydrazine (UDMH) 393, 397
UV, See ultraviolet light

vadose zone 105
valence (oxidation state) 243
Valley of the Drums 198, 215–216
valuation 446–449
value of human life xviii–xx
vapor pressure 308–309
variability xv–xvii
veil of ignorance xx
veliger 290
Vesilind, P. Aarne xxx, 327, 492, 499, 501
Vietnam 14, 404–406
vinclozolin 393–394, 472–473
vinyl chloride 232–234, 448, 527, 537–538, 540–541
viscosity 106–107
vitellogenin 319
VOC, See volatile organic compound
volatile organic compound (VOC) 19
volatilization 85–87, 308–309

Warren County, North Carolina 20, 452–458
Water Resources Act 169
weight of evidence 402, 483
weij-ji 69, 519
West Nile virus 430
wetland 327–328
Will, George 500
World Trade Center, New York 150–152

xenon (Xe) 322

zebra mussel (*Dreissena polymorpha*) 278, 286–287, 290–291
Zeolite 6
zone of saturation 105
zoning, exclusionary 466–468